WOODLAND CONSERVATION AND MANAGEMENT

WOODLAND CONSERVATION AND MANAGEMENT

Second edition

George F. Peterken

CHAPMAN & HALL

London · Glasgow · New York · Tokyo · Melbourne · Madras

Published by Chapman & Hall, 2–6 Boundary Row, London SE1 8HN

Chapman & Hall, 2–6 Boundary Row, London SE1 8HN, UK

Blackie Academic & Professional, Wester Cleddens Road, Bishopbriggs, Glasgow G64 2NZ, UK

Chapman & Hall, One Penn Plaza, 41st Floor, New York NY10119, USA

Chapman & Hall Japan, Thomson Publishing Japan, Hirakawacho Nemoto Building, 6F, 1-7-11 Hirakawa-cho, Chiyoda-ku, Tokyo 102, Japan

Chapman & Hall Australia, Thomas Nelson Australia, 102 Dodds Street, South Melbourne, Victoria 3205, Australia

Chapman & Hall India, R. Seshadri, 32 Second Main Road, CIT East, Madras 600 035, India

First edition 1981
Reprinted 1985, 1987
Second edition 1993

© 1981, 1993 G.F. Peterken

Typeset in 10/12 Bembo by Mews Photosetting, Beckenham, Kent
Printed in Great Britain at the University Press, Cambridge

ISBN 0 412 55730 4

A catalogue record for this book is available from the British Library

Library of Congress Cataloging-in-Publication data available

Contents

Contents

Part 3: Woodland nature conservation

Contents

Part 4:   Management for nature conservation

Contents

Preface to second edition

Many developments have taken place since the first edition of this book was published in 1981. We have learnt more about the history and pre-history of ancient woods and thereby acquired a more realistic picture of their real or imagined links with pre-neolithic forests. We have acquired a greater understanding of natural temperate forests and how they were altered by exploitation and management from North America. The National Vegetation Classification has been published and will henceforward form the standard basis of survey and communication about native woodland. Growing public support for woodland conservation has manifested itself in many ways. These range from profound changes in national forestry policy and practice, the success of the Woodland Trust and the burgeoning interest in new native woodlands to the active acknowledgement of a more spiritual significance of trees and threatened civil disobedience in the face of threats to much-loved woods. The basis for woodland nature conservation has been improved by the completion of the Inventory of Ancient and Semi-natural Woods by the Nature Conservancy Council.

Ordinarily, changes of this magnitude would justify a wholesale revision of the text, but I would never have found the time to undertake this, and in any case the original text still seems to have much to offer to the present generation of ecologists and conservationists. Instead, I have written an additional chapter, which reviews progress since 1981 in the subjects covered in the first edition, and substantially expanded the references, mostly with material from the last decade or so.

One of the pleasures of working in woodland conservation is the generally supportive and constructive relationships between the individuals involved. On this occasion, I would like to thank Keith Kirby, Rob Soutar and Jonathan Humphrey for reviewing a draft of the new chapter: it may be significant that two of these have, like me, recently chosen to leave the statutory nature conservation agencies to work with the Forestry Commission. I would also like to thank the librarians of the former Nature Conservancy Council for being consistently helpful and constructive. My wife, Susan, has shared my interest and helped in many ways over the years.

G.F. Peterken
July 1993

Preface to first edition

Professor John Harper, in his recent *Population Biology of Plants* (1977), made a comment and asked a question which effectively states the theme of this book. Noting that 'one of the consequences of the development of the theory of vegetational climax has been to guide the observer's mind forwards', i.e. that 'vegetation is interpreted as a stage on the way to something', he commented that 'it might be more healthy and scientifically more sound to look more often backwards and search for the explanation of the present in the past, to explain systems in relation to their history rather than their goal'. He went on to contrast the 'disaster theory' of plant succession, which holds that communities are a response to the effects of past disasters, with the 'climax theory', that they are stages in the approach to a climax state, and then asked 'do we account most completely for the characteristics of a population by a knowledge of its history or of its destiny?'

Had this question been put to R.S. Adamson, E.J. Salisbury, A.G. Tansley or A.S. Watt, who are amongst the giants of the first forty years of woodland ecology in Britain, their answer would surely have been that understanding lies in a knowledge of destiny. Whilst not unaware of the historical facts of British woodlands, they were preoccupied with ideas of natural succession and climax, and tended to interpret their observations in these terms. Great strides were made on this basis and many of the papers published before 1940 have lasting value, but since then there has been a shift in emphasis. Now many ecologists try to understand existing woods in terms of their history, relying on events which actually happened in the past, rather than developments which may or may not happen in the future.

This book about existing British woodlands and the conservation of their wild flora, fauna and other natural features, unequivocably adopts the historical approach. Partly, of course, this reflects personal temperament: I belong to that half of humanity which, on confronting something new, first asks how it came to be as it is. (The other half asks how it works.) I accept that woodlands can function in a purely mechanical sense and I derive greater interest and understanding from a knowledge of their history, the events which have taken place in and around them and their natural responses over as long a period as the record will allow. The historical approach, however, is also a particularly suitable basis for nature conservation. We are used to the idea that no wood in Britain is completely natural, that human actions in the past have in varying degrees determined their present characteristics, and conclude that nature conservation must include continuing control of the relationship between man and nature. If we know what impacts woods have sustained in the past and how they have responded, we have a sense of direction in our appreciation of each wood, we can confidently predict how woods will react in the future, and we can identify the relatively undisturbed descendants of primitive woodland, all without recourse to circular reasoning. 'Disaster theory' is an unduly melodramatic label for the perception of ecosystems as natural responses to past events, but it is far more relevant than the remote climax theory to those who manage woods and propose to determine the 'climax' for themselves.

This book is in four parts. Part 1 considers woodland origins, management history and ecological characteristics with an emphasis on the traditional management systems of coppice and wood pasture. Here we owe a considerable debt to Dr Oliver Rackham, whose *Trees and Woodlands in the British Landscape* (1976) not only demonstrates that ecological literature can be popular without being popularised, but also greatly advances our knowledge of that important group the ancient, semi-natural woodlands. Part 2 describes semi-natural woodlands in terms of a classification of stand types which attempts to take management history into account, as well as more conventional characteristics such as composition and soil. This classification is not entirely new, for it derives its inspiration from the early work of C.E. Moss, W.M. Rankin, E.J. Salisbury, A.G. Tansley and A.S. Watt, but it is based entirely on my own data collected since 1970 and it attempts to meet modern practical needs and limitations. Part 3 examines the objectives and priorities of nature conservation in woodlands and some aspects of recording and assessment. Finally, in Part 4, we consider the difficult task of reconciling the material and commercial drives behind timber growing with the restraints needed for effective conservation of wildlife and other natural features. Readers should therefore be warned that this is not a general textbook of woodland ecology, for it says little or nothing about nutrient cycling, trophic levels, primary production, or indeed the woodland fauna. Nor is it a forestry manual, though I hope that foresters will act on some of the points made in Part 4. On the other hand it strays occasionally beyond British woods into the European Continent in order to consider virgin forest stands, high forest management systems, the wider occurrence of British woodland types and the different requirements for nature conservation in at least some continental countries.

Several people have read parts of the text in draft and made numerous useful comments and suggestions — and in one case induced a major revision. These people include Eric Duffey, Margaret Game, Suzanne Goodfellow, Max Hooper, Joanna Martin, Susan Peterken, Ernest Pollard, Derek Ratcliffe, Francis Rose, John Sheail and Paul Stevens. They should not of course be held responsible for the use I have made of their help. Others, whilst having no direct connection with the book, have been major influences on my approach to woodland ecology and nature conservation, though not all will take the view that I have responded to their influence in the manner they would have wished: John Jeffers, Kenneth Mellanby, Duncan Poore, Oliver Rackham, Richard Steele, Colin Tubbs and John Workman. My colleagues in the Nature Conservancy Council, who possess a wonderfully wide range of attitudes to both forestry and nature conservation, have been a constant source of ideas, problems, sustaining enthusiasm and stimulating criticism. As a member of the Institute of Foresters I have enjoyed many valuable discussions with those whose primary purpose with woodlands is to grow timber as a commercial enterprise: I hope that these contacts have balanced my not uncritical comments on forestry and made realistic my approach to the integration of nature conservation and forestry.

I am grateful to those who have allowed me to use their photographs and illustrations. These are acknowledged in the text, but I would particularly like to thank Derek Ratcliffe and Peter Wakely for their photographs and Jan Čeřovský and Jaroslav Řehák for obtaining a copy of Joseph John's original records. Sandra Collinge helped me to prepare many of the figures.

Finally, I make the common but, as I now realise, no less sincere acknowledgement to my wife and sons for their forbearance whilst this book was being written.

George Peterken
April 1980

Note on terminology

Although the terms 'primary' and 'secondary' are commonly used in Britain to classify woods according to their origins, confusion could arise because similar terms have been used for similar, but significantly different, attributes of tropical rain forest. In the tropics 'primary forest' is virgin forest (Richards, 1952), whereas 'secondary forest', or second-growth forest, is that which grows up after primary forest has been destroyed by logging, cultivation, wind or fire. Thus, in the tropics the terms refer to the stand, but in Britain they refer to the site as woodland.

Despite the difficulties, I have retained 'primary woodland' and 'secondary woodland' for several reasons. The distinction is worth making. These terms are familiar and unlikely to be confused by most British readers. And no alternative labels come readily to mind. In practice, 'primary' and 'secondary' will usually be displaced by the more useful 'ancient' and 'recent secondary' (see Section 2.1).

Another difficulty affects 'management'. Foresters use this to encompass administration, supervision, staff matters, equipment, finance and any other matter which relates to forestry as a business. 'Treatment' refers to silvicultural operations, i.e. to a particular aspect of management (Hiley, 1954). Whilst this specialised application of words in common use could have been adopted here – in which case Chapter 9 would have been *Treatment Variants of Stand Types* – I felt that non-foresters might be at least as confused as foresters will be by my lax use of 'management' where 'treatment' would be more appropriate. In any case foresters themselves are not invariably particular in this regard.

The nomenclature of vascular plants follows Clapham, Tutin and Warburg (1962).

PART ONE

Origins, management and ecological characteristics of British woodlands

Chapter 1

Original natural woodland

Millenia of forest clearance, exploitation and management have apparently left Britain without any truly natural woodland. If, contrary to belief, some natural woodland does survive, it is likely to be small and to be confined to remote sites such as ravines, and so to be hardly representative of British woods as a whole. Furthermore, such woodland could not be proved to be natural, but only assumed to be so in the absence of contrary evidence. Nevertheless, the lost primaeval, natural woodlands which once covered most of Britain are not just a tantalising and fascinating subject for ecological historians, but are important as the sources of much of Britain's native fauna and flora, as the direct antecedents of some existing woodlands, and as the yardsticks – or controls in the scientific sense – which enable us to discover how much we have modified our natural environment.

The last wholly natural woodlands in Britain were the forests of the Atlantic period, which flourished before human influence grew and spread with the Neolithic cultivators. Their composition and structure can be reconstructed from the virgin forests which survive elsewhere in Europe, and from fossil remains preserved as pollen, leaves and trunks in the peaty deposits of bogs and fens.

1.1 Virgin forests in the north temperate zone

Virgin stands have survived in remote districts in Central and Eastern Europe, Asia and North America. Most lie in districts which were not settled and intensively exploited until recent times, and many were protected within huge hunting preserves. Few, if any, have entirely escaped some modification by selective felling, by community imbalances induced by the artificially high stock of game and low populations of predators in hunting grounds, by use as pasture and by use as sources of charcoal. Nevertheless, they provide valuable evidence on the structure and reproduction of natural woodland. This section is based mainly on a valuable review by Jones (1945).

Two of the many surviving virgin forests in Czechoslovakia exemplify the character of these stands and the manner of their survival. The Lanžhot forest (Fig. 1.1) is a 31 ha broadleaved stand on alluvium at the confluence of the Morava and Dyje in southern Moravia. It has survived first as a communal forest, then as an aristocrat's hunting preserve, and finally, since 1920, as a nature reserve. The largest trees are *Quercus robur* and *Fraxinus angustifolia*, some of which are over 450 years old, but other trees are present and form a variety of communities: *Acer campestre, Alnus glutinosa, Carpinus betulus, Populus alba, Salix alba, S. fragilis, Tilia cordata, T. platyphyllos* and *Ulmus carpinifolia*. The second forest, the Boubinský Prales, covers 666 ha, but only a core of 47 ha is truly virgin. Lying in the Šumava Hills of southern Bohemia, where permanent settlement did not start until the twelfth century (Rybníčková, 1974), it became a reserve as early as 1858 in order to preserve a

3

Fig. 1.1 Virgin forest near Lanžhot in southern Czechoslovakia, composed mainly of *Quercus robur*, *Tilia cordata*, *Fraxinus excelsior*, *F. angustifolia*, *Acer campestre*, *Carpinus betulus*, *Ulmus laevis*, *U. carpinifolia*, *Populus alba*, *P. tremula* and *P. nigra*, a combination which is not dissimilar to the underwood of East Anglian coppices. Although the intimate mixture of species and age classes shown here apparently developed only in the least disturbed natural forests, it is possible that stands like this were frequent in the primitive forest from which the mixed coppices of lowland England were derived. (Photo. R. Janda)

sample of the remaining untouched forest. Now it is a magnificent stand of *Picea abies*, *Fagus sylvatica* and *Abies alba* up to 58 m tall, with a good representation of younger age classes and much dead wood standing and lying. Since it became a reserve, the standing volume of timber has remained constant, but beech has increased at the expense of spruce and fir. In Poland a 4600 ha portion of Białoweiža Forest, consisting of almost undisturbed stands of *Acer platanoides*, *Carpinus betulus*, *Picea abies*, *Quercus robur*, *Tilia cordata* and *Ulmus glabra*, is protected from structural treatment (Pigott, 1975).

1.1.1 Structural characteristics

The dominant trees of virgin forests usually live for 300 years or more, with many growing to 500 years and only exceptional trees growing for much longer. Individual trees commonly grow to around 30 m in height, rising to over 40 m in favourable sites. They take the characteristic form of high forest stands, the small trunks rarely exceeding 1 m in diameter, and surmounted by narrow crowns.

Virgin stands are often envisaged to be like selection forests (Section 4.3), with an irregular canopy and a more-or-less even spread of age classes of all species throughout the stand, but this structure appears to be rare and to be confined to irregular sites and rigorous climates in mountain districts. In fact, most virgin stands contain some degree of even-agedness, ranging from completely even-aged stands to mixed-age stands in which one or two age classes are over-represented. One example quoted by Jones (1945) was a Bosnian beech forest where the age range was 200 years, but where no stems were less than 90 years old and 70% were between 170 and 200 years. Even-aged stands are quite common in northern coniferous forests, though in most instances there is a narrow spread of ages covering the time taken to close the canopy after the initiation of regeneration. Most mixed-age stands look more even-aged than they actually are, because trees spend only a small proportion of their life growing to full height.

Since age structure is largely determined by the periodicity of regeneration, it follows from the widespread occurrence of even-agedness that opportunities for successful regeneration occur irregularly in natural stands, depending on the death or weakening of dominant trees and simultaneous production of seed. Even-aged stands develop after natural catastrophes, principally fire, wind-throw and inundation. The subsequent stand may last for 300 years or more, but being even-aged it breaks up over a short period and a new, more-or-less even-aged stand again develops, prolonging the effect of one cataclysm into second and subsequent generations. Less catastrophic climatic fluctuations, such as prolonged drought, cause stress. Many trees consequently die in a short period, and a flush of regeneration follows, forming an even-aged element in a mixed stand. Even-aged components can clearly persist in a stand for several hundred years, time enough for even the rarest of catastrophes to strike again and renew the even-aged condition. Nevertheless, there are some areas, such as the White Mountains region of northern New Hampshire, USA, where catastrophes are so infrequent that a temperate broadleaved forest could achieve a steady state (Borman and Likens, 1979) or climax.

Even- and mixed-agedness is a matter of scale. In northern areas, which under natural conditions are particularly susceptible to fire and tempest, extensive even-aged stands develop. Over a large area the forest consists of a large-scale mosaic of even-aged stands, but the stands are of all ages and there is no net change in the general age structure of the forest. Even-agedness can also occur at the scale of an individual tree, the fall of which initiates a patch of even-aged regeneration. Even-agedness seems naturally to be well-defined and on a large scale in northern coniferous stands, but somewhat ill-defined and small scale in broadleaf temperate stands: by implication opportunities for regeneration are least irregular in space and time in broadleaf stands on mesic sites.

1.1.2 Tree species in virgin stands

Jones (1945) divided trees into tolerant and intolerant groups. The 'tolerant' genera – *Abies, Acer, Fagus, Tilia, Ulmus*, etc. – bear shade, and are the ultimate dominants on mesic sites with adequate rain. Dominant individuals are long lived and do not burn easily. In stands mainly composed of tolerant species, catastrophes are rare and on a small scale. Nevertheless, a measure of even-agedness exists, because, in addition to the residual effects of rare catastrophes, seed production does not always coincide with opportunities for regeneration. The 'intolerant' genera, such as *Betula, Castanea, Fraxinus, Larix, Pinus* and *Quercus*, do not bear shade well, but large quantities of seed are regularly produced and are well dispersed. They survive in undisturbed stands of tolerant species as itinerant colonists of canopy gaps. Some, notably oak, are long-lived, large trees which can remain dominant for centuries in undisturbed woodland following an opportunity for regeneration. Intolerant species remain permanently dominant at high altitudes and latitudes, on barren soils in temperate regions, and in warm, arid regions to the south, where catastrophes and climatic stress are more frequent. Being more susceptible to fire, forests of intolerant species were more easily set on fire by lightning and more easily cleared by early man.

Under stable climatic conditions, free from major disturbances, the composition of natural stands still seems to fluctuate in space and time. At any one site it might be changing as a result of the tendency of one species to regenerate better beneath another species than beneath its own canopy (alternation of species), or else one species might increase because good mast years happen to coincide with the death of many canopy trees, but the net change over a large area and a long period might remain zero, i.e. a dynamic stable state. However, major climatic changes have occurred every one or two thousands years, time enough for perhaps only three generations of the longer-lived trees, and as a result, species and communities have migrated across Europe and the stand composition at any given point has changed permanently. Whatever the theoretical ideal, natural forests have in fact had little chance of ever developing a completely stable state.

Jones (1945) summarised his view of natural woodland in the following terms: 'In a mixed stand there will almost certainly be species differing from each other in their natural span of life, the periodicity of their seed years, the exact conditions of site and climate needed for the establishment of regeneration, and other biological traits. Climate itself undergoes cyclical changes quite apart from any secular trends, and a generation of trees initiated by a favourable period in the cycle will persist through many succeeding cycles, and when conditions become ripe for regeneration once more it might well be for regeneration by a different species at a different phase in the cycle. It would not, then, be surprising if the entry of any one species in sufficient quantity to form an important constituent of a stand occurred only at distant intervals in time, as the result of some rare combination of circumstances. Thus a forest which is ever varying in the proportions of its species would arise. On a mountain side in central Europe beech might be dominant for a generation, to be followed in one place by a preponderance of silver fir, in another place perhaps by a preponderance of spruce. The many communities of varying specific composition, which can so often replace each other on the same site as the result of some slight disturbance of balance, may then have equal claim to be considered as part of the climax, and the climax itself would be a mosaic of communities, at least in time, and possibly in space also. This is the kind of picture which seems to be suggested, however vaguely, if the temperate forests which have been accepted as ''virgin'' are in fact the products of immense periods of growth undisturbed by catastrophes or by extraneous interference.'

1.2 Atlantic forests in Britain

The forests of the Atlantic period, which covered most of Britain between 7000 and 5000 years ago, were the last which could be described as wholly natural. We see in them 'for the first (and possibly last) time the major climax woodlands of the British Isles in stable equilibrium, their distributions determined by natural environmental controls' (Godwin, 1975). They had developed by a natural succession starting approximately 10 000 years ago when the remaining ice sheets decayed under a milder climate and the rapid development of closed woodland was permitted (Table 1.1). The most abundant species in the early stages were *Juniperus communis*, *Betula pendula*, *B. pubescens*, *Pinus sylvestris*, *Corylus avellana*, *Populus tremula*, *Sorbus aucuparia*, *S. rupicola* and *Salix* spp. Then, during the Boreal period 9000–7000 years

Table 1.1 Changes in the extent and composition of woodland in Britain during the Flandrian Period (Modified from Godwin, 1975).

Years before the present	Blytt and Sernander Periods	Forest cover
0–2700	Sub-Atlantic	Substantial clearance by Iron Age cultures onwards. Severe reduction in forest cover. Extensive cultivation and soil modification. Alder, birch, oak, beech woodland.
2700–5000	Sub-Boreal	Some forest clearance by Neolithic and Bronze Age cultures, but forest remaining widespread. Alder, oak, lime woodland.
5000–7500	Atlantic	Forest cover complete. Alder, oak, elm, lime woodland with pine, birch in north (see Fig. 1.1).
7500–9500	Boreal	Forest cover complete. Hazel, birch, pine woodland, becoming pine, hazel, elm, oak.
9500–10 500	Pre-Boreal	Some open vegetation remaining from the Weichselian period (last glaciation), but forest cover increasing and becoming almost complete. Birch, pine woodland.

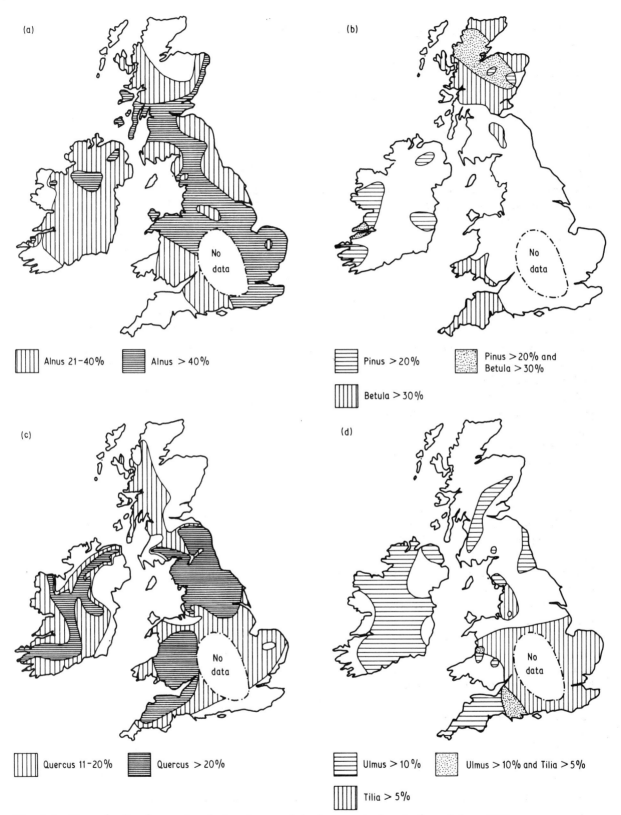

Fig. 1.2 Maps of pollen frequencies of selected tree and shrub genera in deposits formed about 5000 years ago. Modified from Birks *et al.* (1975). (a) *Alnus*, (b) *Betula* and *Pinus*, (c) *Quercus*, (d) *Tilia* and *Ulmus*.

ago, *Ulmus*, *Quercus* and *Alnus glutinosa* became more important in the south and west, whilst pine expanded into the birch forests of the north and west (Godwin, 1975).

The Atlantic period was the warmest since the last glaciation, with a mean summer temperature perhaps 2.5°C higher than it is now, and a longer growing season. Although the forests were fairly stable in composition, they were nevertheless adjusting to the change from the preceding, drier, cooler Boreal period. In particular, *Alnus*, *Tilia*, *Ilex*, *Ulmus* and *Quercus* expanded or became more frequent within the range already established, whilst *Pinus* and *Betula* generally declined, presumably in the face of increased competition.

The broad distribution of species in the Atlantic forests (Godwin, 1975; Birks *et al.*, 1975) was remarkably similar to that of existing semi-natural (Fig. 1.2) woodlands. The main regional contrast was between central and northern Scotland and the rest of Britain and Ireland. *Pinus*, *Betula* and *Alnus* were dominant in the former region, with *Pinus* concentrated into two areas in the Cairngorms and north-west Scotland. The extreme north of Scotland possessed woodland of *Betula*, *Alnus* and *Corylus* without *Pinus*. *Quercus* and *Ulmus* were present, particularly along the western seaboard where they probably formed outliers of the mixed broadleaf woodland further south. Elsewhere the so-called mixed oak forest prevailed, although, outside certain parts of Ireland, southern Scotland and Wales, *Quercus* was probably not dominant. *Ulmus*, *Tilia*, *Quercus*, *Corylus* and *Alnus* were the most abundant genera, with frequent *Betula*, *Ilex*, *Fraxinus*, *Pinus* and some shrubs, e.g. *Rhamnus*, *Viburnum* and *Crataegus*. *Quercus*, presumably both *Q. petraea* and *Q. robur*, was still expanding in the north and west where it achieved its highest frequencies. *Tilia*, most of which was *T. cordata*, was frequent in lowland England, mid-Wales and north-west England, where it must have been an important component of the canopy. *Ulmus*, presumably *U. glabra*, was infrequent in the English lowlands, and more frequent in western Britain. Likewise *Fraxinus* was most frequent in the carboniferous limestone regions of north-west and south-west England, but infrequent elsewhere. *Corylus* and *Ilex* were least frequent in south-east

England, where they may have been suppressed by a more vigorous canopy. *Betula* was most frequent in the north, west and the north-east seaboard of England. Pinewoods survived locally in all areas except south-west England, where this species was never abundant, even in the Boreal.

Local variation in composition, which was superimposed on the broad geographical variation, is difficult to establish from pollen analysis, but macroscopic remains provide some indications. Alternatively, one can reconstruct spatial variation in composition by assuming that the edaphic range of species has not changed, and test this by comparing pollen frequencies in different parts of a site. Thus Pigott and Pigott (1959) suggest that yew, which was undoubtedly present somewhere near Malham Tarn, was on the limestone cliffs. The mixed oak–elm–lime forests of Denmark (Iversen, 1960) were evidently dominated by lime and elm, which can tolerate more shade than oak, but only on soils that did not limit their growth. Soils undergoing podsolisation, on which lime does not thrive, were probably dominated by sessile oak. Pedunculate oak was most likely to be dominant on heavy, poorly drained soils.

The structure and function of Atlantic forests were presumably much the same as in the virgin forest remnants of recent times, but it is difficult to confirm this. Certainly the forests were dominated by shade-tolerant species, whilst shade-intolerant species seem to have remained as initial colonisers of canopy gaps and permanent dominants in the sites with more extreme soils and climates. Pollen profiles in mor humus deposits within woodland can show small-scale events. For example, at one point in an oak-dominated part of Draved Forest in Jutland (Iversen, 1964), an oak fell and was replaced temporarily by *Frangula*, *Salix* and *Pteridium*. Later, another oak died at the same point and was replaced by *Betula* and grasses before oak resumed its dominance.

The Atlantic forests were interrupted by coastal formations, lakes, large rivers, some inland cliff faces and developing peatlands. Here, those species which cannot stand the shade of closed woodland were able to survive the forest maximum. Land at and above the tree line provided extensive refuges in some areas. Within the main wooded lands, however, clearings must have been infrequent, for grassland species such as *Campanula rotundifolia*, *Centaurea nigra*, *Jasione montana*, *Poterium sanguisorba* and *Sanguisorba officinalis* disappeared temporarily from the pollen record. Godwin (1975) assumed that they persisted in lightly wooded areas until grassland again increased with the Neolithic clearances. Nevertheless, natural woodland must have contained some gaps which attracted herbivores in search of more plentiful pasture, and thereby remained open. Certainly Rose and James (1974) were led by their studies of epiphytic lichen communities in the New Forest to believe that natural woodland must have been fairly open.

1.3 Natural woodland since the Atlantic period

The Atlantic period ends with the 'elm decline', a sharp and widespread decrease of *Ulmus* in the pollen profiles which is believed to reflect the first major impact of man on the natural woodlands. Neolithic farmers cleared the woodland, cultivated the ground and depastured stock in the surviving forests, but their impact seems to have been uneven. Clearance was widespread as early as 3700 years ago in south-east England (Godwin, 1962), whereas woodland in much of the Highland Zone was only locally and temporarily cleared until 2500 years ago (Turner, 1965). Pioneer species such as *Betula* increased as a direct consequence of woodland disturbance by man. Other major vegetational developments were less obviously the result of disturbance, notably in southern Britain the expansion of *Carpinus* and *Fagus*, whose rise is so often associated with woodland disturbance that their spread may not be entirely due to natural processes. Rackham (1976) thought that the last approximation to virgin forest – his 'wildwood' – in England was probably in the Forest of Dean around 1150, whereas in Scotland fragments may have survived as late as 1830 in the pine forest of Glen More.

Although essentially natural woodland remained widespread in the Sub-Boreal and Sub-Atlantic periods, and fragments may have survived into historical times, its composition did not remain unchanged. In Draved Forest (Iversen, 1964) local natural changes in the post-Atlantic periods were detected by pollen profiles in mor humus. At one point a succession from lime–oak forest on grey–brown podzolic soil to a pure oak forest on podzol took place well before clearance caused temporary major changes. At a second site, beech partly replaced oak in a mixed forest of lime, oak, alder and beech on mull gley soil. A third site on low-lying ground originally supported oak–alder forest with hazel and lime, but first oak and lime were partly replaced by an increase in alder and ash, and then, as acid peat developed, alder was reduced, lime, oak and hazel were lost, and birch and *Frangula* increased. Changes in the herb pollen support these interpretations: for example, succession at the third site was accompanied by the loss of *Anemone* and the rise of an acidophilous flora of *Melampyrum*, *Dryopteris*, *Osmunda* and *Sphagnum*, which itself later declined. These are all examples of retrogressive succession within natural woodland.

Another cause of retrogressive succession is human interference. Even apparently natural successions have involved man as an initiator and accelerator of retrogressive changes. The lowland heaths appear to have formed after prehistoric man had cut and burned the woodland and temporarily cultivated the soil. The loss of soil fertility was equivalent to a much longer period of natural leaching (Dimbleby, 1962). The western blanket bogs were forming from the Neolithic age onwards, but in many cases the presence of man has been detected in the pollen profile at the point where bog formation started. In the Exmoor uplands, for example, the climate was deteriorating and prolonged leaching of soil nutrients had already taken place when prehistoric man started to clear, cultivate and graze within the woodland. Man's intervention may have tipped the balance towards blanket bog formation rather than continued occupation by natural woodland (Moore, 1974). Nevertheless, some retrogressive successions appear to be entirely natural. For example, in the north-west Highlands beside Loch Sionascaig, the first undoubted traces of man are recorded about 1500 years ago, but retrogressive succession in the pine–birch–alder woods could be detected 4500 years earlier, and blanket bog had replaced woodland over wide areas 4000 years ago (Pennington *et al.*, 1972).

1.4 A bog oak from the fens

Bog oaks preserved beneath the fenland peat are a

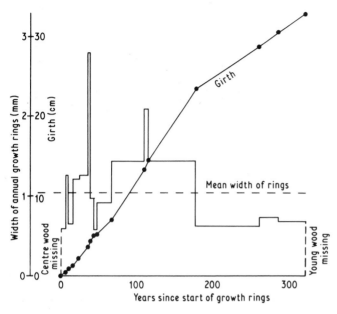

Fig. 1.3 Growth of a bog oak found in fen peat in Cambridgeshire. The long-term increase in girth was relatively steady, but marked short-term fluctuations in ring width (histogram) occurred during the tree's youth.

vivid, direct link with Britain's primaeval, natural woodland. One trunk, found in May 1976 on Eversden Farm, near Ramsey, Cambs, measured one metre in diameter at 6.5 metres from the roots and still contained adult and larval *Cerambyx* beetles entombed in their galleries. This oak grew 5000 years ago on silt by the edge of what later became the fens, apparently in undisturbed, mature woodland. (Its growth rate was very similar to that of modern oaks grown in densely stocked high forest in France.) Assuming that 20 growth rings were missing at the start of the sequence, its growth can be reconstructed in the following terms (Fig. 1.3). Until about 85 years the tree grew at a fluctuating but generally slow rate, shaded no doubt by the mature woodland canopy.

Short bursts of rapid growth were initiated abruptly at 27 and 56 years, presumably as a response to increased light admitted after the fall of a mature tree nearby. From 85 to 197 years growth was fairly rapid and remarkably even, save for a short burst of rapid increment around 130 years. Here, perhaps, we see a mature tree entering the canopy and maintaining its position in competition with neighbours. At 197 years there is an abrupt transition to the very slow, even growth of a mature, aging tree, which lasted until at least 340 years (after which the wood is decayed). Even though the tree was older than the great majority of oaks now seen in Britain, its trunk was still sound. Today it demonstrates the reality of natural woodland in Britain.

Chapter 2

Ancient woodland and traditional management

The primaeval forests were not all cleared away. The small fragments which survived into historical times mostly ceased to be 'waste' ground ripe for exploitation and clearance, and eventually they became valuable assets managed as coppice or wood-pasture. Despite continued clearance and changes in management, many of these woods – the direct descendants of primaeval forest – survive today.

2.1 Primary, secondary, ancient and recent woodland

Existing woods originated in two main ways. Some descended directly from patches of primaeval woodland which were never cleared away, whereas others came into existence on land which, though it may have been wooded in prehistoric times, was at some time completely clear of trees. This distinction has come to be described by the terms 'primary' and 'secondary' (Peterken, 1977a). Definition of these terms is based on the recognition of a period when the original, extensive, primaeval forest was fragmented by clearance into separate patches of woodland. Before this period the forest had covered the landscape, and any other habitats were islands within it, but once clearance had passed a critical point the surviving woods became islands in a sea of unwooded ground. Primary woodlands are those occupying sites which have been continuously wooded since that period, whereas secondary woodlands occupy sites which have not been continuously wooded. The terms refer only to the woodland in relation to specific sites: they say nothing about the type of woodland nor the age of trees.

This distinction has important implications. Secondary woods can and do range in age from recent scrub on an abandoned field to, say, a wood which formed in the Dark Ages and has survived ever since. Correspondingly, primary woods vary in the time which has elapsed since they became habitat islands: clearance and therefore isolation of primary woods occurred much earlier in the southern chalklands than in the clay plateaux of East Anglia. Other regions have never been extensively cleared (e.g. parts of the Weald and the sides of the lower Wye Valley and Dean Forest), and here the distinction between primary and secondary woodland barely applies.

Although the distinction between primary and secondary woodland appears to be sharp and unambiguous, several factors tend to reduce the differences. A small clearing might be cut in a primary wood and then allowed to revert to woodland, e.g. an old ride. This woodland is technically secondary, but it is not isolated and it quickly acquires the appropriate species from the primary woodland all around, eventually becoming undistinguishable from it if the soil has not been disturbed. Clearance itself is sometimes incomplete,

11

for example when a scatter of trees survives in the newly-formed pasture. (Clear-felling does not constitute clearance if the woodland canopy is immediately allowed to grow again without any intervening period of pasturage or cultivation.) Secondary woods originating beside a primary wood can be colonised rapidly by woodland species, but an isolated secondary wood separated by many kilometres from the nearest pre-existing wood will take much longer to be colonised (Section 6.2). The land use following clearance and preceding restoration of (secondary) woodland may be completely hostile to woodland species (e.g. arable) or permit many to survive (e.g. rough pasture, limestone pavement). Thus, the primary wood (which is characterised by minimal soil disturbance and minimal ecological isolation from primaeval woodland) is linked to the isolated secondary wood on former arable land (which has maximum soil disturbance and maximum ecological isolation) by a continuous range of circumstances in which disturbance or isolation (or both) is not so extreme. In fact, the distinction between primary and secondary woodland expresses not a discrete difference but a continuous variable of ecological isolation and disturbance. Correspondingly, the value of distinguishing secondary from primary woodland decreases in those regions where large amounts of primary woodland have survived, for here the secondary woodland is hardly isolated at all.

Primary woodland is a useful hypothetical concept, not an identifiable state (Section 13.2), so it is desirable to employ an alternative concept for practical purposes. A distinction is therefore drawn between 'Ancient' woodland and 'Recent' woodland, defined according to their origin before and after a threshold date, respectively. The most convenient date for this threshold is about the year 1600 (thus the terms are virtually synonymous with 'Medieval' and 'Post-Medieval' woodland), because maps first become available and plantation forestry becomes important from about that time. All primary woods are therefore ancient, but ancient woods include both primary and ancient secondary woods in unknown proportions. Despite its hybrid historical character, ancient woodland has some ecological reality, first because in many areas the great majority of ancient

woods are believed to be primary, and second because the characteristics of ancient secondary woods are likely to resemble those of primary woods. There are several reasons why the latter should be so: (i) ancient secondary woods have had a long time to acquire species and form stable communities; (ii) they formed when primary woods and other semi-natural communities were more abundant than they subsequently became, so their ecological isolation was low; (iii) their soils were not disturbed by modern agriculture and their structure has had several centuries to mature; and (iv) both primary and ancient secondary woods have been managed under the same traditional systems. Thus, whilst it is not strictly correct to use the terms 'primary', 'ancient' and 'medieval' interchangeably, there is some validity and utility in doing so.

2.2 Common woods and the origins of wood pasture

2.2.1 Wood pasture management

Wood pasture is woodland which is permanently available as pasture, or pasture with trees. It developed out of the pre-historic practice of depasturing cattle within the natural woodlands, and survived into historical times on commons and in forests, parks and chases. This long-term maintenance of both trees and pasture on the same ground presented obvious problems: too many trees reduced the herbage, but sustained grazing eventually eliminated the trees. The balance was struck with the help of pollarding, the practice of polling or beheading trees some 2–3 m above the ground. Pollarding, in effect coppicing beyond the reach of deer and other animals, is said to prolong the life of individual trees, and certainly many surviving pollards are 300–500 years old or more. Regeneration from seed was correspondingly less important for maintaining the wooded cover, and wood pasture could remain wooded for centuries without any natural regeneration taking place. Temporary reduction in grazing pressure, even at long intervals, would have been enough to initiate a new generation of trees.

2.2.2 Wood pasture on commons

Woodland, like other kinds of uncultivated ground, was used as pasture in pre-historic times, whilst the need for timber was satisfied by clearance and raids upon the wooded waste. In order to control this exploitation, common rights evolved in 'immemorial antiquity' (Hoskins, 1968). The common rights defined, limited and often rationed or 'stinted' those to whom the commons were available. In some cases the common woods were very extensive and were available to communities over a wide area: Andred's Weald was common to the whole of Kent, and Sherwood (= shire wood) Forest was probably available to all the commoners of Nottinghamshire. Saxon and early Medieval sources indicate that the woodland on the waste was exploited in various ways. For example, around 866 in the common wood of Wolverley (Worcs), 70 swine were allowed to forage, and each year one oak and five wainloads of good brushwood as well as other timber for building and firewood were taken (Finberg, 1972). The oak, beech and ash woods of the Kentish Weald were used for timber, pasture, and as a source of mast for pigs under various rights (with evocative names like estovers or pannage) the usage of which is described in detail by Neilson (1928). The common woods of early historical times remained contiguous with the common grassland and heathland which developed where grazing and exploitation of the trees had been more intensive.

By the later Middle Ages the common woods were enclosed as coppice, incorporated within parks, forests and chases, or mostly devoid of all but scattered large trees and scrub. Many of those which survived as commons were later enclosed, though not all were then cleared of trees and a few remain now in modified form as private woods (Rackham, 1976). By the end of the 18th century, when the Chairmen of General Quarter Sessions of each county were asked 'Are there any commons or commonable woods of considerable extent, and (if so) is the quantity of timber on them considerable?', most answered that there were few or none outside the forests, and that such woodland as remained was poor in timber (Commons Journal, 1792).

Wood pasture survived on commons much later in some places. In the Chilterns, commoners had rights to fuelwood, building timber, herbage and pannage, and the many large common woods 'were generally more open in character than those on private estates – often little or no underwood remained in them and herbage was more abundant' (Roden, 1968). Many were reduced by enclosure after 1550 and the remainder degenerated to heath with a few old trees and patches of scrub. Many are again well wooded as a result of recent natural succession, but very few can have borne woodland continuously: Naphill Common (Bucks) may have done so. In the early Middle Ages the Weald, too, had many wooded commons (Brandon, 1963, 1969 and 1974) which by the 16th century were reduced to open woodland and scrub: the common at Westmeston was 'dispersedly set with oaks and bushes', for example. The demands of ironworks, glass making, and rabbit warrens intensified the continued effects of grazing, fern cutting and heath burning, so that by 1700 most of the remaining woodland on the commons had been wasted and spoiled. Some surviving commons such as the Mens (Tittensor, 1978) and Ebernoe Common have nevertheless been continuously wooded, although the trees dating from the wood-pasture period are now thinly scattered amongst more recent growth.

2.3 Forests, chases and deer parks

Many pasture woods were incorporated within deer parks, forests and chases, where their chances of survival into modern times proved to be better than on commons. Common rights often continued in these places, but the crown and other landowners had a greater incentive to preserve the wooded cover.

2.3.1 Wood pasture in deer parks

The deer park (Fig. 18.5) was essentially a medieval phenomenon which survived in a modified form into modern times. It was originally part of the demesne lands of the Lord of the Manor, for whom it provided both a hunting ground and a source of fresh meat. Typically, it was stocked with red, fallow or roe deer, which were securely enclosed within a substantial earthen bank surmounted by a hedge, paling fence or wall. Some parks contained other semi-wild animals such as hares, swine and white cattle, and others contained domestic stock: a joist taken in 1595 at Cliffe Park (Northants) for example, revealed 258 cows, bullocks, steers and heifers, 101 young beasts and yearling calves, and 101 horses, mares and colts, as well as the deer (B.M. Add. Ms. 34214*). Parks differed from forests and chases by being enclosed and generally smaller, most extending to a few hundred acres or less. They differed too from later landscape and amenity parks, which were created long after the medieval deer parks had declined. Their characteristics have recently been reviewed by Cantor and Hatherly (1979).

Deer parks originated before 1086, when Domesday Book recorded 35 parks scattered through the southern half of Britain (Darby, 1977). The main surge of emparking came in the early Middle Ages in order to satisfy the feudal aristocracy's enthusiasm for hunting and taste for meat, but from about the 15th century onwards more were disparked than newly enclosed. Sussex, for example, had four parks in 1086 and reached a peak of 111 in the Middle Ages (Brandon, 1963 and 1974). Most were disparked in the 16th and 17th centuries or converted into landscape parks, so by the late 19th century only 20 remained as deer parks (Shirley, 1867; Whittaker, 1892). Scotland too had many medieval deer parks; the modern extensive deer forests date only from the early 19th century (Anderson, 1967).

Parks often contained coppices, arable, grassland, and heathland as well as wood pasture (Rackham, 1976; Chapter 8). Sussex parks were mostly enclosed directly from wooded and heathy waste. Such land was probably preferred if it was available, but the supply was very limited in some areas and in later centuries, so enclosed land was also emparked, e.g. Chideock Park (Dorset), which seems to have been little more than a group of fields designated as a park (Cantor and Wilson, 1969).

Wood pasture was therefore common in parks, but not invariably present. The tree cover was partly maintained by allowing established trees to grow to their natural span, prolonged perhaps by pollarding. Very few of the oldest trees in surviving parks are much over 400 years, e.g. Staverton Park (Peterken, 1969a), so some regeneration must also have taken place. Planting, though feasible, was probably not necessary, for many parks were subdivided and deer could be excluded from certain areas for long enough to start a new generation naturally.

Roger Taverner in 1565 (PRO LRRO† 5/39) gives some excellent descriptions of wood pasture in parks. Meere Park (Wilts) had 'old oaks whereof 30 are timber, the rest ruinous and shells the number 600'. The 'divers oaks, ashes and hornbeams' in Little Park (Essex) 'commonly used to be shredd for browse for the deer and there be also growing dispersedly divers other oaks being timber and meet for pole and rail'. In Oakley Park (Shropshire) wood pasture had been formed from already enclosed land: 'in the said park is neither woods nor underwoods except for a few alders, salye and thorn growing dispersedly and certain oaks and ashes growing upon the banks of the ditches which were the partitions of closes before the said ground was imparked of which the oaks and ashes the number of 80 were never topped or shredd, but the rest have been of old time lopped and topped and the lopps and tops thereof yearly to be taken will scarcely suffice for browse for the deer, and the said browsewood will scarcely suffice for the necessary firewood of the keeper there'. The Duchy of Lancaster (DL) Survey, 1587 (PRO DL 42 114), records that Elterwater Park, Furness, was 'for the moste parte Scarres, hilles and craggie rocks' but 400 acres were 'Sclenderlie besett with hollyn, asshes eller and birtche of an old groth . . . 200 acres of small saplinges (oak) . . . and 100 doted oaks.'

Parks provided better circumstances than commons for the survival of wood pasture, but few now remain close to their medieval condition: one example is Staverton Park (Suffolk) (Peterken, 1969), but even here the wood pasture has been reduced and modified. Some are now heathland with scattered trees, e.g. Kentchurch Park (Hereford); many have been incorporated within or converted to landscape parks, e.g. Blenheim Park (Oxon); but most have now

---

* British Museum Additional Manuscript.

† Public Record Office, Land Revenue Record Office.

simply become ordinary farmland. Moccas Park (Hereford) has survived reasonably intact, but even here numerous exotic species were planted from the 18th century onwards in a partial conversion to a landscape park. And like so many parks, Moccas has been tidied: it is now a carefully managed pasture with trees, but in 1876 Francis Kilvert the diarist recorded that 'we came . . . slipping, tearing and sliding through oak and birch and fallow wood of which there seemed to be underfoot an accumulation of several feet, the gathering ruin and decay probably of centuries' (Plomer, 1964). One of the few parks in which pollarding continues is Mersham Hatch Park (Kent).

Many landscape parks, associated with the names of Capability Brown, Humphrey Repton and others, were created *de novo* on agricultural land (Prince, 1958), and can be regarded ecologically as recent, secondary wood pasture.

2.3.2 Wood pasture in forests and chases

The medieval forests also provided favourable circumstances for the survival of wood pasture. These were areas subject to Forest Law (Turner, 1901), the ostensible purpose of which was to preserve the 'beasts of the forest' (hart, hind, hare, boar and wolf) and the 'beasts of the chase' (buck, doe, fox, marten and roe) for the King to hunt. Chases were private forests where the hunting rights were held by a subject but they were often known as forests and treated in a similar fashion. Both the court and the aristocracy fed well on the venison and made good profits. The social, legal and administrative history of individual forests have attracted many writers, and the history of many forests is summarised in the Victoria County Histories (a series of volumes published by Oxford University Press on behalf of the London University Institute of Historical Research).

The rise and fall of the area under Forest Law closely follows that of medieval deer parks. Some areas had been designated as forests by 1086 (Darby, 1977), but the main phase of afforestation followed in the 12th century, by the end of which Forest Law was applied to one-third of England. Bazeley (1921)

reveals the extent of 66 royal forests in the 13th century and notes that over 70 private 'forests' were recorded in addition. During the 13th century Forest Law was forced to retreat, many forests were completely disafforested, others were reduced, and thereafter Forest Law was progressively limited and diluted. Many forests survived into more recent times only to be disafforested under the influence of the enclosure movement, e.g. Rockingham Forest about 1796 and Wychwood Forest in 1857. The survivors were managed by or on behalf of the Office of Woods until 1923, when they were formally transferred to the Forestry Commission (Ryle, 1969).

Although the Crown was initially more concerned with the deer, it was also concerned from at least the Assize of the Forest, 1184, to preserve the woodlands and control the amount and duration of pasturage and pannage (Bagley and Rowley, 1966). This is understandable, for the Royal forests were an important source of timber, much of which was sold outside the forests (Donkin, 1960; Hart, 1966). Management mostly consisted of exploitive felling and a prayer that natural regeneration would follow. Not all woodland was wood pasture, however, even in forests such as New and Dean, because some woods were enclosed as coppice and, in some forests such as Rockingham, the coppices were more extensive than the pasture woods. Moves to improve the growth and supply of timber from the 17th century onwards resulted in the enclosure of considerable tracts of wood pasture and their conversion by planting and sowing to broadleaf high forest. Alice Holt Forest (Hants) exemplifies this transformation. In 1565, according to Taverner, the 'maine wood . . . (was) set with oaks of very great age and with thorne, maple and some beech beside the waste plots', but in 1812 the commoners were bought out and between 1815 and 1825 the whole forest was planted with oak. As in Salcey Forest (Section 4.1), whose preceding management was mostly coppice, the oak plantations have been gradually replaced by conifers in the present century.

The unenclosed woods of the New Forest (Hants) (Fig. 2.1), where the Verderers' Court still preserves a vestige of the medieval administration, remain the best surviving examples of wood pasture in Britain.

Fig. 2.1 Mark Ash Wood, New Forest, Hants, an unenclosed beech–oak wood. The pollard beech to the right (probably over 250 years old) and the maiden oak to the left have both grown in relatively open conditions. The central group of mid 19th Century beech have a slim, narrow-crowned form, and must have regenerated naturally into a gap in the older canopy: their shade has evidently weakened and killed the lower branches of the older trees. In the background an old tree has fallen leaving a gap which will be filled by beech and oak saplings if the holly and other scrub affords enough protection from deer. This is not a natural stand, but its structure has been less determined by man than almost any other woodland type in Britain. (Photo. P. Wakely, Nature Conservancy Council)

Some existing unenclosed pasture woods, such as Ridley Wood, were enclosed and coppiced in the Middle Ages (Tubbs, 1964). Conversely some of the medieval pasture woods have since been enclosed, though only a few of these remain virtually unaltered by planting within the enclosures, e.g. Burley Old Inclosure. The many pasture woods which remained unenclosed developed as the result of a series of natural reactions to changes in management, i.e. to fluctuations in the intensity of grazing, heath burning, turf cutting, timber extraction and planting (Peterken and Tubbs, 1965; Tubbs, 1968; Flower, 1977). The oldest generation of beech, oak and holly dates mainly from 1620–1760, though a few giant oaks survive from earlier periods. The preponderance of beech in some woods is due to selective felling of oak and a limited amount of planting and sowing. After 1760 those woods were not managed, deer populations were at record levels and regeneration was rare. The woods maintained an open, mature canopy of large trees with spreading crowns, many of which had formerly been pollarded. Effective implementation of the Deer Removal Act of 1851 (described in Tubbs, 1968) enabled natural regeneration to proceed apace during the second half of the 19th century. This filled the canopy gaps in the old woodlands with slim, narrow-crowned trees, allowed a dense understorey of holly to develop in some woods, and permitted scrub and emergent woodland to expand on to the heaths. After 1900, regeneration was again slow because grazing pressures rose once more, the gaps in the woodland canopy had been closed, and the spread of scrub over the heaths was checked by increased burning. For a short period during World War II, grazing from commonable stock was greatly reduced and regeneration was again prolific in the canopy gaps recently created by the fall of old beeches and oaks, and on unburnt stretches of heath. Since then grazing pressure has climbed to levels not exceeded since 1850, and regeneration is correspondingly slow.

Surviving stands of unenclosed wood pasture in other forests are smaller and less impressive than those of the New Forest, though their generation structure is often similar. Some, in Sherwood Forest (Notts) and St Leonards Forest (Sussex), have recently been destroyed in the relentless urge to plant conifers, and most of the remainder, like the New Forest, are under damaging pressure from plantation forestry and recreational activities. Epping Forest (Essex) is still an extensive relic of former wood pasture management (Rackham, 1978) thanks to the 19th century efforts of the Commons and Footpaths Preservation Society, who also took an active role in the New Forest (Eversley, 1910). Windsor Forest is large but the wood pasture relic on High Standing Hill (Berks) is small. A small stand of mature beech, oak and holly, which resembled some of the finest New Forest stands, survives at Speech House in the Forest of Dean (Glos), but many of the old trees have been felled recently for fear that branches would fall on

picnickers. The relics of wood pasture in Sherwood Forest (Hopkinson, 1927) are overrun by tourists, attracted *inter alia* by the fame of the Major Oak. The numerous old oaks there may be the survivors of a continuously regenerating medieval population which was 'past maturity' in 1698 and 'fit for cordwood and firewood' in 1790.

2.3.3 Other wood pasture

Holly was once cultivated as a source of winter feed for livestock (Radley, 1961): groups of hollies, often known as 'Hollins', stood in pasture and were pollarded at intervals. The practice was known almost throughout the natural range of holly in England and Wales, but it appeared to have been most frequent in the Pennines, Cumbria and the Welsh Borderland. It originated before the 13th century and was abandoned by the 18th century. The scale of the practice could be huge: when Needwood, an area whose holly and oak foliage had long been used to feed deer (Birrell, 1962), was disafforested in 1802 and enclosed, 148 170 hollies were felled to make bobbins for the Lancashire Cotton Mills (Nicholls, 1972). Hollins survive today on the Stipperstones (Salop) in the Olchon Valley (Hereford), in Needwood Forest (Staffs), the Peak District and the New Forest. Hubberholme Wood in Upper Wharfedale (Yorks) appears to be a hollin modified by invasion by other species.

Since wood pasture was fundamentally a system for reconciling the existence of trees and grazing animals on the same ground, it is legitimate to regard hedges and hedgerow trees and surrounding pasture fields as a form of wood pasture. Hedgerow trees were generally pollarded, and their density often exceeded that of trees in forest wood pasture (Rackham, 1976, 1977b). Like all other forms of wood pasture, hedgerow trees are now declining rapidly.

2.4 Coppice management

The significance of coppice management for ecologists can hardly be exaggerated. Most ancient woods have been managed as coppice for most of the last one thousand years, leaving a legacy of semi-natural woodlands upon which our knowledge of the natural climax vegetation of Britain is founded. And yet, the coppice system hardly rates a mention in modern forestry literature, and the surviving coppice woods are mostly dismissed as 'scrub' by foresters.

2.4.1 The coppice system

The fundamental feature of the coppice system is the method of restocking: each coppice stand grows mainly from shoots which spring from the cut stumps of the previous stand. It yields three main products, namely sticks and brushwood from the underwood (commonly known as coppice), timber from standard trees scattered amongst the underwood, and pasture from the herbaceous field layer beneath the underwood and the grassy rides and clearings.

In theory, the system could be highly systematic. The underwood was cut on a 'y' year rotation (usually 5–25 years) whereby one yth of a wood was cut each year when it had grown for y years. The parcel which was cut in any one year was named a coupe, sale, fell, cant or burrow depending on the region. Standards were grown on a multiple of the underwood rotation, i.e. 'ny' years (usually less than 100 years). One nth of the standards in a coupe where felled at the same time as the underwood was cut, and an equal or greater number of saplings was reserved as 'wavers' and allowed to grow into standard trees. Grazing could not be allowed in freshly cut coupes, but after 'x' years (usually 4–7) the underwood would have grown too tall to be seriously damaged by deer, etc., and grazing animals could be admitted. Thus, any particular part of a wood would alternately be ungrazed for x years and grazed for $(y - x)$ years, and grazing would always be available on one $(y - x)$ x^{-1}th of the wooded area in addition to the pasture which was permanently available on any rides. In spring, just after cutting, a large coppice would contain underwood of all ages from freshly cut to $(y - 1)$ years growth. Standard trees would range from y years to ny years, all age classes being present throughout the wood. In this way the material yield from a coppice and its overall condition could remain constant, whilst the particular distribution of age classes in the underwood changed annually on a pattern which recurred every y years. Coppice management could be more complex than this, for example when the age structure of the standards was manipulated to yield timber of various ages and sizes. On the other hand it could also be simpler. Grazing was often omitted, at least in recent centuries, and standard trees might also be omitted. The latter system is known as 'simple coppice', whereas the system in which timber trees are also grown is known as 'coppice-with-standards'.

In practice, coppice management rarely achieved this theoretical steady state. Any mismanagement could take up to ny years to rectify and any changes in demand which required a change in the rotation of either underwood or standards would also take years to complete. Although mismanagement was common, change in demand was the more important factor, for one of the important features of coppice management was its flexibility. It was not usually required to be systematic, but to provide a diverse resource whose management could be adjusted in response to almost annually changing demands for timber, wood and pasture, and any changes in the labour availability or the personal circumstances of the owner. The example of Gretton Woods, Northants (Fig. 2.2) shows the degree to which a coppice system with pasturage might depart from the theoretical steady state.

Coppices were an integral part of the traditional agricultural economy. They mainly supplied local needs but often produced a surplus for sale to distant communities. For example, the buyers of wood at the 18th and 19th century sales in Newball Wood (Lincs) came almost entirely from local villages (Nottingham University Library Ma ZA 162-167, unpublished) and the wood from the Northamptonshire Forests was sold within 15 miles of its source (Pettit, 1968), but the timber required to build Nonsuch Palace was drawn from woods all over Surrey (Dent, 1962). Much of the produce satisfied bulk needs, notably fuelwood. The lop and top from timber trees and the brushwood from the underwood were bundled into faggots or fascines and used to heat ovens, strengthen the surface of roads, drain fields (buried in deep trenches) and make dead hedges along boundaries which were not permanently fenced. (These included the internal boundaries of coppices in which pasturage was utilised.) A considerable volume of underwood material was converted by skilled craftsmen into a variety of structures, tools and utensils for

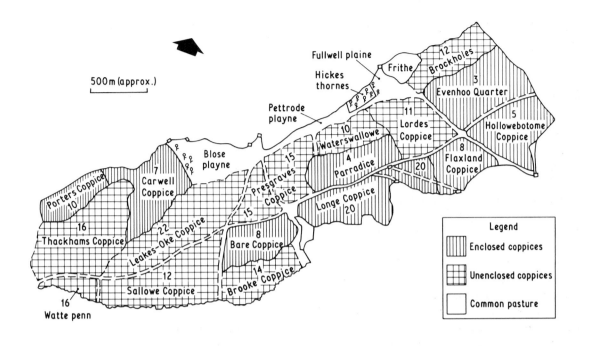

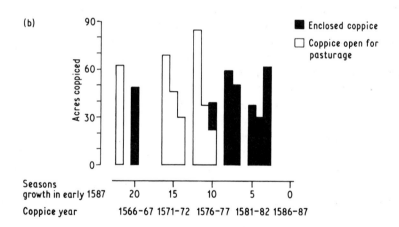

Fig. 2.2 Gretton Woods (Northants) in 1587. Redrawn from the Treswell survey in the Finch–Hatton Manuscripts (Northampton Record Office NRO FH 272 f.7). (a) Map of coppices and unwooded ground, (b) area of different ages of coppice growth in 1587.

The wood consisted of 17 named coppices, grass-covered tracks and rides, almost treeless plains, and Hickes Thornes, apparently a recent intrusion on to the plain. In 1587 the coppices had grown for between 3 and 22 years since they were last cut. Deer had access to the plains, rides and nine coppices: solid lines indicate stock-proof boundaries, whereas broken lines indicate boundaries which were temporarily open to commonable stock and deer. Although designed for a 16-year coppice rotation with deer excluded for 7 years after cutting, the actual practice in Gretton Woods was rather less regular.

domestic, agricultural and industrial use (Fitz-randolph and Hay, 1926; Edlin, 1949; Woods, 1949). These included bowls, chairs, baskets, besoms, spoons and clogs mainly for use in the home; wattle and gate hurdles, rakes, tool handles and various fencing materials mainly for agricultural use; and crates, hoops and other items used in factories. Stakes and wattle were also used in buildings, where they formed part of the panels between the main members of timber-framed buildings. The standards supplied the timber for the frames themselves. Some villages became centres for the coppice crafts, such as Kings Cliffe (Northants), Thatcham (Berks), Kings Somborne (Hants) and Hawkshead (Cumbria). To take one example, the underwood products of Newball Wood (Lincs), in 1868/69 were 806 score of kids (i.e. faggots), 338⅔ hundred of stakes and binders (for hedging), £10 worth of poles, 19½ hundred bean poles, 18⅓ hundred of besom shafts, 10 hundred trevors, a 38 foot length of birch, and some pea rods. In other years there were sales of clothes props, bobbin wood, crate rods, leading, 840 lime poles, stack props, fruit stakes and 'sundries'.

Several detailed historical accounts of individual coppice woods and groups of woods have been published, many in journals and books of local interest. Examples include the Lake District Woods (Fell, 1908; Kipling, 1974), Loch Lomondside (Tittensor, 1970), Leigh Woods near Bristol (Way, 1913), Hayley Woods, Cambs (Rackham, 1975), Monks Wood, Cambs (Hooper, 1973) and Bedford Purlieus, Cambs (Rixon, 1975).

2.4.2 Underwood management

At its simplest, coppicing involves cutting all the woody growth in a compartment, then allowing new shoots to spring from the stumps to form, with any seedling growth, the next underwood crop (Fig. 18.6). In a coppice-with-standards system a proportion of the timber trees is retained into the next rotation and some saplings are spared to grow into standards. Repeated coppicing of individual trees generates a stool, the permanent woody base from which coppice shoots arise.

Coppice stools are mostly low or subterranean, for most underwood was cut close to the ground.

Exceptionally, the underwood was cut much higher so high coppice stools and low-cut pollards were able to develop. Some of the lime coppices on the gorge of the river Wye near Chepstow have been cut at all heights from 3 m downwards, generating a remarkable mixture of pollards, high-cut coppice and low-cut coppice. Coppice stools cut as high as one metre are common in the coppices of heavy soils in the English Midlands and East Anglia, where ash especially but also lime, wych elm, oak, beech and English elm have all been treated in this way. Coppice shoots from low-cut stools are said to be more vigorous and there is less trouble from rotten stools. The advantages of high-cutting are hard to imagine; some high-cut stools marked compartment boundaries in woods which were otherwise cut low. Perhaps they conferred some advantage in the face of rabbits (the high-cut stools at Swanton Novers Great Wood, Norfolk were promoted to this height in the 19th century) or ground frosts, but this hardly explains why some stands were cut at a variety of heights. Individual stools can live for many hundreds of years, growing in this time to a massive size and developing often into a ring of separate stools as they expand and the core slowly decays. Many of the larger coppice stools in surviving coppice woods certainly originated in the Middle Ages, and might have been almost immortal if coppicing had continued unchanged.

Underwood rotations were generally of 5–30 years. In the Middle Ages some East Anglian coppices were cut every three years. On the other hand some rotations reached 40–50 years, as in the 18th century 'springwoods' of Yorkshire (Marshall, 1788), and could be even longer: coppices were sometimes neglected for a century or more as in Rockingham Forest (Peterken, 1976). Underwood was generally cut when it reached usable or marketable size, but many factors interacted to determine the particular rotation in any wood. Growth was faster and thus rotations were shorter in southern Britain and on fertile sites. Specialised markets were a factor. For example, wattle hurdles (Fig. 18.2) can be made from hazel on a 10-year rotation or thereabouts, but not from older hazel, because at that age the thin, weak rods, which are used to bind in the split rods, are generally dead. In Derbyshire, rotations had to be

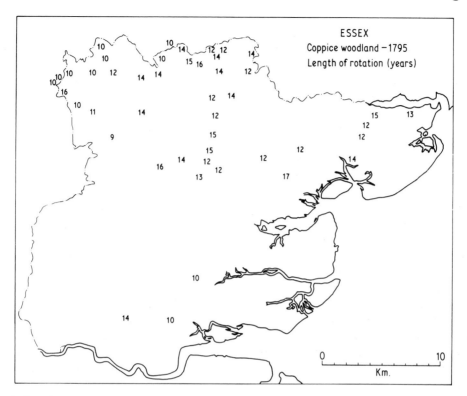

Fig. 2.3 The length in years of coppice rotations in Essex woodlands about 1795 (from Vancouver, 1795). Although most woods were cut on rotations of 10–14 years, there were slight regional variations.

about 25 years to provide puncheons or pit props (Fairey, 1815). Rotations were influenced by the terms of a lease, such that a coppice leased for 21 years might be cut twice on a 10–11-year rotation, even if this was premature (Jones, 1961). Customary rotations were evidently created in some woods and more-or-less fixed by construction of permanent (bank, ditch and hedge) subdivisions into compartments, e.g. Salcey Forest (Northants and Bucks): these rotations must have been inflexible. In any area at any time there was limited variation in rotations, as in Essex in the late 18th century (Vancouver, 1795; see Fig. 2.3).

The system of coppicing was sometimes fairly complex. The species in mixed coppice could be managed on different rotations, as in Cambridgeshire, where some of the ash and maple was cut on a longer rotation than the hazel and sallow (Rackham, 1975). Marshall (1788) describes the 'springwoods' on steep valley sides in Yorkshire. These were managed on a

long, 40–50 year simple coppice system from which thinnings were taken at intervals of 10 years. In parts of Scotland the oak coppices were thinned in mid-rotation (Tittensor, 1970). Selection coppicing ('furetage') of beech was practiced in France and may have occurred in southern England: only a proportion of shoots were cut at any one time from each stool, which therefore bore a mixed-age population of shoots, a practice which helped the weakly coppicing beech to survive as coppice.

Traditionally, the composition of the underwood was not directly influenced by the woodman, who accepted 'the natural growth of the soil'. In Huntingdonshire, for example (Stone, 1793), in the late 18th century the underwood '(was) not carefully selected and planted; the production of it, both in quantity and quality, (was) for the most part left to chance'. The actual mixture with which he was presented was not always ideal, but it could be 'improved' by removing unwanted species or encouraging valuable species. Cleaning – the removal of un-wanted species – was probably commonplace, at least in so far as invasive trees (e.g. birch) and shrubs were controlled in order to maintain the stock of hazel and other coppice species. Unwanted species were removed during the mid-rotation thinning of the oak coppices around Loch Lomond (Tittensor, 1970) and hazel was until recently still cut out as a weed from oak coppices in Cornwall.

Valuable coppice species were encouraged by layering, planting and natural regeneration in gaps created when old stools died or were grubbed out. Around 1800 these practices varied enormously, even within counties (Jones, 1961). In some counties no effort was apparently made to replace worn out stools and in some woods it was actually forbidden under the terms of a lease. Elsewhere old stools were replaced by planting and layering. Plashing – the practice of filling blanks in coppice by layering – was apparently common in Surrey (Stevenson, 1813), where sallow, ash, alder and willow responded well but oak, chestnut, maple and other species did not. Plashing was still common in Surrey, Hampshire, Sussex and Kent in the early years of this century (Nisbet, 1905).

The desirable species had to be present throughout a wood if cleaning and plashing were to succeed. Where they were not, the only feasible method of improvement was planting. Underwood planting appears to have been commonplace only in the improvement era from the late 18th to the late 19th centuries, when many western woods were planted with oak; hazel was planted in south and south-east England; chestnut was planted in south-east and south-west England; and numerous small 'copses' (secondary woodlands) were planted in the newly enclosed cultivated landscape (Chapter 5).

2.4.3 Management of standards

The timber trees of coppice woods were grown on a longer rotation than the underwood and yielded larger timber, which was mostly used in buildings. They competed with the underwood for light, water and nutrients, so timber production tended to be inversely proportional to underwood production and the density of standards was either high, low or nil

according to their value in comparison with under-wood (among other factors). Oak was overwhelm-ingly the main timber tree, so much so that medieval recorders often treated 'timber tree' as synonymous with oak, and oak was still the principal standard species in 90% of all English coppice-with-standards woods in 1947, according to the Forestry Commis-sion census. Even so, other species were allowed to grow into timber, as the following examples recorded in the Calendar of Miscellaneous Inquisitions (HMSO) show. A mixture of 123 oaks, aspens, limes and alders was felled in the Abbot of Chester's woods at Huntingdon and Chauley (? Cheshire) in 1295. Oaks, ash, hornbeam and other trees were felled at Welwyn (Hertfordshire) in 1387 and oak, ash, beech and elm were felled at North Mimms (Middlesex) in 1397. An inquiry of 1377 into alleged waste by Ralph Bisset in the Isle of Axholme (Lincs) found that he had wasted a wood called Countessehage by felling maples, ash and great willows, and assarted part of Mellewood, during which he felled very many great trees called 'lindes'. Houghton's survey of 1564 records the occasional beech and ash among the timber of Rockingham Forest (Northants) (British Museum Add. MS 34214, unpublished). Woodland elm timber was used in the construction of Grundle House, Suffolk, about 1500 (Rackham, 1972).

The history of timber growing in coppice woods is almost synonymous with the history of oak in traditional woodland management (Rackham, 1974). The standards were generally felled when, by modern estimates, they were still small, and only a small proportion were left longer than three times the coppice rotation. In East Anglia most were grown for 25–70 years, yielding a continuous range of size classes up to 45 cm basal diameter. Their trunks were kept clean to a height of about 6 m by the shade of the underwood and occasional trimming of lower branches to provide faggots.

The succession of standards could in theory be so organised that the stock remained almost constant. Equal numbers of each age class might be present if all the standards were felled at the same multiple of the coppice rotation, as described above. Alternatively, the younger age classes might predominate if a greater range of timber sizes were required, as in the example from Liss, Hampshire (Nisbet and Lascelles, 1903;

Table 2.1 A plan, made about 1900, for the management of standard trees in the coppices of an estate at Liss, Hampshire (trees per acre).

	Standards left after each fall		Removed during each fall	
	Age	*Number*	*Age*	*Number*
Underwood	—	—	20	all
Standards				
Young stores	20	40	40	20
Double stores	40	20	60	10
Young trees	60	10	80	5
Old trees	80	5	100	5

Table 2.1). Such strict regularity was, however, rarely achieved, for individual trees were cut irregularly as the need arose, and heavy fellings at long intervals were fairly common in practice. Rackham quotes the example of Hardwick Wood (Cambs), where there were heavy fellings around 1380 and around 1460, but only sporadic felling of a few trees in other years for which the record is available, 1341–1495. In the Chilterns (Roden, 1968) substantial sales of large timber were made only occasionally, for example when cash was needed, when large repairs were planned or when gales had brought down many trees. Inevitably, therefore, the density of standards fluctuated enormously in any given wood, generally within the range of 5–40 per acre (Beevor, 1925; Rackham, 1974). At any given time the density of standards varied from one wood to another. For example, the crown coppices of Rockingham Forest contained anything from 1 to 40 trees per acre in 1564, according to Houghton's survey (1564). Sometimes, when regional or national requirements or factors transcended the needs of individual owners and communities (e.g. the aftermath of the Dissolution, during the Commonwealth and during the two World Wars) there was wholesale simultaneous felling of timber in many woods. Examples include woods in the vicinity of naval dockyards, ironworks and glassworks.

New standards in East Anglian woods were produced by promotion of coppice shoots from the stumps of felled standards and by natural regeneration. Planting was not necessary. Around the year 1800, saplings from acorns were generally preferred to coppice shoots; for instance, at Stratfield Saye, Hants (Vancouver, 1810) the standards from coppice shoots were grown for only 2–3 rotations whereas the standards from acorns were preserved for longer because of their better growth. The importance of promoted coppice shoots as a source of standards, combined with their youth when felled, confuses the distinction between standards and underwood in traditional management: many standards were just long-rotation coppice growing amongst short-rotation coppice.

Apart from timber, the standard oaks yielded bark which was used in tanning. In the Newball Woods, Lincs, bark provided 10–14% of all income from the timber sales between 1700 and 1723. The proportion increased to 22–40% in 1837–54, then fell to about 25% around 1880 and 5–15% in 1904–1910. Bark was also an important product from the oak coppices of western Britain. Vancouver (1813) describes oak coppices in part of Devon run on a 20-year rotation, with a few reserved till 40 years, from which the bark was peeled on every branch 'of an inch or less' in diameter.

2.4.4 Grazing in coppices

Deer, cattle, horses and sheep were commonly allowed to graze in coppices, and hogs were admitted to feed on acorns in autumn. Many coppices were actually enclosed as deer parks, e.g. Monks Park Wood, Suffolk (Rackham, 1971). Others formed part of the woodlands of a forest, where they were subject to grazing by both the King's deer and the commoners' stock; the forest woods in the belt from Stamford to Oxford were mostly coppices. But even ordinary coppices, which were neither parks nor part of a forest, were used as supplementary pastures by the local community. And in some regions the woods were seasonally used by stock introduced from elsewhere: Standish (1616) records an instance of hogs being driven into Shropshire to mast, 'which in former times was a common course, before woods were destroyed, for the champagne countries to feed their hogges in woodland countries'. Some of the pasture was provided by rides and launds (Fig. 2.2) – permanent clearings within the woods – and the remainder came from the field layer beneath the underwood. In the Chilterns (Roden, 1968) the pasturage of common woods was relatively abundant, whereas the pasture of private woods was confined to glades, a clear indication that the former were more open in character and functioned as wood pasture, whilst the latter remained well-wooded as a result of controls on grazing.

Grazing had to be controlled if the underwood was not to be damaged. Animals were excluded from freshly-cut coppices for about 4–7 years by means of hedges, either temporary barriers of dead brushwood or live hedges on banks which formed the permanent boundaries of coppice compartments. Nevertheless, grazing represented a permanent threat to the underwood and occasionally, in periods of mismanagement, destroyed it, as in the medieval coppices of the New Forest (Tubbs, 1964). The dangers were dramatically illustrated by Donaldson (1794) when he described Geddington Chase (Northants): 'All the townships using a commonage in these woods (except one) are in an open field state, and no attention is paid by the occupiers to the description of cattle bred and reared, which are of the most inferior kind, and which, in consequence of the inability of the occupier of an open field farm to procure a sufficiency of food for their support in the winter season, are reduced to an extreme state of leanness and poverty at the time they are turned into the woods, when whole herds of them rush forward like a torrent, and everything that is vegetable and within their reach, inevitably falls a sacrifice to their voracious and devouring appetite.'

'Traditional' grazing within coppices survived into modern times in Cranborne Chase, Dorset and Wilts, where the practice was witnessed in 1930 by E. W. Jones (Jones, 1961), and last occurred in the late 1960s. Common grazing rights still exist in some coppices, but are not exercised, e.g. at Hawkesbury, Glos (M. J. Penistan, personal communication). Many former coppice woods in the north and west of Britain are now continuously grazed by sheep or deer, but this is more a symptom of the decline of coppice management than an integral part of it.

2.5 Origins and trends in coppice management

2.5.1 The origins of coppice management

Coppicing is unlikely to have had a definable origin.

Primitive exploitation of natural woodland by clear-felling large areas and then moving on to another when one area was exhausted would have generated a coppice structure if the intensity of grazing was not too high: any trees which were too large to have been worth cutting would have survived as 'standards', whilst an underwood would have developed from coppice growth and saplings. Once all the available natural woodland had been exploited, a settled community would have been obliged to return to the woodland that had previously been cut over, and a form of coppice would have emerged from woodland exploitation. It could later be refined and made more productive when demand and organising ability increased. Thus, the coppice structure was probably widespread before a coppice system was needed.

The earliest evidence of coppice management comes from the Somerset Levels, where wattle trackways of hazel and ash, which were laid across the marsh, have been dated to the Neolithic, c 2200–2500 BC. The material appears to have come from land set aside for the production of underwood, and to have been cut by 'drawing', so that each coppice stool might bear shoots of various ages (Rackham, 1977a). Another early record comes from Shropshire, where the simultaneous rise of *Corylus* and fall of total tree pollen in the peat of Whixall Moss about 250 BC may indicate coppice management (Turner, 1965). Records of woods in Saxon charters have sometimes been translated as 'copses' even though there was no evidence that they were actually managed as coppice. By 1086, when the Domesday Book recorded numerous woods as *Silva minuta*, *Silva modica* and *Silvulae*, coppice management was evidently widespread throughout the English lowlands and into Devon and Cornwall. The system was still spreading into some remote districts as late as the 18th century, for it was only about 1700 that coppicing started in the woods beside Loch Lomond (Tittensor, 1970) and elsewhere in Scotland. Hitherto these oak woods had been exploited but unmanaged (Anderson, 1967). In England the system may have still been gaining ground during the Middle Ages in the formerly unoccupied wooded waste of the Weald.

2.5.2 Coppice improvement after the Middle Ages

The traditional coppice system described by Rackham

(1976) and summarised above was 'improved' from the 17th century onwards. Some coppices were cleared or changed to other systems, whilst in other districts new coppices were created. Many surviving coppices were re-organised and planting became a significant component of the system.

The conversion of coppices to other management systems only became really significant in the late 19th and 20th centuries, but before then a limited amount of conversion had taken place. The medieval coppices of the New Forest were thrown open to become wood pasture, or converted to high forest by renewed enclosure, sowing and planting (Tubbs, 1964). There, and in other crown woods, former coppices were planted as oak high forest in the early 19th century (Section 4.4), though this did not necessarily preclude continued coppicing of underwood. In the Chilterns (Mansfield, 1952) and other southern chalklands, coppices were either promoted to high forest or planted with beech.

On the other hand, some new coppices were planted, presumably where individual farmers had no ready supply of stakes, etc., or where they found it more economic to grow their own. These planted coppices were generally small, scattered and often planted in places which were inconvenient for other uses, such as wet valleys, steep hillsides, or on field margins. They are particularly common where there were few ancient coppice woods such as the Dorset chalklands, Cambridgeshire, etc., but they were also created where coppices were abundant, such as in Rockingham Forest and parts of central Lincolnshire. Many large oak coppices were planted on the valley sides of western and northern Britain, and in Scotland many of the natural oak woods were coppiced at the same time.

Most of the larger woods which remained as coppice throughout the 17th–19th centuries were re-organised. Hitherto the extent of annual coppice coupes was flexible and their shape was generally irregular like an 'old inclosure' field pattern. Rides were narrow, winding, or merely temporary cuts which gave access to the compartments which were about to be worked. During the 'improvement' era coppices were laid out more systematically; compartments were regularly shaped and cut in an orderly sequence. Straight, wide rides with marginal ditches

were made, often radiating from a central roundel in the manner of a spider's web. This process can be seen at Bedford Purlieus (Peterken and Welch, 1975) and Fulsby Wood (Fig. 2.4). Some woods escaped these changes until recent reforestation (e.g. Chalkney Wood, Essex; Rackham, 1976) and in others the older rides survive with the later additions. Foxley Wood (Norfolk), for instance, has three ages of rides: the modern, wide roads carved straight across the wood by the Forestry Commission; an earlier grid of narrow, straight, grassy rides; and a few narrow, sinuous grassy rides with individual names, which appear to be surviving portions of the earliest ride system.

Traditional coppice management and plantation forestry co-existed from the 17th–20th centuries, but each retained its separate identity enough for the Board of Agriculture Reports of about 1800 (e.g. Fairey, 1815) to distinguish consistently between 'woodlands' (meaning coppices on ancient woodland sites) and 'plantations'. Nevertheless, the two traditions did influence each other, and planting in particular became a feature of coppice management.

Planting permitted wholesale changes in underwood composition, a greater rate of renewal of dead and moribund stools, planting of coppice on un-wooded ground, and planting of standards. It presumably started with the rise of the plantation tradition: certainly Evelyn (1664) was speaking of planting chestnut, sycamore and willow in coppices. By about 1800 planting was widespread, even though traditional methods prevailed and there was considerable variation between areas and between estates within one area. Boys (1794) gives details of 26 groups of woods in Kent, where the 'produce seemingly natural' (i.e. the coppice species present before improvement) was mainly oak (25 sites), ash (20), beech (16) and hazel (15), with some birch (10), hornbeam (5) and aspen (1). No details of 'improvements' were given for one site, but of the 25 sites for which details are given, 13 were either not improved or 'but very little improved', one had been improved with native species, six with chestnut (Fig. 9.1), and five with chestnut and native species together. The technique for improvement was described: 'The plants, whether chestnut, ash, or willow, should be taken up from the nursery, with as much earth to

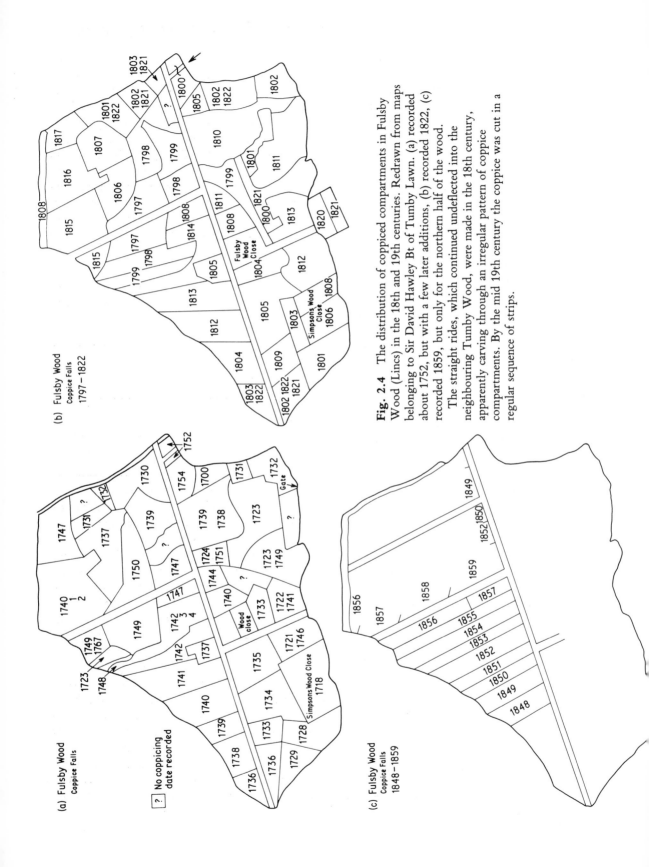

(a) Fulsby Wood
Coppice Falls

[?] No coppicing
date recorded

(b) Fulsby Wood
Coppice Falls
1797–1822

(c) Fulsby Wood
Coppice Falls
1848–1859

Fig. 2.4 The distribution of coppiced compartments in Fulsby Wood (Lincs) in the 18th and 19th centuries. Redrawn from maps belonging to Sir David Hawley Bt of Tumby Lawn. (a) recorded about 1752, but with a few later additions, (b) recorded 1822, (c) recorded 1859, but only for the northern half of the wood.

The straight rides, which continued undeflected into the neighbouring Tumby Wood, were made in the 18th century, apparently carving through an irregular pattern of coppice compartments. By the mid 19th century the coppice was cut in a regular sequence of strips.

their roots as can be conveniently done, and their small roots should be cut as little as possible. Strong plants taken up in this manner, and planted with care, seldom fail: they should be looked over next Spring, to fasten those which the frost may have loosened.' Such improvement was not confined to the southeast. A particularly illuminating statement by Lawe (1798) concerns some woods in Northamptonshire where after a fall, 'the hazels and thorns are mostly stubbed up, and young ashes planted in their stead'. In some years, up to 100 thousand plants were used. Fairey (1815), quoting this example with approval, notes that in Derbyshire this practice 'seems too much neglected'. One hundred years later, around 1900, the Forestry sections of the Victoria County Histories indicate that much the same situation prevailed. Three methods of propagating underwood were mentioned, natural regeneration, layering or plashing, and planting. Layering was 'still commonly practiced' in southern counties. Chestnut was much planted in Essex. Blanks were filled in by planting ash at Alton (Hants) and Cirencester (Glos), whereas at Liss (Hants), in addition to layering, they were filled by sowing seeds of ash, oak, maple, sycamore, chestnut, etc. At Guildford (Surrey) small-scale plantings were made in coppices to make good the ravages of rabbits.

The extent to which standard oaks were planted is difficult to establish. Planting seems to have been relatively unimportant at the time of the Board of Agriculture reports, and around 1900 the VCH Forestry sections (e.g. Nisbet, 1905) show that some estates still preserved natural oak saplings or 'dibbed in' acorns. Planting must have been common and widespread in the 19th century, however. It caused major changes in some woods, such as Overhall Grove (Cambs), where *Ulmus carpinifolia* was planted into an oak over ash coppice-with-standards wood, and minor changes in others, e.g. Clouts Wood (Wilts), where the oak standards from before about 1850 were mostly promoted from coppice, but the younger oak standards are maidens which might have been planted.

2.5.3 The decline of coppice management

Although coppicing had ceased in some woods as early as the 16th century, the main decline of coppice management did not start until the late 19th century. In 1905, coppice woods were still present virtually throughout Britain (Fig. 2.5), but the retreat from Scotland, north Wales and north-east England was already well advanced. Forty years later coppice was largely confined to the English lowlands, Welsh Borderland and the Lake District, and by 1965 it had been completely abandoned over most of Britain and was much reduced in the southern area where coppicing was still practiced. The full extent of the coppice retreat over the post-Medieval period can be judged from three study areas in the East Midlands (Table 2.2).

Table 2.2 Changes in the area (ha) of coppice woodland on ancient woodland sites in three areas in the East Midlands of England (Peterken and Harding, 1975; Peterken, 1976).

Date	Rockingham Forest (Northants)	Central Lincolnshire	West Cambridgeshire
1650	8442	?	?
1817–34	7182	3726	700
1885–87	4200	2670	680
1946	2363	1838	590*
1972	1389*	1014*	367*

* More than 90% of this area no longer cut as coppice.

This trend is well illustrated by the Tumby estate (Lincs) (Fig. 2.6), where wood account books covering the main period of decline still survive. Income from the annual wood sales of Fulsby Wood, Tumby Woods and some small outlying coppices rose to a peak about 1860, but declined erratically from 1870 until about 1895. From 1895 to 1930 income remained constant, but the number of buyers at the auctions gradually decreased. After 1930 income fell heavily and the auctions ceased. These woods have since been partially replanted as conifer high forest.

The main reasons for the decline were economic and social. Coppice woods become less profitable as the price obtainable for coppice products fell. Wages increased and apprentices could not be recruited to the underwood crafts against the competititon of better paid and less arduous work in towns. Rabbits and game preservation increased, damaging the underwood and subordinating silviculture to the pheasant.

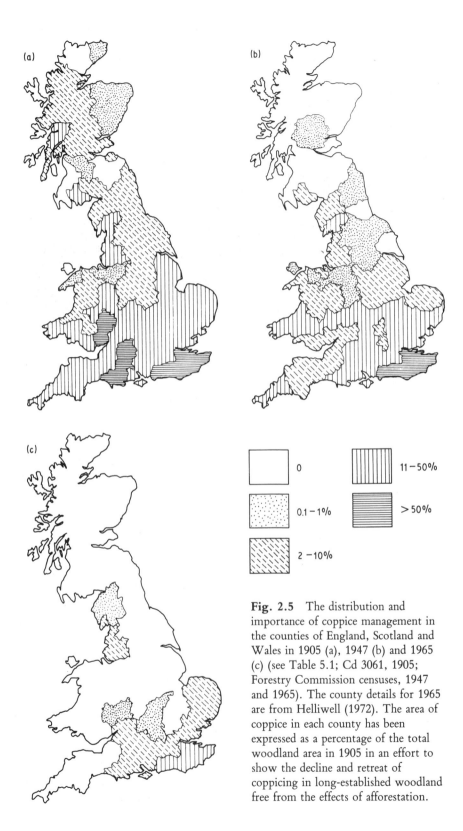

Fig. 2.5 The distribution and importance of coppice management in the counties of England, Scotland and Wales in 1905 (a), 1947 (b) and 1965 (c) (see Table 5.1; Cd 3061, 1905; Forestry Commission censuses, 1947 and 1965). The county details for 1965 are from Helliwell (1972). The area of coppice in each county has been expressed as a percentage of the total woodland area in 1905 in an effort to show the decline and retreat of coppicing in long-established woodland free from the effects of afforestation.

Legend:

0

0.1 – 1%

2 – 10%

11 – 50%

> 50%

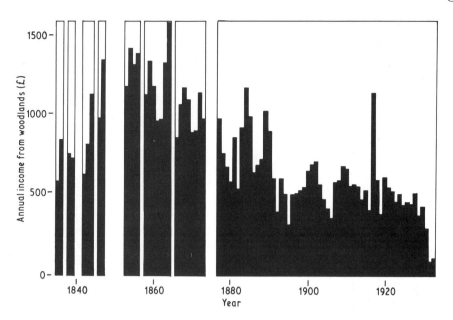

Fig. 2.6 Annual income from the Tumby woodlands (Lincs) 1835–1933. (Lincoln Archive Office 1/Hawley/3B.) These woods were then managed as coppice-with-standards, the timber and underwood being sold at the wood sales held in the village inn in October and December. The record ends at 1933 when, it seems, the wood sales were discontinued.

Local events were also important where, for example, closure of a canal cut off a wood in Berkshire from its traditional market, and brushwood could be obtained cheaply from the now overgrown commons or the thinnings from plantations.

In the early 20th century coppice management was often continued at a loss, and owners seem to have hoped for an upturn in markets for coppice material, but between about 1920 and 1950 numerous woods were coppiced for the last time, and many were gutted of saleable timber, particularly during the two World Wars. Before coppicing was completely abandoned the system appears to have been modified in detail. For example, the succession of oak standards was not maintained, and we now see only old and large standards where we see any at all. In Swanton Novers Woods (Norfolk) numerous oak and ash standards have survived, but almost all originated between 1799 and 1855, according to the few ring counts available. Where annual cuts were reduced, the rotations were inevitably lengthened. Some coppice species were selectively felled, e.g. only the hazel in an ash–hazel mixture might have been cut at

the last coppicing. Weed species such as thorns, birch and sycamore no longer cleaned out as coppicing declined, spread into the woods and competed with coppice.

2.5.4 Coppice woods today

Only 15 550 ha of simple coppice was worked in 1965/67 (Locke, 1970) and most was chestnut (12 900 ha) and hazel (1700 ha). In addition, some of the 11 600 ha of coppice-with-standards and Forestry Commission coppice was also worked. The great majority was in south-east England (Fig. 2.7).

Traditional coppice management survives in few places, notably Bradfield Woods, Suffolk (Rackham, 1971, 1976). Traditional crafts survive where chestnut fencing and hazel wattlehurdles (Fig. 18.2) are made, and a few practitioners of other coppice crafts are still active (Manners, 1974). Faggots have recently been cut for river and sea defence schemes and for at least one specialist bakery. Even so, the total area coppiced for traditional markets is unlikely to exceed 600 ha today. New markets have, however,

29

Fig. 2.7 The distribution of recent coppicing. Each dot indicates that at least one wood or part of a wood in a 10 km square of the national grid was actively managed as coppice in or after 1965 (from Forestry Commission and Nature Conservancy Council).

demand is such that the long decline of coppicing could even be reversed. Another new development which is insignificant quantitatively, is the attempt by many conservation organisations to revive coppicing on a small scale (e.g. Peterken, 1972; Steele and Schofield, 1973; Rackham, 1975). Very recently the fashion for wood-burning stoves has improved the prospects of coppicing for fuel.

Natural changes are slowly taking place in the neglected coppice woods (Figs 8.31 and 18.1). The mixed coppice changes in composition as the taller-growing species compete successfully with the shrubs, e.g. a neglected hornbeam–hazel coppice eventually becomes a pure hornbeam coppice. Remaining standards are deteriorating as their lower branches are shaded and killed by overgrown coppice. Weed species, hitherto cleaned out, compete in the canopy with coppice species and enter long-neglected coppice as a subordinate shrub layer. Ancient hollow stools of ash are disintegrating under the weight of heavy, overgrown trunks. Felled standard oaks have occasionally produced coppice shoots which have been allowed to grow with the coppice.

2.6 Woodland in Scotland

2.6.1 Historical outline

The wood pasture and coppice systems can both be regarded as developments from primitive exploitation of natural woodland which were increasingly systematised before they became obsolescent. Elements of original–natural woodland survived in both regimes, but by the early Middle Ages the great majority of woodlands in England and Wales were used and managed, and therefore far removed from their natural structure. Woodlands in Scotland, and especially those in the Highlands, were somewhat different in character. Lying north of the main area of coppice in sparsely inhabited country remote from large markets, they survived in a near-natural form until much later than the woods further south.

Natural woodland in Scotland was much reduced in prehistoric times by the formation of blanket bog on the coarse, easily-leached soils of the uplands and valley peatlands on the lower ground (Godwin, 1975). Prehistoric man made only trivial inroads, and

opened up, or have been developed from old markets. A few woods are coppiced to provide alder, sycamore or birch to the turnery trade. More importantly, the large demand for hardwood pulpwood, used to make corrugated brown paper for packaging, can be satisfied from mixed coppice, although much of the pulpwood is made from branchwood of timber trees, imported material and coppices being cleared or reforested. The companies involved have tried to encourage continued coppice management, and their even in early historical times there remained extensive woodlands which had scarcely been affected by man. During the Middle Ages the Highland woods were used as wood pasture, providing land for hunting as well as timber, brushwood, pasture and pannage. The 17th and 18th centuries appear to have been exceptionally significant (Anderson, 1967): exploitation of the remaining natural woods was intensified as the demand for timber, charcoal and tanbark increased under more settled conditions and commercial interests were able to penetrate to even the most remote woodlands. Although some 'natural woods' were finally exterminated as a result, many – mainly birch and pine woods in the Highlands – perpetuated themselves as wood pasture, and numerous hitherto unmanaged woods as far north as Sutherland were managed as coppice, thereby restricting the availability of winter pasture (Lindsay, 1977). At the same time plantation forestry became widely established. During the 19th century coppice management declined, and it was largely extinct by the present century except in Perthshire and southwest Scotland. Afforestation became increasingly important and today the great majority of Scottish woodlands are plantations on secondary woodland sites.

The history of woodland in Scotland therefore follows much the same sequence of events as in England, but with significant differences in emphasis and timing. Wood pasture was relatively more important and continued widely into modern times. Coppice management became established several centuries after it was widely practiced in England, but declined and retreated earlier. Some primary woodlands, notably the birch and pinewoods of the Highlands, appear to have retained a high forest structure under wood pasture management, but in England high forest was rare before the advent of planting. Afforestation and plantation forestry started at much the same time, but in the present century they have become overwhelmingly more important in Scotland (so much so that the Forestry Commission moved their headquarters from London to Edinburgh). Scotland was not, however, completely distinct from England, for just as there were deer parks and eventually coppices in parts of Scotland much as in England, so some parts of England followed closely the pattern of events in Scotland. For instance, during the 13th and 14th centuries much of the pasture available in the Vale of York and Yorkshire coalfield was in open woodland, but these wood pastures were largely lost by the 18th century (Sheppard, 1973).

2.6.2 The native pinewoods

The native pinewoods (Fig. 2.8) have attracted more interest in proportion to their extent than any other group of British woodlands. In addition to the classic study of Steven and Carlisle (1959) there have been detailed examinations of particular pinewoods, such as Abernethy (Birks, 1970; O'Sullivan, 1973) and Rannoch (Lindsay, 1974). A symposium volume has recently been published on the present state of the pine woods as a whole, and our knowledge of them (Bunce and Jeffers, 1977). The 'native' pinewoods were defined by Steven and Carlisle (1959) as those which were 'genuinely native, that is descended from one generation to the next by natural means', thus excluding those which were mainly planted, but admitting almost any degree of felling and extraction. The evidence upon which particular pinewoods were identified as 'native' was partly historical. Stands which were described as old and reputedly natural in 16th–18th century records (which excludes early plantations), and which included at least some old trees, were accepted. Unrecorded recent plantations were identified by their narrow range of morphological variation. Steven and Carlisle's list should be extended to include some of the numerous birch stands with natural pine in them, e.g. Crathie Wood, in Deeside. The 35 sites identified by Steven and Carlisle now extend to about 10 900 ha, but only 1300 ha of this is relatively dense pine high forest.

The authentic native pinewoods lie in remote areas where they remained relatively free of human interference throughout the Middle Ages. Inaccessible and virtually unknown outside the Highlands before 1600, they had only to meet local needs, which were mostly small. In the 17th century timber shortages elsewhere, coupled with knowledge of the timber reserves available, stimulated exploitation for the navy, but access remained difficult and transport remained a problem until the flotation technique of

Fig. 2.8 Native pine woodland in Abernethy Forest, situated in the Spey valley on the northern margin of the Cairngorms. Beneath the pines juniper forms an irregular underwood over a carpet of dwarf shrubs, mainly *Calluna vulgaris*, *Vaccinium myrtillus* and *V. vitis-idaea*. The structure must once have been more open, for the older trees (middle distance) have wider crowns than the younger trees (foreground). As in most native pinewoods, timber has been cut but regeneration has been natural. The interaction between man, grazing animals and vegetation has in this instance produced a diverse structure which, in completely natural woodland, was probably less common than a patchy even-agedness. (Photo. D. A. Ratcliffe)

extraction was improved in the 18th century whereupon felling became more extensive. By the 19th century restocking was more difficult, because the woods were exploited as sheep pasture on a greater scale than hitherto, and some were incorporated within deer forests. Sowing and planting, which started by 1600 and occurred on a large scale from the 18th century was occasionally used to perpetuate parts of those woods which are now identified as 'native'.

The structure of existing woods provides ample evidence of developments since the 17th century, and details are given for each wood by Steven and Carlisle. At the start of the 17th century the woods were well stocked, and many trees of about 300 years survive from near that time. The majority of the pines we now see arose naturally between about 1770 and 1820, following the main phase of exploitive fellings, and in some woods there has been very little regeneration since. A complete range of age classes is present at Abernethy, and good recent regeneration occurs at Glentanar, Rothiemurchus and Shieldaig. The pinewoods have occasionally regenerated beyond existing margins, e.g. Abernethy, and would do so more often but for moor-burning. Many native pinewoods have been felled in the last 40 years, and planted with exotic conifers or with introduced strains of Scots pine.

The apparent lack of saplings in most pinewoods has prompted doubts about their ability to perpetuate themselves by natural regeneration. Certainly some pinewoods were replaced by peatland in prehistoric times by natural retrogressive succession (O'Sullivan, 1977); in some woods a high density of red deer ensures that seedling pines are grazed to oblivion (e.g. Mar Forest); and some 'woods' now are just scattered old pines standing in heathlands (e.g. Glen Falloch). Soil developments, notably podzolisation and the growth of thick mor humus and peat over the surface, have also apparently reduced the chances of successful regeneration in what may be a classic natural retrogressive succession which has been accelerated by exploitive fellings, much as blanket bog was initiated by neolithic fellings elsewhere.

2.6.3 Broadleaf woodlands

The records accumulated by Anderson (1967) suggest that almost-virgin broadleaf woodland survived in the Highlands until the early Middle Ages, and some may have remained little altered as late as the 17th century. Records quoted for Stirlingshire seem typical. In about 1142 the Abbey of Holyrood was given permission to take timber to build churches and houses from the Crown woods, as well as to claim pannage for pigs. The monks of Newbattle were granted the right to fuelwood and to pasture of cattle in the wood of Callendar (Perthshire). In 1317 the right of the burgesses of Stirling to common pasture and brushwood in the Torwood was confirmed. Woods, it appears, were discrete and recognisable entities by then, not unbroken expanses of tree-covered land.

Coppice management (Lindsay, 1975 a and b; Tittensor, 1970) was not recorded in Scotland before 1550, when the monks of Coupar Angus were managing an enclosed wood in lowland Perthshire on a 28-year rotation, and it remained very limited throughout the country until after 1700. Coppicing increased during the 18th century, but the main expansion started in the 1790s and progressed so fast that in the early 19th century 'the self-sown deciduous woodland of Scotland was coppiced to such an extent that the terms "coppice" and "natural wood" came to be regarded as interchangeable', according to Lindsay. In some areas, such as Argyll, the main product was charcoal, which could be made from many species and provided an economic use for mixed coppice. Tanbark was an important product too, so oak coppice was particularly valuable and many mixed woods were improved to oak coppice by planting oak and removing other species. Rotations were long, mostly 20–30 years, and most woods were thrown open to stock some 4–7 years after coppicing. Standard oak and ash were reserved in many coppices, but only for two rotations. After the early 19th century expansion, coppice management declined and retreated as fast as it had spread, and by the 20th century it was more or less confined to a few areas in central or south-west Scotland. The abandoned coppices were once again opened to sheep and deer as sheltered pasture, a process which was already well established by 1827 in Argyllshire.

This analysis from historical sources is abundantly confirmed on the ground. Most semi-natural broadleaf stands on ancient woodland sites show some signs of coppicing, even as far north as the ash and hazel of Rassal and the oak at Loch A'Mhuillin. Standard oaks remain in many woods, e.g. the Old Wood of Methven at Almondbank (Perth), where there are some fine standard oaks. Some sessile oak coppices are so pure that planting must be suspected, e.g. the oak over oak coppice-with-standards at Craig Wood by Dunkeld, Tynrioch Wood and Twentyshilling Wood, all in Perthshire. At Loch Lomondside the oak coppices are pure to the south but mixed with other species to the north, where cleaning out of worthless species was less efficient in the remoter woods. Despite these improvements, numerous coppices retained their natural, mixed composition, such as the 'black-woods' of birch, alder and hazel in Argyll, and the woods of oak, wych elm, hazel, birch and oak in the base-rich soils of gorges at, e.g. Airlie (Angus) and Killiecrankie (Perthshire).

Although a large number of semi-natural, broadleaf woods were temporarily managed as coppice, the intensity of management and care of new growth was variable, and many woods appear to have remained mainly as exploited wood pasture, the large trees being cut and sold, and the new growth from seed or stumps surviving or persisting according to the intensity of grazing. Anderson (1967) quotes many descriptions from the 18th and 19th centuries of woods which had been reduced to a scatter of scrub and old trees, whereas others somehow managed to spring again into well-stocked woodland. Two woods in the borders exemplify the results. Cragbank Wood is a mature, more-or-less closed stand of ash and wych elm which arose in the mid-19th century mainly from seedlings, but a spreading 18th century oak and a few ash and elm of coppice origin suggest that managed woodland was present on the site before then. At Airhouse Wood (Berwick) (Gordon, 1911), in contrast, the scattered birch and hazel are all that remains of an ancient wood. The distinction between wood pasture and coppice, which is rarely confused in lowland England, is not at all clear in Scotland, because in numerous woods neither system seems to have moved far beyond primitive exploitation.

Birch woodlands are a particular feature of Scotland. They suffered the same history of wood pasture exploitation as other types of broadleaf woodland, but they show a number of special features. Birch is a short-lived tree which regenerates copiously from seed, so it was rarely managed as coppice (except as part of a mixed stand) and most birchwoods had and have a high forest structure. Birch is also capable of rapidly colonising ungrazed grassland and unburned heathland, so it readily develops as secondary woodland: the birchwood fringes to the oakwoods at Letterewe arose as secondary woodland since the late 19th century, for example. Birch also fails to regenerate within birchwoods, not only because of grazing, but also due to root competition or some other antipathy between birch trees and birch saplings (McVean, 1964), so mature birchwoods tend eventually to die out and the ground returns to an unwooded state. Birchwoods, therefore, like pinewoods, tend to be migratory, developing here and degenerating there in a kaleidoscope of even-aged patches, perpetually maintaining themselves as 'secondary' woodlands. Some birchwoods have had a high proportion of birch for a very long time, though most appear either to have contained other species which are less able to survive the various forms of disturbance (e.g. Morrone Wood at Braemar may once have contained some pine, a tree which is still abundant across the Dee in Mar Forest), or else to contain other species such as hazel as an underwood (e.g. Strathbeag, Sutherland). In the former case the lost species would presumably return if grazing were relaxed and seed sources were reasonably close at hand. At Craigellachie (Speyside), for example, a mature but degenerating birchwood grows on neutral, base-rich soils which appear well capable of supporting ash, wych elm and pedunculate oak, but these species are rare in the surrounding area.

2.7 Distribution of ancient woodland

The amount of ancient woodland in Britain has been tentatively estimated at about 33% of all woodland existing in the late 1960s, or around 574 000 ha (Peterken, 1977a; Chapter 19). Although some is secondary, much of it must be primary. The amount and distribution of ancient woods today has been determined by those factors which influenced the location and extent of woodland clearance.

Despite assertions that particular woods survive because they grow on steep slopes or heavy soils, these factors cannot be the prime determinants of the location of ancient woods: equally steep and heavy sites nearby have actually been cleared. Rather, the pattern of ancient woodland complements the pattern of settlement and cultivation in prehistoric and historic times: as settlements were established and cultivation spread, so the woodland was cleared until it acquired scarcity value (or until some more remote factor, such as forest law, came into operation). Thus the pattern of ancient woods is to some extent the fossilised pattern that was produced by the time it became more economic to retain woodland than clear it.

Woodland clearance has been discussed by many

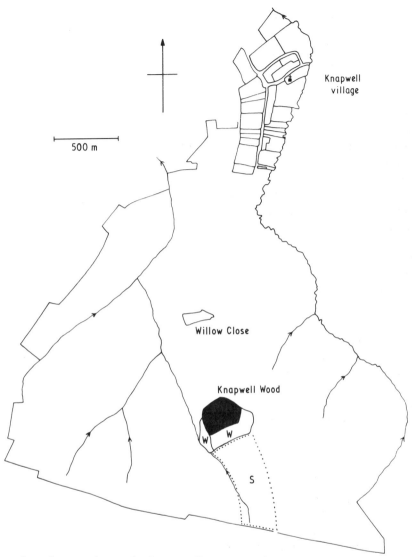

Fig. 2.9 Enclosure boundaries in the parish of Knapwell, Cambs, before enclosure of the open fields in 1775 (Knapwell Enclosure Map, 1775 and Edward Hare's map of the Lordship of Knapwell, 1776, both in Cambridge Record Office; Rackham, 1969).

Apart from Willow Close (which vanished as a separate entity at Enclosure, even though part of its boundary survives today), Knapwell had two discrete groups of old enclosures, centred respectively on the village itself and on Knapwell Wood. The wood is evidently the northern tip of an original, larger wood, whose boundaries once included Stocking Furlong (S), which was incorporated into the open fields, and the two wood closes (W). The eastern side of Stocking Furlong is marked by a wood-relict hedge (Pollard, 1973). See also Fig. 13.1.

writers and can only be outlined here. The southern chalklands were mostly cleared by the Bronze Age (*c* 2000–500 BC), a millenium before clearance was widespread elsewhere (Godwin, 1975; Turner, 1965). During Roman times (Applebaum, 1972) the large

estates centred on villas were mainly restricted to the medium loams and the limestone soils from Gloucester to the Humber, a zone which must have been sparsely wooded. In Essex the first and densest settlements were on the chalk, valley gravels and

35

glacial sands and gravels, leaving the boulder clay zones virtually unoccupied. Settlement in the Highland Zone was sparse and mainly on high ground, and so much woodland survived in the valleys. This Roman landscape was the foundation of the modern landscape in many areas such as Cambridgeshire (Taylor, 1973), where post-Roman settlement consisted either of infills into the established pattern or extensions into hitherto unoccupied areas on the clay plateaux. Elsewhere, perhaps notably on the southern chalk and limestone, the Roman settlements were abandoned on a large scale and the area was recolonised later in a different pattern, e.g. Rockingham Forest (Northants) (Peterken, 1976). Here the population appeared to retreat from higher ground to the main river valleys which were then recolonised by establishing villages along the tributaries. By the time that individual identifiable woods were first recorded in Saxon charters, the modern pattern of woodland had already been created.

Woodland clearance, the development of field patterns and indeed the evolution of the landscape as a whole during historical times have been the subject of numerous studies (Baker and Butlin, 1973; Darby, 1951; Hoskins, 1955 and subsequent volumes on individual counties; Sylvester, 1969). The fragments of surviving woodland were gradually pared back by assarting and by a growing population which required more cultivable ground. The rate of clearance varied, but was especially rapid in the early Middle Ages (until the Black Death), during the Dissolution and its aftermath of rapid land-ownership changes, during the period of agricultural prosperity around 1800 and also during the last 30 years of agricultural intensification. Land cleared in the Middle Ages was not usually incorporated within open fields, but remained within small, hedged closes, and as a result the surviving ancient woods often have 'ancient enclosures' around them. Detailed local studies of South Harting and Rogate parishes in Sussex (Yates, 1960) and Whiteparish, Wilts (Taylor, 1967) show the survival of woodland in an evolving landscape. The distribution of woodland around the Vale of Pickering (Yorks) 700 years ago (Wightman, 1968) can be matched in detail with the existing distribution.

Perhaps the most significant feature for woodland ecology is the existence of 'late woodland areas', where settlement came late and remained sparse. These had a distinctive combination of woodland, common pasture and small fields with characteristically irregular boundaries, formed by piecemeal, unplanned clearance of ancient woodland (Pollard *et al.*, 1974). These contrast with 'champion areas' where virtually all woodland was cleared, the land was intensively cultivated in open fields, and which were eventually enclosed in a planned manner during the 18th and 19th centuries. These distinctions can be seen on several scales. On a national scale, Rackham (1976) has divided lowland England and Wales into two zones, the 'predominantly ancient countryside' of south-east England and the upland margins, and the 'predominantly planned countryside' centred on the Midland clay belt. On a county scale, Warwickshire is divided between Arden (Woodland) and felden regions, and Yelling (1968) was able to divide East Worcestershire into 'woodland' and 'champion' parts. On a local scale, individual parishes can often be divided into a patch of ancient enclosures near the parish margin, which contains any surviving ancient woodland, and planned enclosures nearer the village where the common fields formerly lay (Figs 2.9, 2.10).

Clearance during modern times has proceeded apace, especially during periods of agricultural prosperity. In the East Midlands (Peterken, 1976) the contrast in the amount of woodland from one district to another tended to diminish during the 19th century, clearance being faster in well-wooded districts and afforestation being faster in sparsely-wooded countryside. This equilibration process has been reversed during the mid-20th century, when, in the lowlands at least, clearance has been slight in well-wooded districts where forestry is worthwhile, and rapid in sparsely-wooded areas where woods function mainly as game coverts (see Table 16.1).

During any given period the area of woodland in a district may increase, decrease or remain constant, according to the relative rates of clearance, natural succession to woodland and afforestation by planting. If the area remains constant, the rate of clearance must be balanced by the rate of succession plus afforestation, but the rates may be either slow (low turnover) or fast (high turnover). These rates are influenced by

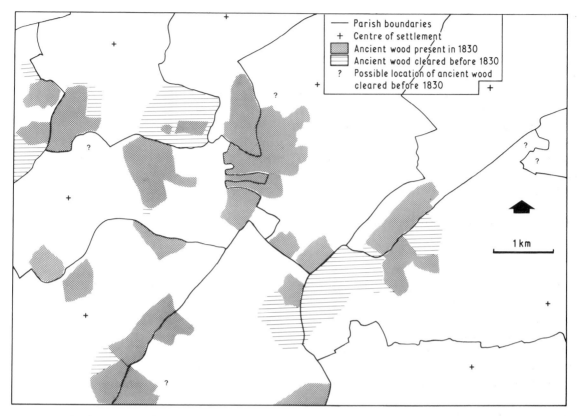

Fig. 2.10 The distribution of ancient woodland in relation to parish boundaries in part of central Lincolnshire near Wragby and Bardney, in 1830. Some ancient woods had been cleared before 1830: their location and extent was determined by evidence from field boundaries, wood relic hedges and extrapolation from woodland earthworks. Since 1830 most of Langton Wood, the large lobster-shaped wood in the centre, has also been cleared for cultivation.

national pressures, events and decisions, but each district responds in a distinctive manner. The sequence of events in Rockingham Forest (Figs. 2.11, 2.12), for example, shows a general long-term reduction in woodland: early extensive clearance reversed by natural succession to woodland in the Dark Ages, periods of ancient woodland clearance in the 19th and 20th centuries, the late rise of afforestation, the current high turnover of wooded ground due to open-caste mining and pit restoration under plantations. Other regions (Fig. 2.13) have different histories, so that some have always been well-wooded (e.g. Lower Wye Valley); some have long been denuded but recently afforested (e.g. Breckland, Galloway); some were partly denuded and re-wooded in prehistoric and early historic times (e.g. the Southern Chalk Plateau and Rockingham Forest);

and others remained well-wooded into historical times but are now predominantly agricultural districts (e.g. west Cambridgeshire and central Lincolnshire).

The present distribution of ancient woodland can be estimated by analysing features shown on Ordnance Survey 1:25 000 maps, which show woods, field patterns, parish boundaries and other significant features from which an historical geographer can deduce some aspects of landscape history. Woods with irregular boundaries lying in a patch of ancient enclosures against the parish boundary are likely to be ancient, especially if they are named after the parish or manor, if they have a name which indicates traditional management (e.g. Spring Wood), if they are linked to the village by 'Wood Lane', or if they possess some other hint of antiquity.

37

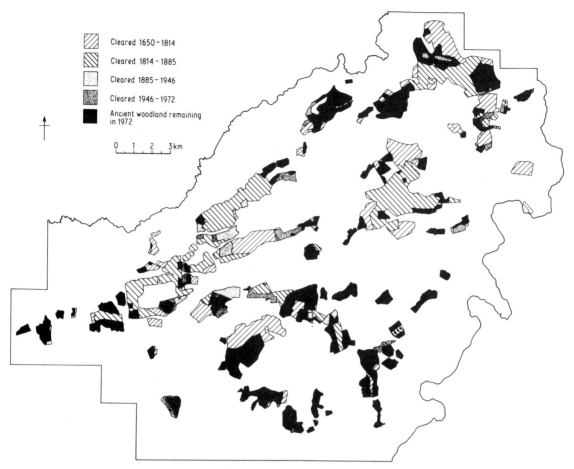

Cleared 1650 – 1814

Cleared 1814 – 1885

Cleared 1885 – 1946

Cleared 1946 – 1972

Ancient woodland remaining in 1972

0   1   2   3 km

Fig. 2.11 The reduction of ancient woodland in Rockingham Forest (Northants) since 1650. Almost all the clearance before 1814 occurred between 1796 and 1814 following the loss of Rockingham's status as a Royal Forest. The surviving ancient woods comprise 42% of those present 330 years ago (Peterken, 1976).

Recent woods, on the other hand, tend to give themselves away by their name (e.g. 'New Plantation' and 'Duke of Plazatoro's Belt'), by their regular shape, their straight sides and their conformist relationship with a local enclosure pattern of recent origin. Using such clues an estimate was made of the extent of ancient woodland in each 5 km square of the national grid. Estimates were checked against the first Ordnance Survey maps, on which ancient woods are easier to identify because at this time there were fewer

(Fig. 2.13 cont'd)

RF, Rockingham Forest (see Fig. 2.10).

WV, The well-wooded district on both sides of the lower Wye Valley, south of Goodrich Castle as far as Chepstow, in Hereford, Gloucester and Gwent.

CL, Central Lincolnshire.

WC, The boulder clay plateau and surroundings west of Cambridge.

SC, The scarp and adjacent plateau areas of the southern chalklands.

G, Galloway, representing much of upland Britain.

B, Breckland (Norfolk, Suffolk and Cambs) representing lowland heaths.

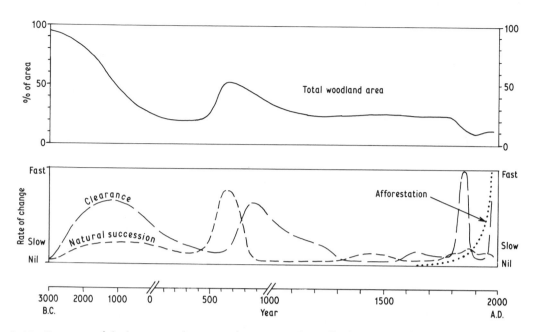

Fig. 2.12 Summary of the long-term changes in the amount of woodland in Rockingham Forest (Northants) and the main components of change. Reconstructed from archaeological, documentary and cartographic evidence.

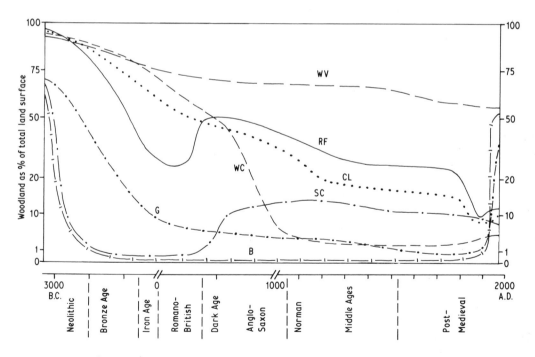

Fig. 2.13 Summary of the long-term changes in the amount of woodland in seven regions of Britain, reconstructed on the same broad and approximate basis as Fig. 2.10, but including the evidence of pollen diagrams where available. The regions were chosen to emphasise the large differences which exist between various parts of Britain. *(cont'd opposite)*

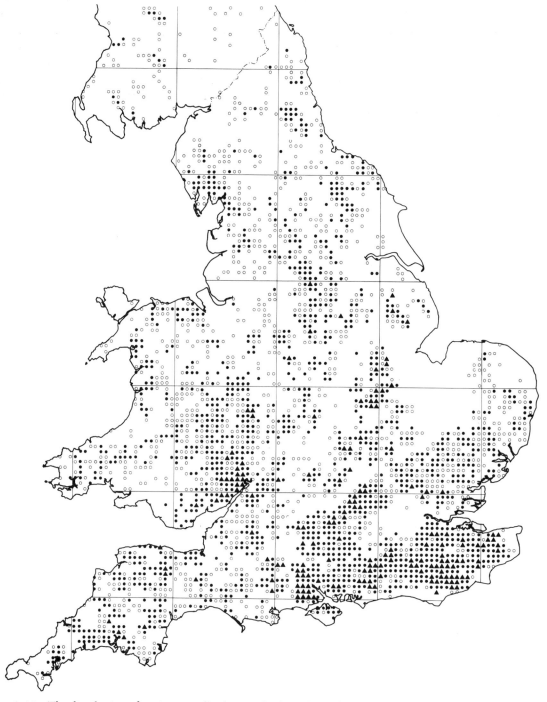

Fig. 2.14 The distribution of ancient woodland in England, Wales and southern Scotland. Each symbol shows the amount of woodland in each 5 km square of the national grid. Open circle: one or two small ancient woods. Triangle: extensive ancient woodland. Closed circle: a moderate amount of ancient woodland, usually distributed as a scatter of separate woods (reconstructed from features on Ordnance Survey 1:25 000 maps).

secondary woods and more ancient woods, and many ancient woods which have since been given straight sides by partial clearance were then larger and still in possession of their traditional, irregular margins. The resulting map (Fig. 2.14) ought to be fairly accurate in the English lowlands but less accurate in the Highland Zone. (Readers should treat individual dots with caution and examine the overall pattern.) It agrees fairly well with the maps of 11th century woodland prepared by Darby (1977) and his collaborators, but differs in style and, I hope, accuracy from similar maps prepared by Wilcox (1933) from literary sources. It shows that ancient woods occur in most regions, but are biased to the south and distinctly clumped into the 'late woodland areas' mentioned above.

Chapter 3

Ecological characteristics of ancient woods

Although many of the ancient woods have descended directly from primaeval forest, they are no longer completely natural. The great majority have been managed, and all have been influenced to some extent by man. This influence is most obvious in the woodland structure, but it is also apparent in the plant and animal communities within woodlands, in the distribution of individual species and in soil characteristics. Nevertheless, despite various changes brought about by exploitation and management, numerous natural features survive.

3.1 Types and degrees of naturalness

A woodland or any individual component of a woodland is 'natural' if its characteristics have not been significantly affected by man. Conversely, a woodland is artificial if its characteristics have been determined by man. Complete naturalness and total artificiality are the extreme states of a continuous variable, the degree of naturalness, the former being 100% natural and the latter 0% natural. Neither extreme exists in practice, for no completely natural woodland survives in Britain and even the most intensively managed plantation of exotic species contains some wildlife. In a sense, therefore, all woodland is semi-natural, owing its present characteristics to both man and nature. However, a distinction needs to be made between relatively natural and relatively artificial woodland: it would be pointless to describe

both the native pinewood at Beinn Eighe and a nearby sitka spruce plantation as 'semi-natural'. Therefore, the term semi-natural is reserved for those woods or features of woods which have a relatively high degree of naturalness.

The boundary between semi-natural woods and others is somewhat arbitrary. Normally it is drawn between predominantly unplanted stands on one hand and plantations on the other. Inevitably many stands occupy the grey area near the boundary. For instance, mature plantations of oak growing on sites which oak would have occupied naturally are technically plantations, but they may well have acquired a natural shrub layer and many ecologists, including Tansley (1939), accept them as semi-natural. At Bedford Purlieus (Peterken and Welch, 1975) a semi-natural, mixed coppice stand became a plantation when it was clear-felled and reforested with oak, but it reverted to semi-natural when some of the former coppice was allowed to grow up with the planted oaks, and has lately become a borderline case following heavy thinning selectively in favour of oak.

Naturalness varies not only in degree but also in kind. It is convenient to recognise four basic kinds of naturalness:

(1) Original-naturalness. The state which existed before man became a significant ecological factor. Original-natural woods were not a single state but a

succession, which continued naturally until the Atlantic period, the last time when all British woods were completely natural.

(2) Present-naturalness. The state which would prevail now if man had not become a significant ecological factor.

(3) Future-naturalness. The state which would develop if man's influence were completely and permanently removed. When grassland reverts naturally to woodland and the succession continues without human influence, the resulting woodland is future-natural at all stages.

(4) Potential-naturalness. The state which would develop if man's influence were completely removed, and the resulting succession were completed in a single instant. Unlike the other forms of naturalness which did, could have or would exist, potential-natural states cannot actually exist. Nevertheless, this is a useful concept which expresses the existing site potential under the prevailing climate, unaffected by any future changes of climate and soil (to which future-naturalness is subject).

The relationships between the various kinds of naturalness are somewhat complex. The main point to recognise is that original- and present-naturalness are linked: together they describe natural features which existed before modification by man. Strictly speaking, no existing natural feature can be described as original-natural: the latest original-natural state is present-natural. Management, however, has somewhat distorted the relationship between the two states. For instance, oak was a relatively important original-natural constituent of lowland woods, but might have declined by now if man had not favoured it in coppice management (Godwin and Deacon, 1974). Lime was a very important component of Atlantic forests and has survived as underwood in many ancient woods, but under a long history of wood pasture management in Epping Forest it has been replaced by beech–oak–hornbeam woodland (Baker *et al.*, 1978). It is useful, therefore, to have a fifth term, 'past-naturalness', to describe existing natural features which have descended directly from original-natural conditions. One need not reach a conclusion on whether man has accelerated or arrested natural changes when pronouncing a feature to be past-natural.

The two forms of naturalness which actually exist in British woods are thus past- and future-naturalness, the former preceding and the latter succeeding human influence. Future-natural features develop as a response to an action by man. The thicket which springs up after a wood has been clear felled, and the scrub growing on abandoned pasture both possess a semi-(future)-natural structure. Their age-structure and the relative importance of coppice and saplings grown from seed will differ, this difference being indirectly a measure of human influence. Past-natural features are those which have survived from a state preceding human influence. They have mostly been modified and some, such as the changing composition of Epping Forest, have changed for partly natural reasons during the period of human activity. In practice past-natural features are usually mixed with future-natural features. A coppiced woodland of lime, oak and hazel growing on light, acid soils in an ancient East Anglian woodland may plausibly be regarded as having a past-natural composition (though its structure is almost wholly artificial), but if coppicing is neglected and birch invades, the composition becomes more future-natural (and so does the structure). Clearly, secondary woods can only contain future-natural woodland features, and past-natural woodland features can only occur in ancient (strictly, primary) woods.

Some ecologists have assumed that natural succession after human disturbance will restore the natural state which preceded human disturbance, i.e. that future-naturalness equals past-naturalness. Watt (1934) described several secondary successions on abandoned pasture in the Chilterns and assumed that they were developing towards a state resembling the mature beechwoods of the district. Subsequent studies by Mansfield (1952) and Roden (1968) showed that the 'climax' beechwoods did indeed have past-natural features, notably the widespread presence of beech, but that their structure and composition had been greatly modified by management. Nevertheless, it was reasonable to assume that past- and future-natural states had indeed been similar in the Chilterns, because many soils had not been irreversibly modified under pasture, and the high density of woodland ensured that all species could eventually recover lost ground. Elsewhere, the assumption of equality is

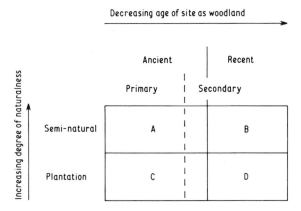

Decreasing age of site as woodland →

Increasing degree of naturalness ↑

	Ancient		Recent
	Primary		Secondary
Semi-natural	A		B
Plantation	C		D

Fig. 3.1 The four main groups of existing woodlands defined in terms of their origin and degree of naturalness. Although the dividing lines between groups A, B, C and D are shown solid, implying that sharp distinctions exist, they are in fact merely arbitrary subdivisions of two continuous variables (shown by arrows).

Examples of each category include:

A, Traditional coppice and wood-pasture stands dating from the Middle Ages or before; native pine woods.

B, Scrub and secondary woodlands developing naturally on formerly unwooded ground.

C, Examples of A which have been clear-felled and replanted with an even-aged stand of broadleaves or conifers.

D, Most recent plantations on lowland heaths and upland peat-lands and moorlands.

unwise. Lime, for example, is a notoriously poor colonist, and it is very unlikely that it will recolonise the widespread sites it occupied in the past. Also, many tree species have been introduced and it is probable that sycamore, for example, will prove to be a common component of future-natural woodland although it was not a component of past-natural stands.

The terms 'ancient', 'primary' and 'semi-natural' are sometimes used interchangeably as if they mean the same thing, but this is unhelpful. We are dealing with two independent variables. One variable, naturalness, describes the degree of human influence on the stand, soils and associated communities, and ranges from 'natural' through 'semi-natural' to 'plantation' (when, in the interests of simplicity, the whole woodland is categorised). The other variable

describes the history of the site as woodland, and is simplified into the categories 'ancient' and 'recent'. The two variables are linked only at the level of kinds of naturalness. If they are combined at the simplest level, naturalness and site history form four major groups, which are illustrated with examples in Fig. 3.1. Since both variables are continuous, some stands must fall on the borderlines: a neglected shelterbelt will be changing from recent plantation (D) to recent semi-natural (B).

3.2 Soils of primary woodland

When we identify a wood as primary we imply that its soil has remained undisturbed by cultivation and major changes in vegetation, and we are entitled to assume that the soil attributes remain close to natural. Primary woodland soils are not, however, completely undisturbed, nor have they necessarily remained unchanged for long periods.

The raw, undifferentiated and often base-rich soils left behind after the last glaciation developed into the mature, stratified and often acid profiles which we see today. The soil-forming processes included leaching and lessivage, whereby dissolved substances and clay particles respectively were washed down the profile and, to some extent, out of the soil altogether. Leaching has tended to have the greatest effects in moist climates, on coarse soils and on land which was inherently poor in bases. It has been counterbalanced where there was a significant inflow of nutrients, such as at the base of slopes and other flushed situations, in oceanic climates where the rainfall is enriched with salts from the sea, and on rocks which weather rapidly or otherwise release large amounts of nutrients into the soil. Some 2–4000 years after the last glaciation – well before the development of the Atlantic Forests – natural woodland soils had matured into a near-stable state, which may have generally been a form of brown earth (Dimbleby, 1965). Subsequent natural changes in undisturbed woodland soils would have been very slow, the rate of soil genesis decreasing logarithmically under stable conditions.

Natural woodland soils were disturbed by burrowing animals and the growth, decay and fall of trees. Similar disturbance continued after such woods had been transformed into coppices, but the character of this disturbance changed. Big trees no longer toppled over, excavating pits on one side whilst throwing up mounds on the other. Instead, more and smaller trees were felled and extracted, and their roots were sometimes excavated for fuel. Ponds were dug, boundary and dividing banks were constructed, and shallow quarries were excavated. Rides and ride-side ditches were carved through the majority of lowland woods and generally linked to a ramifying system of drainage ditches.

The effects of these activities are difficult to estimate, but the general impression is that several centuries of coppice management have not fundamentally changed the character of ancient woodland soils. Certainly, this was assumed by Mackney (1961), who treated the soils of the ancient coppices of Sutton Park (West Midlands) as controls for the study of soil changes under different vegetation outside the wood. Pits, ponds and banks usually disturb only a small fraction of the total area of any given wood. Drainage ditches, whose effects are likely to be all-pervading, have not always been kept open and many seem to be ineffective. For example, the orange and grey mottles which form in the surface water gleys of the ancient woods in the east Midlands reach roughly the same height in the profile, irrespective of whether a ditch is close or distant. (This does not entirely preclude the possibility that the moisture regime near the ditch has been altered.) Disturbance due to timber extraction was, until the advent of modern machinery, probably no greater than disturbance due to the natural fall of great trees. Coppice management sustained for centuries may, however, have reduced the fertility of woodland soils, because extraction of timber and brushwood creates an additional drain on nutrients. Presumably the base-status of woodland soils has been forced to a lower equilibrium. Changes in the major tree species could have altered the soils through the differential effect of various trees on the litter and humus and thence the rate of leaching and lessivage. Simplification of woodland composition may have caused some soil deterioration, leading, for example, to the alleged oak-sickness of the soils beneath the oak coppices of the Wyre Forest (Salop and Worcs).

Soils of ancient woods sometimes contain what

appear to be direct evidence that they have not been disturbed. The lime coppices in Lincolnshire mostly occupy soils with a light, acid surface horizon over a strongly mottled, structureless, neutral or alkaline clay subsoil. In some woods this surface horizon is over one metre deep and the influence of the underlying clay is negligible, but in others the surface sand is so thin that it is barely detectable. Nevertheless it is there, it has clearly survived several centuries of coppice management and it could hardly have survived a single ploughing. The surface horizons of lowland woods on clays are often markedly more acid than the subsoil: this surface acidification apparently formed after centuries of slow leaching and would vanish with a single ploughing.

The soils of ancient woods merit a book on their own. Individual woods vary enormously in complexity: contrast, for example, Hayley Wood (Cambs), whose soils are all referred to the Hanslope Series (Martin and Pigott, 1975), with Bedford Purlieus (Cambs), whose undisturbed areas contain soils of four major groups and numerous sub-groups (Stevens, 1975). The extremes are perhaps the amorphous, alkaline organic ooze beneath alder carrs, the rendzinas beneath beech hangers and the podzols which have developed in coarse, bouldery ground beneath the Highland pinewoods. The conditions which are actually encountered can be illustrated by reference to the 700 profiles examined in my sample plots (Sections 7.3 and 8.1) (Table 3.1), which, although they were taken neither randomly nor systematically, are believed to be reasonably representative of British woodlands. Organic soils are infrequent, developing either on waterlogged, neutral–alkaline conditions (e.g. alder carrs) or freely-drained acid sites (e.g. some pine woods with incipient blanket peat). Ancient woodland mostly occupies mineral soils, with a slight but definite bias to heavy soils, especially in the lowlands. Most profiles in light and medium soils are freely drained, but not surprisingly most heavy soils are poorly drained. Heavy soils cover the full range of soil reaction and profile drainage, with a slight bias towards neutral and alkaline conditions, but light soils are mostly acid. Heavy acid clays may be infrequent, but alkaline sands are almost unknown.

Table 3.1 The frequency of some soil characteristics in the ancient, semi-natural woodlands of Britain. These figures come from 700 samples recorded in ancient, semi-natural woodlands throughout Britain (Chapter 7). The tables show the incidence of soils in four texture classes: heavy, medium, light and organic. Within each texture class the incidence of various combinations of profile drainage and soil reaction is given. Although the samples were taken neither randomly nor systematically, they are believed to be reasonably representative of British woodlands.

*Heavy**	SA	A	N	C			*Medium*	SA	A	N	C	
F	12	8	24	47	91	F		24	27	23	35	109
I	12	10	9	9	40	I		19	11	5	4	39
P	13	28	32	17	90	P		15	17	9	3	44
VP	8	18	15	7	48	VP		5	5	13	1	24
	45	64	80	80	269			63	60	50	43	216

Light	SA	A	N	C			*Organic*	SA	A	N	C	
F	67	32	10	4	113	F		1	9	—	—	10
I	21	8	9	1	39	I		—	—	—	—	0
P	14	10	2	—	26	P		—	—	—	—	0
VP	5	2	3	3	13	VP		1	—	11	2	14
	107	52	24	8	191			2	9	11	2	24

* *Key to classes and abbreviations*

Texture
  Heavy: Clay, silty clay, silty clay loam, clay loam.
  Medium: Silt, silty loam, sandy clay, sandy clay loam, loam.
  Light: Sandy loam, loamy sand, sand, gravel, boulders.
  Organic: No mineral horizon in top 20 cm.

Reaction
  SA Strongly acid, pH below 4.5.
  A   Acid, pH 4.5–5.4.
  N   Neutral, pH 5.5–6.9.
  C   Calcareous or alkaline, pH above 6.9.

Profile drainage
  Depends on height at which mottles or other signs of poor drainage were observed.
  VP Very poorly drained, 10 cm or above.
  P   Poorly drained, 11–20 cm.
  I   Imperfectly drained, below 20 cm.
  F   Freely drained.

3.3 Flora of ancient coppice woodlands

It is possible that the herbaceous communities of ancient woods have descended without much alteration from those of the primaeval, natural woodlands which once occupied the same ground. After all, woodland composed of native species has grown continuously on the site, and historical references contain no evidence that the ground flora was directly affected by traditional management. Management has certainly affected the ground flora indirectly, but in the coppice woods these effects were probably slight. We presume, therefore, that the ground flora of ancient coppice woods is essentially natural, and we are sustained in this belief by the correlations which we repeatedly observe between particular species and identifiable features of their natural environment.

A woodland contains three major plant groups which function almost as independent communities. The tree and shrub layer, or stand, creates the woodland, forming an uppermost canopy and subordinate strata of small trees and shrubs. Epiphyte communities of mainly bryophytes and lichens grow on the trees and shrubs: the stand is both their shelter and substrate, but the epiphytes do not apparently influence the growth and composition of the stand. The field layer of mainly vascular plants and bryophytes itself consists of more-or-less distinct layers of small shrubs (e.g. bramble), herbs and a ground layer of bryophytes. The stand and the field layer are both

Fig. 3.2 Mixed field layer without clear dominants growing along a small stream flowing down the slopes of The Hudnalls, Glos, a fine mature stand of beech, oak, lime, ash, alder, etc., on the Old Red Sandstone of the lower Wye Valley. This community includes *Anemone nemorosa, Cardamine flexuosa, Chrysosplenium alternifolium, C. oppositifolium, Circaea lutetiana, Dryopteris filix-mas, Galeobdolon luteum, Galium odoratum, Mercurialis perennis, Oxalis acetosella* and *Rubus fruticosus*.

rooted in the soil and compete for moisture and nutrients. The stand shades the field layer and to some extent controls its growth and composition, but in a reciprocal manner the field layer can influence the composition of the stand. Tree seedlings are part of the field layer and must compete with it, so it is hardly surprising that, for example, the rate and distribution of ash regeneration in the woods of the Derbyshire limestone is controlled by the growth of *Mercurialis perennis* (Wardle, 1959).

Within any wood one can generally find a range of field layer communities (Figs. 3.2 and 3.3). Many were described in the earlier, descriptive phase of British ecology (Tansley, 1939), but we await the forthcoming National Vegetation Classification (Chapter 7) for a complete coverage. Meanwhile, the vascular plant species found frequently in each woodland type are listed in Chapter 8. The interaction between individual species, communities and the factors which affect them is exemplified by Hayley Wood (Cambs) (Rackham, 1975), where six distinct communities have been recognised, their distributions being related

47

Fig. 3.3 Carpet of dog's mercury, *Mercurialis perennis*, on freely-drained, calcareous loam in High Wood, Horningsham, Wilts, where the field layer is almost monospecific. The alder in the background is part of an outlying example of plateau alder wood (stand type 7C). The standard oak in the foreground must be over 200 years old. (Photo. P. Wakely, Nature Conservancy Council)

to fine variations in slope and drainage (Fig. 13.10) and their composition and structure varying from deep shade to open canopy (Table 3.2). The range of soils and shade to which individual species are limited is determined first by their intrinsic response to these factors, and is then further limited by the degree of competition from other species. Any community recognised in one wood can usually be found in other woods in the vicinity and may also be found in more distant regions, but since no two species have exactly the same distribution, any floristically defined community can form over only a limited geographical range, beyond which it grades into other more or less similar communities on similar soils.

Most herbaceous species have a recognisable habitat range within woods (Fig. 3.4). Thus, for example, *Blechnum spicant* is found on acid soils, *Iris pseudacorus* on wet, base-rich organic and light-textured soils, *Primula elatior* on clay, *Urtica dioica* on soils rich in phosphate and *Viola odorata* on calcareous soils. Some, such as *Melampyrum cristatum* and *Vicia sylvatica* thrive in the half-shade of wood margins. Others grow well when a wood is grazed (e.g. *Deschampsia flexuosa*) whereas others (e.g. *Luzula sylvatica*) growing in the same site are much reduced. The ecological range of some species varies from one part of its range to another: *Convallaria majalis* (Fig. 3.5), for example, is predominantly calcicole in the uplands and calcifuge

Table 3.2 Summary of variation in the composition of the ground vegetation of Hayley Wood, Cambridgeshire in relation to variation in shade and drainage (Rackham, 1975). The species involved are *Carex acutiformis* and *C. riparia* (big sedges), *Primula elatior* (oxlip), *Filipendula ulmaria* (meadowsweet), *Deschampsia caespitosa*, *Endymion nonscriptus* (bluebell), *Rubus caesius*, *Mercurialis perennis* (mercury), *Poa trivialis*, *Hypericum hirsutum*, *Juncus conglomeratus*, *J. effusus*, and *Rubus fruticosus*, sect. *Triviales*.

Zone	Dense shade	Normal uncoppiced canopy	Open canopy	Clearings and ride edges
1	Bare ground	Sparse, non-flowering big sedges	Dense, non-flowering big sedges	Dense flowering big sedges
2	Bare ground or sparse oxlip	Oxlip with sparse meadowsweet	Meadowsweet with oxlip and occasional *Deschampsia*	
3	Sparse oxlip and bluebell	Oxlip and bluebell co-dominant	Meadowsweet and bluebell with oxlip, *Rubus caesius*, occasional *Deschampsia*	Various combinations of *Deschampsia*, *Poa trivialis*, meadowsweet, *Hypericum hirsutum*, *Juncus* spp., *Rubus* sect. *Triviales*, etc.
4	Patchy bluebell	Carpet of bluebell	Bluebell with *Rubus caesius* and *Deschampsia*	
5	Mixture or mosaic of bluebell and mercury		Bluebell and mercury with *Rubus caesius*, *Deschampsia*	
6a	Mercury with sparse bluebell		Mercury with *Rubus caesius* and frequent *Deschampsia*	
6	Carpet of mercury			

Increasing slope and drainage →

in the lowlands. Sometimes the available conditions impose apparent variation on a species. For example, *Endymion non-scriptus* is often abundant on acid and neutral soils in the English lowlands, yet in Bedford Purlieus (Peterborough) it behaves as a calcicole because elsewhere in the wood the neutral and mildly acid soils are all too heavy and poorly-drained and the freely-drained sands are too acid. Despite such apparently clear circumstances, the beautifully intricate and multifarious relationships between woodland species, communities and their environment remain imprecise enough to tantalise ecologists and defy exact explanation.

The ancient woodland flora seem to remain more-or-less constant, adapted to a stable environment and apparently capable of sustaining itself indefinitely. Such records as we have, suggest that the distribution of species and communities remains fairly stable, provided that management is fairly constant. The communities which were described and mapped by Wilson (1911) in some North Kent woods and

Adamson (1912) in Gamlingay Wood (Cambs) are still there, and the vegetation zones of Hayley Wood have not moved significantly in the 30 years since they were first mapped. The oxlips and other species listed by John Ray 320 years ago in Kingston and Madingley Woods (Cambs) (Ewen and Prime, 1975) are mostly still there, and many populations of rare species are known to have survived in one spot for decades.

Several writers have claimed that certain species are restricted to, and thus indicative of, ancient woodland. Sir Hugh Beevor's assertion (1925) about *Endymion non-scriptus* in Norfolk is typical: 'Our original woods may, I believe, be readily identified, because every wood containing the wild hyacinth I take to be such. Outside the wood, bluebells rarely appear in the hedgerow, if so, they proclaim a woodland that has disappeared . . . you may meet a planted wood of a hundred years old, but how much does it fail to charm compared with the primaeval wood, with its glory of bluebells.' Tansley (1939) mentions

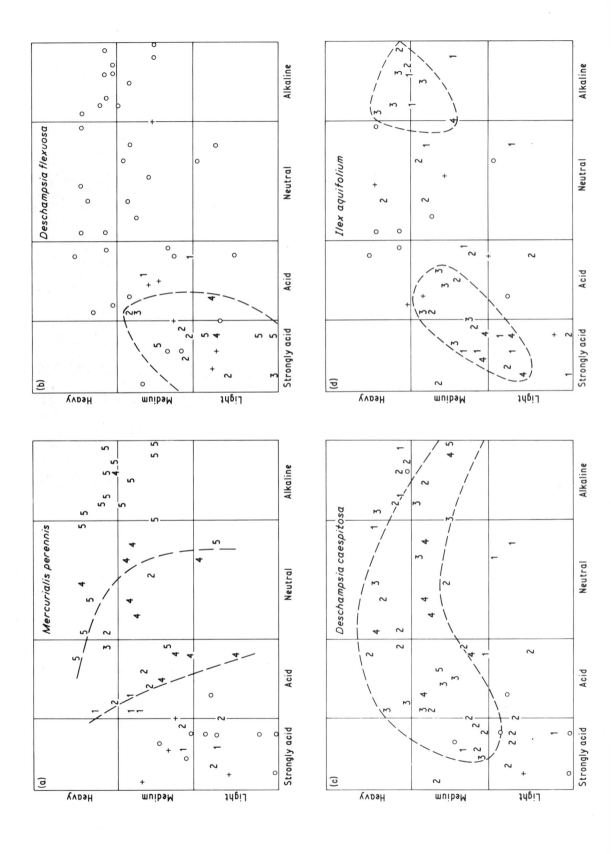

Fig. 3.5 *Convallaria majalis* (Lily of the valley) grows on highly calcareous soils over limestone and strongly acid sands, both of which soils are dry. A good ancient woodland indicator in central Lincolnshire and many other areas. (Photo. P. Wakely)

species which are confined to modified fragments of original forest and which have never entered 'planted woods', and he reports F. B. White's observation that *Corallorhiza trifida*, *Goodyera repens* and *Moneses uniflora* (Fig. 3.6) are confined to 'the old pine forests'. Ratcliffe (1968) considered that concentrations of Atlantic bryophytes found in some western woods would not have survived if the tree cover had ever been removed. Pigott (1969) concluded that the populations of native *Tilia* in the Peak District (Derbyshire) were relict stands which had never been cleared, and suggested that certain herb species (e.g. *Convallaria majalis*, *Melica nutans*), which were strongly associated with *Tilia*, were members of a relict community.

Fig. 3.4 Occurrence of three woodland species in relation to soil texture and reaction, (a) *Mercurialis perennis*, (b) *Deschampsia flexuosa*, (c) *Deschampsia caespitosa* and (d) *Ilex aquifolium*. The frequency of each species in each stand type is shown by superimposition on the array of stand types in Fig. 7.4. Frequencies: (+) 0–10%, (1) 11–20%, (2) 21–40%, (3) 41–60%, (4) 61–80%, (5) 81–100%. Broken lines are drawn to emphasise the main concentrations of high frequency.

Mercurialis is mildly calcicole; *D. flexuosa* is strongly calcifuge; *D. caespitosa* is mildly associated with heavy soils across a wide range of pH; and *Ilex* is found mainly on well-drained soils.

Fig. 3.6 *Moneses uniflora* (One-flowered wintergreen), a rare herb of native pinewoods and other woodland types in north-east Scotland. (Photo. P. Wakely)

My own observations in Lincolnshire confirm that many species are strongly associated with ancient woods and wood-relict hedges (Pollard, 1973) derived from them, though few species are absolutely confined to such sites (Figs. 3.5, 3.7, 3.8, 3.9, 3.10). The list in Table 3.3 is based on species lists from over 300 woods in an area centred on Woodhall Spa, Bardney and Market Rasen, where 85 separate, surviving ancient woods (43% by area of all woodland in the study area) have been identified from historical and archaeological evidence (Sections 13.2, 13.3). This list is true only for the study area, for in other parts of Britain many of the species are known to occur commonly in recent woods and even in unwooded ground. Indeed, very few species appear to be

strongly associated with ancient woods throughout their British range: amongst the most convincing are *Convallaria majalis*, *Dryopteris aemula*, *Primula elatior*, *Sorbus torminalis* and *Tilia cordata* (Fig. 3.7).

3.4 Cyclic effects of coppicing on the flora

The response of ground vegetation to the coppice cycle was described in some of the earliest woodland studies (Wilson, 1911; Adamson, 1912; Salisbury, 1916). When a coppice is cut the light which reaches the field layer suddenly increases from perhaps 5% of full daylight in midsummer to virtually 100%. Under full shade at the end of the 'shade phase' the field layer is generally scattered, poorly-grown and apparently

Table 3.3 Ancient woodland vascular plant species in central Lincolnshire.

LIST A

Species with a strong affinity for ancient woods, showing little or no ability to colonise secondary woodland, and rarely found in other habitats.

Anemone nemorosa	*Luzula sylvatica
†Aquilegia vulgaris	*Lysimachia nemorum
Calamagrostis canescens	†Maianthemum bifolium
Campanula trachelium	*Melampyrum pratense
†Carex laevigata	Melica uniflora
*Carex pallescens	Milium effusum
*Carex pendula	*Neottia nidus-avis
Carex remota	Orchis mascula
†Carex strigosa	Oxalis acetosella
Convallaria majalis	*Paris quadrifolia
*Dipsacus pilosus	Platanthera chlorantha
†Equisetum sylvaticum	Potentilla sterilis
Galeobdolon luteum	Primula vulgaris
*Galium odoratum	Scrophularia nodosa
†Lathraea squamaria	*Sorbus torminalis
†Lathyrus montanus	*Tilia cordata
*Luzula pilosa	†Vicia sylvatica

* Six or more locations, more than 90% of which are in ancient woods.
† Five or fewer locations, all in ancient woods.

Other species (no superscript) occur in six or more locations, 75–90% of which are in ancient woods.

LIST B

Species with a mild affinity for ancient woods, generally with 50–75% of all woodland locations in such woods. Some occur sparingly in other semi-natural habitats.

Adoxa moschatellina	Geum rivale
Agropyron caninum	Hypericum hirsutum
Allium ursinum	Hypericum tetrapterum
Calamagrostis epigejos	Mercurialis perennis
Campanula latifolia	Myosotis sylvatica
Carex acutiformis	Ranunculus auricomus
Carex sylvatica	Stellaria holostea
Chrysosplenium oppositifolium	Valeriana officinalis
Conopodium majus	Veronica montana
Corydalis claviculata	Viola reichenbachiana
Endymion non-scriptus	Viola riviniana
Fragaria vesca	

Table 3.4 Composition and biomass of the ground flora in 450 m^2 plots in chestnut coppice at different stages of the management cycle in Ham Street Wood, Kent (Ford and Newbould, 1977). Biomass is given as the maximum dry weight in kg ha^{-1} achieved in the season.

Year since coppicing	1	2	5	9	15
Shade species					
Anemone nemorosa	34	259	53	21	3
Endymion non-scriptus	1	130	32	174	1
Pteridium aquilinum	69	336	436	20	3
Rubus fruticosus	23	1048	202	49	8
Other woodland species					
Ajuga reptans		×			
Digitalis purpurea		×			
Dryopteris filix-mas				1	
Euphorbia amygdaloides	×	23	10	1	
Fragaria vesca		×		3	
Galeobdolon luteum		13			
Hypericum perforatum		16	4	×	
Juncus sp.	1	2	1	×	
Lonicera periclymenum			×	5	
Lysimachia nemorum		5	7	×	
Luzula pilosa		1	1		
Primula vulgaris	×	2			
Teucrium scorodonia	×	697	109	2	
Veronica officinalis		15	7	×	
Viola riviniana		18	17	3	
Adventive species					
Cerastium sp.		×			
Chamaenerion angustifolium	×	175	31	6	
Cirsium arvense		1			
Galeopsis sp.			1		
Lamium purpureum	×				
Myosotis sp.			2	1	
Potentilla anserina			2		
Sonchus oleraceus		12			
Shrubs and tree seedlings					
Betula pubescens	8	294	127	266	
Castanea sativa					×
Carpinus betulus	×	1	1	×	×
Crataegus monogyna			×	×	
Ilex aquifolium	×	1	×		×
Populus tremula			2		1
Quercus robur	×	2	1	×	
Sarothamnus scoparius			2	×	
Ulex europeaus			3	36	
Number of species	14	23	30	23	8
Ground vegetation biomass	135	2160	956	279	16

× maximum biomass below 0.5 kg ha^{-1}.

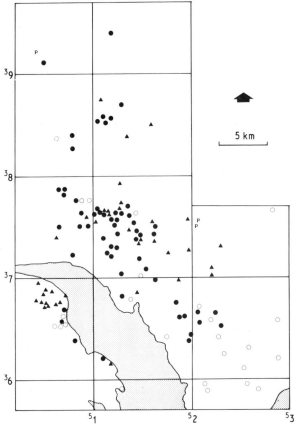

Fig. 3.7 The distribution of small-leaved lime, *Tilia cordata*, in central Lincolnshire. The area shown lies to the east of Lincoln and measures 28 km (or 18 km) by 40 km, extending from the limestone in the west across the Witham Fens (shaded) on to the central clay vale and the chalk wolds in the east. All ancient woods are shown by a circle at their central point.

●, Ancient wood containing *T. cordata*.

▲, *T. cordata* in wood-relict and other mixed hedges.

P, *T. cordata* in secondary woods (invariably planted).

○, Ancient wood containing no *T. cordata*.

impoverished, but in the first season after coppicing – at the start of the 'light phase' – the species already growing beneath the shade grow more vigorously. In the second season the response is tremendous: the shade flora grow and flower with even greater vigour, herb species appear which formerly were not visible and tree and shrub seedlings become established. After two seasons the new coppice shoots are probably 1–2 m tall, and in each

subsequent season their height and productivity increases whilst the productivity of the field layer declines beneath the developing canopy. Eventually the 'shade phase' returns. The cycle is illustrated by the response of the field layer growing in a chestnut coppice on poorly-drained, acid, medium-textured soils in Kent (Table 3.4).

Herbaceous species react to the critical dark phase in five ways (Salisbury, 1924; Rackham, 1975). Some tolerate the shade in a predominantly vegetative form [e.g. *Ajuga reptans*, *Galeobdolon luteum* (Fig. 3.8)]. Others likewise continue to grow, but by starting activity in the early spring they can also flower freely

Fig. 3.8 *Galeobdolon luteum* (Yellow archangel), characteristic of medium-heavy soils mainly in the English lowlands. A very slow colonist of secondary woodland in central Lincolnshire. (Photo. P. Wakely)

Fig. 3.9 *Milium effusum* (Wood millet) in a Somerset hazel coppice. A distinctive grass, confined to woods, which is a good indicator of ancient woodland in central Lincolnshire. (Photo. P. Wakely)

(e.g. *Anemone nemorosa, Ranunculus ficaria*). Most of the species in both these groups are perennials. A third and rather important group survives as dormant seed, buried in the litter and upper mineral horizons of the soil (e.g. *Digitalis purpurea, Juncus conglomeratus*). The dormant seed germinates in response to the environmental change after coppicing. The significance of this group has recently been emphasised by Brown and Oosterhuis (in press). Another group, the 'marginal flora' retreats to ride sides and clearings (e.g. *Cirsium palustre, Calamagrostis epigejos, Melampyrum cristatum*). Most of these species are annuals, biennials or short-lived perennials. Finally, there are casual species with good powers of dispersal and these colonise the bare ground which is available after cutting on cultivated ground outside woods (e.g. *Anagallis arvensis, Cirsium arvense*). Individual species may overlap two groups, e.g. *Rubus fruticosus*, which bears shade, but also has an abundant store of buried seed. *Chamaenerion angustifolium* is an archetypal casual species but can also stand the shade of a closed broadleaf canopy. Thus the coppice system produces a perpetual alternation of light and shade, vigour and decline, appearance and disappearance, flow and ebb, which presumably remains more or less stable from one cycle to the next.

The crucial point in the cycle comes in the two years immediately after coppicing when herbaceous species must colonise, germinate and grow rapidly in order to fruit successfully in the next few years of enhanced competition. Conditions at this time vary – one spring may be dry and the next may be wet – and this influences the growth and competition between species. The effects of chance circumstances in producing local variations in the ground flora was well illustrated in Hayley Wood (Rackham, 1975), where the 1967 coppice plot became dominated by *Juncus* and the adjacent 1968 plot was dominated by *Deschampsia caespitosa*, the change of dominants coinciding with the boundary of the plots. Such effects are likely to persist through at least one coppice cycle, imposing an extra dimension of variation caused by a temporary, vanished combination of circumstances.

Coppice management may be partly responsible for the famed 'vernal aspect' of British woods, generated by carpets of spring-flowering *Anemone nemorosa, Endymion non-scriptus, Primula elatior*, etc. These species flower vigorously in the second spring after coppicing and then survive the subsequent competition with the help of shade from the canopy. The spectacular displays of white, blue and yellow also occur in derelict coppice, where the mature canopy transmits enough light for them to flower, but not enough for their competitors to grow vigorously. Spring-growing species receive some 50% of full daylight before the canopy buds break and thus utilise the absolute peak period for light beneath a temperate deciduous woodland canopy (Fig. 3.11). They derive some benefit from coppicing, but are not unduly

Fig. 3.10 *Primula vulgaris* (Primrose), a widespread and well-known woodland herb which also grows in hedgerows, sea cliffs and ungrazed grassland on, for example, railway embankments. (Photo. P. Wakely)

inconvenienced by the shade phase. Summer-growing species (e.g. *Rubus fruticosus*) receive about 5% of full daylight in their growing season. They benefit greatly from the light phase, but are subdued by the shade phase. Coppice management therefore appears to shift the competitive balance to spring-growers by regularly causing a bottleneck of summer light in the pole stage stand at the end of a normal coppice cycle.

Salisbury (1924) also noticed that coppicing has some effect on the soil and thus indirectly on plant growth. During the light phase the soil surface is liable to dry out in summer, but the subsoil water-table rises because the pumping effect of transpiration from coppice is temporarily reduced. As a result the bulbs of *Endymon non-scriptus* were shallower in the light phase than in the dark phase of the Hertfordshire coppices, and no doubt bluebell was also forced to retreat from the wettest soils it had occupied under the closed canopy. The soils of recently coppiced plots became warm earlier in the season and advanced the onset of flowering. Organic matter decayed faster,

releasing mineral nutrient and nitrates, and slightly acidifying the surface.

3.5 Natural features in existing coppices

The communities of trees and shrubs which form the underwood of ancient, semi-natural coppices have long been managed as a crop, but historical evidence suggests that, though some have certainly been planted, many have not. This raises the possibility that the underwood itself is not generally the creation of past woodmen, but an original-natural woodland community which has survived in ancient woods because the woodland was never cleared and the management was not significantly selective with respect to species. The evidence for this hypothesis is based on the absence of any record of planting in many coppices, and admittedly it is somewhat negative (but this may be all we can expect, for 'natural' means the absence of human influence, and this cannot be strictly proven). More direct and

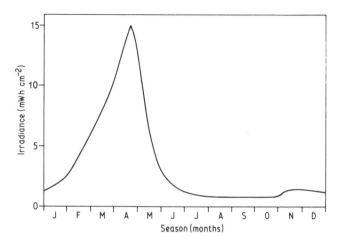

Fig. 3.11 Daily totals of irradiance measured beneath a deciduous canopy in Madingley Wood, Cambs (Anderson, 1964). The graph shows the seasonal trend of light reaching the ground vegetation beneath an ash, oak, elm, hazel canopy. The values shown are mean irradiance: variations from one day to the next were up to *c* 50% of this mean. Light in the open reaches a peak in late June and a trough in late December, but at ground level in the wood the peak comes in late April just before the canopy leaves expand and remains almost uniformly low from late June to mid-January.

positive evidence is available in the form of features which we would expect to observe if the hypothesis is true.

First, we would expect to see a difference between known planted coppices and supposed semi-natural coppices. In central Lincolnshire, where some secondary woods planted between about 1700 and 1850 were managed as coppices, the planted stands are mostly pure or mixtures of two species. The coppice species include aliens to the region, such as sycamore and hornbeam. Stool size is generally small and uniform and straight planting lines can be observed in some sites. In contrast, the coppices of ancient woodland sites contain a mixture of stool sizes and species (Fig. 3.12), and rarely include aliens: sycamore, when present, is usually an obvious recent invader. The underwood of ancient woods is thus unlikely to have been planted after 1700, and before 1700 planting was generally infrequent.

Second, existing coppice and its antecedents on the site should be similar. Pollen in deposits at Old Buckenham Mere, Norfolk, for example (Godwin, 1968), show that the tree species present in the region before the onset of major prehistoric clearance were similar to those present in historic times and still present in ancient, semi-natural coppices, although their proportions have changed. Historical evidence often shows that the composition of the underwood has been stable for centuries. In Rockingham Forest (Peterken, 1976) there have been minor changes in the proportions of species over the last 400 years, mainly an increase of ash and hazel at the expense of

the formerly abundant 'thorn', and some aliens have become established. Elsewhere there have been seemingly greater changes. For example, in 1587 the mixed coppices in Borrowdale were composed of hazel, birch, ash, holly, blackthorn and whitethorn in various combinations with a few small oaks as saplings and pollards (PRO DL 42.114, unpublished), but now the woods in that area are almost all dominated by oak. The ancient coppice hedges provide further evidence, if one accepts that there has been no reason to change their composition over the centuries. Allowing for the fact that they provide drier, better drained and often more calcareous soils than the woods they surround, their composition is usually similar to that of adjacent underwood. Where it is different there is generally a history of planting in the last 200–300 years, as in those plateau beech-woods of the Chilterns which have hornbeam in the hedges.

Third, traditional woodmen regarded the ancient coppices as natural. Reporters for the Board of Agriculture, writing about 1800, would have been made well aware of any significant planting, and some gave details of the species which were being planted into coppices, yet they repeatedly refer to coppice as 'seemingly natural', or they write that its composition was 'left to chance'. Reviewing these, Jones (1961) inferred that the original coppice composition was mixed, particularly on calcareous soils, and that many coppices which had not been improved still remained. The same point is made by Nisbet and Vellacott (1907) when describing 1083 acres of 'old

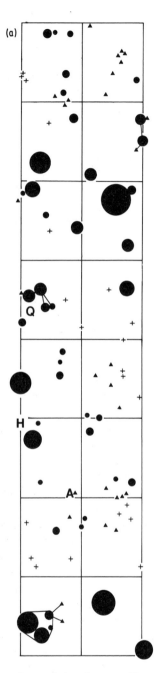

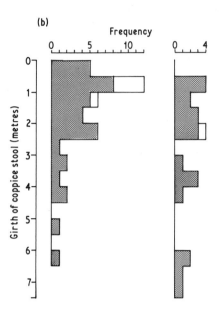

Fig. 3.12 Small-leaved lime, *Tilia cordata*, coppice in Easton Hornstocks (Northants).

(a) Map of the distribution of lime stools (solid circles) and maiden stems (triangles) in an 8 m × 40 m strip. Lime stools are represented by circles, equal in circumference to their girth at ground level at the scale drawn, even though most were irregular in plan. Certain vegetative connections between stools and saplings are indicated by lines. One oak standard (Q), felled at the last coppicing 30 years ago, a hazel stool (H), an ash stool (A) and maiden birch and ash (+) are also shown. Since the oaks nearby were mostly *Quercus petraea* this stand conforms to type 4C.

(b) Girth distribution of lime coppice stools in two 8 m × 40 m sample plots. The left-hand diagram, which represents the area mapped in Fig. 3.12(a), shows an area of rapid stool initiation, mostly from root suckers, and few large stools. The right-hand diagram shows a contrasting area with many large stools – many of which are becoming rings of functionally separate smaller stools – and negligible stool initiation. The two samples were about 200 m apart in an unbroken stand dominated by lime coppice beneath oak standards. The unshaded zone indicates dead stools, killed recently during a phase of coppice neglect.

copsewoods' on the Sedbury Park Estate (Glos): 'the underwoods . . . comprise a mixture of oak, hazel, birch, alder, black withy, lime, maple whitebeam and chestnut growing at random . . . [which] do not appear to have been planted, but seem to be remnants of the ancient woodlands formerly existing in this part of the country'.

Fourth, *Tilia cordata* is a frequent coppice species in eastern England, where it is present in ancient woods, but almost entirely absent from secondary woods and planted coppices (Fig. 3.7). Pollen records show that it was a common constituent of original-natural woodland in the region, and has been present ever since. Did medieval woodmen plant lime? It is difficult to see why they would bother when it was already there, and there is no historical or pollen evidence that they did. The most likely explanation is that this species, together with others in ancient, semi-natural coppice is indeed a direct descendant on the site of original-natural woodland.

Finally, perhaps the most positive and persuasive evidence that the underwood communities of mixed coppice are indeed mainly past-natural communities, directly descended from original-natural woodland, lies in the nature of variation in composition, so well expressed by Rackham (1976): 'Much variation . . . in underwood . . . appears to be almost wholly natural, due to the intrinsic properties of the trees and to their reactions (present or past) to soils and other aspects of the environment. Natural variation is often gradual and subtle, as with the complex underwood changes in the Bradfield Woods. Often the trees and herbaceous plants vary together. . . . Sudden changes in woodland can arise from natural features such as spring-lines; and there are naturally gregarious trees, for example, hornbeam and small-leaved lime, that tend to occur in patches with abrupt edges. Natural transitions, whether sharp or gradual, are irregular, eschew straight lines, and disregard management boundaries'. The causes of variation are not always apparent, as in the rich, complex and somewhat damaged underwood of Bedford Purlieus (Peterken and Welch, 1975). A straight, sharp and obviously artificial boundary within a wood does not necessarily mean that the stands on both sides were planted: the same result is achieved by planting on one side only.

3.6 Flora of ancient wood pasture

Pasture and coppice woods are structurally complementary. Coppice management maintains the underwood and, at least in its traditional form, allows few timber trees to grow to a mature size, whereas in pasture woodlands the underwood is virtually eliminated and most of the trees are mature or over-mature. The coppice field layer is usually dominated by broadleaved, shade-bearing herbs and is largely free of the influence of grazing, but the field layer of pasture woods is generally dominated by heathland and grassland communities. The two systems were not totally distinct, for coppices were much more commonly grazed in the past and several intermediate and transitional states were possible. Some woods were managed as coppice-with-pollards, and vestiges of this arrangement could still be seen recently in the neglected coppices of Whaddon Chase (Bucks). Most coppices had hedgerow pollards on their boundary banks. Some woods changed from coppice to wood pasture, e.g. in the New Forest (Hants) and Hatfield Forest (Essex). Even so, the two systems were reasonably distinct. They provided complementary circumstances in which the main structural components of primaeval woodland could survive into modern times, even though they were teased apart. The mature timber trees and epiphytes survived in wood pasture, and the underwood and field layer communities survived in coppices.

The species of the woodland field layer are obviously capable of bearing shade, but another characteristic, their intolerance of sustained grazing, may be equally significant. Herbaceous communities containing many woodland species are often found on ungrazed mountain ledges (Edgell, 1969), sea cliffs and railway embankments where, for example, *Primula vulgaris*, *Luzula sylvatica* and rarities such as *Pulmonaria longifolia* thrive in the absence of both shade and grazing. Understandably, therefore, the field layer of wood pasture contains relatively few woodland species and, if the canopy is open, is little different from nearby grasslands and heaths without trees. Some species, which are capable of surviving both shade and grazing, are frequently found in wood pasture (e.g. *Conopodium majus*, *Corydalis claviculata*, *Deschampsia flexuosa*, *Pteridium aquilinum*, *Teucrium scorodonia* and

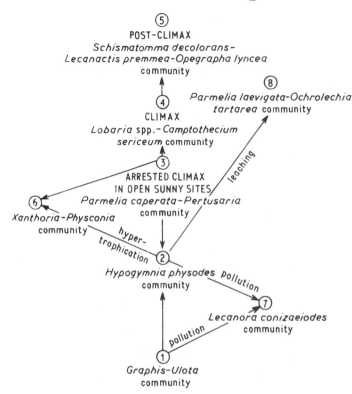

⑤
POST-CLIMAX
Schismatomma decolorans–
Lecanactis premmea–Opegrapha lyncea
community

⑧
Parmelia laevigata–Ochrolechia
tartarea community

④
CLIMAX
Lobaria spp.–*Camptothecium*
sericeum community

leaching

③
ARRESTED CLIMAX
IN OPEN SUNNY SITES
Parmelia caperata–Pertusaria
community

⑥
Xanthoria–Physconia
community

hyper-
trophication

②
Hypogymnia physodes
community

Pollution

⑦
Lecanora conizaeiodes
community

pollution

①
Graphis–Ulota
community

Fig. 3.13 Major epiphytic lichen communities in the New Forest (Rose and James, 1974). The pH of the bark decreases in time from stages 1 to 2, then normally increases up to stage 5. There is a further increase to stage 6 but a sharp decline to stages 7 and 8. *Sorbus aucuparia, Ilex* and *Corylus* support stages 1, 2 and 7; *Fraxinus* 1–4. Stages 1–4, 6 and 7 apply to all tree species with a smooth bark which becomes roughened with ageing. Stage 5 applies to *Quercus* only. Stages 6 and 8 are very poorly developed in the old woodlands; stage 6 requires hypertrophicated conditions characteristic of wayside trees in farming areas and stage 8 requires more consistently moist conditions than is usual in the New Forest.

Vaccinium myrtillus) and most pasture woods contain a few bitten-down patches of *Mercurialis perennis, Primula vulgaris* and other typical woodland herbs, usually in moist or remote ground. Despite the impoverishment, pasture woods may well contain a few woodland herbs which depend on habitat continuity much as they do in coppices: candidate species in the New Forest include *Euphorbia amygdaloides, Ruscus aculeatus* (N.Flower, personal communication), *Anemone nemorosa, Lysimachia nemorum* and *Oxalis acetosella.*

Epiphyte communities have been favoured by wood pasture because the mature trees upon which they thrive have been continuously present. Only in ancient wood pasture is it reasonable to assume that the mature timber habitat is primary. Rose (1974) has recorded 65 species of bryophyte and 303 lichens on oak, and rather fewer species on other native trees (Table 3.5). Some of the richest communities are found in the New Forest (Rose and James, 1974) and in a range of woods close to the Atlantic coast, notably in north-west Scotland. Nine distinct

Table 3.5 Comparison of numbers of lichen taxa recorded as epiphytes on thirteen tree genera or species in the British Isles (Rose, 1974).

Number of lichen taxa	Tree species
303	*Quercus (Q. robur* and *Q. petraea)*
230	*Fraxinus excelsior*
194	*Fagus sylvatica*
171	*Ulmus* spp.
170	*Acer pseudoplatanus*
128	*Salix (S. cinerea* and *S. caprea)*
124	*Corylus avellana*
93	*Betula (B. pendula* and *B. pubescens)*
88	*Acer campestre*
72	*Alnus glutinosa*
68	*Ilex aquifolium*
66	*Tilia* spp.
42	*Carpinus betulus*

communities have been recognised in the New Forest (Fig. 3.13). Some lichen species, which are found mainly in ancient pasture woodlands but not (in the lowlands at least) in secondary wood pasture with

Fig. 3.14 *Lobaria pulmonaria* (L.) Hoffm., one of the most impressive and interesting epiphytic lichens. Away from the western seaboard, this species occurs only where mature trees have been present throughout historical times. (Photo. P. Wakely)

equally elderly trees, appear to require continuity of mature timber for survival (Fig. 3.14). The species in question have been listed by Rose (1974; 1976). The species in question have been listed by Rose (1974; 1976) and used as an 'Index of Ecological Continuity' for individual woods.

The composition of wood pasture stands in forests has changed drastically within relatively recent times. In Epping Forest, for example, the original-natural lime-dominated woodland was transformed by selective forest clearance in the middle Saxon period into the beech and hornbeam woods we see today (Baker *et al.*, 1978). Some of the existing mature beechwoods of the New Forest were dominated by oak until the end of the Middle Ages (Barber, 1975), and previously contained lime. Hazel has been greatly reduced since the Middle Ages. Although wood pasture 'management' has clearly influenced these changes, it is possible that beech and hornbeam would have increased at that time in undisturbed woodlands, which suggests that, unlike coppices, the naturalness in wood-pasture stands is more 'present' than 'original'.

The trees of ancient parklands are mostly far less natural than those of forest wood pasture. Only exceptional parks, such as Staverton (Suffolk) (Peterken, 1969a), have escaped the widespread planting of exotic species during and after the landscape park era. Some parks, such as Moccas (Hereford), contain more-or-less distinct generations, an older semi-natural generation dating from the 17th century or before, and younger generations of planted exotics and native species. Even the older generation is likely to be a careful selection from the naturally available species, made possible and necessary by the care that must have been exercised during the Middle Ages in order to maintain the succession of park trees in the face of perpetual deer grazing. During the last two centuries, neglect of grazing in some parks has permitted natural regeneration around and below the mature park trees, thus initiating a secondary succession within an ancient semi-(past) natural stand (e.g. Staverton Thicks).

3.7 Fauna of ancient woodland

Woodlands contain far more animal species than plant species (e.g. Monks Wood; Table 13.2). The complexity of the relationships between each animal species and its vegetable and physical environment can be appreciated from Elton's (1968) study of Wytham Hill (Oxon), an area of woodland and grassland which contains some ancient woodland, and from which about 5000 species of animal have been recorded. Justice cannot be done here to the woodland fauna, except to consider the evidence that woodland continuity is a significant ecological factor for animals as well as plants.

3.7.1 Black hairstreak butterly (*Strymonidia pruni* L.) (Fig. 3.15)

This is one of the best examples of an invertebrate species which is almost confined to ancient woodland

Fig. 3.15 Black Hairstreak butterfly, *Strymonidia pruni* (L.) (Lepidoptera: Lycaenidae). (Photo. J. L. Mason)

(Thomas, 1974). Its larvae feed on blackthorn (*Prunus spinosa*), a widespread shrub of mainly wet, heavy soils, but in order to be suitable the bushes must have more than 20 years growth, a condition which develops when a wood is relatively neglected. Individual black hairstreak populations are remarkably conservative, persisting in particular parts of a wood but scarcely moving to other, apparently suitable, blackthorn banks in other parts of the same wood.

The black hairstreak is confined in Britain to woods along the Midland clay belt from the fen edge to Oxford, where it is found almost exclusively in ancient coppice woods, though a few populations survive in hedges and two populations thrive in secondary woods close to ancient woods. The Midland clay belt contained a string of medieval forests and chases where there was a high density of semi-natural woodland, long-term stability of woodland sites, relatively long coppice rotations combined with chronically inefficient Crown management, heavy, wet soils conducive to blackthorn and a high proportion of thorn in the coppice [which presumably included blackthorn (Peterken, 1976)]. Each of these features favours the survival of black hairstreak populations, and it was evidently only in this region that they were all continuously present through the era of traditional management. It is reasonable to suggest that the present distribution of the black hairstreak is partly determined by these historical factors.

3.7.2 Terrestrial molluscs

Terrestrial molluscs are sedentary and slow-moving

animals, and it is therefore no surprise to learn that some are slow colonists. Paul (1978) has compared the Mollusc fauna of west Cambridgeshire woods (Table 10.1) in a way which allows historical effects to be detected. Ancient (primary) semi-natural woods contained more species than secondary woods of any age, but secondary woods were rapidly colonised if they lay adjacent to ancient woods (Table 3.6). The ancient woods were not only richer in species, but also contained larger populations of most species. The species of ancient semi-natural woods, which were assumed to be the relicts of the Atlantic forest fauna, had responded in different ways to the formation of secondary woodland and the disturbance of ancient woodland. Some were ubiquitous in other habitats and were clearly capable of rapid colonisation, but others were confined to ancient woods, e.g. *Azecta goodalli* (Férussac), *Carychium tridentatum* (Risso). Two ancient woods, which had been replanted with conifers, lost many species as the plantations reached the pole stage.

Table 3.6 The number of Mollusc species recorded in ancient and recent woodland on calcareous boulder clay in Cambridgeshire (Paul, 1978).

Woodland type	Mean number of species	Number of sites studied
Ancient, semi-natural, probably primary (e.g. Hayley Wood)	24.7	6
Ancient, semi-natural, secondary (Papworth Wood, Overhall Grove)	15.5	2
Recent, broadleaved	15.5	2
Recent, semi-natural, but adjacent to ancient, semi-natural woodland (e.g. Hayley triangle)	23.0	2
Ancient, probably primary, but recently replanted with conifers	18.0	2

Paul's observations follow the early work of Boycott (1934), who was interested in the colonising ability of woodland slugs and snails, and moreover used the modern concept of 'ancient' woodland. No species was absolutely confined to woodland, but many were found mainly in a series of relatively sheltered, undisturbed, semi-natural habitats, amongst which woodland was the most abundant and widespread. The 49 'woodland species' – which also occurred in hedges, scrub, sea cliffs, limestone screes, etc. – formed almost half the total British fauna, and by definition woods were amongst the richest habitats for them; Boycott listed 37 species from Witcombe Wood (Glos), an ancient woodland on the oolite scarp of the Cotswolds. Paul (1975) lists 43 species which are in or by Hayley Wood (Cambs). Several factors tended to favour a large number of species in a particular wood: (i) calcareous soils, (ii) a sheltered, moist environment, and (iii) the fact that a wood was ancient rather than recent. Most of these 'woodland' species were anthropophobic, i.e., they 'do not flourish under any intensive conditions of human occupation', because 'man destroys their characteristic habitats, and if he makes new ones which are suitable it takes a longer time than has yet elapsed for them to be occupied'. (Boycott, 1934).

The colonising ability of the woodland species appears to be generally limited. Boycott noted, for example, that 'the woodland species living in the old forests on the South Downs and the Cotswolds have not spread into the neighbouring beech plantations which were made so freely from about 1750'. Some species would evidently not become established in apparently suitable ground even if they are placed there by hand. Two slugs, *Limax cinereoniger* Wolf and *L. tennelus* Müller occur throughout Britain in woods as different as southern calcareous beechwoods and Highland pine forest, yet they are found only in woods 'which have every appearance of being ancient and which are generally large': the sites he lists are indeed all ancient woodland. Furthermore, 'they do not occur in woods which look like or are known to be plantations, and their discovery is probably as good a piece of evidence as can be had that a wood is on a primaeval site'. Other species which are listed by Boycott and are still believed to be good ancient woodland indicators (E. Pollard, personal communication) are *Acicula fuca* (Montagu), *Acanthinula lamellata* (Jeffreys) and *Ena montana* (Drapernaud).

3.7.3 Coleoptera

Almost 4000 species of beetle have been recorded in Britain, of which approximately 650 are 'woodland species' (Hammond, 1974). Individual woods may harbour well over 1000 species, and the richest

Fig. 3.16 Pupa of *Stenostola ferrea* (Schrank) (Coleoptera: Cerambycidae), a beetle associated with lime and hazel in ancient woodlands of England and Wales. (Photo. F. A. Hunter)

wooded areas, notably the New and Windsor Forests, have each yielded almost half the British total. The composition of the British Coleopterous fauna has changed markedly over the post-glacial period, initially in response to changing climate, but in the last 5000 years clearance of woodland and fragmentation of remaining woods, combined with extension of cultivation and habitation, have been more potent causes of change than changing climate. These developments are described in a valuable review by Hammond (1974), on which this section is largely based.

One long-term change of particular significance is the decline of species dependent on old forests, particularly species associated with the wood of mature or dead trees and with arboreous fungi (Figs. 3.16 and 3.17). Many of these are in the families Cerambycidae, Cleridae, Colydiidae, Cucujidae and Elateridae. They have been found in recent times mainly as small populations scattered in widely separated localities, where they are regarded as relict populations surviving from natural woodland.

If they are indeed relicts, one would expect to find evidence of (i) a former wider distribution, (ii) the tenacity of populations in existing localities, and (iii) that their present populations lie in relict stands of ancient trees. Evidence supporting all these expectations is forthcoming. Buckland and Kenward (1973) examined fossil beetles from Thorne Moor (Yorks), where they had lived in Bronze Age times in the natural, mixed oak forests. They found a number of

Fig. 3.17 Adult *Saperda scalaris* (L.) (Coleoptera: Cerambycidae) beetle on birch, its main food plant in Scotland. Although this species occurs throughout Britain it is commoner in the north, where it associates with ancient broadleaf woodland. (Photo. F. A. Hunter)

species which no longer occur in Britain, and others which are now very rare and confined to old woodland localities, mainly to the south of Thorne Moor. An example quoted is *Teredus cylindricus* (Olivier) which was abundant at Thorne, but is now restricted to parts of Sherwood and Windsor Forests. They ascribe these changes mainly to the loss of suitable woodland: 'many species dependent on mature forest must have been reduced to populations whose small size would render them liable for extinction by any major environmental change, the isolation of each community precluding re-immigration'.

Local extinction of isolated populations of woodland species is continuing. In Essex these species are increasingly confined to the large woodland block of Epping Forest, following progressive local extinctions in the scattered woods elsewhere in the county, whereas in the more extensive woodland of south-east England the woodland beetles are still widespread. In Britain as a whole perhaps 20 old forest species and other species associated with mature trees have become extinct since 1800. Nevertheless, many small populations are known to have survived in favoured localities over a long period, often with gaps of decades separating successive sightings.

The major refuges of the old forest species are the larger surviving ancient wood-pasture stands, e.g. New, Epping, Windsor and Sherwood Forests, and medieval parks at Moccas and Grimsthorpe, etc. These are sites where mature and dead wood is likely to have been present continuously through historical times, so the correlation is consistent with the relict hypothesis. One limiting factor may be the beetles' relatively poor power of dispersal, but this is difficult to prove, because very few sites are available where mature timber is present in a site which is not a survival of medieval wood pasture. Crowson (1962), however, has some evidence that colonising ability is limited. He points out that few woodland beetles have colonised apparently favourable sites in the 19th century broadleaf plantations of Scotland, and many are apparently confined to oak woods on ancient or primary woodland sites, such as the Clyde Valley. Some are evidently old forest species as discussed above, for they are confined to the few broadleaf woods which were not brought into coppice, such as Hamilton High Parks and Dalkeith Old Wood.

Although some species are wingless the majority possess functional wings, but apparently they are rarely induced to undertake large-range dispersal. Crowson suggests that these species will fly only in warm, humid, still air mainly in late spring and early summer, but in Scotland the temperature is high enough only at midday, when beetles in flight are likely to stay within shelter. Crowson (1972) records a similar observation on the Morecambe Bay limestone, ascribing the absence of the 'more restricted (and often flightless) woodland species' from Eaves Wood, Silverdale and their presence at Roudsea Wood to the discontinuity of woodland on the former site. Hammond (1974) points out that, 'as certain species, which are rarely found outside of woodland in the east may be found in the open in the west, the destruction of woodland habitats may be expected to have had a differential effect on such species according to the part of the country'. The parallels with vascular plants (Sections 3.3; 5.5) are close.

Chapter 4

High forest management

'High forest' is not synonymous with 'plantation', although many observers of modern British forestry probably think that it is. Planting, like sowing and natural regeneration, is a method of establishing a stand. 'High forest' describes stands grown mainly from seedlings rather than coppice shoots. In Britain, high forest and plantations are strongly correlated, for most high-forest stands are plantations, whereas an unknown but high proportion of coppices are of natural origin.

Troup (1928) describes the silvicultural systems which have been employed in British and Continental woodlands. The high-forest systems can be grouped as follows:

(A) Clear-cutting system. A stand is cleared in a single felling and replaced by an even-aged stand, usually by planting.

(B) Shelterwood systems. A stand is cleared in two or more successive fellings (known as 'regeneration fellings'). The new stand is established between the first and the last regeneration fellings, often by natural regeneration. Although stands are more-or-less even-aged, a two-aged structure is temporarily created during regeneration.

(C) Selection system. Felling and regeneration are distributed continuously throughout the forest area, unlike the other systems within which at any given time they are concentrated on particular parts of a forest. Regeneration is usually natural. Troup would

have classified the selection system as a special form of shelterwood.

Woods were rarely managed as high forest during the Middle Ages. In Rockingham Forest, for example, coppice and wood pasture predominated until the 19th century, and in 1565 Roger Taverner recorded just one, small high-forest stand in the 127 Crown woods which he surveyed (PRO LRRO 5/39). Many woods in the Highlands possessed a high-forest structure during the Middle Ages, but these were natural woods, which were not systematically managed (Section 2.6). High forest grew in importance after the Middle Ages under the influence of writers such as Standish (1616) and Evelyn (1664). Planting and afforestation became commonplace, and many coppices were eventually converted to a high-forest system. During the present century high forest has replaced coppice as the dominant management system in British woods (Table 4.1), so much so that only high-forest systems are seriously considered now by British foresters.

4.1 Plantation management

British forestry is based on even-aged plantation high forest managed by a clear-cutting system. The plantations are established on 'bare ground' (i.e. land devoid of trees) or woodland which has been 'clear felled' (i.e. reduced to bare ground by felling all the previous stand). Seedling trees, or 'plants', are raised in a nursery, then planted out when they are 2–3 years old. The species selected for planting are determined by commercial judgement and site conditions, tempered somewhat by non-commercial factors such as amenity. In practice, most planting by the Forestry Commission has been coniferous, since conifers are regarded as the only trees which will grow on many of the poor, peaty, upland soils where most of the plantations are established. (In the year ended 31 March 1977, conifers comprised just over 99% of all new planting and restocking by the Forestry Commission, according to their Annual Report.) Most planting is arranged in regularly spaced parallel rows, because this permits the most efficient use of available plants, and the subsequent care is easier and therefore less expensive. Initially, the growth of ground vegetation ('weeds') and any woody growth from the previous crop or self-sown trees ('scrub') must be controlled ('cleaned') by hand, machine or selective herbicides in order to ensure that it does not smother the young plants. Some plants succumb, nevertheless, leaving 'blanks' in the developing plantation which have to be 'beaten up' by planting replacements. Once established and growing well the plantation will generally have very few blanks, but it may contain a few self-sown trees which escaped removal during cleaning.

As the plantation grows the canopy closes and the stand passes through the 'thicket' and 'pole' stages, during which the dense shade effectively controls any remaining competition from weeds and scrub. The original 'stocking' (i.e. density of trees) greatly exceeds the stocking needed to produce the 'final crop' of mature trees. 'Thinning' is therefore undertaken at intervals, when a proportion of the 'standing crop' is removed to provide small poles, an early financial return on the investment in planting, and more space for the remaining trees to develop. Lower branches are killed by the shade from higher branches, and, if the subsequent improvement in timber quality is likely to justify the expense, their remains are removed from the trunk ('brashing') and left to rot on the ground. Eventually, the last, or 'final' thinning is complete and only the trees which form the 'final crop' remain: these may be vigorous trees of good growth spaced evenly throughout the

plantation, or they may simply be all trees in every third row of the original plantation – irrespective of quality of growth – if, for economic reasons, the stand had been 'line-thinned'. The mature plantation is allowed to stand until its growth rate declines, the owner needs the cash it represents, a good market for the timber prevails, the due date in the management plan is reached, or until high winds blow it down. Once the decision is taken, the final crop is clear-felled, leaving bare ground once again available for planting. The full cycle from planting to felling ('rotation') may be as short as 40 years for some conifers, and as much as 120–150 years for some broadleaves, such as oak and beech. The norm for the extensive coniferous plantations is about 50–60 years.

This close control of stand composition and structure is extended to the site itself. Afforestation in the uplands is usually preceded by ploughing, which improves drainage and exposes mineral soil, and fertilisers are often applied. In long-established woods, which have undergone conversion from coppice to plantation high forest, the ride-side ditches have frequently been deepened for the same reason.

The main costs of the clear-cutting system are site preparation, planting, fertilising, and control of competing herbaceous and woody growth, i.e. establishment costs which are incurred at the start of the rotation. Significant benefits accrue only at the end of the rotation when the final crop is felled, though thinnings may yield some early returns. The private grower must therefore commit himself to an investment which endures for at least 40 years and generally much longer, and he will only do this if he is sure that the product can be sold at a reasonable price and that the proceeds will benefit his family. Confidence – both commercial and political – is therefore highly important, and recent lack of confidence is blamed for the recent fall in the rate of planting. Economic calculations are based on compound interest on the initial expenditure over the whole rotation, so the cost of longer rotations becomes disproportionately large, and the exact rates of interest which should be applied to forestry materially affect the outcome: they are a matter for debate (Helliwell, 1974; Price, 1976). Economic considerations tend to shorten rotations and favour fast-growing species (mainly conifers). Broadleaf species are nevertheless grown at least

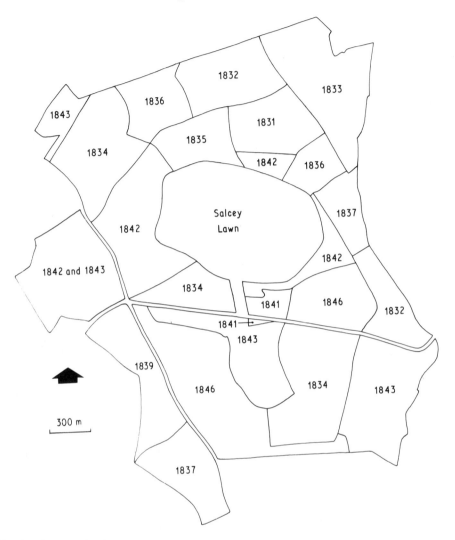

Fig. 4.1 The distribution of pedunculate oak plantations established between 1831 and 1846 in Salcey Forest (Bucks and Northants) (redrawn from an original stock map, Forestry Commission office, Northampton).

partly for non-commercial reasons and latterly with an enhanced subsidy from the State, but here too, economic pressures have generated a search for faster-growing species (such as *Nothofagus*) and systems which can shorten rotations, such as free-growth (Jobling and Pearce, 1977). The economics of private timber growing has been discussed by Lorrain-Smith (1969).

The plantations established on moorland during the 20th century have a well-deserved reputation for monotony. Rapid afforestation in any area inevitably creates plantations with a very limited spread of ages. Most of the expansion of Britain's forests has been achieved with just five species: Sitka Spruce, Scots Pine, Larch, Norway Spruce and Lodgepole Pine. Most stands are monocultures. To take just one small example: between 1950 and 1955, 274 ha of moorland at the southern end of the Long Mynd (Salop) was afforested entirely with conifers, and 174 ha of this was planted with a mixture of Scots Pine and Sitka Spruce. Fortunately, felling of such stands may be spread over a longer period than the original planting, so plantations of the second and subsequent rotations will probably have a wider range of age-classes.

Some long-established high forest plantations have a considerable diversity of species and age-classes, even though most stands are generally monocultures or mixtures of two species. The Inclosures of the New Forest now contain broadleaf stands dating from the 18th–20th centuries, and even coniferous stands ranging up to well over 100 years. Some of the main features and trends in plantation forestry are shown by Salcey Forest (Bucks and Northants), a large but isolated wood which was managed as coppice-with-standards until 1830. Between 1831 and 1846, the entire wood was planted with pedunculate oak (Fig. 4.1). This was thinned between 1867 and 1890. The surviving hazel underwood was still being cut in the 1890s and perhaps later. Salcey Forest must have been fairly monotonous at this time, but from 1901 onwards the oak was slowly replaced, and a more diverse age-structure was created. Planting in the 20th century went in phases (Fig. 4.2). From 1901–1943, mainly broadleaved stands of oak and ash were planted, with just a few mixed stands: there was a marked hiatus in planting during the 1920s, when Salcey was first managed by the Forestry Commission. From 1943 until 1960, most plantings were mixed, consisting mostly of oak with Norway Spruce, and beech (which is probably not native to the area) was increasingly introduced. Then, during the 1960s, only pure conifer stands were planted; some (mainly those in the early 1960s) were planted beneath poorly-growing broadleaf stands established earlier in the 20th century. The result by 1969 (Fig. 4.3) was a remarkably complex pattern of species and age-classes, but the continued replacement of mature oak by conifers after 1969 eroded this diversity and brought forth protests from naturalists. The recent history of forestry in Salcey Forest epitomises the impact of forestry in the traditionally broadleaved woods of the English lowlands, where, from 1945 onwards, the trend in planting was increasingly towards conifers, so much so that by the 1960s most plantings were purely coniferous, and many earlier 20th century broadleaf stands were underplanted with conifers. During the 1970s, fortunately, a more balanced approach has been adopted under pressure from environmentalists, and more broadleaved stands are now retained or planted.

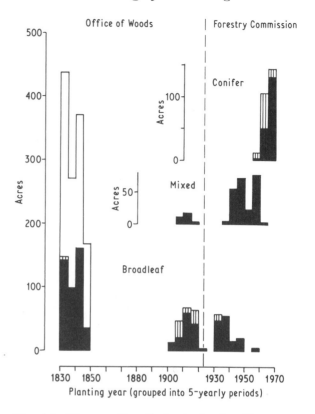

Fig. 4.2 The age of broadleaf, mixed and coniferous plantations in Salcey Forest (Bucks and Northants) in 1969 (simplified from the Forestry Commission stock map. Broadleaves underplanted with conifers are shown in both broadleaf and conifer categories but distinguished by stripes. Broadleaf stands planted between 1831 and 1846 but felled and replaced after 1900 are unshaded. Thus, the diagram summarises the entire history of planting between 1831 and 1969.

4.2 Shelterwood systems

Shelterwood systems are characterised by clearance of the mature stand in two or more successive 'regeneration' fellings and the establishment of the new stand in the shelter of mature trees remaining from the previous felling. The resulting stands are often more-or-less even-aged, though the stand passes through a two-aged condition during regeneration. Compared with the clear-cutting system, shelterwood systems have several advantages. They aim at natural regeneration and therefore avoid the cost of planting. Young

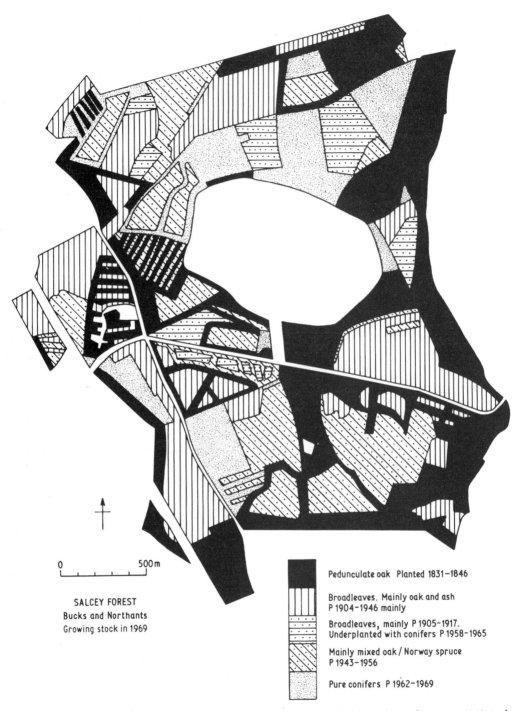

0   500 m

SALCEY FOREST
Bucks and Northants
Growing stock in 1969

Pedunculate oak Planted 1831–1846

Broadleaves. Mainly oak and ash
P 1904–1946 mainly

Broadleaves, mainly P 1905–1917.
Underplanted with conifers P 1958–1965

Mainly mixed oak / Norway spruce
P 1943–1956

Pure conifers P 1962–1969

Fig. 4.3 The distribution of various types of plantation in Salcey Forest (Bucks and Northants) in 1969 (redrawn from the Forestry Commission stock map). Replacement of the early 19th century oak over a prolonged period, covering many changes in forestry policy and practice, had by 1969 produced a diverse wood. Since then more of the mature oak has been felled.

saplings are better protected from frost, drought and winds, and the soil is less likely to dry out or (on steep slopes) be eroded. Storm and snow damage is generally less. The best trees of the old crop are spared for rapid incremental growth at the end of the rotation. These advantages are offset by several disadvantages. Shelterwood systems require more skill than clear-cutting, yet the final felling of the old stand still partly damages young growth. Since the work is dispersed over a larger area, shelterwood systems are less economical. Furthermore, the composition and genetic characteristics of the new crop is strongly determined by the old crop, which can be either an advantage or a disadvantage.

Shelterwood systems have played a minor role in British forestry for several reasons. They are, of course, possible only where a mature high-forest stand exists, and in Britain most long-established woods have been managed as coppice until recently, so modern foresters have not generally wished to retain the coppice species. Hence, virtually the only form of shelterwood in Britain has been underplanting with beech or shade-bearing conifers beneath the shelter of old coppice growth. Where high-forest stands of beech, oak and pine have been due for replacement, the available stock was often considered to be of inferior quality, and the cost of planting after clear-cutting was generally felt to be worthwhile in the face of unreliable natural regeneration, an insignificant gain in protection and general lack of the skills required for shelterwood regeneration. Furthermore, some suitable stands had to be clear-felled during the World Wars. Nevertheless, some oak and pine stands in the New and Dean Forests and some beechwoods on chalk have been successfully regenerated naturally under a shelterwood system. Perhaps such systems would have been more widely applied if British forestry had not been so concerned with afforestation and if taxation and grants for forestry had not been based on the assumption that most stands should be managed on the clear-cutting plantation system.

On the Continent, where natural regeneration is more reliable, where long-established high-forest stands are common and widespread, and where the mountainous terrain creates a greater need for protection measures, shelterwood systems are widely adopted. Some of the most impressive are the oakwoods of France, which are managed by the 'Uniform' system, a form of shelterwood in which the mature stand is thinned uniformly throughout. Thinning is carried out first by a 'seeding' felling, in which the object is to initiate regeneration by admitting more light to the ground and leaving the best trees as seed sources, and then by 'secondary' fellings, in which more light is given to the young seedlings which have appeared. The mother trees are removed at intervals until, in the 'final' felling, they are all removed, leaving a dense, more-or-less even-aged thicket of young oak, beech and other species. Thinning follows at regular intervals when unwanted species are removed, yet the stand remains dense. Oak is the main species in the canopy after a few decades, but beech is permitted or even introduced to grow as an understorey, where it keeps the oak trunks clean, maintains a mull humus, suppresses 'weeds', and provides some extra timber. Periodic thinning with the object of developing straight, clean-limbed oaks with good crowns continues until the stand is mature. On the medium-textured, acid soils of Tronçais and Bercé, the sessile oak is grown on a rotation of about 220 years on a plan which has altered little since the 17th century reformation of French forestry under Colbert. The results, in both the beech and the oak forests, are quite magnificent stands, which cannot fail to make a tremendous impression on the visitor from Britain. Reed (1954) has admirably described some of the more famous forests, together with their social and historical background. Plaisance (1963) provides a detailed inventory of French forests.

Other shelterwood systems are mostly more complex than the uniform system, but are advantageous in some circumstances (Troup, 1928). The 'Group' system is similar, and likewise produces an essentially even-aged stand, but the first fellings are made in groups centred on any pre-existing groups of saplings. These groups are subsequently enlarged until the old stand is completely removed. The system, which during regeneration provides somewhat better protection than the uniform system for both the old and the new stands, has been commonly practised in Central Europe in mixtures of beech, spruce and silver fir. The proportions of these species can be regulated by the size and distribution of group fellings, for spruce demands more light than beech and fir. The 'Irregular Shelterwood' system is also practised on the beech, spruce and silver fir woods of hilly districts in Central Europe, especially in Switzerland where the additional protection it confers is particularly valuable. It involves opening a mature canopy slowly, by light thinning at irregular intervals over perhaps fifty years or more, to produce a distinctly uneven-aged new stand.

The 'Strip' and 'Wedge' systems differ from the uniform, group and irregular systems in that the compartments are regenerated by clear-cutting narrow strips or wedge-shaped patches, followed by planting or natural regeneration. These are still classified as shelterwood systems by Troup (1928), but here the shelter is from the side. They can provide good shelter against the prevailing wind (but are vulnerable to winds from other directions); the new crop grows in full daylight; and mature trees are felled into the old stands, not the new growth. The strips are cut in sequence, so felling and regeneration proceeds as a broad front through any given part of a forest. The strip systems can be seen as part of a continuum from the selection system, the irregular, group and uniform systems, through the strip system to the clear-cutting system: at one end the pattern of felling and replacement is small and continuous, but on the other it is large and discontinuous.

Two other systems can be conveniently described here, even though Troup (1928) did not classify them as Shelterwood. Both result in two-storied high forest which can be regenerated naturally or by planting. 'Two-storeyed High Forest' is formed when a middle-aged stand of a light-foliaged species is heavily thinned and under-planted with a shade-bearing species, and both species are then allowed to grow on together. Beech, for example, may be grown under pine. In 'High Forest with Standards' a few selected trees of one stand are retained through the next rotation in order to provide good-quality large timber.

4.3 Selection system

The selection system (Fig. 4.4) is an extreme form of shelterwood in which selected trees are felled continuously throughout the forest area and natural regeneration springs up in the gaps. This perpetual thinning generates no spatial separation of age-classes, unlike the other shelterwood and clear-cutting systems in which age-classes are separated to a greater (clear-cutting) or lesser (irregular shelterwood) extent. Each part of a selection forest is worked at, say, 10-year intervals; a proportion of the stems of all usable-size classes are removed, especially those of the less-vigorous individuals so that the more-vigorous trees can continue to grow. Regeneration, which is usually but not necessarily natural, generally develops in groups, and hence the age-classes of a mature selection stand also tend to occur in small groups rather than as an even spread throughout the stand. The system works best for shade-bearing species which can regenerate from seed, though Ammon (Reade, 1969) argues that it is suitable for native and naturalised species including light-demanders. Certainly light-demanding species can be given a better chance by felling in small groups, i.e. by the 'Group Selection' system. Selection forestry requires great skill on the part of the forester, both in selecting which trees to fell and in bringing them down without damaging other trees. Operations are dispersed through the forest and thus more expensive. On the other hand, the selection system provides the best protection from soil erosion, drought effects and damage by wind and snow, so it is most appropriate for mountain forests. Felling and regeneration are continuous, so the system makes use of all mast years and provides a constant income even from small woods: hence it is appropriate for small, private woodlands. Furthermore, the external and internal appearance of a selection forest constantly remains a more or less mature forest, so the system is especially appropriate where woods are important in the landscape.

According to Troup, much of the primitive exploitation of European forests took the form of selection fellings, although they were uncontrolled. Selection fellings were common and widespread in the Middle Ages, but the forests tended to deteriorate because trees were selected for immediate needs, not the long-term health of the forest, and uncontrolled grazing retarded natural regeneration. Control of felling was necessarily introduced, but in most forests selection fellings were replaced by the shelterwood

Fig. 4.4 Selection forest, alias Plenterwald, at Emmental, Switzerland. A stand of Norway spruce, silver fir and beech is repeatedly thinned in order to maintain an uneven-aged structure wherein all age classes are mixed together throughout the wood. This structure is not natural. It has been created by sustained, skilled treatment. (Photo reproduced by permission of the director of the Swiss Federal Institute of Forestry Research, Birmensdorf, Zürich)

systems described above. However, selection forests remain frequent in the forests of beech, silver fir, spruce, larch and Cembran pine in the mountainous districts of Europe, especially in Switzerland and France. Some 10% of German forests were managed by the selection system in the 1920s. In Britain, the Chiltern beechwoods were latterly worked as a form of selection forestry, but evidently the woodmen tended to cut too many of the larger and better trees, thus exerting a negative selection pressure. In earlier centuries the common woods must have been selectively felled, but none survive in this form. In the Landes (France), however, at Biscarrosse, a maritime pine forest is maintained under a selection system by the interaction of common rights to timber (whereby trees are generally felled when they reach usable sizes)

and resin (the tapping of which reduces the value of a tree as timber, and ensures its survival into old-age).

Modern selection forests (for instance, those in Emmental, Switzerland) are quite magnificent. Great trees, perhaps 300 years old, tower up into the canopy. Beneath them smaller trees of all sizes and ages develop deep layers of foliage, and groups of saplings of various ages develop in any gaps. The strong vertical lines of the trunks contrast beautifully with the horizontal boughs of foliage, which form from top to bottom of the stand. This is the common conception of a virgin forest, but of course it is quite unnatural, the product of sustained, skilled and highly regulated forestry. Nevertheless, selection forestry has been strongly advocated on the ground that it is the closest approximation to the natural forest (Köstler, 1956; Reade, 1969).

4.4 Conversion of coppice to high forest

Substantial areas of former coppice woodland have been changed to high-forest management in recent decades. Two processes are involved, (i) promotion of the coppice direct to high forest, and (ii) reforestation by planting. The latter is much the more important, as three study areas in the east Midlands show (Table 4.1).

Table 4.1 Restocking of woodlands in the east Midlands between 1946 and 1972/3 (Peterken and Harding, 1975).

	Rockingham Forest	Central Lincolnshire	West Cambridgeshire
Woodland promoted from coppice to high forest (ha)	50	94	7
Woodland clear-felled and replanted (ha)	675	755	123

The nature of the coppice itself limits the potential extent of promotion to high forest. Clearly promotion is out of the question in a hazel coppice, and undesirable in coppices where, even though the coppice species can grow to tree-size, the timber is unsound (e.g. chestnut), the stools are too rotten to support a trunk, or the quality of growth is likely to be poor (e.g. many ash coppices on heavy soils). It is more attractive where the coppice consists of oak, lime, elm, birch, alder or sycamore, and where there is a good stock of maiden stems amongst the coppice. In these circumstances, careful thinning and singling (cutting all but the most promising stem on each stool) can leave a stand which has a high-forest structure. Some former western oak coppices have been treated thus, and in recent years a few attempts have been made to promote lime coppice in Lincolnshire. Simple neglect often achieves the same result in longer time.

Reforestation by planting is a drastic process in which the derelict coppice is removed by felling, scrub bashing and herbicides, and an even-aged high forest is planted in its place. The process differs little from afforestation by planting, but the costs of clearing the ground and controlling the weeds and woody regrowth are greater. Reforestation can be achieved beneath a shelterwood of old coppice but eventually this is removed, leaving an even-aged stand. For a period it was common practice to kill the shelterwood with herbicides and leave the standing dead remains to disintegrate over the years, but the resulting scene of devastation offended too many sensibilities.

Several conversions were made in earlier periods, notably in the Royal Forests of England and in the southern beechwoods. In the New Forest some coppices (Sumner, 1926) were converted to beech or oak high forest by 1700, e.g. Ridley Wood and South Bentley Inclosure. Most of the oak high forest established in Alice Holt, New and Dean Forests was on former wood pasture or heath, but in Hazelborough and Salcey Forests the oak over hazel, ash, or maple coppices were reforested with pedunculate oak in the early 19th century, and in Dymock and Tintern the oak and beech coppices respectively seem to have been mainly promoted to high forest.

Perhaps the most interesting conversion from coppice to high forest in Britain occurred in the southern beechwoods (Fig. 4.5). Beech is such a weak coppicer that it may require selection coppicing to keep it going. Even so, the private Chiltern Woods were managed as coppice for centuries until, in the 18th and 19th centuries, the growth of the furniture trade at High Wycombe (and the loss of the London firewood market to coal) led to wholesale conversion to high forest, apparently by a mixture of planting, natural regeneration and promotion (Mansfield, 1952; Roden, 1968). Thereafter, many of the stands received management which was akin to the selection system, with saplings being cut for chair legs, and larger timber for other uses.

The clear-felling of Hailey Wood at Aston Rowant (Oxon) in 1965 provided an opportunity to reconstruct the recent history of a mature plateau beech stand (type 8D). It appeared to be even-aged, but the canopy beech (and a few oak) ranged in origin from 1760–1876 (median origin of 1804). In the older trees the growth rings originating from before about 1851 were narrow, but at this date there was an abrupt increase before slow growth was resumed in later years. Some trees also showed sharp but small

Fig. 4.5 A beech wood in the Chiltern Hills at Park Wood, Bradenham, Bucks. Characteristically beech dominates the stand to the virtual exclusion of other tree species, but in the past, when these woods were mostly coppiced, ash and oak probably shared dominance with the beech. Natural regeneration of beech in clearings is profuse in neighbouring Bradenham Coppice. With bramble abundant the soil is probably heavy and acid. (Photo. P. Wakely, Nature Conservancy Council)

increases in rings formed in the period 1832–41. Younger trees, showing no abrupt changes, followed the normal pattern of accelerating early growth moving smoothly into slow growth at maturity. Combining such clues with observations on the distribution of age-classes, it appeared that before about 1851 the stand had repeatedly been lightly thinned and groups of beech had regenerated in the gaps. The wood was evidently managed as a selection stand, with mature trees growing in a matrix of roughly-even-aged groups of saplings and subordinate trees: in 1851 three such subordinate groups had been 29–41, c.42 and 51–85 years old. In 1851 a major thinning occurred which promoted the growth of hitherto subordinate trees, and allowed a flush of beech and oak regeneration. This ceased about 1876, after which the canopy was presumably too dense for later seedlings to survive, though some holly succeeded in forming a sparse understorey. The stand remained untouched until 1940, when it was heavily thinned. This permitted a strong flush of ash, beech and birch regeneration, which flourished until the stand was clear-felled in 1965. The apparent absence of ash in 19th century regeneration leaves a mystery: perhaps it was cut out because it was less valuable than beech and oak.

4.5 Flora, fauna and soils of high forest

It is difficult to find ecological characteristics which are due to the high-forest structure as such. Most semi-natural high-forest stands originated recently, either by secondary succession or conversion from coppice and wood pasture, so their flora and fauna are likely to be changing. Furthermore, high-forest stands are mostly plantations, whereas the coppices with which they might be compared are mostly semi-natural. Nevertheless, it is worth considering the changes which take place when coppice is converted to high forest, because this must to some extent reflect the change in structure.

4.5.1 Neglected coppice

The stage at which coppices were normally cut is equivalent to the pole stage of a high-forest rotation, and when coppice is left uncut beyond this stage it increasingly approximates in structure to mature high forest (Figs. 8.34, 18.1 and 18.4). The size of individual trees and trunks gets larger but the number of individuals per unit area decreases through competition and self-thinning. Coppiced individuals are often reduced to one or two large trunks surmounting the stool. Canopy structure becomes more complex as the foliage is distributed over a greater height and subordinate shrub strata develop. As maturity approaches, the vigour of the canopy declines, more light reaches the ground and saplings can again become established, having been excluded during the pole stage. Light phases no longer occur, though they may eventually resume at very long intervals. Amongst the field layer, the marginal and casual species which no longer periodically invade the core of the wood may be seriously reduced or eliminated if rides are not kept open. The bank of buried seed remains largely dormant, infrequently renewed but lasting for decades. Summer-growing

species amongst the shade flora probably increase, but the spring-flowering species decrease in the face of increased competition. As the canopy becomes more heterogeneous, so the field layer becomes more patchy. The effects of repeated disturbance and exposure on the soil surface are reduced, and litter and humus are able to build up.

Fig. 4.6 White Admiral butterfly, *Ladoga camilla* L. (Lepidoptera: Nymphalidae). (Photo. J. L. Mason)

These structural changes undoubtedly influence the fauna. The clearings, rides and multifarious margins of a worked coppice, which evidently provide rich opportunities particularly for insects, are largely replaced by the equable, uniform environment of the mature high forest. Inevitably some species benefit whilst others decline. The White Admiral butterfly (*Lagoda camilla* L.) (Fig. 4.6), for example, has spread from southern England into most of the Midlands, East Anglia and south-west England, mainly between 1930 and 1942, in response both to favourable mid-summer temperatures and the decline of coppice management (Pollard, 1979). The eggs are laid on moderately shaded *Lonicera* lianes which are not available in the light or dark phases of worked coppice, but develop when the underwood is spared to grow towards high forest. On the other hand the dormouse (*Muscardinus avellanarius* L.), a rodent which frequents low bushes, scrub and coppices (Corbet and Southern, 1977), has evidently declined, partly it is thought in response to the decline of coppicing.

4.5.2 Conversion of coppice to high forest

Most of the deliberate conversion of coppice into high

forest (as opposed to the passive neglect of coppices) involves planting, and of late most of the plantations have been coniferous or mixtures of broadleaves and conifers. In three study areas in the east Midlands (Peterken and Harding, 1975) where at least 4791 ha of coppice on ancient woodland sites survived in 1946, 388 ha of the coppice had been converted to broadleaf high forest by 1972 – mostly passively – and 1182 ha had been replanted as conifer and mixed high forest. The destruction of ancient, semi-natural coppice stands and the planting of conifers in their place is an ecologically violent change in which the alteration of structure from coppice to high forest is probably insignificant beside the change from deciduous to evergreen and broadleaf to conifer. The full effects have not been thoroughly investigated, and in any case the change has been so recent in most cases that the effects are not yet complete. Casual observations suggest that there is a close similarity with the changes following upland afforestation (Section 5.4). Initially, the field layer responds as if coppicing had been resumed: spring flowers bloom, herbaceous species grow rapidly, scrub springs up and scrub species, such as the grasshopper warbler, move in. Scrub growth is controlled by herbicides and a variety of cutters, so that eventually the young conifers break free from competition, close their canopy and exclude virtually all natural growth. By the pole stage the mollusc fauna has been greatly reduced (Paul, 1978; see Table 3.6). The field layer species and invasive shrubs, such as *Sambucus nigra*, increase towards the end of the rotation, but the community appears to remain greatly impoverished. Eventually, clear-felling creates a new light phase and disturbs the soil surface, enabling ground vegetation to develop strongly until the canopy closes again, but the community generally seems to be more uniform than its counterpart and predecessor in the coppice rotation, with a reduced woodland shade flora and an increased representation of casual species.

Planting conifers in place of ancient coppice involves the clear and unambiguous loss of the semi-natural stand. The spring-growing shade flora is largely eliminated except around margins, but some other woodland herbs can survive a conifer rotation *in situ* as buried seed or depauperate individuals. Marginal species and casuals are less affected. Under larch these changes are relatively small, for the spring flora only has to cope with a change of structure and an increase in litter and humus at the soil surface, but beneath the most productive conifers the change is almost as great as if the wood had been temporarily cleared away. Rides are kept open; the grassland and the pond-margin communities in their ditches are usually maintained or even enriched, provided that chemicals are not used to control scrub. But other 'advantages' are temporary: the surge of herbs and insects which follows in the seasons after planting is a last awakening before the long night of the conifers descends upon them.

4.5.3 Effects of conifers on woodland soils

The effects of introducing conifer monocultures on to soils which previously supported deciduous, broad-leaved stands have been reviewed by Noirfalise and Vanesse (1975) and summarised by Miles (1978). The most obvious change is the accumulation of organic material as a thick mat on the surface, due not to an increased production of litter but to a reduced rate of decomposition. Litter from resinous trees decomposes slower than that from broadleaved trees because it contains less protein and is therefore more resistant to attack from micro-organisms; phenolic substances are liberated (which can be toxic to fish when they leach into streams); and the conifer canopy and litter keeps the soil surface cooler and drier. The microflora and fauna are also substantially modified. Groups such as the woodlice (*Isopoda*), millipedes (*Diplopoda*) and slugs and snails (*Gastropoda*) decline, whilst the mites and ticks (*Acari*) and springtails (*Collembola*) increase. The biomass of earthworms may be 100 times less under conifers than under broadleaves because conifers acidify the soil to unacceptable levels. This means that the worms are no longer present in sufficient numbers to aerate the soil and stimulate bacterial activity. Bacteria, much reduced, become concentrated at the soil surface and around roots. Fungi generally increase under conifers, though some groups decline.

The chemical and physical properties of the soil are changed. The greater accumulation of litter beneath conifers immobilises a greater quantity of mineral nutrients. Although conifers contain lower concentrations of nutrients in their wood, their productivity is greater and the drain of nutrients in harvested timber is greater than from a broadleaved stand. Soils under Norway Spruce and Scots Pine are more acid by about 0.5 pH units because the conifer litter and decomposition products are more acid, and the increased organic matter increases the exchange capacity. This is not balanced by an increase in nutrient cations, so a greater part of the exchange capacity is occupied by hydrogen ions. On light, acid soils conifers accelerate podzolisation, and on heavy, poorly-drained soils they can aggravate a natural tendency to form gley horizons, possibly through a chain reaction of reduced biological activity, poorer structure and thus poorer surface drainage.

Significant physical and chemical changes can take place in less than 50 years after conifers have replaced broadleaves. Grieve (1978) studied the acid, brown earths of the Neath Series in the Forest of Dean (Glos), and showed that under spruce the soil had been significantly podzolised. Thick mor humus had replaced the mull humus which formerly existed under the broadleaves, and the soil structure had become more open. The changes described by Grieve were likely to result in reduced timber yields and could only be regarded as a deterioration. However, as Noirfalise and Vanesse (1975) emphasise, the effects of conifers on some soils can be reversed by restoring broadleaves or fertilising. The soils which are irreversibly affected by conifers are poor, sandy soils which are podzolised and exhausted of nutrients, and pseudogleys, which are made even less permeable. Conifers are not uniform in their effects: the firs (*Abies*, *Pseudotsuga*) with softer needles have less effect than spruces.

Chapter 5

Recent secondary woodland

Secondary woodland grows on land which was formerly used as pasture, meadow, arable, grouse moors, deer forest, habitation, quarrys, etc., uses which displaced original natural woodland and subsequently kept most of Britain treeless. In the interval between the original clearance and the later restoration of woodland – many hundreds or even thousands of years – the use of most patches of ground has changed many times, as arable cultivation in particular has waxed, waned and moved from one place to another in response to economic pressures and changes in the distribution and size of settlements. Land which ceased to be cultivated tumbled down to pasture, but pasture could readily be reclaimed for cultivation. Most of the lowlands and the low uplands have been cultivated at some time. Exceptionally, however, some unwooded ground has remained uncultivated, being too remote, infertile, steep or irregular to be worth ploughing, or being subject to common pasturage rights and communal haymaking regimes (Duffey, 1974).

5.1 Soils of secondary woodland

The soils bequeathed to secondary woodland almost certainly differ from those which formerly occupied the site before the original woodland was cleared. This is partly because clearance removes the protection against erosion afforded by the cover of trees, shrubs and ground flora. Considerable volumes of soil were eroded from the chalk scarps and the slopes of upland hills, leaving thinner, drier and less fertile soils on the slopes and burying the original soils of low ground beneath colluvium and field wash. Even on gentle slopes soil profiles have been truncated by erosion during cultivation. The second reason why the soils of primary and secondary woodlands on similar sites are likely to differ is that the soil-forming processes, which once progressed slowly and steadily under an unbroken continuity of original-natural woodland, changed abruptly when the woodland was cleared.

Although a soil may exhibit the effects of a particular use long after the use has changed, the effects of later land uses on the characteristics of soils inherited by secondary woodland generally over-ride those of earlier uses. Arable cultivation breaks up any natural mineral stratification in the topsoil, destroys the litter and humus layers of woodland soils, and obliterates any surface acidification and other chemical stratification which had developed under undisturbed, natural woodland. The complex woodland topsoil is replaced by a uniform ploughed horizon. Drainage of heavy soils and flat ground is improved by ditches, under-drains and ridge-and-furrow. Whole catchments are dried out by river improvement schemes and abstraction of water from aquifers. Ploughing, which brings subsoil to the surface, and fertilising ensure that cultivated topsoil is richer in nitrate, phosphate, calcium and other

Soil Series or Complex	ICKNIELD	COOMBE	CHARITY	WINCHESTER	BATCOMBE	BERK-HAMSTED	COW-CROFT	SOUTH-AMPTON
Soil Formation	Rendzina	Brown calcareous	Brown earth	Brown earth	Gleyed brown earth	Gleyed brown earth	Gley	Podzol
Land use	W G G A A	W G A A 2	W W A A	W W G A A	W W A A A A	W G A	W A A	W W

Depth in inches (0 – 40), pH recorded at various depths:

- **ICKNIELD:** 7.9, 7.7 7.5, 8.3, 8.0; 8.1
- **COOMBE:** 7.6, 5.6 7.6, 7.8, 8.5; 6.6, 8.0; 7.7, 7.9, 8.6; 8.2; 7.8, 8.4; 8.4
- **CHARITY:** 4.6 4.5, 7.7, 6.7; 4.5 4.7; 5.0, 7.7, 7.1; 5.3 5.0; 5.4 7.6; 5.8; 8.1 5.5, 7.0, 7.4
- **WINCHESTER:** 4.7 4.8, 7.2 7.5 7.4; 5.3 4.5, 4.2; 4.3, 8.0; 5.3, 7.6, 7.3 7.8; 4.6 7.5; 7.3 7.4 8.2; 5.0; 6.4
- **BATCOMBE:** 4.4 4.0, 7.1 7.0 6.8 6.5; 4.5, 4.2; 4.4, 6.9, 7.1, 7.3; 4.5, 7.5, 7.4; 4.3; 6.8, 4.7; 6.7 7.2; 4.2; 5.0, 7.2 7.4
- **BERK-HAMSTED:** 3.7, 5.5 7.6, 4.5; 5.0 7.9; 4.6 5.0; 5.0, 4.9; 5.4 5.8 7.2
- **COW-CROFT:** 4.1, 4.1 6.2 7.2; 6.6, 4.4 7.2; 6.2; 4.8, 6.8; 5.3
- **SOUTH-AMPTON:** 4.0 3.8, 4.4, 4.1; 4.4, 3.9; 4.3 4.3; 4.4

Land use key: W = Ancient Woodland A = Arable and Ley Grassland
G = Grassland 2 = Secondary Woodland

Fig. 5.1 A comparison between the woodland, grassland and cultivated soils of eight Soil Series or Complexes in the Chiltern Hills (Avery, 1964). The diagram indicates the pH recorded at various depths in 33 profiles. Diagonal bars indicate the depth of the profile as described.

nutrients. Some of the differences are apparent in Avery's (1964) detailed descriptions of soil profiles representing several 'Soil Series' in the Chiltern Hills. His descriptions enable woodland, grassland and cultivated soils of a range of soil formations to be compared (Fig. 5.1). Stratification and surface acidification are apparent in most of the soils under ancient woodland, only the shallow rendzinas being alkaline at the surface (Fig. 5.1). Under grassland and arable, stratification has been reduced, the surface horizons are almost uniformly alkaline, irrespective of the original (woodland) condition, and the weak podzol has lost its identity. In the one example of apparent secondary woodland the pH profile resembles that of grassland and arable soils of the same series, rather than the ancient woodland soil.

Clearance without cultivation leaves the natural mineral stratification undisturbed: indeed, disturbance may diminish because large trees no longer fall and churn up the soil in patches. The cleared land is generally used as pasture or meadow without any form of fertilising, a practice which tends to generate a net outflow of nitrogen, phosphorus and other nutrients and a loss of fertility. These losses have been particularly significant in the soils of the hill lands of the north and west and in sandy soils in the lowlands, most of which were naturally acid and infertile. Before the original clearance, a high proportion of the total mineral nutrients in the system were located in the trees themselves, so removal of the trees caused a substantial immediate loss of fertility which in turn led to accelerated leaching, the formation of mor humus and, in the wetter climate of the north and west, to the development of blanket bog (Tallis and McGuire, 1972). In drier climates and in freely-drained sites the acid brown earths of light soils beneath original woodland were transformed into podzolic soils supporting *Calluna* and other forms of heathland, from which the loss of nutrients was periodically accelerated by burning (Dimbleby, 1962). On the other hand there are other soil types and situations where the loss of nutrients due to hay cropping and pasturage is either insignificant or is made good by cultural and natural processes. Hay meadows, for example, generally occur in valleys where seasonal inundation tends to restore nutrient levels. Even where they are not in valleys, hay meadows generally occupy fertile soils whose fertility is maintained by allowing stock on to the grass after the hay has been taken. The heavy and inherently fertile soils beneath most lowland pastures are similarly buffered against the loss of nutrients.

Secondary woodlands therefore inherit two main types of soil depending on whether the land has been cultivated or long used as an essentially extractive form of pasturage. Compared with primary woodland on the same type of site in the same region the soil of former arable ground will be generally better drained, eutrophicated, less acid, devoid of surface litter accumulations and without chemical and physical stratification of the old ploughed horizon, whereas the soils of former unimproved pasture and heathland will be generally less fertile, leached and possibly either podzolised or covered in blanket peat.

5.1.1 Secondary woodland on former cultivated ground

When secondary woodland develops on former arable land, soil-forming processes such as leaching and humus accumulation can resume free from the chronic disturbance caused by cultivation. Eventually, one might suppose, the soil would revert to its original-natural state and become indistinguishable from the soil of primary woods on the same parent material, but in the east Midlands at least there is little sign of this happening. The enhanced fertility and improved drainage of cultivated soils persists beneath woodland, encouraging earthworm activity, preventing the development of surface litter and humus layers and resisting any trend towards surface acidification, even on light, inherently acid soils. Thus, in central Lincolnshire, where the soils of ancient woods generally have a pH of 4–6 at 10 cm depth, the soils of secondary woods originating in the last 300 years are mostly above pH 6. Secondary woods in eastern England on former cultivated or inhabited sites still have high phosphate levels in their soils and a more friable topsoil than the equivalent ancient woods, even 400 years after they were last enriched. In central Lincolnshire, where the thin surface layer of sand is mixed with heavier subsoil when a site is first ploughed, no length of time free of

subsequent cultivation will return the sand to the surface.

5.1.2 Secondary woodland on former heathland

Soil changes associated with the growth of secondary, semi-natural birch stands on heathland in Scotland have been studied by Miles (1978). The soil under the *Calluna* was a well-developed podzol with a pH of 3.8 near the surface, but under the birch the pH, extractable calcium, rate of nitrogen mineralisation and total phosphorous were all greater. Earthworms, which were virtually absent under heather, built up to a peak under 38-year-old birch and the mor humus was gradually converted to a mull. The bleached horizon gradually disappeared, i.e. the podzol became a brown podzolic soil. However, there were signs under a 90-year-old birch stand that these changes might eventually be reversed, allowing the birch wood to degenerate again into heathland. In the New Forest an analogous succession involving holly, oak and beech also permitted the regeneration of mull humus on former heathland podzols (Dimbleby and Gill, 1955).

Afforestation of heathland and blanket bog is mostly too recent to show the full effects on soils. Any changes observed below older stands cannot readily be extrapolated to younger stands. The younger stands have usually been established by different and hopefully better techniques. Such techniques involve ploughing, draining, fertilising and the choice of more appropriate species, and all of these are intended to release nutrients from peat, encourage deeper rooting and faster growth, and generally increase the amounts of nutrients circulating in the system. However, one should not assume that coniferous afforestation invariably increases soil fertility (Gimmingham, 1972) for the conifers themselves produce considerable amounts of acid litter which accumulates as deep mor humus and maintains the podzol-forming tendency. Furthermore, the eventual removal of a harvest of timber will take nutrients out of the system. Pyatt and Craven (1979) have recently reviewed the changes which take place in ironpan soils, gleys and deep peats under even-aged conifer plantations.

5.2 Natural succession to woodland

Following the definitions by Horn (1974), succession is defined as 'a pattern of changes in specific composition of a community' and secondary succession is 'the process of re-establishment of a reasonable facsimile of the original community after a temporary disturbance'. When the composition becomes constant or merely fluctuates cyclically, the community has reached the climax state. Since much of Britain was formerly wooded, most secondary successions lead to woodland. Seedling trees and shrubs, which had hitherto been killed by ploughing, grazing, etc., survive when cultivation and pasturage cease and develop through scrub and emergent woodland into closed-canopy, semi-natural, secondary woodland. Thus, secondary successions to woodland involve not only changes in composition but also changes in structure.

Several examples of secondary succession to woodland have been described. Indeed, this subject so interested Tansley (1939) and his contemporaries that ideas of succession and climax almost dominated British (and American) ecology for two decades. One of the best known examples is the Broadbalk Wilderness (Tansley, 1939), a small patch of former arable ground which was abandoned to its fate in 1882 and is now a mature mixed broadleaf stand of maple, sycamore, pedunculate oak, hazel and other species. This site, however, is far from exceptional (apart from the record made of its progress); numerous larger and more interesting examples can be found in most parts of Britain. For example, many southern heaths have developed into birch (Fig. 5.2), pine or oak woodland (Summerhayes *et al.*, 1924). The scrub successions leading to beech woodland in the Chilterns have been described in detail by Watt (1934). More recently Merton (1970) reconstructed the history of the Peak District ashwoods (Derbyshire) which were found to be the products of natural succession initiated since 1800. Most of these successions are by-products of agricultural changes, notably the loss of traditional grazing patterns on commons and hilly districts. Other successions have been initiated by drainage in the vicinity of bogs and fens, for example at Woodwalton Fen (Poore, 1956b) and Wicken Fen (Cambs) (Godwin *et al.*, 1974), or by non-agricultural

development: the railway between Oxford and Cambridge cut off the corner of a field which eventually became the triangle of Hayley Wood (Rackham, 1975). Abandoned gravel pits, railway lines and other areas of industrial dereliction revert rapidly to scrub.

The change from traditional to modern farming patterns in the last century undoubtedly initiated a wave of scrub development which still has a substantial effect on the landscape, even though much of the scrub has since been reclaimed for agriculture. Before then one supposes that from the time when the first clearances made secondary succession possible, some scrub was always developing somewhere, though the area affected was probably smaller than of late. Domesday Book, for example, records that many patches of 'spinetum' (presumably secondary thorn scrub) mainly on the Midland clays, as well as some land in north-west Hereford and Radnor, which had lately been converted to woodland (Darby and Terrett, 1954). There is some evidence that the recent burst of extensive and widespread scrub development was preceded by two similar phases. During the Dark Ages, extensive tracts of high ground on the southern chalk and limestone reverted to woodland: Celtic fields, Roman villas and Romano-British settlements are found beneath woodland which undoubtedly existed before the 11th century, e.g. Doles Wood near Andover, Hants (Dewer c. 1926). Earlier still, in Neolithic times, shifting cultivation must have provided the conditions for scrub development on a large scale.

Scrub composition is determined by soil type (Chapter 10), available seed sources in the surrounding countryside, the condition of the land at the time it was abandoned, and chance events such as mast years and pattern of seed dispersal. Inevitably, there is a random element in succession which can affect its course and rate of progress. A good example is the New Forest holly scrub, which has developed in patches on areas of bracken and gorse. Most holly thickets contain oaks, established at the same time as the holly, and these eventually grow to form a canopy over a holly understorey. This oakwood may later be colonised by beech (Dimbleby and Gill, 1955). The succession from heathland to beechwood takes as little as 100 years if the beech gets in early, or perhaps

Fig. 5.2 An example of recent, semi-natural woodland: Birch woods developing naturally on heather at Tuddenham Heath, Suffolk. Birch is usually abundant in the early stages of succession to woodland on heathland, but it is itself eventually displaced by oak (sapling on extreme right) and beech. (Photo D. A. Ratcliffe)

300 years if it does not. Furthermore, if neither oak nor beech colonise in the decade or two before the holly scrub forms a closed canopy, the succession is effectively arrested for as much as 200–300 years at the scrub stage until the holly degenerates.

Successions are described as primary when they are a direct response to natural changes in the environment which make a new patch of ground available for colonisation. Primary succession to woodland was widespread 10 000–12 000 years ago following the retreat of the ice and it continued in the form of changing woodland composition as the climate and soils continued to change. Nowadays primary successions to woodland are rare in Britain – small-scale

infills behind the main Flandrian succession – and much modified by man. The zonation from reedbeds to alder carr and birch–oak woodland at Sweat Mere (Salop) (Tansley, 1939) has been described as a primary hydrosere made possible by the slow silting of the lake, but such examples have been much affected by human activities in and around the sites (Sinker, 1962; Green and Pearson, 1968). Likewise, the succession to holly at Dungeness (Kent), on shingle banks cast up since Roman times, was certainly modified and probably managed (Peterken and Hubbard, 1972). Colonisation of coastal landslips at Lyme Regis (Dorset) by ash woodland has not been directly affected by man, but the seed has arrived from

Fig. 5.3 Conifer plantations on Hopehouse Moor and Whitehill Moor (Northumberland) entirely surround Coom Rigg Moss, a group of raised bogs in a matrix of blanket bog lying on the watershed between the river Irthing and the Chirdon Burn. The almost unbroken expanse of conifers is part of Kielder Forest. (Cambridge University Collection: copyright reserved)

a managed landscape. In other countries it is still possible to see extensive examples of primary succession, for example behind retreating glaciers (Crocker and Major, 1955) and on the extensive, mature dune systems of the Lake Michigan shoreline (Olsen, 1958).

5.3 Afforestation

Many people believe that tree planting and therefore afforestation started with the publication of *Silva* by John Evelyn (1664). However, some planting and sowing are known from earlier periods; Evelyn was significantly preceded by Standish (1616) as a 17th century advocate of planting; and both were part of a general movement towards scientific enquiry and better production and use of timber (Sharp, 1975). The Romans apparently planted trees; several species were probably introduced by them (e.g. walnut, horse chestnut), and a Roman farm at Langton in Yorkshire possessed what appeared to be a windbreak of oak, ash, elder, willow, alder, walnut, sweet chestnut, sycamore and cherry, some at least of which must have been planted (Applebaum, 1972). There are scattered records of plantations in the Middle Ages, e.g. in 1456 the almoner of Peterborough records the planting of 700 ash and two cartloads of willows on the monastic estates (King, 1954). Many of the earlier 'plantations' were probably sown or 'set' with acorns and thorn berries. Most plantings

Table 5.1 A list of national woodland censuses and estimates of total woodland area in England, Scotland and Wales from 1895 onwards. Areas are given to the nearest thousand hectares.

Date	Woodland area (kha)		Source
	Total	Coppice woods*	
1895	1104		Board of Agriculture and Fisheries, Agricultural Statistics (1905), Cd. 3061, Table 7.
1905	1121	234	Board of Agriculture and Fisheries, Agricultural Statistics (1905), Cd. 3061, Table 7.
1913–14	1108	230	Board of Agriculture and Fisheries, Agricultural Statistics (1913), Cd. 7325, Table 11.
			Board of Agriculture for Scotland, Agricultural Statistics (1914), Cd. 7958, Table 7.
1924	1198	214	Forestry Commission Census report. HMSO.
1938	Incomplete		Forestry Commission Census. MS report. PRO F22.
1942	Incomplete		Forestry Commission Census. MS report. PRO F22.
1947–49	1396	142	Forestry Commission Census reports. HMSO.
1965–67	1743	32	Locke (1970), Census of Woodland, 1965–67. HMSO.
1976	1980		The Place of Forestry in England and Wales. Forestry Commission (1978).
1980	2036		Forestry Facts and Figures 1979–80. Forestry Commission (1980).

* Coppice woods are hard to define for census purposes. In general these figures refer to worked coppice and thus omit woods which appeared to be no longer worked by this system.

were probably of native species initially, but by the 18th century nurseries held a wide range of exotic species (Harvey, 1974).

The Board of Agriculture Reports, 1790–1813 (Jones, 1961) record that plantations then were still on a small scale, though the rate of planting had increased towards the end of the 18th century. Most planting evidently took place on recently enclosed heathland in southern England and in timber-deficient areas in northern England, central and eastern Scotland. In the agricultural lowlands the rate of afforestation during the 17th–19th centuries appears to have been fairly even, judging from three study areas in the east Midlands (Peterken, 1976), even though the density of ancient woodland varied greatly. Private landowners afforested not only to produce income from the sale of timber, but also to enhance the landscape, provide cover for foxes and game, shelter for exposed houses and fields, and simply to make some productive use of otherwise unused ground. The interest of individual landowners in tree planting varied enormously, and many estates stand out from their neighbours even now because of the spinneys and shelterbelts planted through the initiative and energy of a particular 18th or 19th century owner, e.g. Limber and surrounding parishes in north Lincs, planted from 1787 by the Earl of Yarborough; and Althorp Park (Northants) where,

remarkably, each plantation from 1567 onwards is recorded with a datestone.

During the present century afforestation has been carried out by the Forestry Commission and by private owners and commercial syndicates with state aid, both acting under a forestry policy directed initially at creating a reserve of timber and latterly at growing utilisable timber as economically as possible. The woodland area of Britain has almost doubled in the last 80 years (Table 5.1). Most of this planting has been on the upland sheep walks of western and northern Britain, where heathland, acid grassland, blanket bogs and valley mires have been afforested on a very large scale indeed (Fig. 5.3), transforming some regions from treeless moorland to virtually unbroken expanses of forest. This process has been described by Ryle (1969), Mather (1971, 1978) and numerous Forestry Commission publications. Almost without exception the plantations have been even-aged mono-cultures or two-species mixtures of conifers, though the proportions of each species planted – spruce, pine, fir, larch, hemlock, etc. – have changed as experience has accumulated and new planting has increasingly been forced on to less productive ground. Some of the earliest and now most productive new forests were planted on the East Anglian heaths in Breckland and the Sandlings. Lowland agricultural districts have not been entirely ignored, for

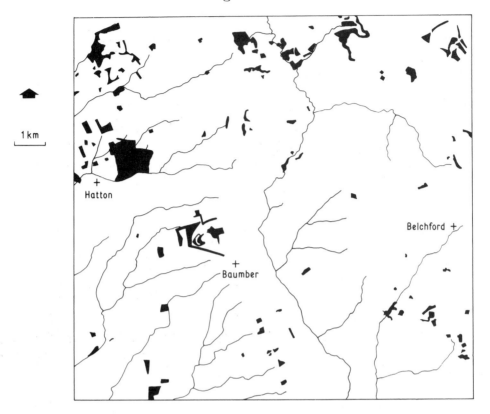

Fig. 5.4 The pattern of post-medieval secondary woodlands in part of the Wolds of central Lincolnshire. The main river in the centre is the Bain. The large wood near Hatton was formed by 20th century infilling around a cluster of small woods. Otherwise the scatter of small woods and parkland belts is of 18th and 19th century origin.

Table 5.2 The rate of afforestation (ha per year) in three study areas in the east Midlands of England (Peterken, 1976).

Period	Rockingham Forest	Central Lincolnshire	West Cambridgeshire
1650–1817	2.8	3.1*	1.5*
1817/34–1885/7	3.9	5.4	4.9
1885/7–1946	16.5	8.0	5.3
1946–1972/3	18.0	13.3	0.4
Extent of study Areas (ha)	41 550	49 950	42 900

* Estimated. Information for the mid-seventeenth century was incomplete.

agricultural land could be acquired fairly cheaply in the 1930s, and until about 1950 many fields were afforested. During this period many lowland woods were extended in an effort to make their management more economic so, apart from the heaths, most lowland afforestation went into relatively well-wooded districts, leaving sparsely-wooded districts such as Cambridgeshire largely untouched (Table 5.2). Nowadays, afforestation is concentrated in the uplands of Scotland, where it is making substantial inroads into the peaty expanses of the extreme north.

The characteristic pattern of lowland afforestation has been small, scattered stands and long, narrow belts, spread more-or-less evenly over the landscape (Fig. 5.4). Before the 20th century afforestation tended to be greater in those areas which lacked ancient woods, for example on the chalk wolds and limestone heath of central Lincolnshire. More recently afforestation has been concentrated into large blocks, because large forests are more efficient to manage than small woods, and remaining farms in a predominantly afforested area tend to become less

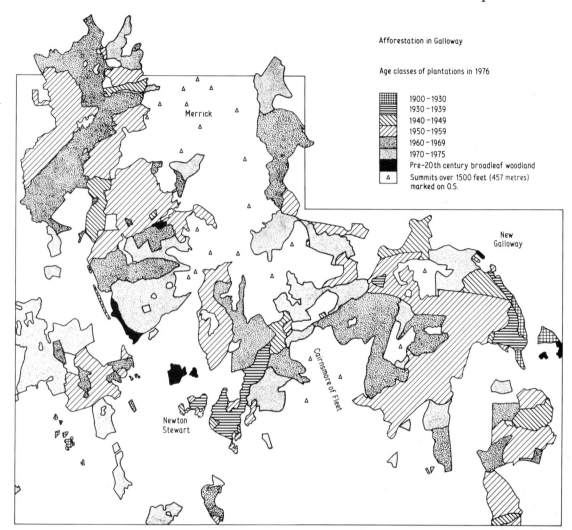

Afforestation in Galloway

Age classes of plantations in 1976

- 1900 – 1930
- 1930 – 1939
- 1940 – 1949
- 1950 – 1959
- 1960 – 1969
- 1970 – 1975
- Pre-20th century broadleaf woodland
- △ Summits over 1500 feet (457 metres) marked on O.S.

Fig. 5.5 Intensive afforestation in Galloway, south-west Scotland (Nature Conservancy Council from information furnished by the Forestry Commission and private forestry groups). This region was virtually devoid of woodland in 1900, the only substantial patches being along the river Cree and its tributaries above Newton Stewart. As late as 1940 the only significant blocks of afforestation lay between Newton Stewart and Bargally Glen in the centre of the area and in the east along the side of Loch Ken south of New Galloway. From these nuclei, however, plantations have since spread rapidly across the landscape, and now cover most of the middle ground between the fertile valleys and the hill tops of Merrick and Cairnsmore of Fleet.

economic (Stewart, 1978). As a result huge forests, such as those at Kielder and in Galloway, have spread over the landscape (Fig. 5.5). Over much of this area afforestation in a general sense restores the woodland which was destroyed in prehistoric times, but this is not always the case: much of the flowe country in northern Scotland which is now being afforested has

not been wooded since the last glaciation (Peglar, 1979).

5.4 Effects of upland afforestation on flora and fauna

When moorland is afforested the sheep and deer are

Fig. 5.6 Vertical ploughing for afforestation, Glen Kinglas, Argyll. The higher ground is not being planted and along the burns semi-natural scrub and narrow strips of moorland have been retained. Apart from the profound ecological changes which will shortly take place on the planted ground, the accelerated run-off will alter the hydrology and ecology of the streams beyond the limits of planting. (Photo. D. A. Ratcliffe)

excluded by a ring fence, the soil is often ploughed (Fig. 5.6) and fertilised, drainage is improved and forest roads are constructed. For perhaps 10–15 years the saplings have little effect on the vegetation, which develops into a leggy form of heathland and grassland. Then the conifer canopy closes, the stand passes through the thicket stage and the ground vegetation is virtually exterminated. Eventually, as the stand matures, some vegetation returns and later, when the stand is felled and the ground replanted, it develops strongly.

The plant succession which follows afforestation has been studied by Hill (1979). One example of succession concerns the changes in Caeo Forest (Dyfed) between 1944 and 1976 (Hill and Jones, 1978). This land, which had been hill pasture with patches of sessile oak coppice, was planted with various conifers between 1932 and 1941, and was therefore in the light phase of plantation growth when it was first recorded. The ground vegetation was mostly dominated by *Calluna vulgaris*, *Pteridium aquilinum* and *Vaccinium myrtillus*, with *Molinia*

Fig. 5.7 Spruce plantations on Loch Ken, Cairn Edward Forest, Galloway. Very few wildlife species inhabit the dark and deeply littered interior of mature spruce stands. (Photo. D. A. Ratcliffe)

caerulea on damper ground. By 1976, when these plantations were around 40 years old, the ground vegetation was fairly well developed below pine, larch, and mixtures containing pine, but in the dense shade below spruce (Fig. 5.7) it remained exceptionally sparse (Table 5.3). Only *Dryopteris dilatata*, *Galium saxatile* and *Vaccinium myrtillus* achieved a frequency of 40% or more in the 200 m² quadrats beneath spruce, whereas five species (*Agrostis canina*, *Deschampsia flexuosa*, *D. dilatata*, *G. saxatile* and *V. myrtillus*) achieved 90% frequency or more in the quadrats beneath other species. Between 1944 and

1976 the number of species per quadrat fell appreciably beneath spruce but had scarcely changed beneath pine and larch. *Dryopteris dilatata* had increased and was almost abundant. Other ferns (*Athyrium filix-femina*, *Blechnum spicant* and *Dryopteris filix-mas*) had also increased together with many bryophytes, such as *Eurynchium praelongum*, *Isopterygium elegans*, *Mnium hornum* and *Plagiothecium undulatum*. These, and the species which showed little change (e.g. *Agrostis canina*, *Deschampsia flexuosa*, *Vaccinium myrtillus*, *Dicranum scoparium* and *Rhytidiadelphus loreus*), are mainly woodland species. The species which declined substantially or disappeared from the quadrats were mainly heath and grassland species, such as *Erica tetralix*, *Festuca ovina*, *Rhytidiadelphus squarrosus*, *Empetrum nigrum*, *Achillea millefolium* and *Polygala serpyllifolia*, but they also included some species of deciduous woodland, such as *Lathyrus montanus*, *Betonica officinalis* and *Viola riviniana*. Hill and Jones concluded that in 32 years there had been a shift from a heath and grassland flora to a woodland flora. However, the flora of the plantations (especially spruce) differed significantly from that of sessile oak woods nearby, primarily because of the perpetual rain of coniferous litter and the deep shade of the thicket stage: bryophytes were more abundant beneath plantations, but *Pteridium* was almost eliminated.

Table 5.3 The cover of ground vegetation beneath approximately 40-year-old plantations of various conifer species in Caeo Forest (Dyfed) in 1976 (Hill and Jones, 1978). Values are mean % cover with standard deviations in parentheses.

Crop species	Number of plots	Vascular ground flora	Bryophyte ground flora
Sitka Spruce	30	0.7 (1.1)	5.0 (6.4)
Norway Spruce	9	5.3 (7.3)	13.0 (7.1)
Sitka Spruce + Scots Pine	2	11	22.5
Norway Spruce + Scots Pine	2	28.5	17
Japanese Larch	7	55.6 (11.6)	24.0 (15.0)
Scots Pine	4	23.0 (17.6)	6.2 (3.7)
Shore Pine	1	75	10

Succession during the first rotation, exemplified by Caeo Forest, is only one aspect of the floristic effects of afforestation. Subsequent rotations start with a woodland flora, which, like the flora of coppices, must have mechanisms for surviving the thicket stage. Few species survive vegetatively beneath evergreen conifer thickets, which cast a deep shade, do not allow a spring flush of growth, and tend to create deep mats of raw humus. Most of the successful species have to rely on buried seed banks or immigration if they are to thrive in the light phase. Thus, when a stable, cyclic state is reached, bryophytes, ferns and such herbs and shrubs as *Chamaenerion angustifolium*, *Calluna vulgaris*, *Deschampsia flexuosa*, *Digitalis purpurea* and *Rubus fruticosus* are likely to be successful, but spring herbs are not. Northern herbs such as *Goodyera repens* and *Trientalis europea* have colonised pine afforestation in eastern Scotland and north-east England and may eventually become widely established. Plantations contain roads, rides and other unplanted patches which to some extent act as refuges for the former moorland plants, but the improved drainage tends to eliminate bog and marshland species. Forest roads, sometimes constructed of rubble brought in from beyond the forest, provide new habitats which are colonised by calcicole herbs. Some moorland outside the plantations, cut off from grazing, will become overgrown. Downstream, the increased run-off of soil particles and nutrients, caused by drainage and fertilisers, changes the aquatic flora and fauna.

Changes in the fauna are equally profound. Moorland birds hang on for some years after afforestation and others move in to take advantage of the enlarged vole populations in the rank herbage (e.g. short-eared owl), but once the plantations reach the thicket stage they are ousted completely. Even large species with huge territories are affected. For example (Marquiss *et al.*, 1978), before 1960 the raven population of southern Scotland and Northumberland was stable, with 123 nest sites, most of which had been occupied year after year, but between 1960 and 1975 the breeding population was reduced to 44% of its former level. This decline was blamed mainly on afforestation which had reduced the food supply – sheep and other carrion, eggs and small animals – and made it harder to find. Significantly, raven populations remained unchanged in the Lake District, a similar area ecologically, where there had been practically no afforestation.

Just as new plantations acquire a woodland flora of sorts during the first rotation, so they also acquire a woodland fauna. Colonising invertebrates include pest species such as the pine beauty moth. Some of the beetles associated with the native pinewoods have colonised pine plantations elsewhere in the Highlands (Hunter, 1977). A particularly interesting record of colonisation comes from Rhum (Wormell, 1977), which was denuded of its original woodland by the late 18th century. About 1900 a small policy wood was planted, and since 1957, when the island became a National Nature Reserve, several patches of

Table 5.4 Densities of song birds (pairs per km²) in plots in coniferous plantations, semi-natural pine and birch–pine woods and a mature mixed woodland in Scotland (Moss, 1978). The species are arranged in order of decreasing affinity with pure coniferous woodland and increasing affinity with woodlands containing broadleaved trees. Where plots were recorded in more than one year, the first year of observation has been quoted.

Origin of stand	PLANTATION											SEMI-NATURAL			?
Species	Sitka and Norway spruce						Larch	Scots Pine				Scots Pine/birch			Mixed*
Planting year	1927	1938	1940	1941	1942	1953	1946	1926	1932	1932	1935				
Siskin	9		5	6		5						6			
Crossbill	9		5	6		9						11	11		
Crested Tit								4	4			17			
Goldcrest	136	372	283	247	417	172	99	22	57	36	34	57	39	21	70
Coal Tit	20	46	29	28	21	41	47	35	26	34	57	46	45	8	70
Chaffinch	102	101	59	125	145	96	82	26	43	30	88	66	45	124	302
Wren	57	52	59	80		25	116	44	4	75	92	57	71	146	186
Robin	5	16	5	23	14	53	47	22	9	20	29	49	79	92	140
Dunnock						9					4	6		4	
Tree Creeper	9	11	10				17		9	16	13	17	5	8	47
Blackbird				6									16	17	47
Willow Warbler							17				23	46	105	292	279
Long-tailed Tit							17						11	4	23
Song Thrush									4			6	5	17	23
Blue Tit													16	25	186
Redstart															23
Garden Warbler													5	8	
Spotted Flycatcher														4	23
Spotted Woodpecker													5		12
Total species in plot	9	6	8	8	4	8	8	6	8	6	8	12	15	18	17
Pairs per km²	351	598	454	520	597	411	444	151	155	208	340	385	463	825	1593

* A mainly broadleaved stand (birch, oak, ash, sycamore) with scattered spruce and pines.

moorland have been afforested with native broadleaves. Now numerous woodland invertebrates are present, having either colonised from the mainland or spread from minute populations on fragments of scrub on Rhum itself. Song-birds such as goldcrest, chaffinch and coal tit become common, but as Moss (1978) has shown, the song-bird populations of pure conifer plantations are neither as large nor as diverse as those in woods containing at least some broadleaves, and moreover some species of broadleaf stands do not seem to become established at all in pure conifer stands (Table 5.4).

5.5 The flora of lowland secondary woods

The upland plantations have only existed for a few decades and we cannot therefore know how their

Table 5.5 Extent of existing secondary woodlands originating at various periods in three east Midland areas (Peterken, 1976).

Period of origin	Area (ha) of woodland present in 1972/3		
	Rockingham Forest	Central Lincolnshire	West Cambridgeshire
Ancient woodland*	3558	2558	598
Secondary woodland			
1650–1817/34	296	222	236
1817/34–1885/7	244	329	198
1885/7–1946	960	443	192
1946–1972/3	467	360	10

* A small fraction of the ancient woods in each area can be proved to be secondary.

flora will develop in the long-term. In the lowlands, however, patches of secondary woodland have been initiated over hundreds of years, and the origin of most can be dated to a particular period, e.g. the Dark Ages. Most areas contain secondary woods originating over several centuries (Table 5.5). These secondary woods of widely differing ages provide an opportunity to study the long-term course of floristic succession in the lowlands, even though interpretations of differences between woods originating at different periods must be cautious: age of site as woodland is confounded with period and circumstances of origin.

5.5.1 Central Lincolnshire secondary woodland flora

Most of my experience of succession in secondary woods comes from central Lincolnshire where, in a study area centred on Market Rasen, Bardney and Woodhall Spa, some 57% of existing woodland (by area) is secondary. Two classes of woodland can be recognised, namely (i) those originating on former arable ground or ley grassland and (ii) those originating on former semi-natural vegetation. Intermediate conditions also have to be recognised, e.g. the wood on former unimproved grassland which had been cultivated in the more distant past. Each wood was assigned to one of these classes on the basis of historical records and evidence on the ground (e.g. ridge-and-furrow). Poolthorn Covert, from which Woodroffe–Peacock (1918) provided one of the fullest descriptions of a secondary woodland flora, lay just to the north of the study area, but is now, alas, once again an arable field.

The species which rapidly colonise secondary woods on former cultivated ground are mostly common and widespread. Some, the 'woodland colonists' (Table 5.6A) are characteristic components of communities in ancient woods (Fig. 5.8), but others, the 'shade-bearing weeds' (Table 5.6B), have their headquarters in hedges and waste ground, and in ancient woods they are mostly restricted to wood margins and disturbed ground (Fig. 5.9). Since many of the woodland colonisers are also common in hedges, one can broadly characterise the secondary woodland flora as an expanded hedgerow flora. Many species in both groups have light seeds or spores, hooked and feathered seeds, berries and other obvious

Fig. 5.8 *Arum maculatum* (Cuckoo-pint), widespread and common in woods and hedges on calcareous and base-rich soils. A rapid colonist of secondary woods in the lowlands. (Photo. P. Wakely)

means of rapid dispersal, and there is a distinct tendency for calcicoles and nitrophilous species to predominate. Few are in any sense secondary woodland indicators, except perhaps *Viola odorata*, a calcicole which Clapham *et al.* (1962), effectively describe as such. Many can withstand the seasonal drought of heavy clay soils.

Secondary woods on former semi-natural habitats in central Lincolnshire grow mostly on former heathland ranging from dry grassland and *Pteridium* brakes to *Calluna* moor and *Molinia* flushes. Many heathland species survive the change to woodland, especially semi-natural birch and oak stands, but others cannot

Table 5.6 Vascular plant species of secondary woodlands on former cultivated ground in central Lincolnshire.

LIST A
Fast-colonising woodland species: most are characteristic members of undisturbed communities in ancient woodlands.

*Arum maculatum	**Moehringia trinervia*
Brachypodium sylvaticum	**Poa trivialis*
Circaea lutetiana	*Rubus caesius*
Dryopteris dilatata	**Rubus fruticosus* agg.
Dryopteris filix-mas	*Rubus ideaus*
Dryopteris spinulosa	**Rumex sanguineus*
Epilobium montanum	*Sanicula europaea*
Festuca gigantea	**Stachys sylvatica*
*Geranium robertianum	*Tamus communis*
*Geum urbanum	*Silene dioica*
*Glechoma hederacea	*Viola odorata*
*Hedera helix	**Urtica dioica*
Listera ovata	*Zerna ramosa*
Lonicera periclymenum	

* Present in more than 50% of secondary woods originating after 1820.

LIST B
Shade-bearing weeds. Species found mainly on waste ground and hedges; infrequent in undisturbed ancient woodland communities.

Aegopodium podagraria	*Lamium album*
Alliaria petiolata	*Lapsana communis*
Anthriscus sylvestris	*Myosotis arvensis*
Arctium lappa	*Poa annua*
Arctium minus	*Potentilla anserina*
Arrhenatherum elatius	*Potentilla reptans*
Bryonia dioica	*Rumex obtusifolius*
Chamaenerion angustifolium	*Solanum dulcamara*
Cirsium arvense	*Stellaria media*
Cirsium vulgare	*Taraxacum officinale*
Dactylis glomerata	*Torilis japonica*
Galeopsis tetrahit	*Veronica hederifolia*
Galium aparine	*Viola arvensis*
Heracleum sphondylium	

tolerate a dense shade and survive mainly in rides and on wood margins (Table 5.7). The same range of shade tolerance is shown by the herbs of unimproved grassland, but most of the fen and marsh species seem to survive the change to woodland without a tremor. The species in Table 5.7, which are in effect woodland species with a wide habitat range in central Lincolnshire, include some which may be wholly confined to long-undisturbed habitats, i.e., they resemble the ancient woodland species of Table 3.2, but enjoy a wider habitat range and can presumably withstand woodland clearance and restoration without moving, provided that the land is not cultivated in the interim. Other species are clearly good colonists, such as *Deschampsia caespitosa* in grassland and *Scrophularia aquatica* beside streams.

Table 5.7 Some shade-bearing vascular plant species of central Lincolnshire which occur in both ancient woodland and heathland, wetlands or unimproved grasslands.

Ajuga reptans	*Holcus lanatus*
Angelica sylvestris	*Iris pseudacorus*
Anthoxanthum odoratum	*Lathyrus pratensis*
Betonica officinalis	*Lychnis flos-cuculi*
Calluna vulgaris	*Molinia caerulea*
Caltha palustris	*Potentilla erecta*
Cirsium palustre	*Primula veris*
Deschampsia caespitosa	*Prunella vulgaris*
Deschampsia flexuosa	*Pteridium aquilinum*
Digitalis purpurea	*Ranunculus ficaria*
Epilobium hirsutum	*Ranunculus repens*
Eupatorium cannabinum	*Scrophularia aquatica*
Filipendula ulmaria	*Succisa pratensis*
Galium palustre	*Teucrium scorodonia*
Galium saxatile	*Veronica chamaedrys*

5.5.2 Origin and structure of field layer communities in secondary woodland

The development of secondary woodland communities depends on two processes, colonisation from outside and survival of woodland species on the site when it was unwooded. Colonisation is overwhelmingly more significant in woods on former arable simply because very few species thrive in both arable and undisturbed woodland. A detailed study of established field layer communities in secondary woodland would probably reveal that the process of colonisation has a lasting effect. The constituent species often appear to be under-dispersed, or clumped into patches apparently round the original site of colonisation: certainly this patchiness can be seen even in woods where the soil is uniform, so edaphic factors do not seem to be responsible. The ground vegetation tends to consist of species-poor patches dominated by one species, although the variety and number of patches may be large and the total number of species in the wood may be comparable with the number in an equivalent ancient wood. One's impression is that

Fig. 5.9 *Heracleum sphondylium* (Hogweed) growing with *Urtica dioica* (Nettles) and *Poa trivialis* (Meadow grass) on the edge of a wood containing *Pteridium aquilinum* (Bracken). The first three species are often found abundantly in secondary woods growing on former arable land. (Photo. P. Wakely)

species become established at random in the early stages of colonisation and then spread from the point of establishment to form single-species patches, which subsequently resist invasion through inertia (the competitive advantage of an established community over a would-be colonist). For example, *Hedera helix* might spread from a marginal hedge and form a dense mat which excludes all other species except *Arum maculatum*. Nearby *Anthriscus sylvestris* arrives first in other vacant territory and forms a community with *Poa trivialis*. Eventually, the two armies meet and

form an armistice line, which is often quite sharply defined. This patchiness contrasts strongly with the ground vegetation of ancient woods, which tends to be an intimate mixture of species whose distribution is closely correlated with soil variation. Obviously some secondary woodland communities are well mixed, and perhaps in the longer-term they will all develop from cliques into communities – this certainly seems to be so in the few medieval secondary woods examined – but in the early stages at least, the ground vegetation of secondary woods assumes a patchy, high entropy state which contrasts with the integrated low entropy state of ancient woodland communities.

Survival of woodland species on unwooded ground depends on their habitat range and the survival of semi-natural habitats. Few species are absolutely confined to woodland in Britain (perhaps *Dentaria bulbifera Neottia nidus-avis* qualify), for even apparently obligate woodland herbs such as *Paris quadrifolia* remain in lightly grazed grassland, formed recently from woodland. Only a few woodland species thrive in heavily grazed grassland (e.g. *Conopodium majus*), but many grow in grassland on railway embankments and in meadows (e.g. *Anemone nemorosa*, *Listera ovata*, *Ranunculus auricomus*), a regime which like coppicing permits spring growth and flowering. Many fen woodland species survive if the trees are cleared, again in a habitat in which grazing is moderated by the terrain. A group of northern pine and birch wood species grow in dune slacks (e.g. *Corallorhiza trifida*, *Pyrola minor*), and otherwise characteristic woodland species inhabit sea cliffs (e.g. *Primula vulgaris*, *Vicia sylvatica*). Certain specialised habitats on mountain ledges, limestone pavement and indeed almost any rock outcrop have a woodland flora protected from grazing. The habitat range of many species varies across their geographical range. For example, on an east–west axis, *Endymion non-scriptus* and *Luzula sylvatica* are both largely confined to woodland in eastern England but spread out over the hillsides of Wales. On a north–south axis, *Iris foetidissima* is almost an obligate woodland species, but not near the Dorset coast, where it also grows in grassland. Some species which are almost confined to woodland in Britain are widespread outside woods on the Continent, e.g. *Primula elatior*. Since semi-natural

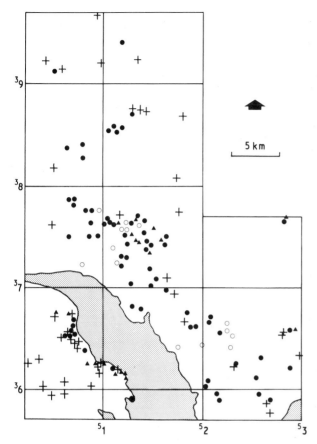

Fig. 5.10 The distribution of bluebell, *Endymion non-scriptus*, in central Lincolnshire (see Fig. 3.7).

●, *Endymion* present in an ancient wood.

▲, Survival site: *Endymion* present in a wood relict hedge and other habitats where it could have survived since that site was occupied by ancient woodland.

+, Colonising site: *Endymion* present in secondary woodland, recent hedge and other sites where it could not have survived since that site was occupied by ancient woodland, and to which it must therefore have colonised from a distant refuge.

○, Ancient wood in which *Endymion* has not been recorded.

Although secondary woods occur throughout the area, long-distance colonisation appears to be restricted to the north-east and south-west corners of the study area. Only 15 of the 42 colonising populations lie more than 2 km from a relict population in an ancient wood or wood-relict hedge.

habitats have survived abundantly in the uplands and the climate is moist and equable outside and inside woods, woodland species are common outside woods and woodland continuity is a relatively insignificant ecological factor. In the lowlands semi-natural habitats are patchy, the climate outside woods is more Continental, the soil beneath secondary woods has more often been modified by cultivation, and the habitat range of woodland species is restricted naturally and by intensive agriculture: here woodland continuity is a significant ecological factor. Hedges to some extent reduce this difference, especially where wood-relict hedges (Pollard, 1973) are common, but not all species can take advantage of them. In central Lincolnshire the calcicoles (e.g. *Mercurialis perennis*, *Melica uniflora*) are generally well suited to hedges, whereas the calcifuge species are not. This is presumably a straightforward response to the soil characteristics of hedges, which are drier and often more alkaline than the surrounding land.

5.5.3 Community relationships

The field layer communities of secondary woodland cannot be properly related to others until the National Vegetation Classification (Chapter 7) is complete but tentative comments are in order. Where the woodland has developed on heath or improved grassland the communities are initially an attenuated form of the previous community in which only shade-bearing species survive, but where the wood has replaced arable and improved grassland, seemingly distinctive communities develop which have much in common with the vegetation of hedgerows, trackways, abandoned cemeteries and waste ground. These communities appear to be very widespread. For example, Kopecký and Hejný (1973) describe an *Alliario-Chaerophylletum temuli* Association, growing in shaded sites on fresh to moderately damp, humus- and nitrogen-rich, neutral-alkaline soils in the area of *Carpinion* woodland communities (Section 11.6) in

Bohemia, which is remarkably similar to communities found in isolated recent secondary woods on former arable in the east Midland claylands. The species which are frequent in both Czechoslovakia and Cambridgeshire include *Alliaria petiolata, Anthriscus sylvestris, Chaerophyllum temulentum, Dactylis glomerata, Galium aparine, Geranium robertianum, Geum urbanum, Glechoma hederacea, Heracleum sphondylium, Lapsana communis, Moehringia trinervia, Poa trivialis, Rubus caesius, Stellaria media, Urtica dioica, Veronica hederifolia* and *Viola odorata*.

These initial communities, which form very rapidly once the land is no longer cultivated or grazed, are eventually colonised by some of the species of undisturbed ancient woods, such as the 'woodland colonists' of Table 5.6, whilst other species scarcely colonise at all (Table 3.3A). This colonising process is influenced by numerous factors, including ecological isolation (Section 6.2). The overall effect of survival in ancient woods and other refuges, and of colonisation into secondary woods of various ages, on the distribution of individual species is illustrated by *Endymion non-scriptus* (Fig. 5.10). Since most of the central Lincolnshire soils are apparently suitable for bluebell, gaps in its occurrence must be due mainly to (a) the absence of ancient woods to provide refuges against clearance, and (b) the failure so far of the species to colonise the more remote secondary woods.

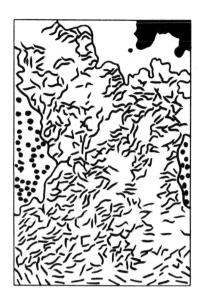

Chapter 6

Long-term changes in the woodland flora

The long-term response of the woodland flora to the broad pattern of woodland change over the last few thousand years can be considered in two contrasting ways, namely the direct observation of these changes made possible by pollen analysis and the application of Island Theory to British woodlands. These approaches combine to identify groups of species on the basis of their response to change, and they eventually establish some of the basic priorities of woodland nature conservation.

6.1 Long-term history of the woodland flora

Only a minority of woodland herbs have left any trace in the sub-fossil record, and very few indeed have left enough pollen, spores, fruits, etc. in peat deposits for a detailed historical analysis, but this is only to be expected. The field layer of closed woodland is usually remote from the bogs and fens in which pollen has been preserved and the chance of pollen reaching them has been minimised by the shelter of the woodland canopy. Godwin (1975) gives details of the sub-fossil finds of about 90 shade-bearing herbs and a commentary on their history, from which the following analysis has been prepared. The species are divided into four groups whose increases and decreases are considered against the background of changes in the post-glacial, or Flandrian, period (Table 1.1).

6.1.1 Rapid colonisers

The first group contains 19 species and genera which are conspicuously rapid colonisers of recent secondary woodland, especially in well-cultivated regions. All are recorded before the Boreal and most appeared by the Late Weichselian. As a group they are relatively frequent in the Late Weichselian, infrequent during the Boreal and Atlantic, but frequent again in the Sub-Boreal and Sub-Atlantic. The group includes *Chamaenerion angustifolium* and *Lapsana communis*, which are well suited to non-woodland habitats and would respond to clearance and cultivation, typical components of scrub (e.g. *Rubus fruticosus*, *R. idaeus*), and species of nutrient-rich soils both inside and outside woodland, e.g. *Circaea lutetiana*, *Galium aparine*, *Heracleum sphondylium*, *Stachys sylvatica*, *Urtica dioica*. The group appears to have been favoured by the open conditions and rich soils of the Late Weichselian, restricted as soils matured, and again favoured once clearance and cultivation had disturbed and enriched the former woodland soils. All these species were evidently present during the forest maximum – 11 species are recorded during the Boreal and/or the Atlantic, including *Dryopteris* agg., *Geum urbanum/rivale* and *Silene dioica* – presumably as members of stable communities in flushed sites and as opportunists in temporarily disturbed areas of forest.

6.1.2 Wetland–woodland species

The shade-bearing components of fens, marshes, bogs, wet grassland, pools and riversides are relatively well represented in the pollen record. Twenty-two of the 32 recorded species appeared by the Late Weichselian and a further seven by the early Boreal. Only *Cardamine flexuosa*, *Chrysosplenium oppositifolium* and *Iris pseudacorus* are not recorded before the Late Boreal. Many species are recorded in all periods after their first appearance, and in sufficient quantities to detect long-term changes in frequency. Some are most frequent in the Late Weichselian (e.g. *Caltha palustris*, *Valeriana officinalis*), others are more frequent in the Atlantic or later periods (e.g. *Ajuga reptans*, *Lythrum salicaria*, *Molinia caerulea*, *Osmunda regalis*, *Phragmites communis*, *Viola palustris*), whilst a few reach peaks of frequency at both times (e.g. *Filipendula ulmaria*, *Montia fontana*). As a group they tend to be most frequent in the Late Weichselian and Pre-Boreal, and also in the Atlantic and later, but are infrequent in the Boreal. Indeed, some species disappear completely from the record during the Boreal (e.g. *Ajuga reptans*, *Carex paniculata*).

Wetland habitats were widespread throughout the Late Weichselian, during which time a large fraction of the present wetland flora appears to have become established. Many species were presumably restricted later by the advance of woodland, but the shade-bearing species, if they were reduced at all, may have been restricted only in flowering. Moist conditions of the Atlantic period and the expansion of wetlands favoured these species, and clearance of later periods accentuated and prolonged the trend.

6.1.3 Heathland–woodland species

Some eight species which now grow in heathland and woodland are recorded from the Late Weichselian and the Boreal onwards, whereas three species, *Digitalis purpurea*, *Teucrium scorodonia* and *Vaccinium vitis-idaea*, are recorded only from the Boreal and later periods. This apparent increase in the number of species is matched by the great expansion in the Atlantic and subsequent periods of the four abundantly recorded species, *Polypodium vulgare*, *Potentilla erecta*, *Pteridium aquilinum* and *Succisa pratensis*. Heath–woodland species were mostly present but not abundant in the Late Weichselian, but declined in the Pre-Boreal as woodland spread. Their later increase may be the result of two mutually-enhancing factors, namely

leaching – leading to the formation of podzols – and forest clearance.

6.1.4 Woodland species

The fourth group – species mainly confined to woodlands and today having only limited powers of reaching secondary woods – contains 20 species which fall into four sub-groups. *Crepis paludosa*, *Linnaea borealis*, *Rubus saxatilis*, *Thelypteris dryopteris* and *T. phegopteris* are recorded only from the Late Weichselian and Pre-Boreal. They all have a strong northerly distribution now, having perhaps been reduced in southern districts by failure to compete with later-arrivals which could respond to the warmer climate. The second group contains 12 species which are not recorded before the Boreal: *Carex pendula*, *C. strigosa*, *Endymion non-scriptus*, *Euphorbia amygdaloides*, *Fragaria vesca*, *Geranium lucidum*, *Lonicera periclymenum*, *Luzula sylvatica*, *Lysimachia nemorum* or *nummularia*, *Oxalis acetosella*, *Potentilla anglica* and *Scirpus sylvatica*. They form a mixed group ecologically and are now mostly widespread. The third sub-group – *Mercurialis perennis*, *Potentilla sterilis* and *Stellaria holostea* – is similar, but those species were recorded in the Late Weichselian as well as later periods. Fourthly, *Rubus caesius* and *Senecio sylvaticus* were recorded only in the Sub-Atlantic. These woodland species were probably well represented in the Boreal and Atlantic forests, but now the early arrivals are understandably found mainly in the north.

The general pattern that emerges, however uncertainly, is that the fourth group of species, now found mainly in ancient woods, represents the characteristic woodland herbs of the natural Boreal and Atlantic forests, where presumably they (and other species which have left no trace in the pollen record) were mixed together in a complex pattern of communities determined by soil and climate, much as they are today. The other groups were also present at the forest maximum, but appear to have been restricted to specialised habitats. Before then the colonists were widespread, and they again expanded when clearance and especially the soil disturbance and eutrophication associated with the agriculture of settled communities provided more open habitats and

Fig. 6.1 Intensively farmed chalklands south-east of Cambridge, looking from Wandlebury towards Balsham, in 1959. The scattered plantations and shelterbelts are habitat islands in a sea of cultivation. (Cambridge University Collection: copyright reserved)

fertile soils. The heath–woodland species were also present before the forest maximum, yet not abundant, but they spread when clearance and shifting agriculture provided open ground and enhanced the natural trend towards mature, infertile soils on light, freely-drained sites. Wetland–woodland species were also well established before the forest maximum and may have been little affected by either the advent of forests or their clearance. Wetland–woodland and heath–woodland species had the advantage of a wide habitat range, but their inability to respond to site disturbance, eutrophication and drainage has now placed them at a disadvantage in the face of intensive agriculture.

6.2 Island theory and British woods

The application of so-called 'island theory' to British woods depends on the analogy drawn between real islands of land surrounded by water (or vice versa) and habitat islands of one habitat surrounded by another (Fig. 6.1). The theory developed from the observation that patches of habitat and islands tend to have more species if they are larger. The relationship generally takes the form

$$S = cA^z,$$

where S = number of species, A = area of island and c and z are constants. The value of z usually lies between 0.18 and 0.35: when $z = 0.3$ a large island has roughly twice as many species as a small island of one-tenth its area.

The species–area relationship has been explained by two main hypotheses. The habitat variety hypothesis (Williams, 1964) states that S is a function of habitat variety: larger islands tend to have a greater variety of habitat and this enables more species to maintain a foothold. The equilibrium hypothesis, commonly called the Island Theory of Biogeography (MacArthur and Wilson, 1967), states that the number of species on an island depends on a dynamic equilibrium between extinction and immigration, both of which are influenced by a number of factors. Extinction rates are proportional to S (if more species are present there is a greater chance that one will become extinct in a fixed period) but inversely proportional to A (larger islands will tend to support larger populations of each species). Immigration rates are inversely proportional to S (the chance that a newly arrived organism will add a species to the island decreases as the number already present increases) and isolation (islands which lie close to continents with a large number of species are more likely to be colonised than islands which lie distant), but proportional to A (larger islands present a larger landing surface for dispersing organisms than small islands). When the two rates are equal, S remains more or less constant but extinctions and immigrations continue: the turnover rate is large in relation to S in small islands and small in relation to S in large islands. The equilibrium hypothesis also makes a useful distinction between oceanic and land-bridge islands. The former originate in the ocean with no species and reach equilibrium S from below. The latter are parts of a continent which has been flooded and reduced to an archipelago of islands: initially these have a high S in relation to area, but with increased isolation S declines to a lower equilibrium, a process known as 'relaxation'.

Species vary in relation to the significant factors of both hypotheses. Not only does each species have its peculiar range of habitats, which may be wide or narrow, but its response to habitat change is distinctive. Some tend to colonise rapidly but then succumb to competition from later arrivals: they lead a fugitive existence. At the other extreme are species which form or indefinitely survive in stable habitats: these tend to be slow colonists.

6.2.1 Island theory applied to woodlands

It is exceedingly tempting to apply the habitat variety and equilibrium hypotheses to British woods as if they were proven theory. They are neatly complementary: the former is static and the latter dynamic, and together they could help our understanding and management of the ecological effects of habitat change. Furthermore, the analogy between oceanic islands and secondary woods on the one hand and land-bridge islands and primary woods on the other is potentially useful. However, although the habitat variety hypothesis is intuitively acceptable, since it is supported by the common experience of every field naturalist, it is not immediately obvious that the fauna and flora of individual woods are in a state of dynamic equilibrium.

Several studies have confirmed that the species–area relationship takes the form $S = cA^z$, e.g. Moore and

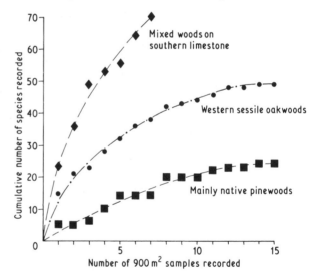

Fig. 6.2 Relationship between the cumulative number of vascular plant species recorded and the number of 900 m² samples in three types of woodland. The types were selected from those generated by an unpublished Indicator Species Analysis of field layer species lists from 700 samples recorded throughout Britain (see p. 112). The field layer types were associated with the following stand types:

♦, Mixed woods on southern limestones, mainly 1A and 4B

●, Western sessile oakwoods, 6A

■, Mainly native pinewoods, 11A

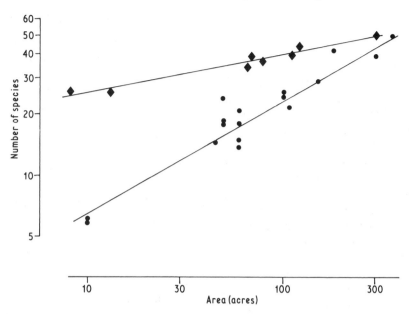

Fig. 6.3 The relationship between species and area in two groups of woods differing mainly in their degree of isolation. Data (unpublished) by kind permission of Dr M. D. Hooper. Each point represents a single wood in either Rockingham Forest (◆) or West Cambridgeshire (●). The species counted were shade-bearing vascular plants.

Both groups of woods grow on heavy, base-rich clays, but those in Rockingham Forest are far less isolated from each other than those in Cambridgeshire. Moreover, until the 19th century, Rockingham was very densely wooded. Hooper interprets the difference in the species–area relationship as a case of relaxation: whilst the woods of Rockingham may have lost some species since they were isolated, the greater and longer isolation of the Cambridgeshire woods has apparently caused about 25 species to be lost in roughly 1000 years from all woods below about 300 acres (121 ha).

Hooper (1975) for birds, Peterken (1974a) and Rackham (1980) for vascular plants. However, woodlands vary enormously in the number of species they contain per unit area (Fig. 6.2), those on carboniferous limestone being rich whereas pinewoods and birch-oakwoods on strongly acid soils with mor humus are conspicuously poor, and this factor can swamp any variation in S due to area. For example, S for an upland oakwood or a lowland hornbeam wood depends not on area but on whether a base-rich stream with its rich associated flora flows through the site.

When groups of woods with a similar habitat range but differing considerably in isolation are compared, the value of z is found to vary. Hooper (1970) compared the ground flora of two groups of ancient woods on heavy calcareous clay in East Anglia (Fig. 6.3). In Rockingham Forest, where woodland had covered at least 25% of the land surface until 1800 and still covers 13%, $z = 0.2$, a value well within the

range characteristic of contiguous habitats and therefore low isolation. In Cambridgeshire, by contrast, where woodland has long covered less than 2% of the land, $z = 0.5$, a value which is said to be indicative of high isolation. Since for a given area the Cambridgeshire woods have fewer species than Rockingham Forest woods, Hooper interpreted this as a case of differential relaxation from a similar original species–area relationship in natural, contiguous woodland. Rackham (1980), however, failed to obtain such clear-cut relationships elsewhere in East Anglia.

One cannot assume that a dynamic equilibrium exists merely because a species–area relationship conforms to the above formula. In Lincolnshire Dr Margaret Game and I found the usual relationship between area and the number of woodland herbs, but correlated this with habitat variety, which was itself correlated with area. If isolation were a factor, we would have expected parts of woods to contain more

101

Fig. 6.4 Panorama of the north Cotswold plateau near Sezincote, Glos, in 1956. The small, scattered woods are linked by a network of hedges containing many mature trees. Hedges are said to be wildlife highways along which species spread across the farmed landscape: they certainly reduce the ecological isolation of small woods, but many woodland species seem incapable of using them as stepping stones to distant woods. (Cambridge University Collection: copyright reserved)

species than whole woods of the same area − the latter would have relaxed but the former would not − but in fact we found no difference. With regard to individual species, we consider that isolation cannot be a factor in the case of, say, *Hedera helix*, which is common in the countryside between woods, and can only be a factor for species which are more or less confined to woods. Amongst the latter it seems that isolation does not influence the survival of species in ancient woods, but does influence the rate of colonisation of secondary woods: certainly this is the case for *Mercurialis perennis*, the only species we have

so far examined in detail. If this finding can be extrapolated we would expect isolation to be a significant factor affecting the floristic richness of secondary woods in the sparsely wooded arable lowlands, but not in those districts which have numerous, ancient hedges (e.g. Dartmoor fringes) and in upland districts where the semi-natural vegetation between woods often contains most of the species found in the woods themselves.

Dr Game and I also compared the species–area relationship of ancient and recent secondary woods in Lincolnshire. Above 3 ha ancient woods had far more species than secondary woods, area for area, due to the inability of many species to colonise from the ancient, presumed primary, woods into the secondary woods (Table 3.3). Secondary woods originating between 1600 and 1820 had the same species–area relationship as those originating between 1820 and 1887 and 1877 to 1947: the flora evidently develops within a few decades then, if it increases at all, does so very slowly indeed.

Clearly we must suspend judgement on the applicability of the equilibrium hypothesis to woods until more evidence is available. When it is we may well find that the hypothesis applies more to some groups, such as birds with their mobility and territorial behaviour, than others. Turnover at equilibrium may be far greater for birds and other animals than for plants. Isolation may not influence survival in ancient woods (or else its influence may be undetectably small beside that of habitat management), but it may prove highly significant in the colonisation of secondary woods. In any case these factors may be completely obscured by variation in habitat diversity within woods and the extent to which habitats between woods harbour woodland species and thus reduce the ecological isolation of the woods themselves (Fig. 6.4). Despite these reservations, some useful generalisations may emerge:

(1) Some districts where large amounts of primary woodland still survive may function as 'continents', in which woodland communities sustain minimal isolation, e.g. lower Wye Valley, Chilterns (see the 'area' approach to nature conservation, Section 15.4).

(2) The fauna of recently reduced primary woods may still be relaxing to a new equilibrium, i.e. species may still become extinct even if management is stable.

(3) Recent secondary woods are unlikely to acquire all the woodland species they are capable of supporting.

(4) Secondary woods close to primary woods are likely to be richer than remote secondary woods, especially if the primary woodland source is large.

6.2.2 Extinction-prone species in British woodlands

One of the more useful ideas to emerge from the attempt to relate 'island theory' to conservation needs is that of 'extinction-prone species'. Terborgh (1974), considering birds, listed six kinds:

(1) Species on the top tropic rung, and the largest members of guilds.

(2) Widespread species with poor dispersal and colonising ability, a group which 'includes a major fraction of the species that inhabited the pristine landscape'.

(3) Continental endemics.

(4) Endemics of oceanic islands.

(5) Species with colonial nesting habits.

(6) Migratory species.

This notion can be applied with modification to British woods. Species on the top trophic rung are not just extinction-prone, they are extinct, e.g. wolves. Likewise the largest members have gone, e.g. bison. We have very few endemic species, but the idea can be extended to include rare species and those represented by very localised populations. Colonial nesters might include species whose populations are locally concentrated even though they are relatively widespread, e.g. wood white butterfly, wood ants, heron. However, the largest and most important group is the slow colonists, many of which are also rare.

6.2.3 Conclusion

Finally, it is worth trying to bring our knowledge of island theory, woodland history and palynology together into a brief summary of the response of woodland species to long-term habitat change. The Atlantic forest maximum provided most woodland species with their chance to expand and occupy all the available ground within their climatic and edaphic tolerances. Later, with clearance after the Atlantic, woodland species responded according to their habitat range, those depending entirely on woodland being reduced whilst others capable of tolerating largely treeless and more disturbed habitats expanded on to cultivated ground and those habitats such as heathland and grassland which we now describe as semi-natural. Such differential responses might have been reversed locally wherever secondary woodland developed, but the capacity of the system to restore itself to its original state declined over the millenia. Initially, complete restoration might have been possible in many areas, those where soils had not been greatly altered by clearance and substantial tracts of original woodland survived to provide sources of seed. Later, restoration was deflected by the progressive alteration of soils outside woodlands and limited by the intensified isolation of new secondary woodland from the dwindling remnants of the original forests. In the last century even the non-woodland, semi-natural habitats, which harboured many woodland species for so long, have been greatly reduced by agricultural intensification. The result in the lowlands is that any secondary woodland formed during the present century stands more isolated than ever before from the places whence its complement of wild species might come. Fortunately for the strict woodland species – exemplified by the last group discussed in Section 6.1 – fragments of the original woodland were never cleared and still survive today. These are the primary woods, distinctive ecologically for their lack of disturbance and continuity (the absence of isolation in time) with the Atlantic forests. Here the original inhabitants of the 'pristine landscape' have survived as relict communities, modified somewhat by traditional management and increasingly isolated from each other by continued clearance of ancient woodland and agricultural intensification, but nevertheless maintaining, we believe, some of the characteristics of the primaeval forests from which they have descended. These are the communities which are difficult or impossible to re-create once they have been destroyed and which therefore need most protection.

PART TWO

Types of semi-natural woodland in Britain

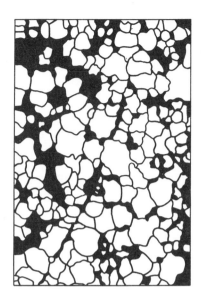

Chapter 7

Woodland classification

7.1 Existing and forthcoming classifications

Any attempt to describe British woodland types is bedevilled by the absence of a stable, well-known and widely accepted classification, for unlike their continental colleagues who have put much effort into descriptive ecology, British ecologists have concentrated on the study of processes and the methodology of classification. For decades, the only available general ecological classification of British woodlands has been the classification contained in the woodland chapters of *The British Islands and their Vegetation* (Tansley, 1939) which developed out of the earliest activities of British ecologists (Moss *et al.*, 1910; Tansley, 1911; Moss, 1913) (Table 7.1), the detailed studies of oak–hornbeam woods by Salisbury (1916, 1918) and the studies of beechwoods and associated types on the southern chalk (Adamson, 1921; Watt, 1923; 1924; 1925; 1926; 1934). It is now about 70 years old in its essential features. Ecologists have long felt that a more up-to-date treatment is needed, and as a result not one but two new systems are in preparation, one by Bunce (1982) at Merlewood, and another by the National Vegetation Classification.

Tansley's system (1939) is set out in Part 4 of his classic work. He described woods in classes which were based on the most abundant species in the tallest stratum: oakwoods, beechwoods, ashwoods, pinewoods, birchwoods, alderwoods and scrub. Mixed woods were recognised in, for example, the oak–ash class, but other mixed stands were included within classes according to which species was most abundant: for example, ash–wych elm woods were included within ash woods because ash is usually more abundant than elm, at least in those woods which had been studied at that time. Some classes were systematically sub-divided, as in the case of beechwoods, which were divided into *Fagetum calcicolum*, *F. rubosum* and *F. ericetosum*.

It would be helpful in many ways if we could continue to use Tansley's classification in perpetuity, if only because it is well-known and has in fact been useful, but there are three main reasons why this is not possible. Most woods consist of mixtures, and it is only in extreme conditions that one species is clearly in the majority. Tansley's system remains useful when dealing with extremes, but is unsatisfactory at the centre of the range of variation: it is a classification with a hollow centre. Dominance is, in any case, an unsatisfactory general basis for classification because the influence of the most abundant species on the whole community varies from one community to another (Poore, 1956a). A second general difficulty is that Tansley's classification is incomplete. Even if we accept that dominance is the most convenient basis for classification, Tansley omits many such types or disguises them within others, notably woods dominated by hornbeam, small-leaved lime and wych elm. The third general problem is that the woods

107

Table 7.1 The woodland associations of Great Britain recognised by Moss *et al.* (1910) and Moss (1913). Definitions are based on dominant tree species and soil characteristics.

ALDER AND WILLOW ASSOCIATIONS on very wet soils

A 1 Acid *Alnus* and *Salix* thickets (lowland moors).
B 2 Basic *Alnus* and *Salix* thickets (East Anglian fens).
C 3 *Alnus* and *Salix* thickets on fresh soils subject to periodical inundations (stream sides).

OAK AND BIRCH ASSOCIATIONS on non-calcareous soils

D On deep clays:
    4 *Quercus robur* woods (English lowlands).
    4a *Quercus robur–Carpinus* woods.

E On dry sands and gravels (south and east England).
    5 *Quercus petraea* and/or *Q. robur* woods.
    6 *Betula* woods (*B. pubescens* with or without *B. pendula*).
    7 *Pinus sylvestris* woods.
    7a Mixed woods of *Quercus, Fagus, Betula, Pinus*.
    8 *Fagus* woods.

F On shallow soils of the older siliceous rocks.
    9 *Quercus petraea* woods (north and west Britain).
    10 *Fraxinus–Q. petraea* woods (Lake District).
    11 *Betula pubescens* woods (high altitude and north Scotland).
    12 *Betula pubescens–Fraxinus* woods (Lake District).
    13 *Betula–Pinus* woods (mid-Scotland).
    14 *Pinus* woods (mid-Scotland).

ASH AND BEECH ASSOCIATIONS

G 15 *Fraxinus–Q. robur* woods on deep marls or calcareous clays (south England).
H On shallow soils over limestone.
    16 *Fraxinus* woods (north and west England, chalklands).
    17 *Betula pubescens–Fraxinus* woods (north England).
I 18 *Fagus* woods on shallow soils over chalk.

---

themselves have changed, thereby undermining the value of the oakwood category. This class may have been useful when so many woods were dominated by oak standards over a coppiced underwood, but since Moss *et al.* (1910) evolved their groups the coppice system has declined almost to extinction, oak standards have been felled without replacement, and the coppice has grown tall with neglect. Woods which were once dominated by oak now consist mainly of ash, maple, lime, hornbeam, wych elm, hazel, etc., and there is no Tansleyan class into which they can be meaningfully placed.

7.1.1 Recent classifications

Bunce's 'Merlewood classification' (1982; Fig. 7.1) is based on Indicator Species Analysis (ISA) (Hill *et al.*, 1975) of 1648 200 m^2 samples taken from 103 woods. These woods formed a stratified random sample of the 2500 British woods which had been recorded in a national survey carried out by the Nature Conservancy. The ISA, which took into account vascular plants and bryophytes, generated 32 'plot types' at what was considered to be the most useful level of detail. These plot types will be characterised in terms of their composition, structure, soil types, distribution and many other features.

The National Vegetation Classification will adopt a more continental methodology and presentation. It will construct, among other things, a national classification of woodland types which should bear some resemblance to the classes now widely and successfully used by phytosociologists elsewhere in Europe. This will, in some ways, complete the work started by local studies (McVean and Ratcliffe, 1962; Birks, 1973) and the ambitious attempt by Klötzli (1970) to classify oak–ash, ash and alderwoods on the basis of published lists and sites seen on a rapid tour of Britain. The associations recognised and related to continental woods in this remarkably interesting paper were:

Blechno–Quercetum } in hyper-oceanic
Dryopterido–Fraxinetum } districts

Querco–Betuletum
Querco–Fraxinetum } in oceanic districts
Hyperico–Fraxinetum

Osmundo–Alnetum
Pellio–Alnetum

These new classifications cannot be assessed until they are published and have been in use for some time. If theoretical rectitude and industry are any guide then they will be valuable.

7.2 Principles and approaches to woodland classification

Gilmour and Walters (1964) made three important points about classification. The term 'classification'

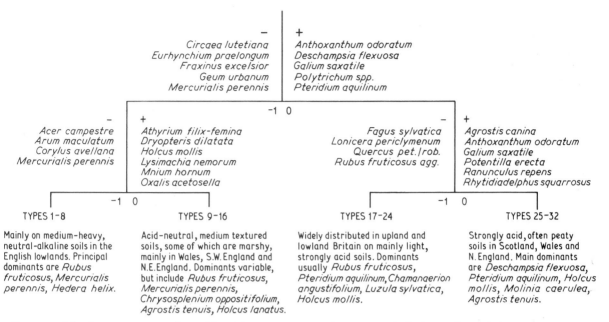

Fig. 7.1 The four main groups formed by Indicator Species Analysis upon which the Merlewood National Classification of British Woodlands is based (Bunce, 1982).

describes the act of grouping objects into classes because of certain attributes which they have in common. This is man's basic method of dealing with the multiplicity of individual objects in his surroundings. The actual classifications made are therefore determined by man's desires and purposes in relation to his environment, and the suitability of any particular classification can only be judged by its usefulness. Gilmour and Walters distinguished 'general-purpose' from 'special-purpose' classifications, although these are linked by intermediates. The classes of general-purpose classifications have a large number of attributes in common, whereas those of special-purpose classifications have few. The former can be made only when the objects being classified are influenced by a powerful factor, which causes a number of attributes to be highly correlated in their occurrence, and if no such factor exists then only special-purpose classifications are possible. Their third point was to recognise two approaches to classification. Under the 'typological' approach the classifier defines a type or node and individual objects are said to resemble it to a greater or lesser degree. In contrast, the 'definitional' approach requires rigid definitions

which are applied precisely and each object must be pigeon-holed into one compartment of the classification.

Phytosociological classifications, as Gilmour and Walters (1964) pointed out, suffer from a number of disadvantages when compared with taxonomic classifications. Amongst these is the problem of selecting the basic unit, i.e. size of sample, which is comparable with the individual plant as a basis for taxonomy. Likewise, phytosociologists lack the equivalent of morphological features as an agreed basis for classification, and have tended to resort to either floristic or physiognomic-habitat classifications. Furthermore, in the absence of any single all-powerful ecological factor comparable with inheritance in taxonomy, general-purpose phytosociological classifications are difficult to achieve.

Classification of woodlands must start with an appreciation of the character of variation in composition and the factors which significantly affect it. The environmental factors associated with variation in semi- (past and future) natural stands can be divided into three groups: site, time and management. Site factors include climate, which operates on a regional

109

scale, and geology, topography, drainage and other edaphic factors, which cause variation on a local scale. Together these factors determine the natural limits of composition on any particular site, subject to any residual instability from past changes in the factors. Time factors revolve round the age and development of woodland, i.e. the contrast between primary and secondary, the time since a secondary wood originated, and the past and present ecological isolation of a wood. Management factors cover past and present human influence which, in semi-natural woods, is largely concerned with traditional management, i.e. the direct and indirect effects of coppicing, etc. on composition, and the effects of domestic grazing and browsing.

Webb (1954) summed up the experience of most field ecologists when he noted that variation in the composition of vegetation tended to 'fluctuate tantalisingly between the continuous and discontinuous'. This exactly reflects my own experience of British woodlands. Variation in stand composition appears to be a multi-dimensional continuum, all points of which would occur equally if the factors influencing composition were to occur equally and randomly in space and time. In fact, they are far from random. Certain soil types are common and others are rare. Even common soil types do not occur throughout Britain but are confined to particular regions and climates. Ancient woods are common as are recent secondary woods, but secondary woods of pre-18th century origin are relatively rare or are restricted to certain regions. In this way particular points in the continuum are over-represented, and thereby become convenient nodal points (Poore, 1955) around which to define classes.

Other factors can be locally important. Under constant environmental conditions communities can fluctuate in composition by, for example, the natural cycle of species (Section 1.1.2) or chance fluctuations in rare species. Thus, within the terms of any classification based on composition, a site may be capable of supporting more than one woodland type, and natural fluctuations may determine the type which is actually present at any one time. Such effects would cancel each other out above the local scale. Another factor concerns the effects of isolation on very long-established communities. For example, a minute

outcrop of limestone in a predominantly acid area may never acquire the species which are characteristic of the site-type elsewhere because they are not present within colonising distance. Conversely, the species of the acid environment may, in the absence of competition from calcicoles, spread beyond their normal edaphic limits. A third related factor is proximity and population pressure: for example, a calcicole like *Viburnum lantana* is most likely to occur within an acid stand type if both calcareous and acid types are present within the same wood, and the calcareous type has a large, vigorous population of this species.

Another factor which especially affects woodland classifications is structural complexity, which, in my opinion, justifies three separate, independent classifications for the tree and shrub layer (stand), the field layer and the communities of epiphytes. These are variously the environment and substrate for each other. That the epiflora remains separate seems to be generally agreed, but the stand and field layer are usually merged. Separation of the stand and the field layer is justified for three reasons:

(1) Field layer communities often vary at a scale far below that of individual trees and generally respond as communities to microtopographical features to which tree and shrub species can only respond as individuals. Any quadrats which are large enough to contain a representative sample of the stand may include several distinct field layer communities.

(2) Stand composition may be naturally independent of field layer composition. Admittedly, they are both responding to the same soils and tree seedlings start as members of the field layer, so a strong correlation is likely. Indeed, field ecologists agree that these correlations do in fact exist, but it seems dangerous to assume that they do in all instances.

(3) Irrespective of (2), the correlations may be destroyed by management, especially planting. Whilst it is usually safe to assume that the field layer has not been directly and selectively modified by management – even though some people dig up primroses whilst others plant them into other woods – this can never be assumed for the stand. In any case, it is convenient to have a field layer classification which can be used in plantations.

Viewed against this background, both the Merlewood and the NVC are general-purpose classifications based on floristics which combine the stand and field layers and exclude epiphytes, but the former is definitional and the latter more typological. They also differ significantly in their sampling, as Merlewood uses a stratified systematic approach which is often described as objective and the NVC selects apparently uniform samples by subjective decisions. They differ too in their treatment of species, Merlewood merging difficult species pairs (e.g. *Quercus petraea* and *Q. robur*) and confining attention amongst cryptogams to the readily-identified bryophytes, whereas the NVC will record all species. The appearance of the final classes will differ. Whereas the type 27 of the Merlewood classification is the '*Calluna vulgaris–Pteridium aquilinum* type' defined by five sets of indicator species, the NVC equivalent will probably be a form of *Pinetum*, defined broadly on a canopy dominant and on certain constant and faithful species.

7.3 A new classification of semi-natural stand types

The description of British woodlands in this volume requires a classification, but none of those available or in preparation is completely suitable. The Tansleyan approach comes closest to my need to concentrate on the stand, management and the dynamic aspects of woodland ecology, but it has several disadvantages which are outlined above. On the other hand the new classifications are neither published nor tested, nor is their structure appropriate to my purpose. Therefore I have prepared yet another classification, which refers only to the stand composition and the factors of site management and time which influence it.

This classification was evolved as a framework for communication at national and regional levels (i.e. much as Tansley's classification has been used) and as a means of describing and mapping individual woods. It is broadly Tansleyan in character in so far as it is based on (a) the composition of selected elements in the stand; (b) soil reaction, texture and drainage; (c) stand history and development; and (d) regional variation in these. Greater attention is, however, paid to mixed stands and to the influence of management

on the composition of 'climax' woodland. Moreover, the time factors and successional relationships are treated differently. On the other hand, this classification is, like Tansley's, robust in the face of incomplete data and the hands of inexperienced observers, for stand types can be recognised from as little as two or three species and a scant knowledge of the site. It could even be construed as an extension of the approach to woodland classification which was adopted by the pioneers of British ecology.

The objective has been a general-purpose classification, but in view of the difficulties outlined above, its success as such can only be limited. It attempts to combine the advantages of both the typological and definitional approaches. Thus, the major units, the Stand Groups, have the clarity and artificiality which tends to be associated with definitional classifications: the units have features (species) which are invariably present and others which are invariably absent. However, the sub-divisions of these major units, the Stand Types and Sub-types, are more typological (although some are clearly definitional, such as the types within the hornbeam group), thereby accepting the natural fact that variation within semi-natural stands is indeed continuous and that it is more convenient to regard some stands as intermediate between two types than to force all stands into moulds. The typological character of the units is emphasised in practice, for the definitional character of the stand group is diluted when one uses the classification to describe real woods and their gradations of composition (e.g. Section 13.6).

7.3.1 Strategy of classification

The strategy of this classification may be briefly summarised. Thirty nine stand types are defined which collectively cover the entire range of the site dimension of variation. This is achieved by holding the time and management dimensions constant in so far as this is possible, by defining the stand types from coppice stands on ancient woodland sites. Variation in those dimensions is introduced subsequently by considering, respectively, successional variants of each stand type and by comparing the coppice and wood-pasture stands assignable to particular stand types. In theory the time and management dimensions could have been held constant in other ways. The time dimension could, for example, have been held constant by basing stand types on recent secondary woodland, and the management dimension could have been held constant by considering only wood-pasture woods. However, recent secondary woods of the necessarily narrow age range are probably not available on the full range of site conditions, and their composition is subject to many factors of dispersal and chance which are not related to site. Wood-pasture woods are of very limited occurrence, whereas coppice woods (which include those which have been coppiced in the past) on ancient woodland sites still occur in most regions and on most soil types. Coppice composition has been affected by management, but there is much evidence that it is still substantially determined by site factors, and that it embodies past-natural features (Section 3.5).

7.3.2 Methods of recording

Approximately 700 sample stands were recorded in semi-natural woods throughout Britain (Fig. 7.2). They were neither evenly nor randomly distributed, for they were recorded opportunistically during site visits for other purposes: any attempt at an objective, random or regular sample programme would have been prohibitively expensive for the number of stands and geographical area covered. All the samples were in semi-natural stands, but a few were in recent secondary woodland and wood-pasture woodlands, and were not therefore available for analysis as coppice and former coppice on ancient woodland sites. Ancient woodland was identified by methods which will be discussed in Sections 13.2 and 13.3. Stands were accepted as semi-natural if no positive evidence of planting could be seen. Planted coppices occur in recent secondary woodland and their origin is easily detected by the purity of the stand, the even stool size and the planting lines. Similar signs of planting occur in ancient coppices of native species, especially of hazel in southern counties, and these were also rejected. Likewise, coppices dominated by chestnut and long-established sycamore were not recorded. The remaining stands were accepted as semi-natural even if some standards appeared to have been planted.

Most woods within which sample stands were to

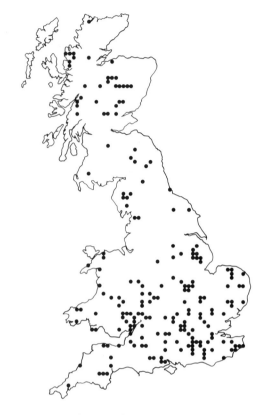

Fig. 7.2 Distribution of the sample stands upon which the classification of stand types (Chapter 8) is based. Each dot represents one or more samples in a 10 km square of the national grid.

be recorded were surveyed so that the major directions of variation and 'woodland types' were identified. For example a wood which consisted of ash–hazel coppice on clay at one end and graded through to birch–oak coppice on sand at the other was judged to include two woodland types. Sample stands for detailed recording were then carefully selected to represent each woodland type, any significant variants of these and extensive transition zones. Thus, in the example, one sample was recorded from the birch–oak coppice, two were recorded in the ash–hazel coppice to represent the main type and the variant characterised by aspen, and a fourth sample was recorded in the transition zone. Sample stands were therefore selected by reference to variation within individual sites and not to any pre-existing formal or informal national or regional classification. The exact location of a sample within the type it represented was carefully chosen to include the essential characteristics and exclude unusual features. Apart from a few exceptional cases, no more than seven samples were recorded in any one wood.

Most samples were 30 × 30 m or its aerial equivalent, a size which was felt to be the best compromise between the need for a large sample which included all the species in the woodland type, and the need for a small sample which would have been quicker to record and more homogeneous. Within each sample the following records were made:

(1) A list of tree and shrub species; the cover of each species judged in relation to a Domin-type scale (Table 8.1); and the structural types (e.g. standards, coppice, shrubs, saplings) in which each species occurred.

(2) A list of vascular plants, noting those which were abundant or common, but otherwise not quantified.

(3) Soil strata as seen in a soil pit about 30 cm deep, dug in the centre of each sample; pH at 10 cm; and texture into classes adopted by the Soil Survey for field identification.

(4) Brief notes which stated the development of the present stand structure (amplifying (1)); estimated height and cover of each stratum; the distribution and general character of the woodland type which the sample represented; and the relationship to other woodland types within the site.

7.3.3 Analysis

The object when analysing the recorded stands was to define stand types which (i) were reasonably homogeneous with respect to composition and site factors, (ii) each possessed certain features which were invariably present and invariably absent from others, and (iii) collectively covered all semi-natural woodland on ancient woodland sites.

Pursuit of these ideals seems inevitably to involve a compromise between the clarity and artificiality of the definitional approach and the naturalness of the typological approach (see above). The compromise has been sought in the present classification by defining

12 stand groups artificially, simply on the presence or absence of 13 tree and shrub species, and then to define ecologically homogeneous types and sub-types within each group by a form of successive approximation (Poore, 1955; 1962) based more on the full range of composition and site characteristics. The artificial stand groups have been made as natural as possible (see below).

The strategy in this classification comprised four stages:

(1) To generate a set of narrowly defined types or 'organising noda' based on correlations in the occurrence of certain tree and shrub species. These accommodated about 170 of the 450 samples available when the analysis started, and eventually became the stand groups.

(2) To sub-divide those organising noda within which there were correlated discontinuities of composition and site features. These subdivisions eventually became the stand types and sub-types.

(3) To expand each organising node and sub-node by adding samples which were closely related to one of them, and no other, and which possessed the minimum definitional qualifications. This process increased the number of samples which were assigned to a particular type and reduced the number which remained unassigned. It continued until the optimum balance was achieved between the desirable but conflicting needs to produce broad types but retain homogeneity at sub-node level within each unit.

(4) To identify gaps left by the previous stages and to fill them by various methods. By the end of this stage about 90% of all samples recorded could be accommodated within a defined stand type, leaving a small residue of intermediate and unclassified stands.

These stages were undertaken in parallel, but the emphasis was initially on (1) and moved progressively through (2) and (3) to (4): this is the essence of the method of successive approximations.

The organising noda (Stage 1) were generated by the following method. Thirteen species were present as coppice in more than 20 of the 450 samples available when the analysis started, the most frequent being hazel in 331 samples. The coppice component was chosen because it is more natural and stable than

113

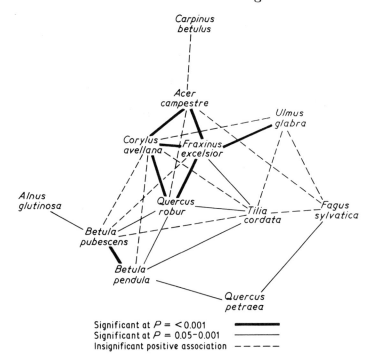

Significant at $P = <0.001$ ▬▬▬▬
Significant at $P = 0.05-0.001$ ────────
Insignificant positive association ── ── ──

Fig. 7.3 Constellation of positive associations (chi-square), significant at $P = 0.05$, between the main coppice species of ancient, semi-natural woodland.

either the standards (Rackham, 1974) or the self-sown trees, shrubs and species which are present only as saplings and seedlings. Associations between each pair of species were tested by chi-squares, and the positive associations were displayed as a constellation diagram (Fig. 7.3). Only hornbeam and sycamore had no significant positive associations but hornbeam could be linked to the constellation through its one insignificant positive association with maple. Sycamore could not be related, for its insignificant positive associations were with alder, pedunculate oak, ash and wych elm, species which were spread throughout the diagram.

The core of variation in coppice composition is the strong association between ash and hazel, and their separate associations with field maple and pedunculate oak (as coppice). Small-leaved lime and the birch species constitute an extension of the core which is linked to ash and pedunculate oak, whilst wych elm forms another core extension through its strong association with ash. Alder, sessile oak, beech and hornbeam remain as satellites. The organising noda were formed from these positive associations, and are listed below with the Groups (1–9) into which they eventually developed:

ash–wych elm (Group 1)
ash–maple–hazel (Group 2)
ash–hazel–pedunculate oak (Group 3)
ash–lime–pedunculate oak (Group 4)
lime–pedunculate oak–birches (Group 5)
sessile oak–birches (Group 6)
alder–birches (Group 7)
sessile oak–beech (Group 8)
hornbeam (Group 9)

The gaps which had to be filled in Stage 4 were of three overlapping kinds:

(1) Disguised correlations. Two coppice species might be positively correlated on one site type, and therefore justify the creation of a stand type, but this might be masked in Stage 1 by negative association on other site types. The beech–pedunculate oak association on deep, calcareous clay loams, beech–ash on shallow calcareous soils and sessile oak–ash on shallow limestone soils were all obscured by stronger negative associations on other soils.

(2) Species occurring mainly as high forest. Some species are infrequent as coppice but widespread as maiden trees because they coppice weakly (beech), are short-lived and regenerate well from seed (birch),

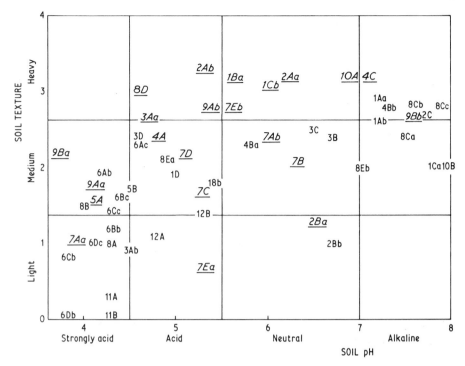

Fig. 7.4 The distribution of stand types and sub-types in relation to soil texture and reaction. The position of each type indicates its median pH and mean texture (calculated by the procedure described in Section 8.1.2). Types characteristically associated with wet and poorly-drained soils are underlined.

Stand types on freely-drained soils fall into distinct acid and alkaline groups, separated by a cluster of types on poorly-drained soils in the middle pH range.

grow beyond the range of coppice management (Group 12, Birch woods), or are relatively recent arrivals in many woods (suckering elms). These therefore occur mainly as high forest stands and as maiden or standard trees within coppice. Definition of stand types in Group 8 beechwoods therefore includes samples from high forest stands, and Groups 10 and 12 were created at Stage 4 to accommodate respectively suckering elm stands and birch stands.

(3) Types beyond coppice range. The stand types based on coppice came so close to being a complete classification of semi-natural woodlands, that it seemed worth adding a group to accommodate the pine woods (Group 11).

7.3.4 Structure of the classification

The foundation of the classification is the 12 *Stand Groups*, which collectively cover the site dimension of variation within ancient, semi- (past) natural coppice

and high forest woods. These stand groups may be arranged in the following order of precedence:

Group 7	Alderwoods	*Alnus glutinosa*
Group 8	Beechwoods	*Fagus sylvatica*
Group 9	Hornbeam woods	*Carpinus betulus*
Group 10	Suckering elm woods	*Ulmus carpinifolia, U. procera*
Group 11	Pinewoods	*Pinus sylvestris*
Group 1	Ash–wych elm woods	*Ulmus glabra*
Groups 4 and 5	Ash–limewoods Acid oak–limewoods	*Tilia cordata*
Group 2	Ash–maple woods	*Acer campestre*
Group 3	Hazel–ash woods	*Fraxinus excelsior*
Group 6	Birch–oakwoods	*Quercus petraea, Q. robur*
Group 12	Birchwoods	*Betula pendula, B. pubescens*

115

The groups above the line are based on the presence of five 'species' which rarely occur together, and which occupy well-defined, limited edaphic and geographical ranges. Those below the line form an attenuating sequence whereby, for example, *Fraxinus* is present in Group 3, and may be present in any group above it in the table, but must be absent from the groups below (i.e. 6 and 12). Thus, the mixed deciduous woods below the line are split successively on the presence of species with an increasingly wide range, i.e. from the relatively narrow *Ulmus glabra* and *Tilia cordata* to the relatively wide *Betula* and *Corylus* (which is used to sub-divide group 12).

The value of this structure can be appreciated by considering the alternatives. A classification based on minor species, such as *Crataegus*, *Ilex*, and *Ribes*, would be interesting, but it would be difficult to use and would lack credibility as a classification of woodlands. A classification which was based on the species which have in fact been used, but in the reverse order of precedence, would be quite impractical. Most stands would be 'birchwoods' and most of the remainder would be 'ashwoods'.

Each stand group is each divided into *stand types*. The stand types within any group are defined independently of those in other groups. Nevertheless there are some links between types in different groups. Thus, both acid beechwoods, hornbeam woods and oak–limewoods are divided on the basis of oak species, forming analogous types 5B, 8A and 9B. The dry, calcareous types in several groups have the same suffix: 1C, 2C, 3C, 4C, 8C.

Some stand types are divided into *stand sub-types* for three closely related reasons. A stand type can include two edaphically distinct forms (9B). More often, a stand type covers a wide geographical (e.g. 1A, 4B, 7E) or edaphic (e.g. 3A, 7A, 9A) range which can usefully be sub-divided to show significant variation. Alternatively, some sub-types are defined on composition because this seems to be ecologically significant (e.g. 2A, 6A, 6B, 6C, 6D, 8C).

The groups, types and sub-types of the basic classification, which refers to semi-natural coppice and naturally regenerating high forest stands on ancient woodland sites, are described in Chapter 8. *Successional variants* of each type can be recognised which describe the seral stages leading to or from each type. *Management variants* of several kinds can also be described for each type. These attempt to describe the major modifications wrought on each type by wood-pasture management, and by the more intensive forms of the coppice system and high forest stands. These are all described in Chapters 9 and 10.

Chapter 8

Types of ancient semi-natural woodland

8.1 Introduction

8.1.1 Descriptions of stand groups and stand types

The descriptions of stand groups and types are presented as a catalogue of more-or-less standardised accounts. Stand groups are defined and described in terms of their composition, distribution, edaphic range, structure and management. The basis of their division into stand groups is explained, and their phytosociological relationships and some published descriptions of them are briefly mentioned. Stand types and sub-types are similarly described, but with a greater emphasis on detail. A brief account of the vascular flora associated with each type and sub-type is given by mentioning the characteristic dominants and listing these species which have been recorded in more than 20% of the recorded samples. (Those which were recorded in more than 60% of the samples are marked with an asterisk.)

For ease of reference all stand types and sub-types are listed at the end of this chapter (p. 173).

8.1.2 Tables

Many of the important details are given in Tables 8.2–8.11, which summarise the composition and site conditions of each type. Since they are derived solely from my samples these tables suffer from certain limitations, namely the variable and often limited number of samples available and the possibility that the samples are not representative of the type. Furthermore, they inevitably obscure local variations in the features of a type, and these variations may be quite significant. For each type and sub-type the tables show the following features:

(a) the number of samples recorded.
(b) constancy of each tree and shrub species.
(c) median pH.
(d) an index of soil texture.
(e) the number of samples in each of four profile drainage classes.
(f) characteristic site drainage classes.

The constancy of tree and shrub species expresses the proportion of samples in which each species occurs and is classed as follows: I 20% or less, II 21–40%, III 41–60%, IV 61–80% and V 81–100%. For example, *Betula pendula* was recorded in 6 of the 14 samples (43%) in 6Cc, and is therefore shown in Constancy Class III. This gives an accurate picture of each type except in Constancy Class I, which contains both characteristic but sparse members of the community, as well as rare species (some of which are strays from another community). Therefore, when 10 or more samples are available, Constancy Class I is divided into two parts, +(10% or less) and I (11–20%). The symbol ' + ' is also used to denote species which have been recorded in a type but not in a

117

Table 8.1   Floristic table for stand type 2C, dry ash–maple woods. The proportion of the sample covered by each species is recorded by the following modified Domin scale:

10	100%	3	One large plant
9	Above 75%	2	One small plant
8	50–75%	1	Saplings only
7	33–50%	×	Seedlings only
6	15–33%	D	Dead
5	5–15%	+	Recorded close to the sample, but not within it
4	Below 5%		

	Stand number											Frequency	Constancy class
	223	240	370	371	443	447	486	487	553	680	684		
Acer campestre	6	4	4	4	2	4	6	7	3	5	4	11	V
Betula pubescens									6			1	+
Corylus avellana	6	8	8	9	9	5	9	7	8	8	8	11	V
*Fagus sylvatica**	2										1	2	*
Fraxinus excelsior	7	8	9	7	+	9		6	6	7	5	9	V
Quercus robur	6	6	7		6	5	2	8	+	3		8	IV
*Ulmus glabra**									1			1	*
Acer pseudoplatanus										5	4	2	I
Clematis vitalba											4	1	+
Crataegus monogyna	3	5	4	1	1	4		4	2	4	4	10	V
Crataegus oxyacanthoides					2							1	+
Euonymus europaeus			4		4			1		4	1	5	III
Hedera helix		4	×	4		4		×		4	5	7	IV
Ilex aquifolium		4		1							1	3	II
Ligustrum vulgare				1							4	2	I
Lonicera periclymenum			4	×				×		4		4	II
Malus sylvestris							2					1	+
Prunus avium		6								6		2	I
Prunus spinosa	D	4	4		2							4	II
Rosa arvensis			2	2					2		4	4	II
Rosa canina									2	4		2	I
Salix caprea									2		4	2	I
Sambucus nigra			2									1	+
Sorbus aria			3		3							2	I
Thelycrania sanguinea			2	2	4	2		2		2	4	7	IV
Viburnum lantana											4	1	+
Viburnum opulus										2	2	2	I

* Although two samples contained *Fagus* and another contained *Ulmus glabra*, these samples are not placed in stand Groups 8 and 1 respectively because only recently arrived saplings and small plants were present.

sample of that type. In most stand types there are at least as many species in Constancy Class V as in Class IV and this is regarded as a test of a true Association. Very few vascular plant species reached Constancy Class V, some reached Class IV and III, and most were in II and I – which indicates that these were not true Associations. The floristic table for stand type 2C (Table 8.1) provides an example of the information from which the summaries in Tables 8.2 to 8.11 have been prepared.

The importance of chemical composition of the soil as an ecological factor is repeatedly demonstrated by experienced ecologists who can estimate, for example, pH from a list of plant species. More rigorously, perhaps, studies based on systematic sampling and numerical analysis show that the principal components of variation within sites are related to this factor (Barkham and Norris, 1970), and that units in a national classification can be characterised as acid or alkaline (e.g. the Merlewood types). Soil chemistry is indicated here only by the median pH, although if the data had been available, values for nitrate and phosphate would have been equally useful.

Soil texture classes were recorded for each sample. A mean texture index was calculated for each type and sub-type by counting 0 for sand and loamy sand; 1 for sandy loam; 2 for sandy clay loam, loam, sandy clay and silty loam; 3 for silt, clay loam and silty clay loam; 4 for silty clay and clay; and then averaging the scores. These figures express in simple terms the ratio of sand and clay in each soil and enable the soil texture of each stand type to be more accurately expressed than a simple 'light', 'medium', or 'heavy' description. Organic soils were excluded from the calculation.

Profile drainage was determined from a soil pit dug to 30 cm depth, in which signs of drainage impedance and soil structure were recorded. Impervious or poorly-drained horizons generally have a compact structure, grey hue and orange mottles. Profile drainage classes depended on the depth at which such a horizon was found: very poor, 0–10 cm; poor, 11–20 cm; imperfect, below 20 cm; free, no signs of impeded drainage. Although these are not the thresholds used by the Soil Survey, they are certainly significant in woodland soils, and can be fairly readily predicted from the ground flora before a soil pit is dug.

The site drainage classes express the characteristic location of each stand type and sub-type in relation to relief: v = valley; r = receiving sites at the base of slopes; s = slopes, including shedding sites at the top of slopes; p = plateau. When a type is strongly associated with a particular topographical position (or where, in the case of sloping sites, the gradient is often medium or steep) this is shown by a capital letter.

8.1.3 Distribution maps

The distribution of most stand types is shown by a dot map indicating presence in 20 × 20 km squares of the national grid. These maps are very incomplete but they do at least give some indication of each type's range and frequency. Apart from my own direct observations, they include my interpretation of some published descriptions and the unpublished results of surveys which happened to be available. The last source notably includes the records collected by a national woodland survey organised in the late 1960s by the Nature Conservancy and County Naturalist Trusts, which are stored at Monks Wood Experimental Station, near Huntingdon. Dr F. Rose kindly supplied many records and is almost solely responsible for the valley alderwood (7B) maps. My estimate of the completeness of each map is indicated by a four-point scale: A = almost complete; B = fairly good coverage; C = many gaps, but the main range is apparent; D = very incomplete.

8.2 Ash–wych elm woodland (Group 1)

Stands with *Ulmus glabra*, but not *Alnus*, *Fagus*, *Carpinus*, *Ulmus procera*, *U. carpinifolia* or *Pinus sylvestris*. Almost all stands contain *Fraxinus*, *Corylus* and *Crataegus monogyna*. *Fraxinus* is often far more abundant than *Ulmus glabra*, and indeed many stands have been described as 'ashwoods' in the past. *Betula* is characteristically infrequent.

Wych elm is a weak calcicole which can grow on mildly acid soils, down to about pH 4.7. Although it occurs throughout Britain, it is infrequent in southern England and north Scotland, and most abundant in Wales and west and north England. Wych elm has a pronounced preference for deep, flushed, moist soils and appears to be moderately nitrophilous. Ash–wych elm woods, however, rarely occur on valley bottoms because alder (and, exceptionally, other elms) is generally present, but they are particularly characteristic of western limestone slopes and the base of steep, upland valley sides. Wych elm is an occasional component of ash–beech woods in the English lowlands, and is rare in hornbeam and suckering elm woods. Like other species of moist, base-rich soils (e.g. ash), wych elm can grow on dry, limestone sites but it does so only rarely, and then usually as a constituent of ash–beech woods. Most ash–wych elm woods have been managed as coppice, either as simple coppice, or with standards of

119

oak (generally), ash, wych elm and lime. Almost all the surviving stands have not been cut since 1940 or much earlier.

Stand types (Table 8.2) are difficult to define because this group covers a wide edaphic and geographical range within which there are no sharp discontinuities. Three distinct and fairly common forms which can be recognised with confidence are types 1A, 1B and 1D. Type 1A is the characteristic form on limestone slopes in the north and west on deep, freely-drained, heavy, calcareous soils. Type 1D is the form associated with coarse, neutral-acid, flushed soils also in the north and west. Both types are fairly common from south-west England to the Scottish Highlands, becoming rather less distinct from each other towards the northern end of this range. Together they are the 'mixed deciduous' woodland of most of upland Britain. Another well-defined type on poorly-drained, heavy soils occurs mainly in the English lowlands, and, together with an infrequent form on low-lying, light soils in the same area, forms type 1B. The associated oaks are almost invariably *Q. robur* in types 1A and 1B and *Q. petraea* in type 1D. Confusion occurs in a minority of ash–wych elm stands which are found on or near limestone, for they combine the features of types 1A, 1B and 1D, whilst possessing some distinctive characters. These are provisionally placed in type 1C, a 'dustbin' category which will almost certainly need revision after more detailed investigations.

Ash–wych elm woods fall mainly within types 1–3 and 4–6 of the Merlewood classification, although type 1D is biased more to their types 5, 10, 12 and 14. Their position in the Klötzli (1970) classification is also rather spread, but 1A falls clearly into the *Dryopteridio–Fraxinetum* and 1Ba into the *Querco–Fraxinetum*. Parts of type 1C approach his *Hyperico–Fraxinetum*, whilst 1D appears to be on the border between *Dryopteridio–Fraxinetum* and *Blechno–Quercetum*. Tansley (1939) would have described most stands as oak–ash woodland or *Fraxinetum calcicolum*, but type 1D is included within the descriptions of sessile oakwoods.

8.2.1 Stand type 1A. Calcareous ash–wych elm woods

Ash–wych elm woods on medium-heavy, freely-drained, calcareous soils, usually clay loams with a pH

Table 8.2 Composition and site features of ash–wych elm woodland (Group 1).

	1Aa	1Ab	1Ba	1Bb	1Ca	1Cb	1D
Number of samples	20	16	11	6	5	5	21
Acer campestre	IV	I	III	II	III	IV	+
Betula pendula	+	+	II	I	I	III	+
Betula pubescens	II	I	I	I	I	I	II
Corylus avellana	V	IV	V	V	IV	V	V
Fraxinus excelsior	V	V	V	V	V	V	V
Quercus petraea	+	+				IV	III
Quercus robur	III	II	IV	V	IV	+	+
Tilia cordata	II	I	+	II	I	I	II
Ulmus glabra	V	V	V	V	V	V	V
Acer pseudoplatanus	I	II	II	III	III	III	III
Castanea sativa							+
Clematis vitalba	II			IV			
Crataegus monogyna	IV	IV	IV	IV	V	V	III
Crataegus oxyacanthoides	+		III		III		
Daphne laureola	I	+				I	
Euonymus europaeus	I		II		III		+
Hedera helix	IV	I	II	II	IV	IV	II
Ilex aquifolium	III	I		I	I	II	II
Ligustrum vulgare	I		I		III		
Lonicera periclymenum	+	I	IV	III	I	III	III
Malus sylvestris	+		II	I	I		+
Populus tremula	+		+				
Prunus avium	II			+		II	+
Prunus padus		+					I
Prunus spinosa			+		II		+
Ribes sylvestre	+					I	
Ribes uva-crispa	+	+					+
Rosa arvensis	II	I	II		I	I	+
Rosa canina	+	I	I	II	II		I
Rosa villosa		I					+
Salix aurita							+
Salix caprea	+	+	II		I		+
Salix cinerea							+
Sambucus nigra	II	III	+	I			I
Sorbus aria	+						
Sorbus aucuparia		II				I	III
Sorbus torminalis			+		II		
Taxus baccata	+	I					
Thelycrania sanguinea	+	+	II	I		I	
Tilia platyphyllos	+	+				I	
Tilia vulgaris				+			
Viburnum lantana					I		
Viburnum opulus	+	+	I	I		II	
Median pH	7.2	7.2	5.6	5.4	7.8	6.0	5.0
Mean texture	2.9	2.6	3.2	1.8	2.0	3.1	1.9
Drainage free	15	15		2	5	2	17
imperfect	4	1	2	3		1	3
poor	1		3	1		4	
very poor			6				1
Site	Sr	SR	P	sr	s	Sr	sR

120

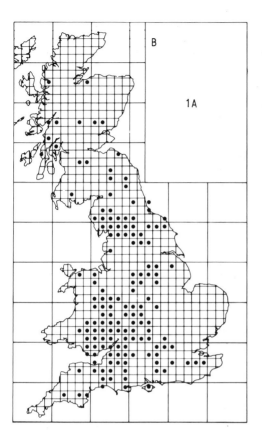

Fig. 8.1 Distribution of stand type 1A, calcareous ash–wych elm woods.

of 7.0–8.0. The oak, when present, is usually *Q. robur*, and only rarely *Q. petraea*. *Sambucus* is frequent, but *Crataegus oxyacanthoides* and *Lonicera* are rare. These stands are generally on sloping ground associated with limestone. They are known north to Moray, but are especially characteristic of woods on carboniferous limestone in England and Wales (Fig. 8.1). They are divided into a southern sub-type (1Aa) and a northern sub-type (1Ab) which is poorer in tree and shrub species. This type is closely related to 1C, which occupies drier, thinner soils on limestone. Type 1A seems to be the core of Klötzli's *Dryopteridio–Fraxinetum* (1970).

(a) *Sub-type 1Aa. Southern calcareous ash–wych elm woods*

The ash–wych elm woods of the south-west limestones and parts of the southern chalk are amongst the richest woodlands in Britain. They generally occur as mixed coppice of *Fraxinus, Corylus, Ulmus glabra, Acer campestre* and *Tilia cordata*, with a low stocking of *Quercus robur* standards, and a tendency for *Prunus avium* to accompany *Fraxinus* and *Acer pseudoplatanus* as the tree species which invades after coppicing. Although 1Aa is defined geographically, it is clearly distinguished from 1Ab by the presence or much higher frequency of *Acer campestre, Clematis, Hedera, Ilex, Ligustrum* and *Prunus avium*: indeed, almost the same sub-type would have been formed on the basis of the presence of *Acer campestre* and/or *Clematis*. The sub-type is most frequent in the Welsh borderland and the limestones of south-west England and south Wales, where it tends to occupy the moister, deeper soils towards the base of slopes. It is known also from the Chilterns in similar situations, and the chalk of south-east England, although here *U. glabra* is infrequent and *Fagus* is much more likely to be present. This sub-type is closely related to 2C, 3B, 4Bb, 8Ca and 8Cb, and often occurs with them in the southern Welsh Borderland.

The field layer is usually dominated by *Mercurialis perennis*, or, less often, *Endymion non-scriptus, Hedera helix* and *Allium ursinum*. Ferns are often abundant. Frequent species are *Allium ursinum, Anemone nemorosa*, Arum maculatum*, Brachypodium sylvaticum, Carex sylvatica, Circaea lutetiana, Conopodium majus, Deschampsia caespitosa, Dryopteris dilatata, D. filix-mas*, Endymion non-scriptus*, Euphorbia amygdaloides, Galeobdolon luteum, Geum urbanum, Hedera helix, Mercurialis perennis*, Phyllitis scolopendrium, Poa trivialis, Polystichum aculeatum, P. setiferum, Primula vulgaris, Ranunculus ficaria, Rubus fruticosus*, Taraxacum officinale, Urtica dioica, Veronica montana* and *Viola riviniana*. Other characteristic species include *Carex strigosa, Galium odoratum, Paris quadrifolia* and, in southern England, *Polygonatum multiflorum*.

(b) *Sub-type 1Ab. Northern calcareous ash–wych elm woods*

Essentially similar woods occur on the limestone and base-rich flush sites of north England and south and central Scotland, but, since they lie at or beyond the geographical range of many calcicole tree and shrub species, they are poor by comparison with 1Aa. *Acer*

121

campestre and *Tilia cordata* are absent, except in a few woods in north England, but *Sorbus aucuparia* occurs sparingly. Some characteristic associates are less abundant than in 1Aa, notably *Corylus* and *Quercus robur*. The soil, though similar, is more often a loam than a clay loam, and rocks are commonly exposed within the stand. This sub-type grades into 1D, but the two sub-types may be distinguished by the species of oak and by other floristic differences. Stands of 1Ab type have been recorded on Magnesian limestone, chalk and oolite in north-east England, and from Shropshire northwards on the western side. The Clyde valley woods and the few ancient woods of south-east Scotland are mostly of this type. The northernmost examples appear to be in the gorge woods of Perth, Angus and Moray. Most stands have not been coppiced since 1910 and it is now often difficult to determine whether they were formerly simple coppice or coppice-with-standards, but both forms can still be found. This sub-type lies between 3C and 1D in the range of variation.

The field layer may be dominated by *Mercurialis perennis* and, rarely, *Allium ursinum*, but generally it is a rich mixture of species without clear dominants. Frequent species are *Allium ursinum*, *Anemone nemorosa*, *Arum maculatum*, *Brachypodium sylvaticum**, *Circaea lutetiana*, *Deschampsia caespitosa*, *Dryopteris dilatata*, *D. filix-mas**, *Endymion non-scriptus*, *Epilobium montanum*, *Fragaria vesca*, *Galium aparine*, *Geranium robertianum*, *Geum rivale*, *Geum urbanum*, *Hedera helix*, *Holcus mollis*, *Mercurialis perennis**, *Oxalis acetosella*, *Phyllitis scolopendrium*, *Poa trivialis*, *Polystichum aculeatum*, *Potentilla sterilis*, *Primula vulgaris*, *Ranunculus ficaria*, *Rubus fruticosus*, *R. ideaus*, *Sanicula europea*, *Taraxacum officinale*, *Urtica dioica*, *Viola riviniana*. Some northern herbs occur at low frequency: *Crepis paludosa*, *Geranium sylvaticum*, *Myrrhis odorata*, *Polygonatum verticillatum*, and *Rubus saxatilis*. Grasses are much more frequent than in 1Aa, and may include *Arrhenatherum elatius* and *Anthoxanthum odoratum*. *Orchis mascula* and *Paris quadrifolia* occur rarely in both sub-types.

8.2.2 Stand type 1B. Wet ash–wych elm woods

Ash–wych elm woods on heavy, poorly-drained soils

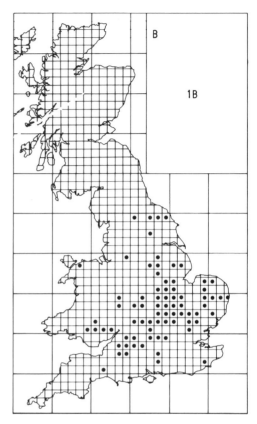

Fig. 8.2 Distribution of stand type 1B, wet ash–wych elm woods.

or medium-textured, flushed soils, usually with a neutral pH. *Quercus robur*, which is commoner in this type than other ash–wych elm types, is the only oak present. Two sub-types are recognised, both of which are virtually confined to the English lowlands (Fig. 8.2).

(a) Sub-type 1Ba. Heavy soil form

Pedunculate oak–ash–wych elm woods on heavy, poorly-drained soils. *Crataegus oxyacanthoides*, *Lonicera* and *Salix caprea* are relatively frequent, but *Ilex* and *Tilia cordata* are rarer than in other ash–wych elm woods. These stands are most frequent in the Midlands and East Anglia, where they normally occur as small patches in woodlands of the closely related types 2A and 10A, and are also present in Hampshire, west Wiltshire, Shropshire and the north-east Midlands.

Like most of the ancient woods in these areas, the type occurs mainly on plateau sites. All recorded stands have been managed as coppice with oak (or, rarely, ash or elm) standards. This sub-type forms part of Klötzli's *Querco–Fraxinetum*.

The field layer is generally rather limited for ash–wych elm woods but can be rich, e.g. 49 species were recorded in a recently coppiced part of Great Sorrells Copse (Hants). Dominants are usually *Deschampsia caespitosa*, *Mercurialis perennis*, *Poa trivialis* or *Rubus fruticosus*. Frequent species are *Ajuga reptans*, *Anemone nemorosa*, *Arum maculatum*, *Brachypodium sylvaticum*, *Carex sylvatica*, *Circaea lutetiana*, *Deschampsia caespitosa*, *Dryopteris filix-mas*, *Endymion non-scriptus**, *Galeobdolon luteum*, *Galium aparine*, *G. odoratum*, *Geum urbanum*, *Hedera helix*, *Lonicera periclymenum*, *Luzula pilosa*, *Mercurialis perennis**, *Poa trivialis**, *Potentilla sterilis*, *Primula vulgaris*, *Ranunculus ficaria*, *Rubus fruticosus**, *Veronica chamaedrys*, *V. montana* and *Viola riviniana*.

(b) Sub-type 1Bb. Light soil form

Pedunculate oak–ash–wych elm woods on light-medium textured soils which are freely or imperfectly drained. They differ from 1Ba in lacking the species of calcareous clays, i.e. *Crataegus oxyacanthoides*, *Euonymus europaeus*, *Rosa arvensis* and *Salix caprea*. They occur sparingly on light drift in Norfolk and Lincolnshire, with outlying examples in Worcestershire and at Coed Gors-wen in north Wales. In each case they occupy more-or-less flat ground which is slightly too dry for *Alnus*. All recorded sites have been managed as coppice with oak (rarely elm) standards.

The field layer is generally dominated by *Mercurialis perennis*, *Rubus fruticosus* or *Endymion non-scriptus*, three species which are also present in almost all stands. Frequent species are, in addition, *Anemone nemorosa*, *Circaea lutetiana*, *Deschampsia caespitosa*, *Filipendula ulmaria*, *Galeobdolon luteum*, *Geum rivale*, *G. urbanum*, *Hedera helix*, *Lysimachia nemorum*, *Poa trivialis**, *Potentilla sterilis*, *Primula vulgaris*, *Prunella vulgaris*, *Pteridium aquilinum*, *Sanicula europea*, *Stachys sylvatica*, *Tamus communis*, *Urtica dioica*, *Veronica chamaedrys* and *Viola riviniana*.

8.2.3 Stand type 1C. Calcareous ash–wych elm woods on dry and/or heavy soils

This stand type contains those ash–wych elm stands

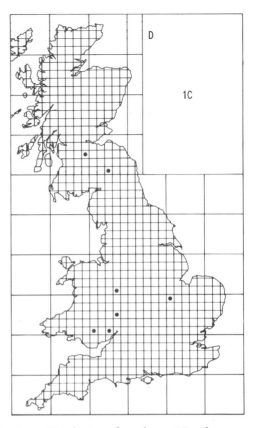

Fig. 8.3 Distribution of stand type 1C, calcareous ash–wych elm woods on dry and/or heavy soils.

which cannot conveniently be placed in other categories. They are found on thin, freely-drained soils, shedding sites or poorly-drained clays. The only apparent link between the diverse stands within type 1C is that, for various reasons, they are all liable to far greater seasonal moisture stress than other ash–wych stands. *Hedera* is characteristically present.

(a) Sub-type 1Ca. Eastern calcareous ash–wych elm woods

Ash–wych elm stands on medium-textured freely-drained, highly calcareous soils in eastern England. The only extensive occurrence of this sub-type is in Bedford Purlieus (Cambs), but the form probably occurs elsewhere on the oolite of Kesteven and the Soke of Peterborough (Fig. 8.3). *Quercus robur* is the main oak, and some calcicole species are more frequent here than in other types: *Clematis*, *Euonymus*,

Ligustrum and *Viburnum lantana. Sorbus torminalis* and *Prunus spinosa* are also found, and *Crataegus oxyacanthoides* is frequent as in sub-type 1Ba, but *Thelycrania* is surprisingly absent from the stands recorded. All examples known have been managed as oak standards over mixed coppice. This sub-type can be regarded as a form of type 1A in a more continental climate, which cannot develop south of the East Midlands because *Fagus* is present.

Frequent field layer species are *Anemone nemorosa**, *Arum maculatum**, *Campanula trachelium*, *Deschampsia caespitosa**, *Endymion non-scriptus**, *Galeobdolon luteum**, *Geum urbanum*, *Hedera helix**, *Heracleum sphondylium*, *Listera ovata*, *Melica uniflora*, *Mercurialis perennis**, *Orchis mascula*, *Poa trivialis*, *Primula vulgaris**, *Rubus fruticosus*, *Tamus communis* and *Viola riviniana*.

(b) *Sub-type 1Cb. Sessile oak–ash–wych elm woods on heavy soil*

Ash–wych elm stands on heavy, poorly-drained soils with *Quercus petraea* as the associated oak. These are known only from silurian limestones in the west Midlands, and are especially frequent on Wenlock Edge in Shropshire (Fig. 8.3). Their features combine those of types 1Aa, 1Ba and 1D, but the combination of sessile oak and heavy, poorly-drained soils is distinctive. Type 1Cb differs from 1Aa in the frequency of *Betula pendula* and the apparent absence of *Sambucus*; from 1Ba in the absence of *Crataegus oxyacanthoides*, *Euonymus* and *Ligustrum*; and from 1D by the presence of *Acer campestre* and *Betula pendula*, and the rarity of *Sorbus aucuparia*. Good examples are known from Halesend Wood (Hereford) and Tick Wood (Salop), which, like other examples, have been managed as mixed coppice, with occasional oak, ash, elm or lime standards. This sub-type is closely related to stand type 4C, with which it occasionally occurs.

Field layer is usually dominated by *Hedera helix*, *Mercurialis perennis* or *Rubus fruticosus*. Frequent species are *Allium ursinum*, *Anemone nemorosa*, *Brachypodium sylvaticum*, *Carex sylvatica**, *Circaea lutetiana*, *Deschampsia caespitosa**, *Dryopteris dilatata**, *D. borreri*, *D. filix-mas**, *Endymion non-scriptus**, *Fragaria vesca*, *Galeobdolon luteum**, *Geum urbanum**, *Hedera helix**, *Lonicera periclymenum*, *Mercurialis perennis**, *Poa*

trivialis, *Potentilla sterilis**, *Primula vulgaris*, *Rubus fruticosus**, *Sanicula europea**, *Tamus communis*, *Taraxacum officinale*, *Veronica montana* and *Viola riviniana**.

(c) *Other stands within type 1C*

Two other recorded ash–wych elm stands fall within 1C, but cannot be accommodated within the sub-types described. They may eventually prove to be separate sub-types. A stand at the top edge of Lady Park Wood (Gwent) on a thin woodland rendzina soil was dominated by *Fraxinus*, *Corylus*, *Acer campestre* and *Betula pendula*, with various calcicole shrubs. The field layer was dominated by *Mercurialis perennis* with other calcicoles, but also contained *Luzula pilosa*, *Pteridium aquilinum* and *Digitalis purpurea*. *Ulmus glabra* was present, so the stand fell within group 1, but *Fagus* had been present and the stand had formerly been in type 8C. This was clearly an example of ash–wych elm woodland at the driest end of its range, which cannot be placed in 1A, but since the soil was freely drained and the oak present was *Q. robur*, the stand could not be referred to 1Cb. Likewise, a stand on the upper levels of the steeply sloping Cragbank Wood (Roxburgh) was an excessively drained, neutral loam forming a matrix for boulders, supporting ash–wych elm woodland containing *Q. petraea*, and *Betula pendula*. The field layer was rich in calcicole species, but had been heavily grazed and was consequently dominated by grasses, mainly *Brachypodium sylvaticum* and *Anthoxanthum odoratum*.

8.2.4 Stand type 1D. Western valley ash–wych elm woods

Ash–wych elm stands on light–medium-textured, freely-drained, acid–neutral soils, with *Quercus petraea* as the associated oak. These normally occur in the dry and wet flush zones along streams and at the base of slopes in 'western' birch–oakwoods, and can be regarded as transitional to these woods. *Ilex* and *Sorbus aucuparia* are commonly present but the southern and lowland calcicole species are absent or rare, e.g. *Acer campestre*, *Clematis*, *Ligustrum*, *Thelycrania*. The type occurs throughout western and northern Britain from the tip of Cornwall to the

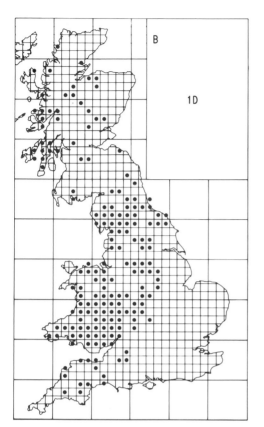

Fig. 8.4 Distribution of stand type 1D, western valley ash–wych elm woods.

odoratum, Athyrium filix-femina, Blechnum spicant, Brachypodium sylvaticum, Cardamine flexuosa, Chrysosplenium oppositifolium, Circaea lutetiana, Conopodium majus, Deschampsia caespitosa, Digitalis purpurea, Dryopteris dilatata, D. borreri, D. filix-mas*, Epilobium montanum, Filipendula ulmaria, Fragaria vesca, Galium aparine, G. odoratum, Geranium robertianum*, Geum urbanum, Hedera helix, Holcus mollis, Lonicera periclymenum, Lysimachia nemorum, Melampyrum pratense, Melica uniflora*, Mercurialis perennis*, Oxalis acetosella*, Poa nemoralis, Polypodium vulgare, Potentilla sterilis, Primula vulgaris, Prunella vulgaris, Pteridium aquilinum, Ranunculus ficaria, Rubus fruticosus*, R. idaeus, Sanicula europea, Silene dioica, Stachys sylvatica, Stellaria holostea, Teucrium scorodonia, Veronica chamaedrys, V. montana and *Viola riviniana*. Some rare species occur in this sub-type, e.g. *Circaea alpina, Festuca altissima* and *Impatiens noli-tangere*.

8.3 Ash–maple woodland (Group 2)

Stands containing *Acer campestre*, but not *Alnus, Carpinus, Fagus, Pinus, Tilia* or *Ulmus*. They usually contain *Crataegus monogyna, Corylus, Fraxinus* and *Quercus robur*. Several minor species occur commonly: *Crataegus oxyacanthoides, Lonicera, Prunus spinosa* and *Sambucus. Betula* is characteristically infrequent and the species of light or dry sites such as *Quercus petraea, Sorbus aucuparia, Taxus*, are absent or very rare. Some species, which are frequent in related stand groups, have only rarely been found in ash–maple woodland: *Clematis, Daphne laureola* and *Prunus avium*.

Maple is a moderate calcicole which is largely confined to the English lowlands and nearby parts of the Highland zone. It flourishes on the heaviest soils, which carries it beyond the edaphic range of beech, lime and pine, and does not appear to require the soils of high nutrient status to which wych elm is attracted. Nevertheless, it is a common and characteristic component of ash–wych elm woodland, calcareous ash–beech woodland and some types of lime woodland. Hornbeam and maple both grow on base-rich clays, so the occurrence of ash–maple woodland is somewhat restricted in south-east England. Ash–maple woodland is therefore fairly abundant only on the heavy soils of the English Midlands, and the drier, calcareous soils of the south-west and

northern Highlands, and is especially characteristic of the acid uplands of England and Wales (Fig. 8.4). Most stands have probably been coppiced in the past, but very few have been cut since 1920, and many are now mature high forest. This sub-type tends to merge with 1Ab in northern Britain. It is closely related to type 3D, and often occurs in the flushed zone between valley alderwoods and stands of type 3D or 6A. It probably just falls within the *Dryopteridio–Fraxinetum* of Klötzli (1970).

The list of species recorded from the field layer is longer than the list for any other stand type, and indeed most stands are rich. The sites are usually rugged and often contain streams, so edaphic conditions within each stand are very heterogeneous, and the flora consists of both calcifuge, calcicole, nitrophilous and wet soil species mixed together. Frequent species are *Anemone nemorosa, Anthoxanthum* north-east England, beyond the main edaphic and geographical rangs of beech, hornbeam and wych elm. It can also occur on light, flushed soils in east and south-east England where the soil is perhaps insufficiently moist for alder, and wych elm is generally infrequent.

Most ash–maple woods have been managed as coppice ash, hazel and maple with standards of pedunculate oak and, less often, ash. Ash and hazel are generally more abundant than maple. Now that most coppices are neglected, the canopy is usually dominated by oak and ash.

Stand types (Table 8.3) within the group are based on soil properties. The few stands on light soils (i.e. loamy sand to the lighter loams) are separated as type 2B. The great majority of stands which occur on heavier soils, are divided into a type 2A on poorly-drained ground and type 2C on freely-drained sites.

Ash–maple woodland is the core of Tansley's (1939) damp oakwoods and oak–ash woods. Klötzli followed Tansley in placing most stands in his *Querco-Fraxinetum*. Stand types 2A and 2B correspond with Merlewood types 1, 3, 5 and 7, but type 2C equals Merlewood types 5–7. Ash–maple woodland is infrequent on the Continent, where the closely related pedunculate oak–hornbeam woodland is prevalent, except in the most oceanic climates. Within the range of beech, both on the Continent and in south-east England, ash–maple woodland is associated with coppice management, and may be replaced by beech woodland under other systems.

8.3.1 Stand type 2A. Wet ash–maple woods

Ash–maple woodland on heavy, poorly-drained soils. *Crataegus oxyacanthoides* and *Lonicera* are present in most stands, and rare or absent in other stand types within the group. Spinose shrubs are well represented, and include *Prunus spinosa*, *Rosa arvensis*, *R. canina* and the *Crataegus* species. The soils are mostly clay or clay loams with a recorded pH range of 4.6–7.8. Typically this type occurs on flat, plateau sites or gently sloping ground nearby, though occasionally it develops on receiving sites, notably in the east Midlands where *Alnus* is rare. Wet ash–maple woods are especially characteristic of the Midland clay belt from Lincolnshire and Norfolk

Table 8.3 Composition and site features of ash–maple woods (Group 2).

	2Aa VP	2Aa P	2Aa I	2Ab	2Ba	2Bb	2C
Number of samples	18	15	11	14	7	4	11
Acer campestre	V	V	V	V	V	4	V
Betula pendula	II	II	I	I	I		
Betula pubescens	II	I	II	+	I		+
Corylus avellana	V	V	V	V	V	4	V
Fraxinus excelsior	V	V	V	II	V	4	V
Quercus robur	V	V	IV	V	V	3	IV
Acer pseudoplatanus	I	+		+		1	I
Castanea sativa			+				
Clematis vitalba		+		+			+
Crataegus monogyna	IV	III	III	IV	III	2	V
Crataegus oxyacanthoides	IV	III	II	III	I		+
Euonymus europaeus	I	+	I	+			III
Frangula alnus		+					
Hedera helix	II	I	+		II	1	IV
Ilex aquifolium	+	I	+			1	II
Ligustrum vulgare	I	I	I				I
Lonicera periclymenum	IV	III	III	III	II		II
Malus sylvestris	I	I	III	I	I		+
Populus tremula	II	II	+	+	+		
Prunus avium	+	+	I				I
Prunus spinosa	II	II	II	V	II		II
Rhamnus catharticus		+					
Ribes sylvestre	+	I		I		1	
Ribes uva-crispa			+				
Rosa arvensis	III	III	+	+	I		II
Rosa canina	I	II	II	II	I		I
Salix alba		+					
Salix caprea	I	I		II	III	1	
Salix cinerea		II	II	+			
Salix viminalis		+					
Sambucus nigra	+	II	III	III	IV	4	+
Sorbus aria							I
Sorbus aucuparia	+				I		
Sorbus torminalis	+			+			
Taxus baccata	+						
Thelycrania sanguinea	III	IV	I	+	III	1	IV
Viburnum lantana		+	+				+
Viburnum opulus	+	+	I	I	I		I
Median pH	6.2	6.2	5.6	5.3	6.5	6.7	7.7
Mean texture	3.5	3.0	3.5	3.3	1.3	1.0	2.7
Drainage free						4	11
imperfect			11	4	3		
poor		15		6	2		
very poor	18			4	2		
Site	Pr	Pv	ps	ps	ps	sr	s

126

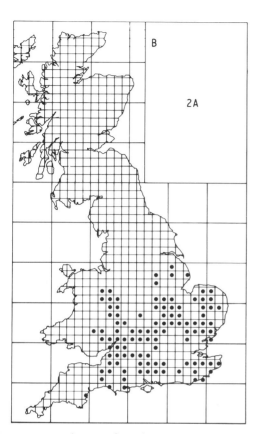

Fig. 8.5 Distribution of stand type 2A, wet ash–maple woods.

south-west into Somerset and Dorset, but they are also frequent on gault clay and heavy deposits over the chalk plateau in the south and south-east and on Lias clay in the west Midlands (Fig. 8.5).

Two sub-types are recognised in order to separate the typical stands from those which lack ash, or have done so until recently.

(a) Sub-type 2Aa. 'Typical' wet ash–maple woods

Wet ash–maple woods in which *Fraxinus* is a long-established component. This is by far the commoner sub-type, and enough stands have been recorded to show how the composition varies with drainage. *Malus* and *Sambucus* are more frequent where the surface horizons are freely drained, but more species are more frequent on the poor and very poorly-drained soils, notably *Crataegus, Hedera, Populus*

tremula, Rosa arvensis and *Thelycrania*. Ash is generally the dominant tree, but in many woods aspen has suckered to form dense groves. Apart from the Midland and East Anglian clays where this is by far the commonest semi-natural stand type, good examples have been recorded on the chalk at Clouts Wood (Wilts) and the Alkam valley (Kent), on the gault clay at Asholt Wood (Kent), and on a clay plateau over carboniferous limestone at Salisbury Wood (Gwent). An exceptional coastal stand at Fairlight Glen (Sussex) is rich in *Ilex*. This sub-type is commonly invaded by suckering elms to form stand type 10A. It is often also found with types 3A and 9A, which, however, are generally on rather lighter or more acid soils. Detailed published descriptions of wet ash–maple woodland in the centre of its distribution in Cambridgeshire are available for Gamlingay Wood (Adamson, 1912), Monks Wood (Steele and Welch, 1973) and Hayley Wood (Rackham, 1975).

The composition of the field layer varies according to soil wetness, as described for Hayley Wood. *Mercurialis perennis* and *Endymion non-scriptus* are most likely to dominate, especially in the drier stands, but *Deschampsia caespitosa, Rubus fruticosus, Filipendula ulmaria* and *Poa trivialis* are also often dominant. *Primula elatior* is especially characteristic of this type in East Anglia and is often very abundant. Frequent species are *Ajuga reptans, Anemone nemorosa, Arum maculatum, Brachypodium sylvaticum, Carex sylvatica, Circaea lutetiana, Deschampsia caespitosa, Dryopteris filix-mas, Endymion non-scriptis, Filipendula ulmaria, Galeobdolon luteum, Geum urbanum, Glechoma hederacea, Lonicera periclymenum, Mercurialis perennis*, Poa trivialis*, Potentilla sterilis, Primula vulgaris, Ranunculus ficaria, Rubus fruticosus*, Urtica dioica,* and *Viola riviniana* (and *reichenbachiana*). *Angelica sylvestris* and *Cirsium palustre* are also frequent on the wetter sites, where *Carex acutiformis* may be found. Characteristic, but infrequent species include *Iris foetidissima, Orchis mascula, Rubus caesius, Sanicula europea* and *Veronica chamaedrys*.

(b) Sub-type 2Ab. Wet maple woods

Stands conforming to the specifications of 2Aa, but in which *Fraxinus* is rare, absent or represented only by recently arrived saplings, occur sparingly from the

East Midlands to Oxfordshire and Berkshire. They are otherwise distinguished from 2Aa by the high frequency of *Prunus spinosa*, the rarity of *Thelycrania* and the apparent absence of *Ligustrum*. They tend to occupy slightly more acid soils on average, though the range (pH 3.8–7.4) is great, but in other respects their soils and sites are much the same as 2Aa. Most stands have hazel and maple as the main coppice species, but pedunculate oak is frequently coppiced as well as being virtually the only standard tree. The type has been recorded in various woods in Rockingham Forest (Northants) and in Salcey Forest (Northants and Bucks) and Great Park Wood (Berks). Ash may be rare because the other coppice species were preferred, in which case the sub-type is no more than an artefact of management. Even if this is the explanation there is some evidence that the ash-free state of some Rockingham Forest woods was established before the 16th century (Peterken, 1976).

The field layer is usually dominated by *Mercurialis perennis*, but most of the other species which can dominate 2Aa are occasionally abundant in 2Ab. Frequent species are much the same as those in 2Aa, but nitrophilous species are better represented: *Ajuga reptans*, *Anemone nemorosa*, *Arctium minus*, *Arum maculatum*, *Brachypodium sylvaticum*, *Carex sylvatica*, *Circaea lutetiana*, *Deschampsia caespitosa*, *Dryopteris filix-mas*, *Endymion non-scriptus*, *Geum urbanum*, *Glechoma hederacea*, *Mercurialis perennis**, *Oxalis acetosella*, *Poa trivialis**, *Primula vulgaris*, *Rubus caesius*, *R. fruticosus*, *Stellaria media*, *Urtica dioica** and *Viola riviniana*.

8.3.2 Stand type 2B. Ash–maple woods on light soils

Ash–maple woodland on sand and light loamy soils; *Sambucus* is characteristically abundant but *Crataegus oxyacanthoides* and *Rosa* species are rare. Although they occur on light soils, most (possibly all) stands either have a heavy sub-soil or occupy receiving sites and flush lines. In both circumstances base status is apparently maintained by flushing which, however, is insufficiently permanent to support alder. This type is thus closely related to sub-type 1Bb, wet ash–wych elm woodland on light soil, and occurs in much the same places (Fig. 8.6). Two sub-types are described (but on insufficient evidence) in order to demonstrate

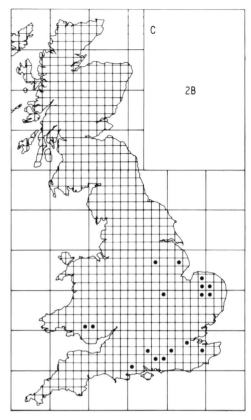

Fig. 8.6 Distribution of stand type 2B, ash–maple woods on light soils.

the greater richness of the stands on wetter soils (cf. 2Aa).

(a) *Sub-type 2Ba. Variant on poorly-drained soils*

Ash–maple woodland on light soils with imperfect–very poor drainage. These are similar to 2Aa, but tend to have more *Salix caprea* and *Sambucus*, and less *Crataegus oxyacanthoides* and *Rosa* species. Hazel and ash are normally the most abundant coppice species. The sub-type occurs mainly on plateau sites of the north-east Midlands and Norfolk, where light drift overlies clay sub-soil, but has also been found on Thanet Sand in north Kent. These are much the same conditions as permit 7C, plateau alderwood, to develop and indeed Foxley and Wayland Woods (Norfolk) have good examples of both, the ash–maple woodland occupying apparently drier ground. In other sites the ash—maple type lies between valley alderwood and birch—oak woodland.

Nitrophilous species are well represented in the field layer, which can be dominated by many different species. Frequent species are: *Ajuga reptans, Anemone nemorosa, Brachypodium sylvaticum, Carex sylvatica, Circaea lutetiana, Endymion non-scriptus, Festuca gigantea, Filipendula ulmaria, Galeobdolon luteum, Geum rivale, G. urbanum, Glechoma hederacea, Hedera helix, Heracleum sphondylium, Lonicera periclymenum, Mercurialis perennis*, Milium effusum, Moehringia trinervia, Poa trivialis*, Potentilla sterilis, Primula vulgaris, Rubus caesius, R. fruticosus, Rumex sanguineus, Sanicula europea, Silene dioica, Stachys sylvatica, Urtica dioica*, Veronica chamaedrys* and *Viola riviniana.*

(b) Sub-type 2Bb. Variant on freely-drained soils

Ash—maple woodland on light, freely-drained soils. This sub-type is known only from Holme and Shire Woods (Lincs) and Colyers Hanger (Surrey), on sandy loams flushed from the chalk and limestone. Apart from the general characteristics of type 2B, this sub-type is mainly notable for the poverty of associated species. Together with 1Bb and 2Ba, this sub-type seems to be an eastern analogue of type 1D in an area where wych elm is rare and the soils are coarse but deep enough to support pedunculate oak: this is especially obvious at Colyers Hanger, where 2Bb forms a zone on the lower slopes between types 6Dc and 7B.

Allium ursinum, Endymion non-scriptus, Galeobdolon luteum and *Mercurialis perennis* can all dominate in the field layer, but only six species were recorded in more than one of the four samples: *Circaea lutetiana, Geum urbanum, Mercurialis perennis, Poa trivialis, Ranunculus ficaria* and *Urtica dioica.*

8.3.3 Stand type 2C. Dry ash—maple woods

Ash—maple woodland on medium-heavy, freely-drained soils. These are almost invariably loam or clay loam in texture with a pH above 7.0. Compared with 2A, these stands have little or no *Crataegus oxyacanthoides, Populus tremula* and *Sambucus,* but *Euonymus, Hedera* and *Ilex* are frequent, and many other calcicole shrubs occur sparingly, notably *Clematis, Sorbus aria,*

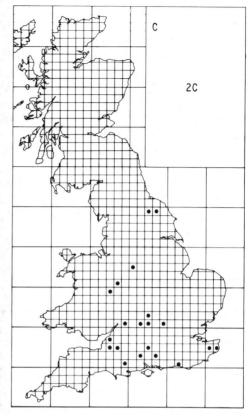

Fig. 8.7 Distribution of stand type 2C, dry ash—maple woods.

and *Viburnum lantana.* Pedunculate oak is usually present only as standards, whilst hazel and ash are normally more abundant than maple in the coppice. This type occurs mainly on chalk and limestone, especially on deeper soils which are free of flushing. This is just the soil on which beech will grow well, so the dry ash—maple woods are found mainly beyond its range. Fine examples are found (Fig. 8.7) on the Mendips (e.g. Asham Wood, Rodney Stoke Wood), and in north Somerset (e.g. Weston Big Wood). The type is rare in the Wye Valley, but was common in the north Cotswolds, just north of the main beech area, until the best examples at Guiting Wood and Chatcombe Wood (Glos) were ruined by replanting. Dry ash—maple woods have also been recorded elsewhere on the oolite in Lincolnshire and the North Yorkshire Moors. They are found frequently in the south and south-east in coppice woods, e.g. Cranborne

Chase (Wilts and Dorset), Sladden Wood and Yockletts Bank (Kent). Type 2C seems to occupy slightly deeper soils than types 1C, 4C and 8C and somewhat drier, better drained soils than types 1A, 4B and 9Bb. All these types are closely related, and often occur with type 2C. Tansley (1939) describes some examples as ash woods. Klötzli (1970) would place them between his *Dryopterido–Fraxinetum* and *Hyperico–Fraxinetum*.

The field layer is generally dominated by *Mercurialis perennis*, *Hedera helix*, *Endymion non-scriptus*, or *Anemone nemorosa*. Frequent species are *Ajuga reptans*, *Anemone nemorosa*, *Arum maculatum**, *Brachypodium sylvaticum*, *Campanula trachelium*, *Carex sylvatica*, *Circaea lutetiana*, *Conopodium majus*, *Dactylorchis fuchsii*, *Deschampsia caespitosa*, *Dryopteris filix-mas*, *Endymion non-scriptus**, *Euphorbia amygdaloides*, *Fragaria vesca*, *Galeobdolon luteum**, *Galium aparine*, *G. odoratum*, *Geranium robertianum*, *Geum urbanum**, *Glechoma hederacea*, *Hedera helix**, *Hypericum hirsutum*, *Listera ovata*, *Lonicera periclymenum*, *Mercurialis perennis**, *Orchis mascula*, *Paris quadrifolia*, *Poa trivialis**, *Potentilla sterilis*, *Primula vulgaris**, *Rubus fruticosus**, *Sanicula europea*, *Taraxacum officinale*, *Veronica chamaedrys* and *Viola riviniana**. Some other strict calcicoles occur, including *Cephalanthera damasonium*, *Colchicum autumnale*, *Ophrys insectifera*, *Phyllitis scolopendrium*, *Polygonatum multiflorum* and *Viola hirta*. Some species which are normally associated with more acid woods also occur rarely, including *Campanula rotundifolia*, *Holcus mollis*, *Lysimachia nemorum*, *Melampyrum pratense*, *Pteridium aquilinum* and *Teucrium scorodonia*. In Kent this stand type is notable for the presence of *Orchis purpurea*, *Pimpinella major* and *Veronica montana*.

8.4 Hazel–ash woodland (Group 3)

Stands containing *Fraxinus*, but not *Acer campestre*, *Alnus*, *Carpinus*, *Fagus*, *Pinus sylvestris*, *Tilia* or *Ulmus*. Almost all these stands contain *Corylus*, which is often more abundant than *Fraxinus*. Most stands contain *Betula*, *Crataegus monogyna*, *Lonicera* and *Quercus*.

Both hazel and ash are catholic species found almost throughout Britain and growing on a wide range of soils. They are present in most other stand groups and abundant in some, but hazel–ash woodland

Table 8.4 Composition and site features of hazel–ash woodland (Group 3).

	3Aa G	3Aa W	3Ab	3B	3C	3D
Number of samples	19	5	9	8	8	14
Betula pendula	II		II		IV	I
Betula pubescens	III	IV	III	I	IV	IV
Corylus avellana	V	V	V	V	V	IV
Fraxinus excelsior	V	V	V	V	V	V
Quercus petraea					III	V
Quercus robur	V	V	V	IV		
Acer pseudoplatanus	+	II	II	II	II	II
Castanea sativa	+			I		
Clematis vitalba	+			I		
Crataegus monogyna	IV	IV	IV	IV	IV	III
Crataegus oxyacanthoides	II	I				
Euonymus europaeus		I		II	II	
Frangula alnus			I			
Hedera helix	+	IV		III	III	II
Ilex aquifolium		V	I	I	II	III
Ligustrum vulgare				I	I	
Lonicera periclymenum	V	V	IV	III	II	IV
Malus sylvestris	I		I	I	I	
Populus tremula	II		I	II		
Prunus avium	I			I	+	+
Prunus padus						+
Prunus spinosa	II		III	II	II	
Rhododendron ponticum			I			
Ribes sylvestre			I			
Ribes uva-crispa						+
Rosa arvensis	II	I	I	II		I
Rosa canina	III			I	III	II
Rosa villosa					I	
Salix aurita					I	
Salix caprea	I	I		II	I	I
Salix cinerea	II	I				+
Sambucus nigra	I		III	I	I	
Sorbus aucuparia		I	I		IV	II
Sorbus torminalis	+					
Taxus baccata				I	IV	II
Thelycrania sanguinea	II		I	II		+
Viburnum lantana				II		
Viburnum opulus	II	III	I	II	+	+
Median pH	4.7	4.7	4.5	6.7	6.5	4.6
Mean texture	2.6	3.0	0.9	2.4	2.5	2.4
Drainage free			5	4	7	9
imperfect	7		1	2		3
poor	8	4	1	1	1	1
very poor	4	1	2	1		1
Site	ps	s	s	Sp	s	Sr

(i.e. without the species characteristic of other groups) occurs mainly in the north and west beyond the ranges of beech, hornbeam, maple and lime and in the southern lowlands on heavy, acid soils beyond the edaphic ranges of wych elm and maple. The main coppice species are ash and hazel, but oak is also commonly found as coppice as well as standards.

The four stand types (Table 8.4) are based on associated oak species. The pedunculate oak–hazel–ash woods divide conveniently into types on acid, heavy, poorly-drained soils (3A) and neutral–calcareous, freely-drained soils (3B), with few intermediates. The sessile oak–hazel–ash woods also divide into an acid type (3D) and a calcareous type (3C), but both are mainly found on freely-drained soils, and intermediate stands are relatively common. Each stand type can be regarded as an attenuated version of stand types in groups 1, 2 and 4.

8.4.1 Stand type 3A. Acid pedunculate oak–hazel–ash woods

Hazel–ash woodland containing *Quercus robur* growing on acid soils. Most stands contain *Crataegus monogyna*, *Lonicera* and one or other *Betula* species. *Q. petraea* is rarely present (but see discussion of 3D). Most stands have been managed as hazel–ash coppice with oak standards, but birch and oak are commonly found as coppice (unlike the related type 2A, where oak is rare as coppice). The type occurs mainly on heavy, poorly-drained, acid (pH 4–5) soils, but some stands grow on markedly lighter and often well-drained soils. This difference is the basis of a division into sub-types on heavy (3Aa) and light (3Ab) soils. The former is closely related to type 2A, whereas the light-soil sub-type is a counterpart of type 2B: both tend to occur on sites which are too acid for maple. Both sub-types are found throughout lowland England and on heavy valley soils in adjacent upland districts, but the type is very rare in Scotland (Fig. 8.8).

(a) *Sub-type 3Aa. Heavy soil form*

Acid pedunculate oak–hazel–ash woods on loams, silt and clay (Fig. 18.2). The soils are invariably poorly drained and most have a pH between 4.4 and

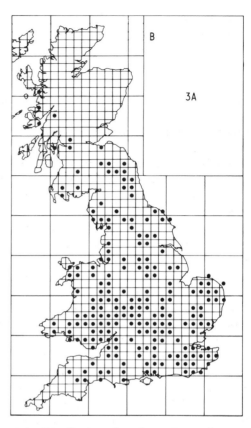

Fig. 8.8 Distribution of stand type 3A, acid pedunculate oak–hazel–ash woods.

5.4. This sub-type is widespread in lowland England and Wales, and is especially characteristic of acid clays in the Midlands, East Anglia and the Weald. Typically, towards the east and south, it is distinguished from 3Ab by the greater frequency of *Crataegus oxyacanthoides*, *Rosa* spp., *Salix* spp. and *Thelycrania*, and the rarity of *Sambucus*. Good examples occur commonly in the Midland clay belt, on boulder clay in eastern counties, on Wadhurst Clay in east Sussex and on plateau deposits on the chalk. In most of these areas the woods are on flat or gently-sloping ground near a plateau, and the associated stand types are 2A, 6Dc, 9A and (in eastern England) 4A. Salcey Forest (Bucks and Northants) is a good example which was planted as oak high forest in the early 19th century. Some of the most western examples are so distinct in floristic terms that they may merit a separate sub-type, in which *Hedera* and

Ilex are common; *Populus tremula, Prunus spinosa, Rosa canina* and *Thelycrania* are absent or rare; and the only birch recorded is *B. pubescens*. Good examples occur at Ashen Copse (Somerset), Pen-Gelly Forest (Pembs) and in the Tywi valley near Llandeilo (Carms), where they occur on low-lying ground close to valley alderwoods. In the English lowlands, type 3Aa can be regarded as an attenuated form of 2A, 4A and 9A which is transitional to type 6Da, but in the west it seems to be transitional between 7A or 7B on wetter sites and 6Bc or 3D on drier sites. Stands of type 3A are widely scattered across a range of Merlewood types, falling mostly between types 5 and 12 inclusive. Klötzli (1970) would include them in his *Querco–Fraxinetum*.

The field layer is often dominated by *Rubus fruticosus*, but many other species may be abundant, including *Deschampsia caespitosa* and *Filipendula ulmaria* on wetter soils and *Holcus mollis* on medium-textured soils. Frequent species are *Ajuga reptans, Anemone nemorosa, Arum maculatum, Carex sylvatica, Circaea lutetiana, Deschampsia caespitosa, Dryopteris filix-mas, Endymion non-scriptus, Galium aparine, Geum urbanum, Glechoma hederacea, Holcus mollis, Lonicera periclymenum**, *Luzula pilosa, Mercurialis perennis, Oxalis acetosella, Poa trivialis, Potentilla sterilis, Primula vulgaris, Rubus fruticosus**, *Stellaria holostea, Veronica chamaedrys* and *Viola riviniana**. In western examples *Dryopteris dilatata, Carex remota* and *Veronica montana* are also frequent.

(b) *Sub-type 3Ab. Light soil form*

Acid pedunculate oak–hazel–ash woods on light soils, generally sandy loam and loamy sand. *Sambucus* is frequent, but spinose shrubs are rare. This sub-type appears to be a heterogeneous collection of stands. In central Lincolnshire, where it is common with stand type 4A, 3Ab seems to be loosely associated with wetter soils and ancient, secondary woodland. In Norfolk, where it is also common, it occurs as a transition between 7A, 7Ea and 2B in or near a valley and 6D or 5A on a plateau. In south-east England 3Ab is again associated with 7A and 6D, but also with 9A in the London Basin and Weald, and it has been recorded from freely-drained, light deposits over chalk near Andover (Hants) where it can be regarded

as an acid form of 3B. In an oakwood to the south of Exmoor (Castle Close Wood, Devon) 3Ab occurs on a flush zone in a wood of type 6B, and is thus analogous to type 3D in a wood of 6A. Throughout these areas 3Ab occurs on freely-drained as well as poorly-drained soils. Ecologically it appears to be marginal to birch–oakwoods, occurring on soils which are richer in bases, irrespective of whether the soil is dry, flushed or waterlogged. Not surprisingly, 3Ab stands fall within a wide range of Merlewood types, mostly 1–12.

Anemone nemorosa, Endymion non-scriptus, and *Rubus fruticosus* are most likely to dominate the field layer, but many other species may be abundant. Correspondingly, perhaps the variety of species tends to be rather limited. Frequent species are *Ajuga reptans, Anemone nemorosa, Circaea lutetiana, Conopodium majus, Convallaria majalis, Deschampsia caespitosa, Dryopteris dilatata, D. filix-mas, D. spinulosa, Endymion non-scriptus, Filipendula ulmaria, Galeobdolon luteum, Galium aparine, Geranium robertianum, Geum urbanum, Glechoma hederacea, Holcus mollis, Lonicera periclymenum, Luzula pilosa, Mercurialis perennis, Moehringia trinervia, Myosotis sylvatica, Oxalis acetosella, Poa trivialis, Primula vulgaris, Pteridium aquilinum, Rubus fruticosus**, *Silene dioica, Stellaria holostea, Urtica dioica, Veronica chamaedrys, V. montana*, and *Viola riviniana*. Eight of the ten species recorded in the sample from the Devon example were in this list, but the most abundant species were *Phyllitis scolopendrium* and *Polystichum aculeatum*.

8.4.2 Stand type 3B. Southern calcareous hazel–ash woods

Hazel–ash woodland on calcareous soils or associated with limestone in which *Quercus robur* is the associated oak. These stands have only been recorded from carboniferous limestones (South Wales, Somerset), oolite (Glos, Northants), gault clay (Kent) and Thanet Sand over Chalk (Kent) in southern Britain (Fig. 8.9). Like 3C, stand type 3B has many calcicole shrubs, but differs in the near-absence of *Betula*. The ecological status of 3B varies across its range. Eastwards from the Cotswolds it is a form of 2A, 2C or 8C which happens to lack maple and beech. In Somerset and the southern Welsh borderland it is associated

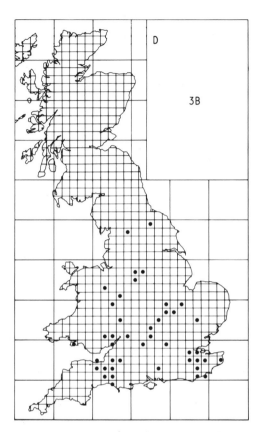

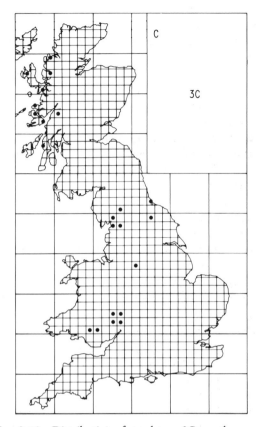

Fig. 8.9 Distribution of stand type 3B, southern calcareous hazel–ash woods.

Fig. 8.10 Distribution of stand type 3C, northern calcareous hazel–ash woods.

with 1A, 2C, 4Bb and 8C. Throughout this range 3B occurs only in small patches and must be regarded as a chance or temporary phenomenon like stand type 12B in southern Britain. However, at Nicholaston Wood on the Gower (and perhaps elsewhere in west and north Wales), beyond the range of most maple and beech, pedunculate oak–hazel–ash woodland appears to be characteristic of dry slopes in a zone above type 1A, which is on the flushed slopes below. This type falls mainly within Merlewood types 5–7.

The field layer of most of the samples recorded resembles that of 2A and 2C closely. At Nicholaston Wood, where the type has greater validity, the field layer was dominated by *Hedera helix*, and contained a mixture of calcicoles (*Arum maculatum*, *Mercurialis perennis*, *Viola reichenbachiana*), species of oceanic climates (*Asplenium adiantum-nigrum*, *Iris foetidissima*, *Phyllitis scolopendrium*, *Rubia peregrina*), species of

lighter, drier soils (*Melica uniflora*, *Teucrium scorodonia*) and more catholic species (*Endymion non-scriptus*, *Lonicera periclymenum*, *Rubus fruticosus*, *Viola riviniana*).

8.4.3 Stand type 3C. Northern calcareous hazel–ash woods

Hazel–ash woodland on calcareous soils or in association with limestone in which *Quercus petraea* is the associated oak. These stands have been recorded only from northern England northwards, with good examples at Gaitbarrows (Lancs) on carboniferous limestone, Castle Eden Dene (Durham) on Magnesian limestone, and at Rassal in north-west Scotland on Durness limestone (Fig. 8.10). They are distinguished from type 3B by the high frequency of *Betula* (both species), *Sorbus aucuparia* and *Taxus*. Oak is generally infrequent. The dominant species are normally ash and hazel, but yew may also be abundant, notably in Castle Eden Dene, where the unstable slopes maintain the stand as perpetual scrub. The soils are generally shallow, often with some surface acidification and a marked accumulation of humus, notably on those stands growing over limestone pavement. Type 3C can be regarded as an attenuated form of type 4C growing mostly beyond the range of *Tilia*. Likewise the hazel scrub on extreme oceanic limestone sites can be regarded as an attenuated form of 3C. Both 3C and 4C occur together around Morecambe Bay, and perhaps in the Peak District, but further south it seems likely that any 3C stands are small, temporary derivatives from stand types 1C, 4C and 8C, for example, the ash thickets which spring up when a gap occurs in a limestone beechwood. The stands recorded fall mostly into Merlewood types 1, 8 and 30. Klötzli (1970) places them in his *Hyperico−Fraxinetum*.

The field layer is rich, but without clear dominants. Grassland species are well represented, for the canopy is rarely vigorous and some stands are actually grazed. Rassal Ashwood, one of the more famous limestone woods, has been heavily grazed for decades, but the woodland flora and structure has been restored remarkably well by excluding sheep from a sample area. Frequent species are *Agrostis tenuis, Alchemilla glabra, Anemone nemorosa, Anthoxanthum odoratum, Athyrium filix-femina, Brachypodium sylvaticum**, *Carex sylvatica, Conopodium majus, Crepis paludosa, Deschampsia caespitosa, Dryopteris dilatata, D. borreri, D. filix-mas**, *Endymion non-scriptus**, *Epilobium montanum, Filipendula ulmaria, Fragaria vesca**, *Geum rivale, Hedera helix**, *Holcus mollis, Lonicera periclymenum, Lysimachia nemorum, Melica nutans, M. uniflora, Mercurialis perennis**, *Oxalis acetosella, Phyllitis scolopendrium, Poa nemoralis, Polypodium vulgare, Polystichum aculeatum, Potentilla sterilis, Primula vulgaris**, *Prunella vulgaris, Pteridium aquilinum**, *Ranunculus acris, Rubus caesius, Rubus fruticosus**, *Senecio jacobea, Taraxacum officinale, Teucrium scorodonia, Thelypteris limbosperma, Veronica chamaedrys, V. montana* and *Viola riviniana*. Other characteristic species include *Convallaria majalis, Rubus saxatilis* and *Viola hirta*. Many heathland species occur infrequently within the type, rooted in deeper pockets of soil.

8.4.4 Stand type 3D. Acid sessile oak−hazel−ash woods

Hazel−ash woodland on acid soils containing *Quercus petraea*. Such stands generally contain *Betula pubescens* and some *Ilex* and *Sorbus aucuparia*. Type 3D occupies both dry and moist flushes in a strongly acid environment, where the soil is too base-poor for the closely related stand type 1D to develop: pH 4.7 is more-or-less the dividing line between the two types. Ash streaks within western sessile oakwoods often contain a core of stand type 1D on richer soils where wych elm prospers, and a marginal zone of 3D, which is transitional to the surrounding type 6A. Type 3D is linked to 3C by the presence of *Q. petraea, Sorbus aucuparia* and *Taxus*, but oak is generally far more abundant in 3D and *Betula pendula* is rare. Both 3D and 3Ab grow on light, acid soils, but the former consistently lacks *Prunus spinosa, Quercus robur* and *Sambucus*. The western form of 3Aa is very similar, but the oaks differ and there seems to be a consistent difference in the soils they occupy. Type 3D has been recorded from south-west England, west Wales, west Midlands, Lake District and west Scotland, with good examples on Exmoor, the Rheidol Valley and Borrowdale (especially Seatoller Woods) (Fig. 8.11). It could also occur in south-east England but the only instances so far encountered have been converted into chestnut coppices. Samples of type 3D are almost randomly scattered through the Merlewood types. An anomalous stand on very poorly drained acid clay at Saplins Wood (Salop) was effectively an acid version of the equally anomalous type 1Cb, which occurs in the same region. At Coed Gors-wen (Gwynedd) a form which is intermediate between 2A and 3D occurs on gently sloping ground on neutral, imperfectly drained loam.

Ferns and grasses dominate the field layer of most examples, but where sheep are absent *Rubus fruticosus* is often abundant. As in the related type 1D, numerous species are frequent: *Agrostis tenuis, Anemone nemorosa, Anthoxanthum odoratum, Athyrium filix-femina, Blechnum spicant, Brachypodium sylvaticum, Carex sylvatica, Circaea lutetiana, Conopodium majus, Deschampsia caespitosa, D. flexuosa, Digitalis purpurea, Dryopteris dilatata, D. filix-mas**, *Endymion non-scriptus, Festuca ovina, Galeobdolon luteum, Galium aparine,*

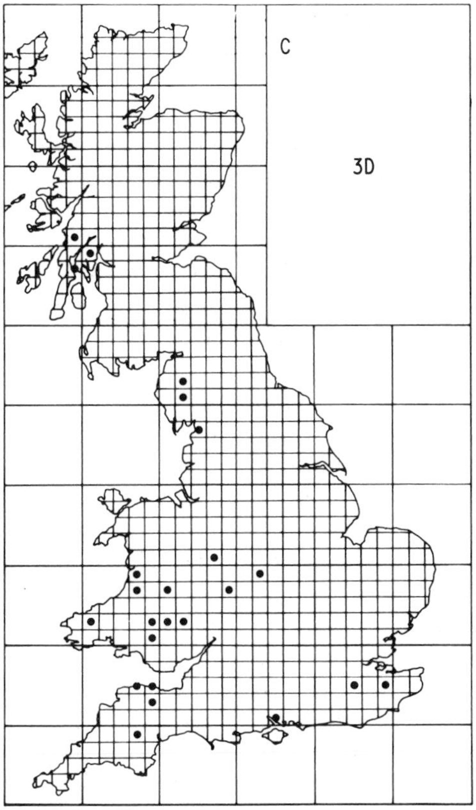

Fig. 8.11 Distribution of stand type 3D, acid sessile oak–hazel–ash woods.

G. saxatile, Geranium robertianum, Geum urbanum, Hedera helix, Holcus mollis, Lonicera periclymenum, Luzula pilosa, Lysimachia nemorum, Oxalis acetosella*, Potentilla erecta, Potentilla sterilis, Pteridium aquilinum*, Rubus fruticosus*, Stellaria holostea, Teucrium scorodonia, Veronica chamaedrys, V. montana and Viola riviniana*.

8.5 Ash–lime woodland (Group 4)

Stands with Tilia and Fraxinus, but not Alnus, Carpinus, Fagus, Pinus sylvestris, Ulmus carpinifolia, U. glabra or U. procera. Almost all stands contain Betula, Corylus, Crataegus monogyna, Quercus and Lonicera, whilst Rosa arvensis and R. canina are frequent. Sambucus and Sorbus aucuparia are both rare. Almost all the Tilia can be referred to T. cordata, but both T. platyphyllos and T. vulgaris occur rarely.

Tilia was abundant and widespread in the Atlantic

Table 8.5 Composition and site features of ash–lime (Group 4) and oak–lime (Group 5) woodland.

	4A	4Ba	4Bb	4C	5A	5B
Number of samples	37	13	4	10	14	10
Acer campestre		V	4	IV	I	+
Betula pendula	II	III	2	III	II	IV
Betula pubescens	IV	III		+	IV	III
Corylus avellana	V	V	4	V	IV	IV
Fraxinus excelsior	V	V	4	V		
Quercus petraea				IV	+	V
Quercus robur	V	V	2	+	V	
Tilia cordata	V	V	4	V	V	V
Acer pseudoplatanus	+		1	II	+	II
Castanea sativa	+				I	
Clematis vitalba	+		2	I		
Crataegus monogyna	V	V	3	IV	III	III
Crataegus oxyacanthoides	I	II				
Daphne laureola			1	III		
Euonymus europaeus	+	+	2	I		
Frangula alnus	+				+	
Hedera helix	+		4	IV		II
Ilex aquifolium	+		2	III	I	III
Ligustrum vulgare	I	II	2	III		
Lonicera periclymenum	V	IV	3	IV	V	IV
Malus sylvestris	II	II	1	+		
Populus tremula	II	II		+	II	
Prunus avium	+	+	2	I	+	I
Prunus spinosa	+	II	1	II		
Rhododendron ponticum			1			
Ribes sylvestre				+		
Rosa arvensis	II	II	2	II	+	I
Rosa canina	II	II	2	II	+	
Salix caprea	+	II		I	+	
Salix cinerea	+	II			II	
Sambucus nigra	+	I	1		II	+
Sarothamnus scoparius						+
Sorbus aria			1	+		
Sorbus aucuparia	+			+	II	I
Sorbus torminalis		+		I		I
Taxus baccata				III		I
Thelycrania sanguinea	+	II	1	III		
Tilia vulgaris	+					
Viburnum lantana			1			
Viburnum opulus	I			II	+	I
Median pH	4.8	5.8	7.3	7.1	4.1	4.5
Mean texture	2.4	2.3	2.8	3.2	1.6	1.7
Drainage free		2	4	3	5	7
imperfect	3	1		1	3	
poor	18	7		3	2	2
very poor	16	3		3	4	1
Site	Pr	P	s	S	Ps	s

forests, but is now very rare in south-east England and does not occur north of Cumbria except as a planted tree. It occurs sparingly in western Britain as a constituent of valley ash–wych elm woodland (1D) and is commonly found in the southern calcareous ash–wych elm woods (1Aa). Ash–lime woodland and oak–lime woodland (Group 5) are therefore uncommon and found mainly in England between the Chepstow–London and Morecambe Bay–Humber lines.

In East Anglia and the east and north Midlands lime (almost invariably *T. cordata*) occurs mainly on light, acid soils; its edaphic range overlaps that of ash only where such soils are poorly drained due to the flat topography and an impermeable sub-soil. Although such sites might also support *Alnus* or *Carpinus*, which would place the stand in groups 7 or 9, there is in fact very little overlap with these groups. Coppices with both lime and hornbeam are found in Essex (e.g. Chalkney Wood) and very rarely in Bedfordshire and Kent. Lime is also characteristic of limestone soils, especially in northern and western England and the Welsh borderland, but here a range of forms can occur, including *T. platyphyllos* (Pigott, 1969). Ash is also characteristic of such soils, but since wych elm is also common on the deeper, flushed soils (and rarely on the dry, shallow soils) many stands containing both ash and lime are placed in Group 1. In the Wye Valley area beech is also present in many stands, which are accordingly placed in Group 8.

Soil features form the basis of stand types within ash–lime woodland (Table 8.5). Stands on acid soils with *Q. robur* as the associated oak form type 4A, acid birch–ash–lime woods. Those on calcareous soils or associated with limestone in which *Q. petraea* is the associated oak form type 4C, the sessile oak–ash–limewoods. Two other forms are distinguished as type 4B, maple–ash–lime woods one of which, 4Ba, is a form of 4A on neutral and mildly acid soils. The other, 4Bb, is often an artefact of sampling but may nevertheless be a valid type (see below).

8.5.1 Stand type 4A. Acid birch–ash–lime woods

Ash–lime woodland lacking *Acer campestre*. These stands occur on poorly-drained, mildly acid soils.

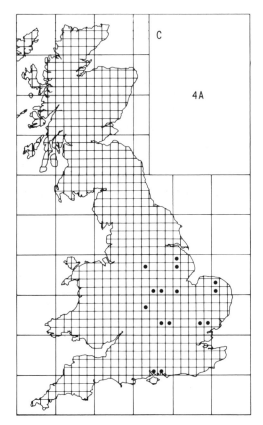

Fig. 8.12 Distribution of stand type 4A, acid birch–ash–lime woods.

Betula (mostly *B. pubescens*), *Corylus*, *Crataegus monogyna*, *Lonicera* and *Quercus robur* are present in almost all stands, but most other tree and shrub species are rather sparse. *Quercus petraea* is absent. Most stands were managed as oak standards over mixed coppice of lime, ash and hazel, with occasional birch and oak in the coppice, but all known examples are now neglected as coppice, and most of the oak standards have long since been removed, so that lime, ash and birch (a recent entrant after the last coppicing) now dominate most stands. The type occurs on flat or gently sloping ground, usually where a light top soil covers a heavy sub-soil. The depth of the sandy top soil varies from a thin smear to deep, slightly leached sands (where, however, types 5A, 6Db and even 12A are more likely to develop). This type is concentrated in central Lincolnshire (Fig. 8.12), where, with some stands of types 4Ba, 5A and 5B, it is the main component of the 'Lincolnshire Limewoods': good examples are found in Stainfield, Potterhanworth, Hatton, Newball and Great West Woods. Elsewhere, 4A has been sparingly recorded in Bedfordshire, Leicestershire, Soke of Peterborough and East Anglia. It lies between types 3A and 5A in the range of variation and close to type 7C. Apart from the stand in Bedford Purlieus (Peterken and Welch, 1975), the acid birch–ash–lime woods have never apparently been noticed in the ecological literature. They fall mainly within types 5, 7, 11 and 12 of the Merlewood classifications and the *Querco–Fraxinetum* of Klötzli (1970).

The field layer is usually a mixture of species in which any one of a number of species may be abundant. Frequent species are *Ajuga reptans**, *Anemone nemorosa*, *Brachypodium sylvaticum*, *Carex sylvatica**, *Circaea lutetiana*, *Deschampsia caespitosa**, *Dryopteris filix-mas*, *Filipendula ulmaria*, *Geum rivale*, *Holcus mollis*, *Lonicera periclymenum**, *Luzula pilosa**, *Poa trivialis*, *Potentilla sterilis*, *Primula vulgaris*, *Rubus fruticosus**, *Veronica chamaedrys* and *Viola riviniana*. The less-frequent species include *Convallaria majalis* and *Platanthera chlorantha*.

8.5.2 Stand type 4B. Maple–ash–lime woods

Ash–lime woodland containing *Acer campestre*, but not *Quercus petraea*. These stands usually also contain *Betula*, *Corylus*, *Crataegus monogyna*, *Lonicera*, *Quercus robur* and either *Rosa arvensis* or *R. canina*. Two subtypes, which are distinguished on floristic and edaphic features, occur in eastern and western districts (Fig. 8.13).

(a) *Sub-type 4Ba. Lowland maple–ash–lime woods*

The main concentration of ash–lime woodland with *Q. robur* as the associated oak, which is found in Lincolnshire and neighbouring regions, extends on to neutral and mildly calcareous soils. This end of the range has rather more tree and shrub species, including mildly calcicole species, and is distinguished as a separate stand type, defined by the presence of *Acer campestre*. The constant associates of 4A are still constant here, but *Crataegus oxyacanthoides*, *Ligustrum vulgare*, *Prunus spinosa*, *Salix caprea*, *S. cinerea* and

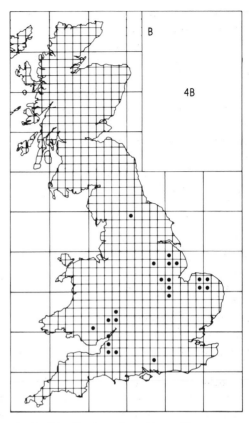

Fig. 8.13 Distribution of stand type 4B, maple–ash–lime woods.

Thelycrania sanguinea are all more frequent. Stand structure is also similar to 4A, except that maple is also present in the coppice, and lime is generally less abundant than ash. Most stands are found on neutral soils, pH 4.8–6.9, which are otherwise similar to those of type 4A. Good examples occur in Wickenby, Newball and Goslings Corner woods in Central Lincolnshire. Sub-type 4Ba links types 5A and 2A in the range of variation. It falls within type 5 in the Merlewood classification.

The main field layer dominants are *Geum rivale* on the wetter sites and *Mercurialis perennis*, *Endymion non-scriptus* or *Anemone nemorosa* on soils with a freely-drained surface horizon. Frequent species are *Ajuga reptans*, *Anemone nemorosa**, *Angelica sylvestris*, *Arum maculatum**, *Brachypodium sylvaticum*, *Carex sylvatica**, *Circaea lutetiana*, *Cirsium palustre*, *Conopodium majus*, *Deschampsia caespitosa**, *Endymion non-scriptus*,

Filipendula ulmaria, Galeobdolon luteum, Geum rivale, G. urbanum, Glechoma hederacea, Heracleum sphondylium, Lonicera periclymenum, Luzula pilosa, Mercurialis perennis*, Myosotis sylvatica, Poa trivialis*, Potentilla sterilis*, Primula vulgaris, Prunella vulgaris, Ranunculus ficaria, Rubus caesius, R. fruticosus*, Veronica chamaedrys and *Viola riviniana.*

(b) Sub-type 4Bb. Western maple–ash–lime woods

Four stands recorded within the rich carboniferous limestone woods near Caerwent (Mon) and Weston-in-Gordano (Som) fell technically within type 4B, but grew on freely-drained, calcareous soils, and were also distinguished from 4Ba by the presence of *Clematis, Daphne laureola, Hedera, Ilex, Prunus avium* and calcicoles such as *Sorbus aria* and *Viburnum lantana.* Although they were clearly related to type 4C, they lacked *Quercus petraea* and *Taxus baccata.* The recorded stands occurred as part of a mosaic with types 1Aa, 2C and 8Cb, within which 4Bb may rarely be extensive enough to justify recognition, except as an artefact of the recording scale, i.e. sub-type 4Bb may be a form of 1Aa in which *Ulmus glabra* happened to be absent or sparse. The only extensive examples found are in Salisbury Wood (Gwent) and a number of woods in the Mendips (Som), notably Asham Wood and Cheddar Wood, where 4Bb is scattered amongst 2C and 1Aa. As a distinct type, 4Bb lies between 1Aa, 2C and 4C in the range of variation. Three of the four samples fell within type 7 of the Merlewood classification.

The field layer, which is dominated by *Mercurialis perennis* and *Endymion non-scriptus* or *Hedera helix*, is rich and very similar to that of types 1Aa and 2C.

8.5.3 Stand type 4C. Sessile oak–ash–lime woods

Ash–lime woodland containing *Quercus petraea* as the main associated oak. These stands have only been recorded on limestone (Fig. 8.14) and they are characteristically rich communities: indeed, a sample recorded in Halesend Wood (Hereford), which contained 20 tree and shrub species, is the richest sample recorded in any type. *Acer campestre, Corylus, Crateagus monogyna, Hedera* and *Lonicera* are present in most stands, and several calcicole species are frequent,

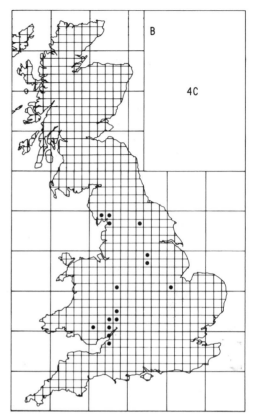

Fig. 8.14 Distribution of stand type 4C, sessile oak–ash–lime woods.

e.g. *Daphne laureola, Ligustrum vulgare* and *Thelycrania.* The characteristic birch is *B. pendula*, and both *Ilex* and *Taxus* are frequent. Most stands have been managed as simple coppice containing a rich mixture of species in which oak and lime usually predominate, but a few stands have also contained oak standards. The soils are mainly calcareous, usually strongly so, or rarely neutral or even mildly acid in the surface horizons. They are generally heavy and are mostly poorly structured, mottled and apparently poorly drained, although it is difficult to believe that this is so when such features are observed on shedding sites over limestone. Their main feature appears to be dryness, caused by the shedding or steeply sloping site, the heavy, poorly structured, often thin soil, or by a combination of these features. Good examples have been recorded on carboniferous limestone in the Lower Wye Valley and Eaves Wood (Lancs); on

Amnystry limestone at Halesend Wood (Hereford); on Wenlock limestone at Tick Wood (Salop); on oolite at Collyweston Great Wood and Easton Hornstocks (Northants); and on Magnesian limestone at Edlington Wood (Yorks). The type falls within the *Hyperico–Fraxinetum* of Klötzli (1970), and was previously described by Salisbury and Tansley (1921). It is an extreme type which is most closely related to type 3C in Britain, and forms a link with the mixed woodlands of dry and limestone soils in southern Europe, such as the *Quercion pubescenti-petraeae*. Most stands fall within types 7 and 8 of the Merlewood classification.

The field layer, which is rather poor when compared with the richness of the stand, is most often dominated by *Hedera helix* or *Rubus fruticosus*. Frequent species are *Ajuga reptans, Anemone nemorosa, Arum maculatum, Brachypodium sylvaticum, Carex sylvatica, Deschampsia caespitosa, Dryopteris dilatata, D. filix-mas, Endymion non-scriptus**, *Fragaria vesca, Galeobdolon luteum, Hedera helix**, *Lonicera periclymenum, Mercurialis perennis**, *Phyllitis scolopendrium, Potentilla sterilis, Primula vulgaris, Rubus fruticosus**, *Sanicula europea* and *Viola riviniana*. Amongst the less-frequent species recorded are *Convallaria majalis* and *Melica uniflora*.

8.6 Oak–lime woodland (Group 5)

Stands with *Tilia cordata*, but not *Alnus, Carpinus, Fagus, Fraxinus, Pinus sylvestris, Ulmus carpinifolia, U. glabra* or *U. procera*. Within these stands *Betula, Corylus, Lonicera* and *Quercus* are common. They are formally distinguished from the ash–lime woodlands of Group 4 by the absence of *Fraxinus*, but there are many other differences. *Acer campestre, Euonymus, Ligustrum, Malus, Prunus spinosa, Rosa arvensis, R. canina, Salix caprea* and *Thelycrania* are absent or much rarer. *Corylus* and *Crataegus monogyna*, though common, are less frequent than in ash–lime woods, and only *Sorbus aucuparia* seems to be commoner in oak–lime woods.

Oak–lime woodland occurs on the only soil which is within the range of lime and beyond the range of ash, an acid (pH 3.6–5.0), generally light soil, at least at the surface. Most stands occur as mixed coppice of lime and hazel with some oak and birch below

standard oaks, but birch has become dominant in many stands.

Stand types (Table 8.5) are based on the two oak species, as in groups 6, 8 (acid beechwoods) and 9. The two resulting types are floristically and geographically distinct, with few truly intermediate stands. They occur scattered through East Anglia, Lincolnshire and the east Midlands, continuing in an arc through the north Midlands down through the Welsh borderland into Devon. At the eastern end of this range the pedunculate oak type (5A) is found to the virtual exclusion of the sessile oak type (5B), but from Lincolnshire westwards the latter is increasingly found to the eventual exclusion of the former.

Oak–lime woods are more-or-less transitional between groups 4 and 6. They fall within the Merlewood types 17–24 together with the birch-oakwoods and acid beechwoods. Neither Tansley nor Klötzli include these stands, but they broadly constitute Rackham's (1971) 'light-soil' type of East Anglian woods.

8.6.1 Stand type 5A. Acid pedunculate oak–lime woods

Oak–lime woods containing *Quercus robur*, but little or no *Q. petraea*. *Betula* is constantly present, with *B. pubescens* as the commoner species. *Populus tremula* and *Salix cinerea* have been recorded in 5A but not 5B. All recorded stands have been managed as mixed coppice of lime, hazel, birch and, rarely, either oak with standards of oak, or else lime and birch. Most stands grow on the plateaux or valleys where a freely-drained, light sandy or loam layer overlies a poorly-drained, often heavier sub-soil. This type has been recorded from Essex, Suffolk, Norfolk, Lincolnshire, Soke of Peterborough and north Gloucestershire, but it probably occurs also in the north Midlands (Fig. 8.15). Good examples are recorded at Stainfield and Bracken Woods (Lincs), Hockering Wood (Norfolk) and Chalkney Wood (Essex): at the last site *Castanea* is common. The type is clearly transitional between 4A, 4Ba and 6D, and generally occurs with these types. In East Anglia type 5A is often associated with the deeper deposits of acid, sandy material, which survive in ancient woods composed mainly of heavy soil types such as 2A and 3A.

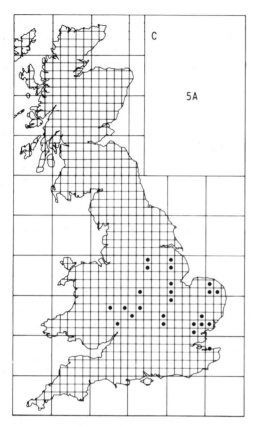

Fig. 8.15 Distribution of stand type 5A, acid pedunculate oak–lime woods

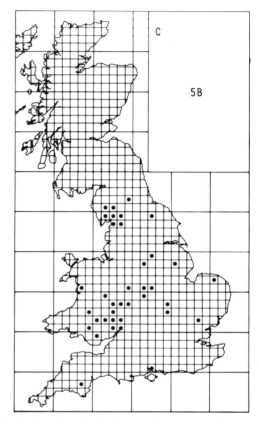

Fig. 8.16 Distribution of stand type 5B, acid sessile oak–lime woods.

Many of the following frequent field layer species may be dominant: *Anemone nemorosa, Chamaenerion angustifolium, Convallaria majalis, Deschampsia caespitosa, Dryopteris dilatata, D. filix-mas, Endymion nonscriptus, Galeobdolon luteum, Lonicera periclymenum*, Luzula pilosa, L. sylvatica, Oxalis acetosella, Poa trivialis, Pteridium aquilinum, Rubus fruticosus*, Teucrium scorodonia* and *Viola riviniana*.

8.6.2 Stand type 5B. Acid sessile oak–lime woods

Oak–lime woods containing *Quercus petraea* but little or no *Q. robur*. *Betula* is constantly present, with *B. pendula* as the commoner species. *Hedera, Ilex, Sorbus torminalis* and *Taxus* seem to be characteristic of 5B but rarely found in 5A. Most stands have been managed as oak standards over mixed lime, hazel, oak

and, rarely, birch coppice. They grow on light, acid, mostly freely-drained soils, though the type has been recorded on very poorly-drained silty clay loam. These stands are scattered from Norfolk and Lincolnshire to the west Midlands, with isolated stands on the fringe of Dartmoor and on the Hamble estuary (Hants) (Fig. 8.16). They are most frequent, but still rare, in the central Welsh borderland south to north Gloucestershire. Good examples occur at Collinpark Wood, (Glos), Shrawley Wood (Worcs), Swithland Wood (Leics) and Swanton Novers Great Wood (Norfolk). In the few woods where both 5A and 5B occur, the former occupies lower, somewhat damper ground. Type 5B lies between 6A, 6C and 4C in the range of variation, and often occurs with birch–sessile oak woodland. Like type 4C, type 5B has relatively high frequencies of *Betula pendula, Acer pseudoplatanus, Sorbus torminalis* and *Taxus baccata*.

The field layer may be dominated by *Endymion non-scriptus*, *Pteridium aquilinum*, *Holcus mollis*, *Rubus fruticosus*, *Luzula sylvatica* or even *Maianthemum bifolium*. Frequent species are *Anemone nemorosa Deschampsia caespitosa*, *Dryopteris filix-mas*, *Endymion nonscriptus*, *Euphorbia amygdaloides*, *Galeobdolon luteum*, *Hedera helix*, *Holcus mollis*, *Lonicera periclymenum**, *Luzula pilosa*, *L. sylvatica*, *Oxalis acetosella*, *Pteridium aquilinum*, *Rubus fruticosus**, *Teucrium scorodonia* and *Viola riviniana*.

8.7 Birch–oak woodland (Group 6)

Stands containing *Quercus petraea* and/or *Q. robur*, but not *Acer campestre*, *Alnus*, *Carpinus*, *Fagus*, *Fraxinus*, *Pinus sylvestris*, *Tilia* or *Ulmus*. *Fagus* and *Pinus* are often present, but only as a result of planting in or near the stand. (For classificatory purposes all occurrences of *Fagus* and *Pinus* beyond their native range in Britain are regarded as introductions.) *Acer campestre* may also be present under exceptional circumstances (see 6C). *Betula pendula*, *B. pubescens*, *Ilex*, *Lonicera* and *Sorbus aucuparia* occur in all stand types, mostly at medium to high frequency. *Betula* species are present in virtually all stands and in 82 of the 102 recorded samples.

Birch–oak woodland occurs on acid (mostly pH 3.8–5.0), light–medium-textured soils, most of which are freely drained. These range from leached soils covered with accumulations of mor humus, to fertile, acid, brown earth with mull humus. Such conditions are mostly beyond the edaphic range of *Acer campestre*, *Alnus*, *Carpinus*, *Fraxinus* and *Ulmus*, but well within the capabilities of *Fagus*, *Pinus* and *Tilia cordata*. Oak–lime woodland (Group 5), oak–beech woodland (Types 8A and 8B) and pine woodland (Group 11) are therefore closely related to birch–oak woodland. Pine woodland in the Highlands seems to be restricted to the most acid and infertile sites, leaving the acid, brown earths to birch–oak woodland. Oak–beech woodland certainly occupies the same range of sites as birch–oak woodland in southern Britain and indeed many stands of the former are known to have developed from the latter. In this region the reluctance of beech to coppice vigorously may be important and birch–oak woodland may be favoured by coppice management.

Oak–lime woodland in the Midlands and East Anglia seems to occupy unleached acid, brown earths, whereas the birch–oak woodland grows mainly on drier, less fertile soils. *Tilia*, however, was more abundant in primaeval woodland than in existing ancient, semi-natural woodland, and is a poor coloniser after disturbance. Quite possibly, therefore, many birch–oakwoods in that region have been derived from oak–lime woods following disturbance in prehistoric and historic times. In Scotland, where maple, lime and hornbeam are absent, birch–oak woodland extends on to apparently more fertile soils than in England and Wales.

Most birch–oak woodlands have been managed as coppice oak or hazel. Those with hazel coppice mostly have oak standards, but many of the oak coppices in the north and west lack standards. Birch is usually present as short-lived coppice stools or, more often, as maiden trees dating from the last coppicing. Almost all these coppices are neglected now, and may have grown into a high forest form. Many birch–oakwoods in Hereford and south-east England were converted to chestnut coppices in the early 19th century and are still coppiced. Wood-pasture forms of birch–oak woodland survive in, for example, Staverton Park (Suffolk) and Sherwood Forest (Notts).

Birch–oak woodlands occur virtually throughout Britain except in the extreme north. In parts of the west and north they are far more abundant than any other stand type, and have in consequence attracted the popular title of 'western oakwoods'. They are rare in the Midlands and eastern England, where the soils are generally heavy and poorly drained, but occur frequently in south-east England on sandy deposits.

The birch–oak woodlands form a distinctive and fairly homogeneous portion of the total range of variation falling within the *Quercion robori-petraeae*. They can be subdivided to form stand types in several ways. The popular contrast is between 'western' sessile oakwoods in oceanic climates and the dry (pedunculate) oakwoods of less oceanic climates in the lowlands, but this is an over-simplification, based on the fact that *Q. petraea* tends to predominate in the north and west, whilst *Q. robur* is the most abundant lowland species. Klötzli (1970) divided the range into the oceanic *Blechno–Quercetum* and the lowland

Querceto–Betuletum, a division which is justified on floristic grounds. Since this division is also correlated with trends in the composition and management of the stands it is accepted here as the basic division into upland (6A, 6B) and lowland (6C, 6D) stand types (Table 8.6). The division is placed arbitrarily along the boundary of palaeozoic rocks. *Hedera* is more frequent in upland types, and many species occur only in lowland types, albeit rarely (e.g. *Thelycrania sanguinea*). Upland oakwoods tend to be on less acid, heavier soils and on steeper ground than their lowland counterparts.

The local distribution of oak species has been disturbed by planting, mainly of *Q. robur* on to sites that might naturally have supported *Q. petraea*, but even so, much of the patchiness in the distribution of the two species appears to be natural (Jones, 1959). The factors which determine their occurrence are far from clear, though there is a tendency for *Q. petraea* to occupy shallow, light soils (including thin soils on limestone) and for *Q. robur* to grow on deeper moister soils: this trend is not obvious within birch–oakwoods. Despite these difficulties, the oak species have been chosen as a second basis for sub-dividing the total range into sessile oakwoods (6A, 6C) and pedunculate oakwoods (6B, 6D). Some stands contain both oak species in an intimate mixture, but rather than create a separate class of mixed oakwoods, these are placed with the minority geographical type; thus mixed oakwoods in the uplands are placed in 6B, and those in the lowlands are placed in 6C. *Ilex* is associated with *Q. petraea*, especially in the lowlands.

Covering such a wide edaphic and geographical range, the field layer of birch–oakwoods is understandably very variable. Some of the most infertile soils are carpeted with *Calluna vulgaris*, *Deschampsia flexuosa* or *Vaccinium myrtillus*, whilst others on more fertile brown earths have a richly mixed herbaceous flora. This has prompted a sub-division of the *Quercion robori-petraeae* into a *Vaccinio–Quercion* on base-poor soils and a *Violo–Quercion* on somewhat more fertile soils (Doing, 1975). This is a useful division and is reflected here in the sub-division of all stand types into two sub-types, respectively the 'b' form in which *Corylus* is absent and *Betula* is the main associate, and the 'c' form in which *Corylus* is present. The former, known as birch–oakwoods, occupy the

Table 8.6 Composition and site features of birch–oak woodland (Group 6).

	6Ab	6Ac	6Bb	6Bc	6Cb	6Cc	6Db	6Dc
Number of samples	18	15	7	8	10	14	6	16
Betula pendula	+	I	II	I	III	III	II	I
Betula pubescens	IV	III	II	IV	V	IV	IV	V
Corylus avellana		V		V		V		V
Quercus petraea	V	V	II	II	V	V		
Quercus robur			V	V	II	III	V	V
Acer pseudoplatanus	+	I		II		II		I
Castanea sativa					I	I		+
Crataegus monogyna		II		II	I	II		II
Euonymus europaeus								+
Frangula alnus							+	+
Hedera helix	I	III	I	II	+	+		I
Ilex aquifolium	III	II	I	II	IV	IV	I	I
Juniperus communis	+							
Lonicera periclymenum	II	III	II	IV	II	V	II	V
Malus sylvestris				II			I	+
Populus tremula				+		II		+
Prunus avium					I	I		
Prunus spinosa				+		I		+
Rhododendron ponticum	+				I	+	I	
Ribes nigrum						+		
Ribes sylvestre								+
Rosa arvensis		+			I	I		+
Rosa canina						+		I
Rosa villosa		+						
Salix caprea						+		+
Salix cinerea		+				+		+
Sambucus nigra					I	I	+	I
Sarothamnus scoparius		+	I		I	+		
Sorbus aucuparia	IV	IV	V	IV	III	III	III	II
Sorbus torminalis	+					+		
Taxus baccata						I		
Thelycrania sanguinea						+		I
Median pH	4.2	4.6	4.3	4.4	3.8	4.3	3.8	4.1
Mean texture	1.9	2.3	1.2	1.6	0.8	1.4	0	1.0
Drainage free	17	13	7	7	7	5	3	6
imperfect	1	1				3	3	5
poor				1		3		4
very poor						3		1
Site	S	S	S	S	s	psr	P	psr

most acid, shallow, leached and degraded soils, and the latter, known as hazel–oakwoods, link these to other stand groups. Some birch–oakwoods have been formed by cutting or grazing hazel out of hazel–oakwoods, but on the whole the occurrence of hazel seems to be mainly determined by natural, edaphic factors. The birch–oakwoods are all remarkably poor in tree and shrub species, but the hazel–oakwoods are moderately rich and characterised by the presence or greater frequency of several species which are found more abundantly in other stand groups: *Acer pseudoplatanus*, *Crataegus monogyna*, *Lonicera*, *Populus tremula*, *Prunus spinosa*, *Rosa* spp. and *Salix cinerea*.

Eight sub-types have therefore been formed on the basis of three features, viz. geographical position, oak species and the presence or absence of hazel. Each feature has more-or-less equal ecological significance. This arrangement is summarised thus:

	Q. petraea Upland	*Q. robur* + mixed Upland	*Q. petraea* + mixed Lowland	*Q. robur* Lowland
birch-form	6Ab	6Bb	6Cb	6Db
hazel-form	6Ac	6Bc	6Cc	6Dc

Birch–oak woodland falls mainly within types 17–20 and 22–25 of the Merlewood classification. Many examples are described in chapters 13–17 of Tansley (1939), including the exceptionally oceanic stands in western Ireland which were described later by Braun-Blanquet and Tüxen (1952). Some Scottish examples have recently been described as a *Galio saxatilis–Quercetum* association by Birse and Robertson (1976). Jones (1959) described four broad classes of oakwood, two of which appear to be equivalent to my birch–oak woodland, namely (a) those on nutrient-deficient soils, usually podzols and gley-podzols on highly siliceous rocks, and (b) those on base-deficient soils better supplied with nutrients than (a), i.e. low-base status brown forest soils: Jones gives floristic lists for both types on dry, damp and wet soils. Klötzli (1970) also reviews British birch–oak woodland and points out that similar soils in most of Europe are mainly occupied by beech woodland. The ecology and composition of European birch–oak woodland and acid beech woodland is

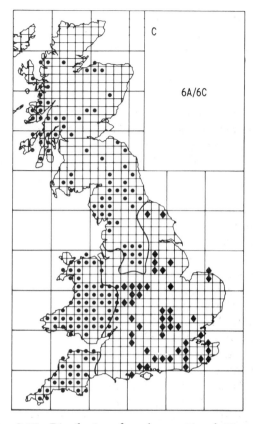

Fig. 8.17 Distribution of stand types 6A and 6C, upland and lowland sessile oakwoods.

described in great detail in a symposium volume edited by Gehu (1975).

8.7.1 Stand type 6A. Upland sessile oakwoods

Birch–oakwoods in upland Britain (Fig. 8.17) containing *Quercus petraea*, but not *Q. robur*. Such stands usually contain *Betula pubescens*, *Ilex*, *Lonicera* and *Sorbus aucuparia*.

(a) Sub-type 6Ab. Upland birch–sessile oakwoods

Upland sessile oakwoods without *Corylus* (Fig. 8.18). Such stands almost invariably contain *Betula*, the great majority of which is *B. pubescens*. Oak is dominant in most stands, and the sole constituent of the canopy in some. Most stands have been managed as simple coppice to supply fuelwood, charcoal and tanbark,

Fig. 8.18 Heavily grazed sessile oak woodland (stand type 6Ab), typical of hillsides in western upland Britain. Once coppiced for tanbark and charcoal (note the two-trunked oaks on the right), these woods in Borrowdale (Cumbria) have merely sheltered sheep for decades. Now they look decrepit and are scarcely regenerating even in clearings. Mats of bryophytes cover the ground and the tree trunks and branches support a rich flora of epiphytic lichens. (Photo. D. A. Ratcliffe)

but the system is now worked in only a handful of sites. Some have not been cut since the mid-19th century and now resemble high forest stands, but others were cut less than 30 years ago. Exceptionally, stands may never have been coppiced, e.g. Coed y Rhygen (Gwynedd). Most are now grazed enough to prevent natural regeneration so that birch, rowan and holly may be present only as seedlings, though normally there are a few mature individuals in the canopy. Characteristically these stands grow on the steeper valley slopes where the soil is strongly acid, medium–light-textured, freely drained and often surfaced by accumulations of mor humus. They occur throughout the west and north, except in the extreme north of Scotland and north-east Scotland, and are especially common in west Wales and Cumbria.

Field layer communities are dominated by grasses (chiefly *Anthoxanthum odoratum* and *Deschampsia flexuosa*), *Vaccinium myrtillus*, *Pteridium aquilinum*, or even *Calluna vulgaris* where the soil, and therefore the canopy, is exceptionally thin. Frequent species are *Agrostis tenuis*, *Anthoxanthum odoratum*, *Calluna vulgaris*, *Deschampsia flexuosa**, *Dryopteris dilatata*, *D. filix-mas*, *Festuca ovina*, *Galium saxatile**, *Holcus mollis*,

Lonicera periclymenum, Luzula campestris, L. pilosa, Melampyrum pratense, Oxalis acetosella, Polypodium vulgare, Pteridium aquilinum*, Teucrium scorodonia, Vaccinium myrtillus*, Veronica officinalis and *Viola riviniana*. These woods are especially rich in bryophytes, which often completely carpet exposed rocks and boulders.

(b) *Sub-type 6Ac. Upland hazel–sessile oakwoods*

Upland sessile oakwoods containing *Corylus*. Most stands also contain *Betula* and *Sorbus aucuparia*. *Crataegus monogyna* and *Hedera* are more frequent here than in 6Ab. Most stands have been managed as simple coppice in which oak is far more abundant than hazel – which was often cut out as a weed in the past – but a few stands were managed as oak standards over hazel coppice. Both rowan and birch may have been included within the coppice, although they occur more often as self-sown trees and shrubs. Most stands have been neglected for decades and used as pasture, a treatment which is gradually exterminating the hazel. These stands occupy much the same range of sites as 6Ab, but the soil tends to be slightly heavier and less acid. Management has somewhat confused the distinction between 6Ab and 6Ac by eliminating hazel from some of the latter: the distinction between the two is often confused on the ground where hazel is thinly and patchily scattered. Nevertheless, it seems that 6Ac is the less extreme form on richer, but still base-poor, soils. The upland hazel–sessile oakwoods occupy the same area as the birch sub-type, but seem to be much rarer in west Wales and the Lake District.

The field layer is generally richer and has a wider range of dominants than in 6Ab, including *Holcus mollis, Endymion non-scriptus, Luzula sylvatica* and *Rubus fruticosus* which rarely dominate in 6Ab. Frequent species are *Agrostis tenuis, Anthoxanthum odoratum, Blechnum spicant*, Circaea lutetiana, Deschampsia flexuosa, Digitalis purpurea, Dryopteris dilatata, Endymion non-scriptus, Hedera helix, Holcus mollis, Lonicera periclymenum, Luzula sylvatica, Lysimachia nemorum, Oxalis acetosella, Polypodium vulgare, Potentilla sterilis, Primula vulgaris, Pteridium aquilinum*, Ranunculus ficaria, Rubus fruticosus, Stellaria holostea, Teucrium scorodonia, Vaccinium*

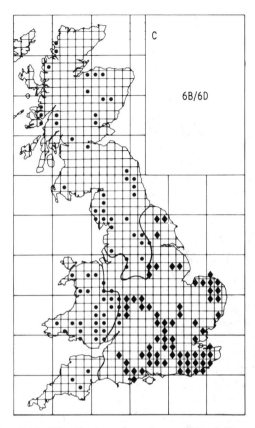

Fig. 8.19 Distribution of stand types 6B and 6D, upland and lowland pedunculate oakwoods.

myrtillus and *Viola riviniana*. Many of the species which are frequent in ash–wych elm woodland occur sparsely in 6Ac.

8.7.2 Stand type 6B. Upland pedunculate oakwoods

Birch–oakwoods in upland Britain containing *Quercus robur*. Stands containing mixed populations of oak are placed in this type. *Betula, Lonicera* and *Sorbus aucuparia* are almost always present, but *Ilex* seems to be less frequent than in 6A. These stands have been recorded throughout upland Britain from Devon to Sutherland and Aberdeen (Fig. 8.19). They are frequent in Devon, rare in Wales outside parts of Pembroke, Carmarthen and eastern Wales, and frequent in the Highlands. Tansley (1939) mentions that *Q. robur* is frequent and undoubtedly native in the far north, where it may be more characteristic of river sides and deep valley soils.

(a) Sub-type 6Bb. Upland birch–pedunculate oakwoods

Upland pedunculate (and mixed) oakwoods without *Corylus*. Oak is dominant in almost all stands, most of which have been managed as coppice. They are similar to sub-type 6Ab in most features, but tend to grow on deeper, coarser soils, such as block scree or deep sands and loamy sands. Wistman's Wood (Devon) and other high-altitude oakwoods on Dartmoor are the most famous examples, but fine stands also occur at Holystone Oaks (Northumberland), Dinnet Wood (Aberdeen) and in the western Highlands north at least to Loch A'Mhuillin (Sutherland).

The field layer is often grazed and therefore dominated by *Pteridium aquilinum* or grasses (*Deschampsia flexuosa*, *Holcus mollis*). Frequent species are *Anemone nemorosa*, *Anthoxanthum odoratum**, *Blechnum spicant*, *Calluna vulgaris*, *Campanula rotundifolia*, *Carex pilulifera*, *Conopodium majus*, *Deschampsia flexuosa**, *Endymion non-scriptus*, *Galium saxatile*, *Holcus mollis**, *Lonicera periclymenum*, *Luzula multiflora*, *L. pilosa*, *Melampyrum pratense*, *Oxalis acetosella**, *Potentilla erecta*, *Pteridium aquilinum*, *Stellaria holostea*, *Trientalis europea*, *Vaccinium myrtillus**, *Veronica chamaedrys*, *V. officinalis* and *Viola riviniana*.

(b) Sub-type 6Bc. Upland hazel–pedunculate oakwoods

Upland pedunculate (and mixed) oakwoods containing *Corylus*. *Betula pubescens* is the main birch species present. Several species which are absent from 6Bb occur in this sub-type, e.g. *Crataegus monogyna*. Most stands have a mixed coppice of oak and hazel, usually with oak standards, thus distinguishing them from 6Ac stands, most of which were simple coppice. Another distinguishing feature is that most stands are not intensively grazed. The soil is usually a deep loam or silty loam, unleached, freely drained and with no more than 2–5 cm of surface litter and humus. Good examples occur south of Exmoor, and in the Old Wood of Methven at Almondbank (Perth), but the sub-type has also been recorded in south-west and south-east Wales, Galloway and Deeside.

The field layer is comparatively rich, containing communities and species which are characteristic of ash–maple woodland and hazel–ash woodland in the lowlands, and no one species is characteristically dominant. Frequent species are *Anemone nemorosa**, *Anthoxanthum odoratum*, *Athyrium filix-femina*, *Brachypodium sylvaticum*, *Conopodium majus*, *Dactylis glomerata*, *Deschampsia caespitosa*, *D. flexuosa*, *Digitalis purpurea*, *Dryopteris dilatata*, *D. filix-mas**, *Endymion non-scriptus*, *Fragaria vesca*, *Galeobdolon luteum*, *Galium aparine*, *Geum urbanum*, *Hedera helix*, *Holcus lanatus*, *H. mollis*, *Lathyrus montanus*, *Lonicera periclymenum*, *Luzula pilosa*, *Lysimachia nemorum*, *Melica uniflora*, *Mercurialis perennis*, *Oxalis acetosella**, *Potentilla sterilis*, *Primula vulgaris*, *Pteridium aquilinum**, *Rubus fruticosus**, *Scrophularia nodosa*, *Silene dioica*, *Stellaria holostea*, *Urtica dioica*, *Veronica chamaedrys* and *Viola riviniana**.

8.7.3 Stand type 6C. Lowland sessile oakwoods

Birch–oakwoods in lowland Britain (Fig. 8. 17) containing Q. petraea. Stands containing both oak species are included here. *Betula* in general and *B. pendula* in particular seems to be more abundant than in other acid oakwood types. *Ilex* is almost always present and *Prunus avium* is recorded only in this type. Many stands which would naturally conform to this type have been converted to *Castanea* coppice, especially in Kent. Throughout the lowlands sessile oak tends to occur in 'massifs' (Jones, 1959), i.e. concentrations where Q. robur is correspondingly infrequent, which are likely to be descendants of natural populations, e.g. north Kent, Wyre Forest (Worcs and Salop). Examples in north Kent have been described by Wilson (1911).

(a) Sub-type 6Cb. Lowland birch–sessile oakwoods

Lowland sessile (and mixed) oakwoods without *Corylus*. Either birches or oaks are dominant. Both *Sorbus aria* and *S. aucuparia* may be present. Most stands recorded were simple oak coppices, but stands in Norfolk and Kent were oak coppice with oak standards. Recorded examples are on strongly acid (pH 3.4–4.2), light-textured soils, freely drained at least in the surface horizons, and sometimes with a deep accumulation of litter and humus. Good examples occur: (a) in west and north Kent; (b) in association with hornbeam woods in Hertfordshire, e.g. Sherrards Park Wood (Herts); (c) in Ruislip Woods (formerly Middlesex); (d) in Swanton Novers Wood (Norfolk), Kings Wood (Bedford) and (e) on pre-Cambrian rocks in Charnwood Forest and at Bentley Park Wood (Warwicks).

The field layer is usually dominated by *Pteridium aquilinum* or *Rubus fruticosus*, and is often extremely poor in species. Frequent species are *Agrostis tenuis*, *Calluna vulgaris*, *Deschampsia flexuosa*, *Holcus mollis*, *Lonicera periclymenum*, *Luzula pilosa*, *Pteridium aquilinum** , *Rubus fruticosus** and *Teucrium scorodonia*. Infrequent but characteristic species include *Convallaria majalis*, *Hypericum humifusum*, *H. pulchrum*, *Poa nemoralis* and *Pyrola minor*. *Luzula sylvatica* and *Vaccinium myrtillus* are locally abundant in west Sussex and west Kent on the Lower Greensand.

(b) Sub-type 6Cc. Lowland hazel–sessile oakwoods

Lowland sessile (and mixed) oakwoods containing *Corylus*. *Betula*, *Ilex* and *Lonicera* are present in almost all stands. Most stands have been managed as oak standards over mixed coppice in which oak, hazel, birch and occasionally wild service are present. Recorded examples were on acid soils below pH 5.1 with a generally medium–light texture, but profile drainage conditions were surprisingly variable, with half the samples on poorly-drained soils. The poorly-drained stands occasionally contained *Populus tremula*, *Prunus spinosa*, *Rosa* spp. and *Salix* spp. Although there is a wide overlap, 6Cc stands appear to be on moister, heavier soils than 6Cb. Examples occur in the same general areas as sub-type 6Cb, but good examples occur also at Collinpark Wood (Glos), and Ashberry Wood (N. Yorks).

The field layer may be dominated by any one of a number of species from *Endymion non-scriptus* and *Pteridium aquilinum* to *Luzula sylvatica* and *Vaccinium myrtillus*. Frequent species are *Anemone nemorosa*, *Deschampsia caespitosa*, *D. flexuosa*, *Dryopteris dilatata*, *D. filix-mas*, *Endymion non-scriptus*, *Holcus lanatus*, *H. mollis*, *Lonicera periclymenum** , *Luzula pilosa*, *L. sylvatica*, *Milium effusum*, *Oxalis acetosella*, *Pteridium aquilinum** , *Rubus fruticosus** , *Teucrium scorodonia*,

Vaccinium myrtillus and *Viola riviniana*. Several infrequent but characteristic species are recorded, including *Betonica officinalis*, *Convallaria majalis*, *Dryopteris spinulosa*, *Melampyrum pratense* and *Stellaria holostea*. The wide variation in the field layer is principally related to variation in drainage conditions.

(c) Lowland sessile oakwoods in north Kent

Extensive stands at Darenth Wood and nearby Stone Thrift, which cover the full range of the stand type from freely to very poorly-drained, sandy soils, have possibly been influenced by a rain of dust from nearby chalk quarries. The soils have a pH range from 4.1–8.0 at 10 cm depth and are rather more acid in the sub-soil. Furthermore, the stand contains many calcicole species, present as saplings and young shrubs, including *Acer campestre*, *Euonymus*, *Ligustrum*, *Thelycrania* and *Viburnum lantana*, which appear to be recent colonisers. The field layer is relatively rich, even in the surviving strongly acid parts, where *Pteridium aquilinum*, *Rubus fruticosus* and *Convallaria majalis* are abundant. Calcicole species such as *Mercurialis perennis* are present, possibly having recently spread in from neighbouring, naturally calcareous sites.

8.7.4 Stand type 6D. Lowland pedunculate oakwoods

Birch–oak woodland in lowland Britain (Fig. 8.19) containing *Quercus robur* but not *Q. petraea*. *Betula pubescens* occurs in almost all stands, but *Ilex* is infrequent. Although pedunculate oak is abundant in the lowlands, this type is infrequent, occurring only on those sites which are too acid or too dry for ash.

(a) Sub-type 6Db. Lowland birch–pedunculate oakwoods

Lowland pedunculate oakwoods without *Corylus*. *Betula*, mostly *B. pubescens*, is almost always present. This is one of the poorest woodland types in the lowlands for, apart from scattered *Sorbus aucuparia* and *Ilex*, very few other species ever occur in it. *Frangula* occasionally occurs at low density. All recorded stands were managed as oak coppice, with or without oak standards, and are still dominated by oak. They occur only on strongly acid (pH 3.2–4.2) sands, with up to 13 cm of surface litter and peaty humus. Apart from one stand, all recorded examples have unleached soils. Examples have been recorded on the fen edge in Lincolnshire, Maidensgrove and Nettlebed in the Chilterns, Addlestone Spinney (Surrey) and Nap Wood (Sussex). Wood-pasture forms of this type occur at Staverton Park (Suffolk), the New Forest (Hants) and elsewhere. These woods correspond with the dry oakwoods and *Quercetum ericetosum* of Tansley (1939). Some examples are secondary semi-natural stands which have subsequently been coppiced.

The field layer, which is outstandingly poor, is most often dominated by *Pteridium aquilinum*, and on the driest sites may be confined to this species. Frequent species are *Convallaria majalis*, *Corydalis claviculata*, *Deschampsia flexuosa*, *Dryopteris dilatata*, *Lonicera periclymenum*, *Pteridium aquilinum** and *Rubus fruticosus**. *Vaccinium myrtillus* is present in southern examples.

(b) *Sub-type 6Dc. Lowland hazel–pedunculate oakwoods*

Lowland pedunculate oakwoods containing *Corylus*. These stands usually contain *Betula pubescens* and *Lonicera*. Most stands are predominantly hazel coppice beneath oak standards, but some are closely related to 6Db with oak as the main coppice species. Like its sessile oak counterpart, 6Cc, 6Dc occurs on strongly acid (pH 3.2–5.0), light–medium-textured soils across the full range of drainage conditions from free to very poor. Furthermore, sub-types 6Cc and 6Dc both contain many species which are not found in the birch sub-types. In fact, there is no obvious explanation in site conditions for the occurrence of one sub-type or the other. However, one repeats Jones' observation that certain districts with extensive light soils are characterised by *Q. petraea*, whilst other districts, usually without extensive tracts of light soil, are characterised by *Q. robur* without *Q. petraea*: possibly 6Cb and 6Cc are ecologically equivalent to 6Db and 6Dc respectively, occupying the same range of sites, differing little in associated species, but with differences in the oaks which appear to have a broader biogeographical explanation. Lowland hazel–pedunculate oakwoods are most frequent in the Midlands (e.g. Sheephouse Wood, Bucks and Waterperry Wood, Oxon), East Anglia and the Weald, but

occur in most districts. This sub-type is closely related to type 3A.

Rubus fruticosus and *Pteridium aquilinum* are most likely to dominate, but the field layer is often a mixture of herbs without clear dominants. In one wood *Convallaria majalis* and *Maianthemum bifolium* are the most abundant species, as in many related woods in France. Frequent species are *Ajuga reptans*, *Anemone nemorosa*, *Carex sylvatica*, *Cirsium palustre*, *Deschampsia caespitosa*, *Dryopteris dilatata*, *D. filix-mas*, *Endymion non-scriptus**, *Galeobdolon luteum*, *Geum urbanum*, *Hedera helix*, *Holcus lanatus*, *H. mollis*, *Lonicera periclymenum**, *Luzula pilosa*, *Lysimachia nemorum*, *Milium effusum*, *Oxalis acetosella*, *Potentilla sterilis*, *Primula vulgaris*, *Pteridium aquilinum*, *Rubus fruticosus**, *Stellaria holostea* and *Viola riviniana*.

8.8 Alder woodland (Group 7)

Stands containing *Alnus glutinosa*. Most of such stands contain *Corylus* and *Fraxinus*. Amongst various 'species-pairs', *Betula pubescens*, *Quercus robur* and *Salix cinerea* are associated with alderwood, whereas *B. pendula*, *Q. petraea* and *S. caprea* are rarely found. Species such as *Clematis*, *Ligustrum*, *Sorbus aria*, *Taxus*, *Tilia cordata* and *Viburnum lantana*, which are characteristic of dry soils, are rare or absent. *Alnus* and *Fagus* hardly ever grow together, but there is a slight overlap with *Carpinus*, *Pinus sylvestris* and suckering *Ulmus* to form stands which are intermediate between group 7 and groups 9–11.

Alder grows throughout Britain on permanently and seasonally wet soils, provided that the water is not stagnant. Thus, alder is frequent on light and medium-textured soils, but infrequent on the clays of the English Midlands. Within these limits alder has a wide tolerance, thriving on both mineral soils with a freely-drained surface horizon and on amorphous, organic ooze in which the watertable is permanently high. Alder occurs on strongly acid and alkaline ground (pH 3.4–7.2 in the samples recorded), but appears to tolerate strongly acid peat better in eastern England. In upland Britain alder is largely a species of base-rich flushes, having abandoned the base-poor sites to bog communities. Suitable soils are, of course, found mainly in valleys: alderwoods are the characteristic woodlands of streamsides, depressions and swampy ground throughout Britain. Permanently moist soils also occur on flat, plateau sites and even on sloping ground in oceanic regions, and here too alderwoods develop. Alder is often coppiced and has frequently been planted on wet ground.

Since the occurrence of alder woodland is so clearly related to topography it seems reasonable to base stand types on this factor (Table 8.7). Two relatively distinct forms are thus immediately separated, the plateau alderwoods (7C) and slope alderwoods (7D), but the more abundant, widespread and variable valley alderwoods justify division into several stand types. These divisions of valley alderwoods are somewhat arbitrary, since there are no sharp ecological discontinuities within them, but they are based on the factors which McVean (1956) regarded as important, i.e. water regime and base status. A typical transect into alder woodland passes through a zone on moist mineral soil in which alder is part of a mixture with the components of the adjacent stand type. The composition of these alderwoods, which can be regarded as wet-soil variants of adjacent stand types, varies according to the characteristics of the mineral soil, but they are all placed within stand type 7A. Because alder is often sparse, they have generally not been described as alderwoods by other ecologists. Not all stand types have an alderwood equivalent. For example, *Quercus petraea* and *Fagus* rarely grow with alder, so the alderwood equivalents of stand types 6A and 6C are the same as those of 6B and 6D, and there are no alderwood equivalents of beech woodland. Conversely, the wet soil variant of stand type 2A on heavy soils in eastern England may be a mixture of ash, suckering elms and poplars without alder. Passing further along the transect on to wetter ground, where the watertable is permanently at or near the surface, one reaches beyond the edaphic range of maple, birch, wych elm, oak, etc.; alder is correspondingly more abundant and generally dominant; and a few species which are especially characteristic of wet soils are found. Those wet alderwoods form stand type 7B. The differences in composition between relatively dry (7A) and relatively wet (7B) alderwoods are clear enough on the ground, but they tend to be masked in the records by the occurrence in wet alderwoods of islands (often coppice alder stools and the surrounding root plate) of

Table 8.7 Composition and site features of alderwoods (Group 7).

	7Aa	7Ab	7B	7C	7D	7Ea	7Eb
Number of samples	10	17	20	12	9	6	5
Acer campestre		II		III			
Alnus glutinosa	V	V	V	V	V	IV	V
Betula pendula	+	+	+	+			III
Betula pubescens	IV	III	I	V	III	III	II
Corylus avellana	V	V	II	V	V	V	III
Fraxinus excelsior	III	IV	IV	V	V	V	IV
Prunus padus					II	V	V
Quercus petraea	+	+		+	I		
Quercus robur	III	III	+	IV	I	V	
Tilia cordata				I			
Ulmus glabra		II			II	+	II
Acer pseudoplatanus	II	II	I	I	II		II
Crataegus monogyna	I	II	I	III	III	I	I
Crataegus oxyacanthoides		+		I			
Euonymus europaeus			+	I			
Frangula alnus	+		+				
Hedera helix	+	II	II	+		I	III
Ilex aquifolium	II	II	+	II	III	II	
Lonicera periclymenum	V	II	II	III	I	V	II
Populus tremula				I			
Prunus avium		I	+				
Prunus spinosa	+	I	I	II	I	I	
Ribes nigrum			I				
Ribes sylvestre		+	+	+		I	II
Ribes uva-crispa		+	+				
Rosa arvensis		I	+	+			
Rosa canina		+	+	+		I	
Salix aurita	+				II		
Salix caprea	+	+		I		I	
Salix cinerea	+	II	III	IV		IV	II
Salix fragilis			+				
Sambucus nigra	II	III	II	+	I	II	I
Sorbus aucuparia	I	I	II	+	III	III	
Thelycrania sanguinea		I	+	I			
Viburnum opulus	+	I	+	II		I	II
Median pH	3.9	6.0	6.3	5.3	5.1	5.3	5.6
Mean texture	1.1	2.4	2.1	1.7	2.2	0.7	2.8
Drainage free	6	6			5	3	
imperfect	3	8	2	8	2	1	
poor		2	1	2		2	
very poor	1	1	17	2	2		5
Site	V	V	V	P	Sr	V	V

drier ground upon which species of dry stands (such as holly or rowan) obtain a roothold. In general, alderwoods achieve a separate identity only on wetter soils, whilst the drier alderwoods on mineral soils are merely variants of surrounding stand types. Exceptionally, however, where *Prunus padus* is present, the drier alderwoods are also distinctive: these bird cherry–alderwoods are recognised as a third stand type (7E) within the valley alderwoods. Alderwoods are spread across a wide range of Merlewood types, with a slight concentration in types 9–14, except for plateau alderwoods (7C) which fall within types 5 and 7.

8.8.1 Stand type 7A. Valley alderwoods on mineral soil

Valley alderwoods growing on mineral soil in which the watertable lies in the sub-soil. Such stands usually contain *Betula pubescens*, *Corylus* and *Quercus robur*, whilst *Acer pseudoplatanus*, *Ilex*, *Lonicera*, *Salix cinerea* and *Sambucus* are frequent. Stands containing *Prunus padus* are placed in type 7E. Characteristically, these stands occur in shallow depressions, on river terraces and around the margins of the wet alderwoods (type 7B), where alder is generally a minor component of a mixture, and alder woodland grades upslope into another stand type on drier soils (Fig. 8.20). Whilst most stands occupy flat ground, valley alderwood may extend a few metres up adjacent steep slopes. The soils are generally light–medium-textured, but cover the wide range from strongly acid to alkaline conditions. Since several species grow only (or more abundantly) at one end of this range, two sub-types are recognised. The division, at soils of pH 5.0, is perhaps an arbitrary cut into a continuum, but it is the lower limit at which calcicole species occur in valley alderwoods.

(a) Sub-type 7Aa. Acid valley alderwoods

Valley alderwoods on relatively dry soils with a pH below 5.0. *Lonicera* (especially) and *Betula pubescens* are more abundant in this sub-type, and *Frangula*, though rare, may be confined to it. The coppice consists mainly of hazel, alder and birch, occasionally with oak standards, and often with many self-sown

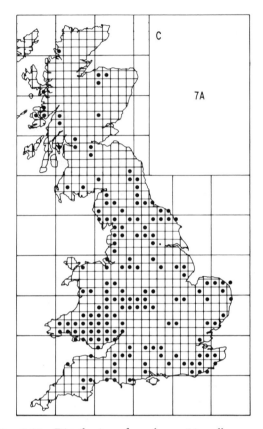

Fig. 8.20 Distribution of stand type 7A, valley alderwoods on mineral soils.

birch dating from the last coppicing. The stands on strongly acid soils are effectively birch–alderwoods, whereas those on mildly acid soils have more ash and hazel. Being acid and at least seasonally moist, considerable quantities of peat and humus have accumulated in many sites, but the watertable generally remains more than 30 cm below the surface. This sub-type occurs throughout Britain in districts possessing light, acid soils, but except on the fen margin woods of central Lincolnshire (e.g. Tumby Woods) they are rarely extensive, presumably because the soil is readily cultivated. In upland Britain this sub-type is virtually absent from steep streamsides, where woodland lacking alder grades sharply into wet alderwood by the stream, and must have been greatly restricted by the development of bogs. Sub-type 7Aa is most likely to be found as small patches in woods containing stands in groups 5, 6, 8 (A and B), 11 and 12.

Klötzli (1970) placed these stands in his *Pellio– Alnetum*.

The field layer is generally dominated by *Rubus fruticosus*, *Pteridium aquilinum* or *Dryopteris* spp., but *Holcus* spp. may be abundant in grazed woods. The following list of frequent species is biased by the prevalence of samples from Lincolnshire: *Ajuga reptans*, *Anemone nemorosa*, *Convallaria majalis*, *Corydalis claviculata*, *Deschampsia caespitosa*, *Dryopteris dilatata*, *D. filix-mas*, *D. spinulosa*, *Endymion non-scriptus*, *Holcus mollis*, *Lonicera periclymenum*, *Oxalis acetosella*, *Poa trivialis*, *Prunella vulgaris*, *Primula vulgaris*, *Pteridium aquilinum*, *Ranunculus repens*, *Rubus fruticosus** and *Viola riviniana*. McVean (1956) adds *Agrostis* spp. (including *A. stolonifera*), *Anthoxanthum odoratum*, *Galium saxatile* and *Luzula sylvatica* as characteristic of relatively dry, base-poor stands.

(b) Sub-type 7Ab. Valley alderwoods on neutral–alkaline soils

Valley alderwoods on relatively dry, mineral soils with a pH of 5.0 or more. Compared with 7Aa, this sub-type contains less *Betula pubescens* and *Lonicera*, more *Fraxinus* and a number of calcicole and clay-soil species, such as *Acer campestre*, *Rosa arvensis* and *Ulmus glabra*, which are not found in 7Aa. *Frangula* and *Salix aurita* appear to be absent, but in general these stands are richer. The coppice consists mainly of ash, hazel and alder, with some oak and ash standards. The soils are medium-textured, usually freely drained in the upper 20–30 cm and free of humus accumulations, except in the sites with a pH below 6.0. This sub-type appears to be most frequent in the mixed coppices of England and Wales, where good examples are recorded in Coed Gors-wen (Gwynedd), Swithland Wood (Leics), Asholt Wood (Kent) and Dobbin Wood (Cumbria). Although 7Ab is strictly the alderwood counterpart of types 1A, 2B, 3A, 3B and 6D, it also occurs on moist ground adjacent to quite different types, such as 1D, 4A, 6A, 8C and 9A.

The field layer is generally very rich, and often lacks dominants, although *Allium ursinum* may be abundant on alkaline soils. Frequent species are *Ajuga reptans*, *Allium ursinum*, *Anemone nemorosa*, *Arum maculatum*, *Athyrium filix-femina*, *Brachypodium sylvaticum*, *Cardamine flexuosa*, *Carex pendula*, *C. remota*, *C. sylvatica*,

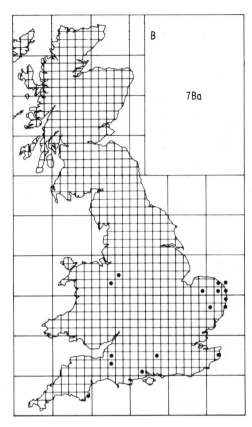

Fig. 8.21 Distribution of stand sub-type 7Ba, sump alderwoods (Dr F. Rose, personal communication).

Circaea lutetiana, *Cirsium palustre*, *Deschampsia caespitosa**, *Dryopteris dilatata*, *D. filix-mas*, *D. spinulosa*, *Endymion non-scriptus*, *Filipendula ulmaria**, *Galium aparine*, *G. palustre*, *Geranium robertianum*, *Geum rivale*, *G. urbanum**, *Glechoma hederacea*, *Hedera helix*, *Heracleum sphondylium*, *Holcus mollis*, *Lonicera periclymenum*, *Lysimachia nemorum*, *Mercurialis perennis**, *Oxalis acetosella*, *Poa trivialis*, *Potentilla sterilis*, *Primula vulgaris*, *Ranunculus ficaria*, *R. repens*, *Rubus fruticosus**, *Stachys sylvatica*, *Stellaria holostea*, *Urtica dioica*, *Valeriana officinalis*, *Veronica montana* and *Viola riviniana*.

8.8.2 Stand type 7B. Wet valley alderwoods

Valley alderwoods with the watertable permanently or seasonally at or near the soil surface. Compared with sub-type 7Ab, which also grows on neutral or

151

Fig. 8.22 Alder carr (stand type 7Ba) at Bure Marshes, Norfolk, with a field layer of *Iris pseudacorus*. Grading into open marsh beside Ranworth and Hoveton Great Broads, this is woodland at one of its edaphic limits. (Photo. Nature Conservancy Council)

alkaline soils, *Salix cinerea* is commoner and often abundant; *Ribes nigrum* is characteristic but rare; but other species are much reduced. *Betula pubescens*, *Corylus* and *Quercus robur* are sparse; *Acer campestre* and *Ulmus glabra* are absent; and even *Fraxinus* is usually reduced to scattered saplings and depauperate individuals. Alder is almost always dominant. Few wet alderwoods are acid: they are confined to catchments which are wholly deficient in bases and often contain *Myrica gale*. The majority have a pH above 5.5 and might properly be termed 'fen alderwoods'. The wetter sites possess permanent pools and an amorphous, organic soil, but stands transitional to type 7A have a shallow, freely-drained surface horizon. The alder carrs around the Norfolk Broads are perhaps the largest and most famous examples, but swampy alderwoods occur throughout Britain on the fringes of lakes, streams and in depressions. They represent an extreme type at the junction between woodland and reedswamp or bog. Klötzli (1970) would place most examples in his *Osmundo–Alnetum*.

The field layer is essentially shaded marsh vegetation containing many species virtually confined, at least in woodland, to this type. Frequent

152

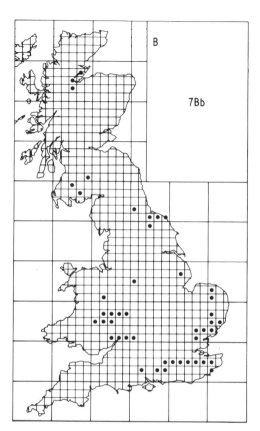

Fig. 8.23 Distribution of stand sub-type 7Bb, base-rich springline alderwoods (Dr F. Rose, personal communication).

vulgaris, *Mentha aquatica*, *Peucedanum palustre*, *Phragmites communis*, *Stellaria alsine*, *Thelypteris palustris*, *Valeriana officinalis* and *V. dioica*. Wet, base-poor alderwoods are characterised by *Eriophorum angustifolium*, *Molinia caerulea*, *Narthecium ossifragum*, *Pinguicula vulgaris* and *Potamogeton polygonifolius* as well as various *Sphagnum* species (McVean, 1956).

Dr F. Rose (personal communication) considers that the following three sub-types can be recognised:

(1) 7Ba. Sump alderwoods in depressions where water movement is mainly up and down within the ooze. The alder carrs of the Norfolk Broads are typical (Figs. 8.21 and 8.22).

(2) 7Bb. Base-rich springline alderwoods of small stream valleys and gentle slopes below springs in which water movement is mainly lateral. This sub-type usually has a pH above 6.0 and characteristically contains *Cardamine amara*, *Carex acutiformis*, *C. strigosa*, *Chrysosplenium alternifolium*, *C. oppositifolium* and *Valeriana dioica*. It is common in the Weald, south-west England and parts of East Anglia (Fig. 8.23).

(3) 7Bc. Base-poor springline alderwoods which differ from the previous sub-type in having a pH about 5.5 and characteristically supporting *Carex laevigata*, *Chrysosplenium oppositifolium* and *Sphagnum palustre*. This is common on the Wealden Sands (Fig. 8.24)

8.8.3 Stand type 7C. Plateau alderwoods

Alderwoods on flat, plateau positions. *Betula pubescens*, *Corylus*, *Fraxinus*, *Quercus robur* and *Salix cinerea* are in most stands, and *Acer campestre*, *Crataegus monogyna* and *Lonicera* are also common, but *Sambucus* is rare. The coppice is usually a mixture of alder, hazel and ash in sub-equal amounts, with some birch, maple and oak coppice and standard oak, ash or birch. These stands generally occur where the soil is heavy enough and the plateau extensive and flat enough to maintain a seasonally high watertable, but not so heavy that drainage in the top 20 cm or so is impeded. Such conditions are found mainly in the gently undulating landscape of East Anglia, where many soils have a light surface horizon over a heavy sub-soil – a fine example forms much of the Bradfield Woods (Suffolk) – but outlying examples are known

species are *Adoxa moschatellina*, *Ajuga reptans*, *Angelica sylvestris*, *Athyrium filix-femina*, *Caltha palustris*, *Cardamine flexuosa**, *Carex acutiformis*, *C. paniculata*, *C. remota*, *Chrysosplenium oppositifolium*, *Circaea lutetiana*, *Deschampsia caespitosa*, *Dryopteris dilatata**, *D. spinulosa*, *Filipendula ulmaria**, *Galium aparine*, *G. palustre*, *Geranium robertianum*, *Geum urbanum*, *Hedera helix*, *Iris pseudacorus*, *Juncus effusus*, *Lysimachia nemorum*, *Mercurialis perennis*, *Oenanthe crocata*, *Oxalis acetosella*, *Poa trivialis**, *Primula vulgaris*, *Ranunculus ficaria**, *R. repens**, *Rubus fruticosus*, *Silene dioica*, *Solanum dulcamara*, *Urtica dioica*, *Veronica montana* and *Viola riviniana*. Other characteristic species include *Calamagrostis canescens*, *Carex strigosa*, *Chrysosplenium alternifolium*, *Cirsium palustre*, *Crepis paludosa*, *Epilobium hirsutum*, *E. palustre*, *Eupatorium cannabinum*, *Galium uliginosum*, *Humulus lupulus*, *Lysimachia*

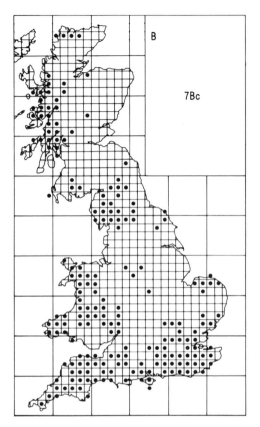

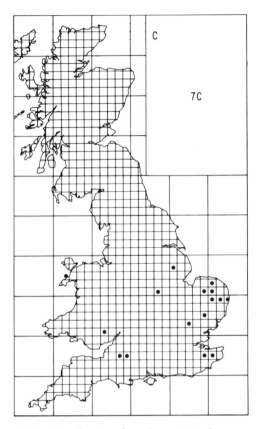

Fig. 8.24 Distribution of stand sub-type 7Bc, base-poor springline alderwoods (Dr F. Rose, personal communication).

Fig. 8.25 Distribution of stand type 7C, plateau alderwoods.

from near Longleat (Somerset) (Fig. 3.3), Coed Elernion on the Lleyn Peninsula and formerly from Woolwick Wood (E. Kent) (Fig. 8.25). Soils vary in texture from loamy sand to clay loam, pH 4.0–7.3, and have mull humus. Adjacent stand types are generally 2A, 3A or 6D, but exceptionally 4A, 5A, 6A and 6C.

The field layer lacks most of the species of wet, nutrient-rich soils which characterise valley alder-woods, and may be dominated by many of the following frequent species: *Ajuga reptans, Angelica sylvestris, Arum maculatum, Brachypodium sylvaticum, Carex sylvatica, Circaea lutetiana, Deschampsia caespitosa*, Dryopteris dilatata, D. filix-mas, D. spinulosa, Endymion non-scriptus, Filipendula ulmaria, Geum rivale, G. urbanum, Holcus mollis, Lonicera periclymenum, Luzula pilosa, Lysimachia nemorum, Mercurialis*

perennis*, Oxalis acetosella, Poa trivialis, Potentilla sterilis, Primula vulgaris, Ranunculus ficaria, Rubus caesius, R. fruticosus*, Valeriana officinalis and *Viola riviniana*. In addition, *Primula elatior* replaces *P. vulgaris* in this type in Suffolk.

8.8.4 Stand type 7D. Slope alderwoods

Valley alderwoods of type 7Ab commonly extend a few metres up nearby slopes, presumably in response to regular flushing in the sub-soil, but occasionally this development is exaggerated to the extent that 'hanging alderwoods' are formed, i.e. alder wood-land on sloping ground which carries no streams. (The strips of alderwood beside hillside streams are valley alderwoods.) This stand type has been recorded mainly in the north west (Fig. 8.26), where sloping ground is presumably more likely to remain perpetu-ally moist in the oceanic climate. Flushing must,

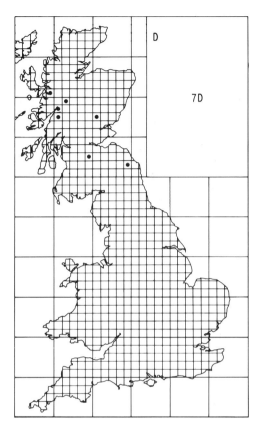

Fig. 8.26 Distribution of stand type 7D, slope alderwoods.

however, remain a factor, for slope alderwoods occur mainly on middle and lower slopes, leaving the presumably less-irrigated upper slopes to other stand types. Most examples contain *Corylus* and *Fraxinus*, whereas *Hedera*, *Lonicera* and *Salix cinerea* are rare. Otherwise, their composition reflects their predominantly north-western occurrence in that *Prunus padus*, *Salix aurita*, *Sorbus aucuparia* and *Ulmus glabra* are present and *Quercus robur* is rare. The type grows mainly on medium-textured, mildly acid–neutral (pH 3.8–6.1 is recorded) soils with a mull humus. Good examples occur at Glasdrum and Ballochuilish (Argyll) and in the Middle Clyde valley. Most examples were managed as simple coppice in which hazel, alder, ash and birch were the main constituents.

The field layer is usually an intimate mixture of many species without clear dominants, which has been fairly heavily grazed. Frequent species are *Agrostis tenuis*, *Ajuga reptans**, *Anemone nemorosa*, *Angelica sylvestris*, *Anthoxanthum odoratum*, *Athyrium filix-femina**, *Blechnum spicant*, *Brachypodium sylvaticum*, *Cardamine flexuosa*, *Carex remota*, *C. sylvatica*, *Cirsium palustre*, *Conopodium majus*, *Crepis paludosa*, *Deschampsia caespitosa**, *Digitalis purpurea*, *Dryopteris austriaca*, *D. borreri*, *D. filix-mas*, *Endymion non-scriptus*, *Filipendula ulmaria*, *Geranium robertianum*, *Geum rivale*, *Juncus effusus*, *Lysimachia nemorum**, *Mercurialis perennis*, *Poa trivialis*, *Potentilla erecta*, *P. sterilis*, *Primula vulgaris**, *Prunella vulgaris*, *Ranunculus acris*, *R. ficaria*, *R. repens*, *Rubus fruticosus**, *R. ideaus*, *Succisa pratensis*, *Taraxacum officinale*, *Teucrium scorodonia*, *Thelypteris oreopteris*, *T. phegopteris*, *Veronica chamaedrys*, *V. montana*, *V. officinalis* and *Viola riviniana*.

8.8.5 Stand type 7E. Bird cherry–alderwoods

Valley alderwoods containing *Prunus padus*. These are scattered through upland Britain (but not south-west England) and Norfolk (Fig. 8.27). The Norfolk examples are distinct enough to be recognised as a separate sub-type.

(a) Sub-type 7Ea. Lowland bird cherry–alderwoods

Bird cherry–alderwoods in lowland England. *Betula pubescens*, *Corylus*, *Fraxinus*, *Lonicera*, *Quercus robur* and *Salix cinerea* are present in most stands, whilst *Sorbus aucuparia* is frequent. This sub-type is virtually confined to Norfolk, where it is characteristic of shallow valleys in sandy soils with a wide pH range (3.7–6.6 recorded), freely-drained surface horizons and a mull humus. Its composition resembles that of valley alderwoods on acid, sandy soils (7Aa) by the high frequency of *Lonicera*, but differs in the examples seen by the abundance of *Salix cinerea*. Most stands have been managed as oak–ash standards over alder–bird cherry–hazel coppice. Good examples occur in central Norfolk at Swanton Novers, Barney and Hindolveston. Bird cherry often extends on to slightly drier ground than alder to form a zone of sub-type 7Ea which actually lacks alder. This zone is notably extensive in Wayland Wood (Norfolk), where much of the coppice is a mixture of hazel and bird cherry. The adjacent stand types in these woods include types 2B, 5B and 6C.

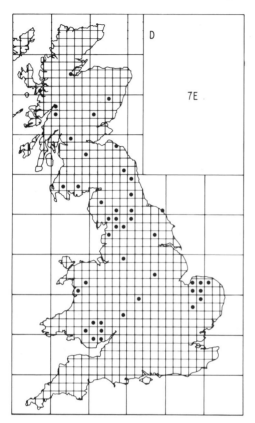

Fig. 8.27 Distribution of stand type 7E, bird cherry–alderwoods.

Many of the following frequent species may be abundant in the field layer: *Ajuga reptans**, *Allium ursinum*, *Anemone nemorosa*, *Angelica sylvestris*, *Arum maculatum*, *Chrysosplenium oppositifolium*, *Deschampsia caespitosa*, *Dryopteris dilatata*, *Endymion non-scriptus**, *Filipendula ulmaria*, *Galeobdolon luteum**, *Geum rivale*, *Glechoma hederacea*, *Lysimachia nemorum*, *Mercurialis perennis**, *Poa trivialis*, *Primula vulgaris*, *Rubus fruticosus**, *R. ideaus*, *Silene dioica*, *Teucrium scorodonia* and *Urtica dioica*.

(b) Sub-type 7Eb. Upland bird cherry–alderwoods

Bird cherry–alderwoods in the uplands of Britain are closely related to types 7A, 7B and those examples of 7D which occur on gently sloping ground. Further investigation may show that 7Eb is not worth recognising as a distinct form. However, the few

examples seen were all on flat ground where the soil is permanently waterlogged and contains a high proportion of organic matter. The examples recorded were on medium-textured, neutral soils. These conditions clearly resemble those of type 7Ab in the uplands, but *Betula pendula* was commoner than *B. pubescens*, *Ulmus glabra* was present and *Sorbus aucuparia* was absent. Stands of 7Eb occur mainly as derelict alder coppice in small patches within upland birch–oakwoods (6A and 6B), and can be regarded as an alderwood counterpart of types 1D, 3D, 6A and 6B.

Frequent species in the field layer of recorded examples were *Ajuga reptans*, *Allium ursinum*, *Athyrium filix-femina**, *Cardamine flexuosa*, *Carex remota*, *Chrysosplenium oppositifolium*, *Cirsium arvense*, *Deschampsia caespitosa*, *Dryopteris dilatata**, *D. filix-mas*, *D. spinulosa*, *Filipendula ulmaria**, *Galium aparine*, *G. palustre*, *Geranium robertianum*, *Geum rivale*, *Hedera helix*, *Juncus conglomeratus/effusus*, *Lysimachia nemorum*, *Mentha aquatica*, *Mercurialis perennis*, *Oxalis acetosella*, *Poa trivialis*, *Ranunculus repens**, *Rubus ideaus*, *Silene dioica*, *Urtica dioica*, *Valeriana officinalis* and *Veronica montana*.

8.9 Beech woodland (Group 8)

Stands containing *Fagus sylvatica*. *Fagus* is usually associated with *Ilex* and *Quercus*, but does not normally grow with *Alnus*, *Populus tremula*, *Ulmus carpinifolia* and *U. procera*. In Epping Forest, the north-western end of the Chilterns and south-east England *Fagus* occasionally occurs with *Carpinus* in stands which are intermediate between beechwoods and hornbeam woods.

Within its native range, in south England and Wales, beech occurs on both acid and calcareous soils, most of which are light to medium in texture. Occasionally it occurs on heavy soils or poorly-drained sites, but it is remarkably rare on neutral soils: only 6 of the 89 samples containing beech had a pH between 5.0 and 6.5. Well-defined concentrations of beechwoods occur on the chalk and limestones of the South Downs, North Downs, Chilterns, Cotswolds, Lower Wye Valley and south Wales, but not all the stands in these areas are on calcareous soils. Indeed, in the Chilterns, where most of the beechwoods grow on the clay-with-flints of the dip slope, the majority of beechwoods are on strongly acid soils. Acid beechwoods are more diffusely scattered through the Weald and on the recent deposits in the Hampshire and London Basins, but here too there are well-marked concentrations in old forest areas such as Epping and New Forests. Further west, acid beechwoods are found in the Forest of Dean and scattered through south-west England to Cornwall and south Wales. Beech has been widely planted beyond its apparent native range and has, like sycamore, become naturalised especially in the acid oakwoods and valley ash–wych elm woods of northern and western Britain.

The occurrence of beech has been substantially affected by management. Beech produces weak coppice shoots and is now infrequent in coppice woods, except in the Cotswolds and further west. On the other hand beech appears to be favoured by wood-pasture management, which eliminates many of its competitors, for it can withstand pollarding, live almost as long as oak and regenerate well under shade whenever the grazing pressure is sufficiently reduced. Beech is also a valuable timber tree which is grown as high forest throughout its native range. Many of these high-forest stands are plantations, but some at least occupy ancient woodland sites where beech was once a component of the preceding coppices.

Tansley (1939) divided beechwoods into *Fagetum ericetosum*, *F. rubosum* and *F. calcicolum*, respectively beechwoods on sands, loams and calcareous soils. These types were based on the work of Adamson (1921) and Watt (1924; 1925; 1934). Watt in particular had recognised various types within these broad categories which were based on associated differences in soils, flora and performance of beech.

The stand types recognised here (Table 8.8) broadly follow Tansley's scheme. The main division is into oak–beechwoods and ash–beechwoods, respectively the beechwoods of acid, base-poor soils, which are equivalent to *Fagetum ericetosum*, and the beechwoods of calcareous and base-rich soils, which are equivalent to *F. calcicolum* and parts of *F. rubosum*. The acid oak–beechwoods are divided into sessile oak (8A) and pedunculate oak (8B) stand types, even though the differences between them are small. The ash–beechwoods are rather more complex, but the

Table 8.8 Composition and site features of beech woodland (Group 8).

	8A	8B	8Ca	8Cb	8Cc	8D	8Ea	8Eb
Number of samples	8	4	9	5	8	1	10	4
Acer campestre			IV	II	IV	1	I	1
Betula pendula	II	1	II		I		II	
Betula pubescens	I	2	II				III	1
Corylus avellana	II	3	V	III	IV	1	III	3
Fagus sylvatica	V	4	V	V	V	1	V	4
Fraxinus excelsior			V	V	IV	1	V	4
Quercus petraea	V						V	4
Quercus robur	+	4	IV	I	IV	1	II	
Tilia cordata			III	I			III	
Ulmus glabra			IV	V		1	II	4
Acer pseudoplatanus			III	IV	II		I	
Buxus sempervirens					+			
Castanea sativa	II						+	
Clematis vitalba			III	I	I		II	2
Crataegus monogyna		1	IV	III	IV	1	III	1
Crataegus oxyacanthoides			I					
Daphne laureola			III	+			II	2
Euonymus europaeus			II	I	I			
Frangula alnus	I							
Hedera helix	III	1	IV	II	III	1	IV	3
Ilex aquifolium	IV	3	III	III	II	1	III	3
Juniperus communis			I		II			
Ligustrum vulgare			II		I			1
Lonicera periclymenum	IV	2	III	I	I	1	III	2
Malus sylvestris		+	I	II	I			
Prunus avium	I		I	I		1	+	
Prunus spinosa			I	I			+	
Rhododendron ponticum		1						
Ribes sylvestre				I				
Ribes uva-crispa						1		
Rosa arvensis			III	II		1	II	1
Rosa canina			I		I		+	
Salix caprea			II		I		I	
Salix cinerea					I		+	
Sambucus nigra			I	IV	II			1
Sorbus aria	I		II		II		+	2
Sorbus aucuparia	IV	2	II	I				
Sorbus torminalis								1
Taxus baccata		1	III	II	I		I	3
Thelycrania sanguinea	I		III	II	I			
Tilia platyphyllos							+	3
Viburnum lantana					+			
Viburnum opulus				I	I			
Median pH	4.3	4.0	7.5	7.6	7.9	4.6	4.9	7.0
Mean texture	1.0	1.5	2.4	2.8	2.8	3.0	2.1	2.0
Drainage free	7	3	9	3	8		6	4
imperfect		1		2			2	
poor							2	
very poor	1					1		
Site	sr	s	S	R	S	p	s	S

157

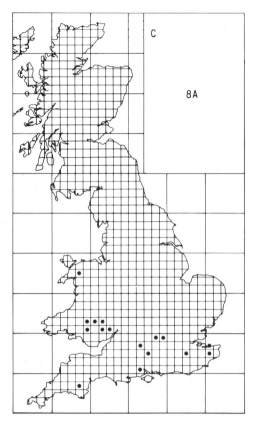

Fig. 8.28 Distribution of stand type 8A, acid sessile oak–beechwoods.

aucuparia are common. *Corylus*, if present, is usually sparse and poorly grown, and other species are rare, or absent. *Frangula* is occasionally found, and in south and south-east England *Sorbus aria* is a rare but characteristic component. This type is virtually confined to light, freely-drained, strongly acid soils which have generally developed a mull or moder humus which is 5–10 cm thick. Beech and oak are generally dominant, but birch will become abundant if a stand is disturbed. Many coppice stands contain *Castanea* and indeed some almost pure chestnut coppices (e.g. in and around the Blean woods of north-east Kent) have probably replaced this stand type. Acid sessile oak–beechwoods are scattered through south-west England, south Wales and south-east England (Fig. 8.28), but are rare in the Chilterns. Good examples occur at Holne Chase on the fringes of Dartmoor, the Hudnalls on the Lower Wye Valley and Coed y Person near Abergavenny. Wood-pasture forms are found in the New Forest. This natural range of acid sessile oak–beechwoods has been greatly extended by planting and seeding of beech into acid birch–sessile oakwoods throughout south-west England, Wales, northern England and central Scotland. Ecologically this sub-type is most closely related to 6Ab, 6Cb and 8B.

Under wood-pasture stands the field layer is exceptionally poor, but beneath coppice it is reasonably well developed, often with *Pteridium aquilinum* as the most abundant species. Frequent species are *Anemone nemorosa*, *Anthoxanthum odoratum*, *Athyrium filix-femina* (western area only), *Blechnum spicant**, *Deschampsia caespitosa*, *D. flexuosa**, *Digitalis purpurea*, *Dryopteris dilatata*, *D. borreri*, *D. filix-mas*, *Endymion non-scriptus*, *Hedera helix*, *Holcus mollis*, *Lonicera periclymenum*, *Luzula pilosa*, *L. sylvatica*, *Melampyrum pratense*, *Oxalis acetosella*, *Pteridium aquilinum**, *Rubus fruticosus*, *Solidago virgaurea*, *Teucrium scorodonia* and *Vaccinium myrtillus*. *Ruscus aculeatus* is found in some stands, including wood-pasture examples.

most satisfactory division is into sessile oak–ash–beechwoods (8E) and pedunculate oak–ash–beechwoods, the latter being further sub-divided into calcareous (8C) and acid (8D) types. The sessile oak–ash–beechwoods are also divided into acid (8Ea) and calcareous (8Eb) forms, which are retained as sub-types rather than types because they tend to occur together without sharp edaphic or floristic discontinuities between them. The acid oak–beechwoods fall mainly within Merlewood types 17, 18, 22 and 23, whereas the calcareous ash–beechwoods fall within types 1, 5 and 7.

8.9.1 Stand type 8A. Acid sessile oak–beechwoods

Beech woodland containing *Quercus petraea* but no *Fraxinus*. *Betula*, *Hedera*, *Ilex*, *Lonicera* and *Sorbus*

8.9.2 Stand type 8B. Acid pedunculate oak–beechwoods

Beech woodland containing *Quercus robur*, but not *Fraxinus* or *Q. petraea*. As in type 8A, the only common associates are *Betula*, *Corylus*, *Ilex*, *Lonicera*

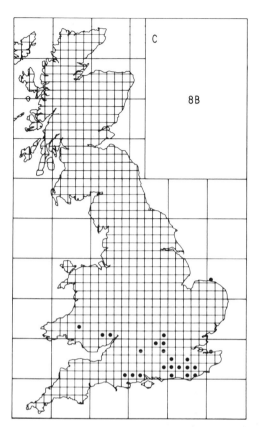

Fig. 8.29 Distribution of stand type 8B, acid pedunculate oak–beechwoods

and *Milium effusum*. Other characteristic species include *Blechnum spicant*, *Oxalis acetosella* and *Vaccinium myrtillus*. Watt (1934) describes the flora of the plateau Chiltern beechwoods in considerable detail.

8.9.3 Stand type 8C. Calcareous pedunculate oak–ash–beechwoods

Beechwoods on calcareous soils containing *Fraxinus* and *Quercus robur*, but not *Q. petraea*. *Hedera* and *Ilex* are common, as in all types of beechwood, but *Lonicera* is infrequent. Several species are commoner here than in other types: *Acer campestre*, *A. pseudoplatanus*, *Euonymus*, *Malus*, *Sambucus*, *Thelycrania* and perhaps *Crataegus monogyna*. *Betula* is relatively rare. Numerous calcicole shrubs are associated with these woods, though they may be rare and confined to margins and canopy gaps in beech-dominated stands. Soils are invariably freely drained.

Type 8C is especially associated with the chalk and limestone scarps of southern England, though it extends into south Wales (Fig. 8.30). Stands in the east of this range are almost all managed as high forest. Coppice stands remain a small minority in the Chilterns, but are fairly common in the Cotswolds, and southern Welsh borderland. Three ill-defined sub-types are recognised. Stands containing *Ulmus glabra*, which grow on deep soils at the foot of slopes (8Cb), are distinguished from those on essentially dry sites. The latter are split into a predominantly western form (8Ca) containing *Tilia* or *Ulmus glabra* and a predominantly English form which lacks these species.

(a) Sub-type 8Ca. Dry lime–wych elm variant

Calcareous pedunculate oak–ash–beechwoods containing *Tilia* or *Ulmus glabra* and growing on essentially dry sites. In all recorded examples the lime was *T. cordata*. Numerous species occur commonly: indeed, the coppice stands of this sub-type are amongst the richest arboreal communities in Britain. Several species occur here which are equally characteristic of beechwoods containing *Q. petraea* (*Betula pendula*, *Sorbus aucuparia* and *Taxus*), or of dry beechwoods generally (*Sorbus aria*). Compared with the maple variant (8Cc), this sub-type contains more

and *Sorbus aucuparia*, but *Hedera* is rare. This type is confined to strongly acid, light soils with a deep mor or moder humus on soils which are either freely or poorly drained. They are rarely found as coppice (e.g. Pishill Bank in the Chilterns and Nap Wood in the Sussex Weald) but are widespread as high forest and wood pasture in the English lowlands (Fig. 8.29). Most of the plateau beechwoods of the Chilterns fall within this type (Fig. 4.5), though many are probably improved versions of type 8D. Beech and oak are often co-dominant, but most high-forest stands are dominated by beech. Ecologically this type is most closely related to type 6D.

The field layer, which is often extremely poor, may be dominated by *Rubus fruticosus*, *Pteridium aquilinum*, *Holcus mollis*, *Deschampsia caespitosa* or *D. flexuosa*. Frequent species also include *Carex pilulifera*, *Dryopteris dilatata*, *Lonicera periclymenum*, *Luzula pilosa*

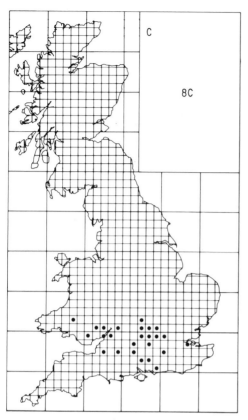

C

8C

Fig. 8.30 Distribution of stand type 8C, calcareous pedunculate oak–ash–beechwoods.

Clematis, Daphne laureola, Sorbus aucuparia and *Taxus*, and less *Rosa arvensis*, as well as the lime and wych elm which are present by definition. Soils may be shallow, but more often are deep with numerous limestone fragments throughout the profile and a high proportion of organic matter. This sub-type is mostly found in the Lower Wye Valley and south Wales where good examples occur at Coombe Woods (Gwent) and James Thorn (Glos) but it is also recorded from the Cotswolds scarp near Birdlip, one locality in the southern Chilterns and the upper slopes of Wealden edge hangers in Hampshire. Sub-type 8Ca is a beech counterpart of type 1A.

Hedera helix and *Mercurialis perennis* generally dominate the field layer. Frequent species are *Allium ursinum, Anemone nemorosa, Arum maculatum*, *Brachypodium sylvaticum, Carex sylvatica, Deschampsia caespitosa, Dryopteris dilatata, D. filix-mas, Endymion*

non-scriptus*, Galeobdolon luteum, Galium odoratum, Geum urbanum, Hedera helix*, Lonicera periclymenum, Mercurialis perennis*, Phyllitis scolopendrium, Polypodium interjectum, Polystichum aculeatum, Potentilla sterilis, Primula vulgaris, Pteridium aquilinum, Rubus fruticosus*, Sanicula europea, Taraxacum officinale, Viola reichenbachiana and *Viola riviniana**. Several other calcicoles occur sparingly including *Carex flacca, Convallaria majalis, Hypericum hirsutum, Inula conyza* and *Viola hirta*.

(b) Sub-type 8Cb. Moist wych elm variant

Calcareous pedunculate oak–ash–beechwoods containing *Ulmus glabra* and growing on deep, freely-drained loams in receiving sites at the base of chalk and limestone slopes. *Acer pseudoplatanus* and *Sambucus* are common, and some species which are characteristic of heavy soils are relatively well represented (*Malus, Rosa arvensis*), but *Q. robur* was rare in the examples recorded. Species such as *Betula pendula, Clematis, Juniperus* and *Sorbus aria*, which are characteristic of dry soils, are rare or absent. In most examples only hazel, ash and wych elm occurs as maiden trees. This sub-type is scattered through the Chilterns and Cotswolds, but appears to be rare in both the South and North Downs and in the Welsh borderland. Examples have been recorded at Norbury Park (Surrey), Shillingridge Wood (Oxon), Buckholt Woods (Glos), Coombe Woods (Gwent) and the lower slopes of the Wealdon edge hangers (Hants). It is the beech counterpart of type 1A.

The field layer is usually dominated by *Mercurialis perennis*, sometimes accompanied by *Allium ursinum* or *Urtica dioica*. Frequent species are *Allium ursinum, Anemone nemorosa, Arum maculatum*, Brachypodium sylvaticum, Carex sylvatica, Deschampsia caespitosa, Dryopteris filix-mas, Endymion non-scriptus, Galeobdolon luteum*, Galium odoratum, Geranium robertianum, Hedera helix, Melica uniflora, Mercurialis perennis*, Rubus fruticosus, Sanicula europea, Urtica dioica* and *Viola riviniana*.

(c) Sub-type 8Cc. Maple variant

Calcareous pedunculate oak–ash–beechwoods without *Tilia* or *Ulmus glabra*. These stands usually contain

Acer campestre, *Corylus* and *Crataegus monogyna*, but *Ilex* is marginally less frequent than in other beech-wood types. *Taxus* is infrequent in coppice stands, but often abundant in high-forest stands. *Buxus* appears to be confined to this type in Britain though it does not occur in any recorded coppice stands. These beechwoods often cover the slopes on the escarpments of the chalk and Cotswold oolite, where the soil commonly takes the form of a rendzina. Beech is usually the most abundant species and is generally completely dominant, but where the beech is more scattered there is a rich association of calcicole trees and shrubs. This is by far the commonest form of ash–beechwood in the Chilterns and Downs, but in all these areas most stands have been promoted to high forest in which beech is dominant. Good examples of coppice stands remain at Hillock Woods (Bucks), Maidensgrove Scrubs (Oxon), Conygre Wood and Chatcombe Wood (Glos). This sub-type is the beechwood counterpart of types 2C and 3B. Tansley's (1939) *Fagetum calcicolum* largely coincides with this sub-type.

The field layer may be very sparse beneath a vigorous beech canopy, but otherwise it is dominated by *Mercurialis perennis*, *Rubus fruticosus* or *Hedera helix*. Frequent species are *Anemone nemorosa*, *Arum maculatum*, *Brachypodium sylvaticum*, *Dryopteris filixmas*, *Endymion non-scriptus*, *Euphorbia amygdaloides*, *Fragaria vesca*, *Galeobdolon luteum*, *Galium aparine*, *G. odoratum*, *Hedera helix*, *Mercurialis perennis**, *Rubus fruticosus**, *Sanicula europea*, *Urtica dioica* and *Viola riviniana*. Other characteristic species include *Arctium minus*, *Carex sylvatica*, *Cephalanthera damasonium*, *Deschampsia caespitosa*, *Eupatorium cannabinum*, *Geum urbanum*, *Melica uniflora*, *Mycelis muralis*, *Potentilla sterilis*, *Viola hirta* and *Zerna ramosa*. Some rare species are found in this type: *Cephalanthera rubra*, *Monotropa hypophagea*, *Stachys alpina*.

8.9.4 Stand type 8D. Acid pedunculate oak–ash–beechwoods

Beechwoods on acid soils containing *Fraxinus* and *Quercus robur*, but not *Q. petraea*. Only one coppice stand of this type has been found, at Commonhill Wood, Bucks, where it grew on a very poorly

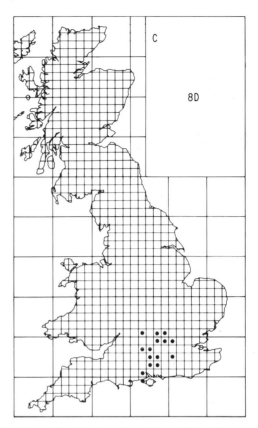

Fig. 8.31 Distribution of stand type 8D, acid pedunculate oak–ash–beechwoods.

drained clay loam of pH 4.6. The main species were *Fagus*, *Fraxinus*, *Ulmus glabra*, *Quercus robur*, *Corylus* and *Prunus avium*. This stand conformed to type A of Watt (1934), and suggests that many of the high forest beech stands of the Chilterns plateau on base-rich soils are best considered to be examples of this type (Fig. 8.31) even though ash is rare or present only as saplings. Ecologically this type is the beech counterpart of types 1B, 2A and 3A.

The field layer seldom has clear dominants. Frequent species are *Brachypodium sylvaticum*, *Carex remota*, *C. sylvatica*, *Deschampsia caespitosa*, *Festuca gigantea*, *Galeobdolon luteum*, *Galium odoratum*, *Hedera helix*, *Hordelymus europeaus*, *Luzula pilosa*, *Melica uniflora*, *Milium effusum**, *Moehringia trinervia*, *Oxalis acetosella*, *Rubus fruticosus** and *Viola riviniana*. Other characteristic species include *Lysimachia nemorum*, *Poa nemoralis*, *Ranunculus ficaria* and *Stellaria holostea*.

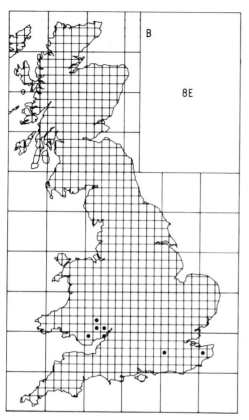

Fig. 8.32 Distribution of stand type 8E, sessile oak–ash–beechwoods.

8.9.5 Stand type 8E. Sessile oak–ash–beechwoods

Beech woodland containing *Fraxinus* and *Quercus petraea*. *Corylus*, *Hedera*, *Ilex*, *Tilia* and *Ulmus glabra* are frequent, but *Acer campestre* is rare. This type resembles both the oak–beechwoods and the lime–wych elm variant of the calcareous pedunculate oak–ash–beechwoods in its relatively high frequency of *Betula* and *Lonicera*, but differs from both in the absence of *Sorbus aucuparia*. The type occurs mostly on light, freely-drained soil and is virtually confined to south Wales and the Wye Valley area (Fig. 8.32). Two merging sub-types are recognised on respectively acid (8Ea) and calcareous (8Eb) soils.

(a) *Sub-type 8Ea. Acid sessile oak–ash–beechwoods*

Sessile oak–ash–beechwoods on acid soils, most of

which lie in the range pH 4.5–5.5. Recorded stands contained *Betula pendula*, *Quercus robur* and *Tilia cordata*, which were not found in 8Eb. Some species are present which are normally regarded as calcicole: *Clematis*, *Daphne laureola*. The coppice species are usually beech, oak, hazel and lime, but not ash. Most examples occurred on light, freely-drained soils where superficial acid deposits covered underlying limestone, as at Garth Wood (Glam) and Ashcombe Wood on the North Downs (Surrey). Alternatively, they occur on sandstone in the vicinity of limestone, as at the Hudnalls and James Thorns (Glos) at the lower end of the Wye Gorge. One of the best examples lies at middle levels in Lady Park Wood, which is also on steep ground overlooking the Wye. This sub-type is transitional between 8Eb and 8Ca on the calcareous side and 8A on the strongly acid side. It is the beechwood counterpart of types 1D, 3D and 5B.

The most likely dominants in the field layer are *Hedera helix*, *Deschampsia caespitosa* and *Luzula sylvatica*. Frequent species are *Blechnum spicant*, *Brachypodium sylvaticum*, *Carex sylvatica*, *Deschampsia caespitosa*, *Dryopteris dilatata*, *D. filix-mas**, *Endymion non-scriptus**, *Galeobdolon luteum*, *Hedera helix*, *Lonicera periclymenum*, *Luzula pilosa*, *L. sylvatica*, *Mercurialis perennis*, *Polystichum setiferum*, *Pteridium aquilinum* and *Rubus fruticosus**. Other characteristic species include *Athyrium filix-femina*, *Circaea lutetiana*, *Oxalis acetosella*, *Tamus communis* and *Viola riviniana*.

(b) *Sub-type 8Eb. Calcareous sessile oak–ash–beechwoods*

Sessile oak–ash–beechwoods growing on limestone. The only examples recorded lie along the Wye Gorge from Symonds Yat and Lady Park Wood to the Blackcliff and Wyndcliff, invariably on precipitous limestone crags where the humus-rich soil collects only in pockets and cracks in the rocks. Characteristically these stands consist of a mature, irregular canopy, but most appear to have been coppiced. *Taxus* and *Ulmus glabra* are common, and *Tilia platyphyllos* tends to be the characteristic lime. *Sorbus aria* is normally present, together with some rare endemic *Sorbi* and *S. torminalis*. *Corylus*, if present, is often infrequent. This is the beechwood counterpart of types 1Cb, 3C and 4C.

The field layer, which is sometimes dominated by *Hedera helix*, is an extraordinary mixture of calcicole and calcifuge species, such as *Deschampsia flexuosa*, *Digitalis purpurea* and *Solidago virgaurea* (which are rare) and *Melica uniflora, Mercurialis perennis, Mycelis muralis* and *Phyllitis scolopendrium* (which are frequent). Species which are characteristic of rock faces are found, such as *Asplenium trichomanes* and *Umbilicus rupestris*. Other frequent species are *Carex sylvatica, Deschampsia caespitosa, Dryopteris filix-mas, Endymion non-scriptus, Galeobdolon luteum, Hypericum perforatum, Primula vulgaris, Pteridium aquilinum, Rubus fruticosus, Urtica dioica* and *Viola riviniana*. Rare species such as *Aquilegia vulgaris*, and *Hordelymus europeaus* are occasionally found.

8.10 Hornbeam woodland (Group 9)

Stands containing *Carpinus betulus*. The most frequent associates are *Betula pubescens, Corylus, Lonicera* and *Quercus*. If *Alnus, Fagus, Ulmus carpinifolia, U. procera* or *Pinus* are present, the stand is intermediate between group 9 and another group, but such intermediates are infrequent. *Alnus* and *Carpinus* occasionally grow together on the margins of valleys and on plateaus where the soil is light, notably in Norfolk. Likewise, *Fagus* and *Carpinus* grow together in Epping Forest (Essex), the north-west end of the Chilterns and scattered locations in south-east England.

Hornbeam coppices vigorously but casts a dense shade. It is therefore infrequent as a standard and occurs mainly as a coppiced underwood in which it often so dominates the community that other arboreal species are thinly scattered and the field layer is absent or sparse. Most hornbeam occurs in south-east England and East Anglia, with notable concentrations in a belt to the north of London from north-west Middlesex into Essex and in the eastern Weald (Christy, 1924). Hornbeam is also scattered through the Midlands west to the Welsh borders and even occurs in Somerset and north to Lancashire in stands which have not obviously been planted. This native range resembles that of beech, but the edaphic ranges of the two species are largely complementary, with beech concentrated upon well-drained soils whilst hornbeam grows mainly on poorly-drained soils across a wide range of texture and base status. Like

Table 8.9 Composition and site features of hornbeam woodland (Group 9).

	9Aa	9Ab	9Ba	9Bb
Number of samples	15	18	13	6
Acer campestre		IV	I	V
Betula pendula	II	+		IV
Betula pubescens	IV	III	IV	III
Carpinus betulus	V	V	V	V
Corylus avellana	III	IV	II	III
Fraxinus excelsior		V	II	III
Quercus petraea			V	V
Quercus robur	V	V	+	
Tilia cordata	+	I		
Ulmus glabra		I		+
Acer pseudoplatanus			I	
Castanea sativa	+	+	+	I
Clematis vitalba		+		I
Crataegus monogyna	I	II	I	IV
Crataegus oxyacanthoides	I	II	I	I
Euonymus europaeus		I		III
Hedera helix	+	+	II	III
Ilex aquifolium	I		II	I
Ligustrum vulgare		+		I
Lonicera periclymenum	III	II	IV	IV
Malus sylvestris			I	
Populus tremula	+	II		
Prunus avium			I	I
Prunus spinosa		+		
Pyrus communis		+		
Rosa arvensis		II		I
Rosa canina	+	II		
Salix caprea	I	+		
Salix cinerea	+	II		
Sambucus nigra	II	II	I	V
Sarothamnus scoparius	+			I
Sorbus aria				I
Sorbus aucuparia			II	
Sorbus torminalis	+			
Taxus baccata			I	
Thelycrania sanguinea		II	I	
Viburnum lantana				III
Viburnum opulus	+	I	I	I
Median pH	4.1	5.4	3.7	7.6
Mean texture	1.8	2.8	2.2	2.7
Drainage free	4	1	1	1
imperfect	5	3	3	3
poor	2	7	4	2
very poor	3	6	5	
Site	ps	psr	ps	sr

Fig. 8.33 Acid pedunculate oak–hornbeam woodland (stand type 9Aa) in Mad Bess Wood, Ruislip (formerly Middlesex), growing characteristically on very acid soils supporting a poor field layer. Originally managed as coppice-with-standards oak over hornbeam, the coppice has not been cut for perhaps 40 years. Now, like so many semi-natural, former coppice woods, the stand is a mixture of 19th Century standards, coppice last cut in the 20th Century, and young trees grown from seed contemporary with the coppice.

beech the range of hornbeam has been somewhat disturbed by planting and the exact limits of its natural geographic and edaphic ranges are somewhat obscured. The boundaries of hornbeam zones within mixed coppices are more sharply defined than those of other species, but there is no evidence that this is a consequence of planting.

Hornbeam woods, unlike beech woods, do not occur in neat ecological clusters which assist in the definition of stand types, and, in the absence of anything better, the distinction made by Salisbury (1916; 1918) between *Quercus robur–Carpinus* woods and *Q. petraea–Carpinus* woods is adopted here (Table 8.9). This appeared to be a natural difference, in so far as the two types were associated with different soils, but planting of *Q. robur* had somewhat disturbed the

natural distribution. Hornbeam woods occur in a wide range of Merlewood types, the more acid stands of 9Aa and 9Ba falling mainly in types 17–24, and the remainder mainly in types 1–8.

8.10.1 Stand type 9A. Pedunculate oak–hornbeam woods

Hornbeam woodland containing *Quercus robur*, but not *Q. petraea* (Fig. 8.33). *Betula* (mostly *B. pubescens*), *Corylus* and *Lonicera* are found in most stands. *Populus tremula*, *Rosa canina*, *Salix caprea*, *S. cinerea* and *Tilia cordata* were recorded in this type, albeit rarely, but not in the sessile oak–hornbeam woods. Type 9A occurs throughout the range of hornbeam woodland in the English lowlands (Fig. 8.34), and is the only

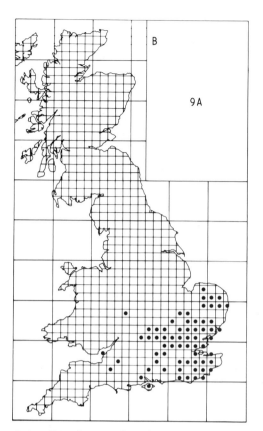

Fig. 8.34 Distribution of stand type 9A, pedunculate oak–hornbeam woods.

type of hornbeam woodland in many areas. Two sub-types are recognised on the basis of the presence and absence of *Acer campestre*, *Fraxinus* and *Ulmus glabra*, even though they merge imperceptibly into each other in general and in particular woods.

(a) *Sub-type 9Aa. Birch–hazel variant*

Pedunculate oak–hornbeam woods lacking *Acer campestre*, *Fraxinus* and *Ulmus glabra*. Such stands usually contain *Betula* (either species), *Corylus* and *Lonicera*, but other tree and shrub species are sparse or absent. Most stands occur as oak standards over coppice of hornbeam and hazel, but hornbeam is usually the most abundant species. In many neglected coppices the hazel is being killed out by the taller hornbeam. Those stands grow on light–medium-textured, strongly acid (pH 3.8–4.6) soils, most of

which are poorly drained, though the surface horizons may be well drained and covered by up to 5 cm of moder humus. Although this sub-type is found in south-east England, it seems to be more frequent in East Anglia, especially at the northern extremity of the hornbeam range in Norfolk. It is the hornbeam counterpart of stand types 5A and 6D, but it only seems to occur with lime in Essex (e.g. Chalkney Wood) and Bedfordshire (Kings Wood). Generally it occurs with types 6D, 7A and 9Ab, but in Hertfordshire it is often found in patches on plateau sites within sessile oak–hornbeam stands of sub-type 9Ba.

The field layer, which is rather poor, may be dominated by *Pteridium aquilinum*, *Rubus fruticosus*, *Endymion non-scriptus*, *Holcus mollis* or *Anemone nemorosa*. Few species are frequent: *Anemone nemorosa*, *Endymion non-scriptus*, *Holcus mollis*, *Lonicera periclymenum*, *Luzula pilosa*, *Moehringia trinervia*, *Pteridium aquilinum*, *Rubus fruticosus**, *Teucrium scorodonia* and *Viola riviniana*. In the Weald *Betonica officinalis* and *Solidago virgaurea* are frequent.

(b) *Sub-type 9Ab. Ash–maple variant*

Pedunculate oak–hornbeam woods containing *Acer campestre*, *Fraxinus* or *Ulmus glabra*. Most stands contain *A. campestre*, *Corylus* and *Fraxinus*, but *U. glabra* is rare. Calcicole species such as *Euonymus* and *Thelycrania* occur here but not in 9Aa, and many species of heavy soils are more abundant, e.g. *Crataegus monogyna*, *Populus tremula*, *Rosa* spp. Most stands have oak standards over a mixed coppice of hornbeam, ash, hazel and maple, but exceptionally lime and wych elm are present. The soils are heavier, cover a wider range of pH (4.2–7.0) and are wetter than those beneath sub-type 9Aa, and there is a significantly greater chance that this sub-type will be found in valleys and depressions. Most of the few patches of 9A in the Midlands are examples of this sub-type, which is frequent on the clay soils of the chalky boulder clay of East Anglia, the Wealden clays and on the clay-with-flints over parts of the chalk-lands. The sub-type is the hornbeam counterpart of stand types 1B, 2A, 3A and 4A, but it mostly occurs with types 2A, 3A, 7B and 9Aa, and on the wettest ground in woods composed mainly of type 9Ba.

The field layer may be dominated by *Mercurialis perennis* on calcareous soils, *Rubus fruticosus* on heavy soils, *Anemone nemorosa* or *Endymion non-scriptus* on light soils, and *Carex pendula*, *Filipendula ulmaria* or *Primula elatior* on wet soils. Clearly, the stand type covers a range of field layer types, in which the frequent species are *Ajuga reptans*, *Carex remota*, *C. sylvatica**, *Circaea lutetiana*, *Deschampsia caespitosa*, *Endymion non-scriptus*, *Epilobium montanum*, *Fragaria vesca*, *Galeobdolon luteum*, *Geum urbanum*, *Glechoma hederacea*, *Lonicera periclymenum*, *Mercurialis perennis*, *Poa trivialis**, *Potentilla sterilis*, *Primula vulgaris*, *Rubus fruticosus**, *Sanicula europea*, *Veronica chamaedrys* and *Viola riviniana**.

8.10.2 Stand type 9B. Sessile oak–hornbeam woods

Hornbeam woodland containing *Quercus petraea*. *Betula pubescens* is common, and *Corylus*, *Hedera* and *Lonicera* are frequent. This type is virtually confined to either strongly acid or alkaline soils: all but two of the 19 samples recorded had a pH either below 4.7 or above 7.3. This striking bimodality is the basis of two sub-types.

(a) *Sub-type 9Ba. Acid sessile oak–hornbeam woods*

Sessile oak–hornbeam woods on acid soils. *Betula pubescens* is common but *B. pendula* was not recorded. Apart from *Lonicera*, other species are rare or absent, but *Sorbus aucuparia* and *Taxus* were recorded only from this form of hornbeam woodland. Most stands have oak standards over coppice, which is generally dominated by hornbeam, but also contains scattered hazel and oak. The light–medium-textured soils are usually exceptionally acid (pH 3.4–3.8) and poorly drained with up to 15 cm of surface litter and mor-moder humus. The best examples are in Hertford-shire, where good stands survive at Sherrards Park Wood and Wormley Wood, but the sub-type is also found occasionally in the Weald (Fig. 8.35). Some of the chestnut coppices of Kent may have been formed in place of this type. This sub-type usually grades into type 6Cb on drier, lighter soils and 9A on heavier, wetter soils. Ecologically, it is the hornbeam equivalent of 6C. One outlying stand in Lords Grove

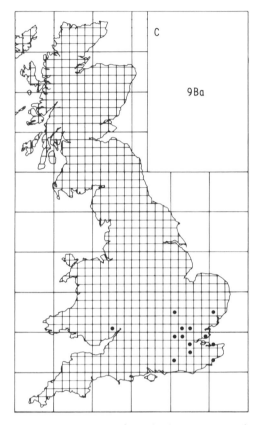

Fig. 8.35 Distribution of stand sub-type 9Ba, acid sessile oak–hornbeam woods.

(Gwent) on the Wye Valley sandstones was exceptional in occupying a steeply-sloping, freely-drained site, but floristically it was similar to stands in south-east England. Another fragment of this type in the same region, at Chaddesley Wood (Worcs), hung on to a steep streamside bank.

The field layer may be almost absent under dense hornbeam and only becomes rich beside streams. *Pteridium aquilinum* and *Rubus fruticosus* are the most likely dominants. Frequent species are *Anemone nemorosa*, *Carex sylvatica*, *Circaea lutetiana*, *Deschampsia caespitosa*, *Dryopteris dilatata*, *D. filix-mas*, *Hedera helix*, *Holcus mollis*, *Lonicera periclymenum*, *Luzula pilosa*, *L. sylvatica*, *Pteridium aquilinum* and *Rubus fruticosus**.

(b) *Sub-type 9Bb. Calcareous sessile oak–hornbeam woods*

Sessile oak–hornbeam woods on alkaline soils. *Acer*

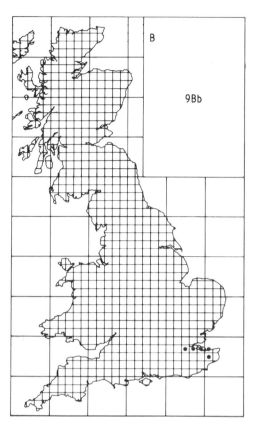

Fig. 8.36 Distribution of stand sub-type 9Bb, calcareous sessile oak–hornbeam woods.

campestre and *Betula pendula* are common in this sub-type but absent from 9Ba, and several other species are more abundant in this sub-type, e.g. *Crataegus monogyna*, *Euonymus*, *Fraxinus*, *Sambucus* and *Viburnum lantana*. Several calcicole shrubs occur sparingly within the stand and more abundantly on the margins. Most stands have oak standards over mixed coppice containing much hornbeam and scattered maple, ash, hazel, oak and birch. This type has only been recorded from north Kent (Fig. 8.36), where it appears to be restricted to the junction between the Thanet Sand and exposures of chalk at the base of slopes. Here the loamy soil contains chalk fragments almost to the surface, but is generally compact and shows signs of poor drainage in the sub-soil. The only known extensive stand is in Darenth Wood, where this sub-type grades through a zone of mildly acid 9Ba to 6Cb on the plateau. Chestnut, which

has been extensively planted in north Kent, may have replaced other stands in Perry Woods and Farningham Wood. Wilson (1911) describes this sub-type, but omits hornbeam from his description. (Re-examination of some of his sites confirms that it is present.) This type perhaps represents a transition between 9Ba and types 2A and the more calcareous form of 9Ab.

Mercurialis perennis dominates the field layer in Darenth Wood but is entirely absent from Perry Wood. Frequent species are *Ajuga reptans*, *Anemone nemorosa*, *Circaea lutetiana*, *Endymion non-scriptus**, *Euphorbia amygdaloides*, *Hedera helix*, *Iris foetidissima*, *Lonicera periclymenum*, *Mercurialis perennis**, *Poa nemoralis*, *Rubus fruticosus**, *Teucrium scorodonia*, *Urtica dioica*, *Veronica chamaedrys* and *Viola riviniana**.

8.11 Suckering elm woodland (Group 10)

Stands containing either *Ulmus carpinifolia* or *U. procera*. Both species and *U. hollandica*, which is believed to be a hybrid between *U. glabra* and *U. carpinifolia*, produce numerous suckers, whereas *U. glabra* does not. Suckering elms rarely reproduce by seed, but they have been widely planted and have spread into many woods by suckers. Numerous local races have been recognised within the suckering elms (Richens, 1958), but their taxonomy remains confusing, many populations are hard to identify, and even the boundary between *U. glabra* and suckering elms is unclear. Suckering ability is itself a variable character and may even be absent from elms which can only be referred to either *U. carpinifolia* or *U. procera*. Conversely, some apparent *U. glabra* populations may exhibit a weak suckering ability. Suckering elms grow occasionally with alder or hornbeam to produce stands which are intermediate between groups 7 or 9 and 10, but scarcely ever grow with beech or pine. Elms in general are mostly found on moist, medium–heavy nutrient-rich soils.

Two stand types are recognised on somewhat uncertain grounds (Table 8.10). The great majority of suckering elm woods fall within the omnibus category of stand type 10A. Suckering elm woods are concentrated into types 2, 3, 5 and 6 of the Merlewood classification.

Table 8.10 Composition and site features of suckering elm (Group 10), pine (Group 11) and birch (Group 12) woodlands.

	10A*	10B	11A	11B	12A	12B
Number of samples	13	2	16	3	10	6
Acer campestre	IV	I				
Betula pendula		1	II	2	II	I
Betula pubescens	II		III		IV	V
Corylus avellana	IV	2				V
Fraxinus excelsior	III	2				
Pinus sylvestris			V	3		
Quercus petraea				2		
Quercus robur	III			2		
Tilia cordata		2				
Ulmus carpinifolia	IV	2				
Ulmus procera	III	2				
Ulmus glabra	+					
Acer pseudoplatanus	II	2				I
Clematis vitalba		1				I
Crataegus monogyna	IV	2				III
Crataegus oxyacanthoides	III	1				
Euonymus europaeus	II					
Hedera helix	II				+	
Ilex aquifolium			+	1		+
Juniperus communis			II	2	III	
Ligustrum vulgare	+	1				
Lonicera periclymenum	II					II
Malus sylvestris	+	1				
Populus tremula	+			1	II	I
Prunus avium						+
Prunus padus						II
Prunus spinosa	+	2				
Ribes sylvestre		1				
Rosa arvensis	II					
Rosa canina				1		II
Salix aurita					+	
Salix caprea	+					
Salix cinerea				+		II
Sambucus nigra	IV					
Sarothamnus scoparius				+	2	
Sorbus aucuparia			II	2	II	III
Thelycrania sanguinea	II	1				
Ulex europaeus				1		I
Median pH	6.9	8.0	4.3	4.3	4.8	5.3
Mean texture	3.2	2.0	0.3	0	1.1	1.4
Drainage free	1	2	14	3	9	3
imperfect	2		2		1	2
poor	6					
very poor	4					
Site	p	v	S	s	S	S

* Only 10A stands which are associated with type 2A.

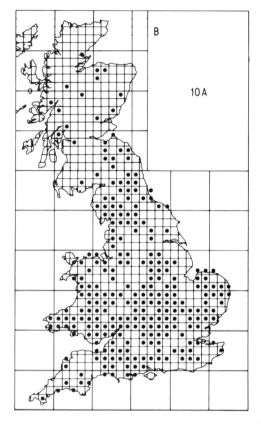

Fig. 8.37 Distribution of stand type 10A, invasive elm woods.

8.11.1 Stand type 10A. Invasive elm woods

Suckering elm stands include many which have undoubtedly developed by invasion of elm into another stand type. This process has been described by Rackham (1975) for Hayley Wood (Cambs), where type 2A has been invaded by *U. carpinifolia*. Such invasions often commence from the wood margin, where elms have often been planted as hedgerow trees, or, less often, from individuals which have been planted as standards or coppice within a wood (e.g. Overhall Grove, Cambs). Most suckering elm stands are completely dominated by elm. In some stands the even-aged thicket of elm poles, showing little or no sign of having been coppiced, presumably developed by invasion after the last coppicing. Elsewhere (e.g. Short Wood, Northants), large and small elm stools are present in a thicket of uncoppiced elm suckers, showing that the elm has been present and spreading within an actively coppiced stand. Whilst most suckering elm stands are or have been invasive, a minority may be long-established and stable. Suckering elms are present almost throughout Britain, but suckering elm woods are more frequent in southern England and Wales, and fairly abundant on the Midland clays (Fig. 8.37). Some stands are extensive, but most are rather small and confined to the woodland margins. Dutch elm disease has killed many adult elm trees within woods, but often the roots survive to produce a new crop of suckers, and it remains to be seen whether elm invasions will be arrested or reversed.

Suckering elms can invade a wide range of stand types, including 1B, 2A, 2B, 3A, 4A, 4B, 6D, 7A, 9A and 12B, but the type which appears to be most susceptible appears to be 2A. A range of sub-types within 10A could be defined by reference to the replaced stand types, but insufficient records are available to describe them or to decide how they might be grouped. Sufficient records are, however, available to describe 10A (2A) and compare it with 2A. Most of the species found in 2A occur, but are somewhat thinly scattered amongst the elm and reduced in frequency at the 30 × 30 m scale of recording. The one exception is *Sambucus*, which is markedly more frequent in 10A (2A). The field layers of 10A (2A) and 2A are similar, but *Galium aparine*, *Heracleum sphondylium*, *Poa trivialis* and *Urtica dioica* are more frequent and occasionally abundant in 10A (2A). These differences indicate that the nutrient content (e.g. phosphate) of the soil beneath 10A (2A) is higher, but we do not know if elms invaded eutrophicated clay soils, or whether the eutrophication has taken place since the elm invaded.

8.11.2 Stand type 10B. Valley elm woods

A stand of suckering elm woodland in Bedford Purlieus (Cambs) grows along the central valley on seasonally wet, freely-drained, highly calcareous loam (Peterken and Welch, 1975). This stand is believed to justify the status of a separate type because (i) it consists of an intimate mixture of elm forms and not the usual patchwork of distinct clones, (ii) it is not obviously the product of recent invasion, (iii) the

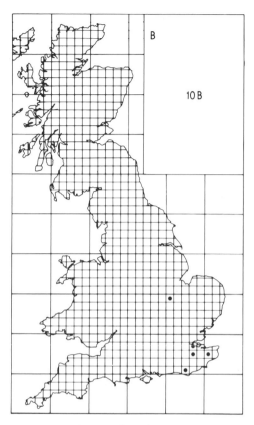

Fig. 8.38 Distribution of stand type 10B, valley elm woods.

constituent elms appear to be regenerating from seed as well as suckers, and (iv) the stand is strikingly confined to the valley. Ash, sycamore, lime and birch are also present in the canopy. The field layer contains abundant *Mercurialis perennis* and *Poa trivialis* with patches of *Allium ursinum* and *Deschampsia caespitosa*, and the following additional species occurred in both recorded samples: *Anemone nemorosa*, *Arum maculatum*, *Carex sylvatica*, *Filipendula ulmaria*, *Galeobdolon luteum*, *Glechoma hederacea* and *Viola riviniana*.

Valley elm woods are widespread in continental Europe, but rare in Britain. Formerly, as Dr E. W. Jones (personal communication) has suggested, valley elm woods may have been common, but were cleared when the valleys were settled in the earlier stages of man's transformation of the English lowlands. Perhaps some village elms were the descendants of these elm woods and, later, the sources of the elms which were planted into hedges and subsequently invaded the ancient woodlands on the plateau. Dr F. Rose (personal communication) regards some valley elm stands in Kent and Sussex as (original−) natural (Fig. 8.38).

8.12 Pine woodland (Group 11)

Stands containing *Pinus sylvestris* as a long-established, native species, i.e. where pine is naturally present in the past-natural sense. *Betula* (both species) is generally present, although many stands are overwhelmingly dominated by *Pinus*. The only other common associates are *Sorbus aucuparia* and *Juniperus communis* (especially in Speyside and Deeside), but *Ilex* is frequent in some western examples.

Pine is regarded as a calcifuge species in Britain, and is indeed generally found on light, freely-drained, strongly acid soils, but like beech, sessile oak and many other species, it is also capable of thriving on calcareous soils in Britain: it can still be found on relatively base-rich, dry soils with a moderately calcicole flora in Ryvoan Pass, between Abernethy and Glen More Forests. Mostly, however, pinewoods are found on podzols over which the deposits of peat may be more than 30 cm deep. The (past−) natural pinewoods are confined to the Highlands of Scotland (Fig. 8.39), where, perhaps because of their rarity, distinctive character and imposing setting, they have attracted ecological attention out of all proportion to their size. Most examples are described in detail as 'Native Pinewoods' by Steven and Carlisle (1959) and various aspects of their ecology and management have been described more recently by the contributors to Bunce and Jeffers (1977). As with so much of British ecology, attention has been focused on an extreme, the pine-dominated woods, leaving those with more birch and less pine (e.g. Crathie Wood, Deeside) largely unstudied.

Although several types of pinewood based on the field layer composition have been recognised by different authors, the relative poverty and uniformity of the tree and shrub layers makes it meaningless to split the pinewoods into stand types to correspond to them. Nevertheless, there is some justification for two stand types differentiated by the presence of oak (either species), for McVean and Ratcliffe (1962) map

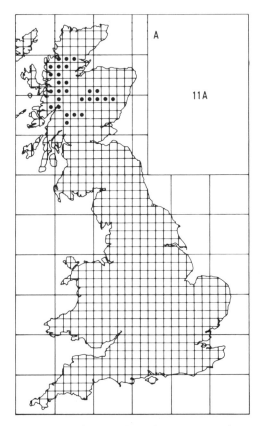

Fig. 8.39 Distribution of stand type 11A, acid birch–pinewoods.

an extensive area as potentially pine–oak woodland. This makes some sense in the extensive secondary woods on heathland in southern England, where oak is also a frequent coloniser, and secondary pine–oak woodland is common. A third type, calcareous pinewood, might also be recognised as part of the natural range of British types, even though it is now extinct in ancient woodland (Table 8.10).

8.12.1 Stand type 11A. Acid birch–pinewoods

Pinewoods on acid soils containing no oak. Such stands usually contain *Betula pendula* and *B. pubescens*, although these may be very sparse. They occur only as high forest in the highlands, with the most extensive stands occurring in Deeside (Glentanar, Ballochbuie, Mar), Speyside [Abernethy (Fig. 2.8),

Rothiemurchus], Glen Affric and Glen Strathfarrar. Other important examples occur further west at Loch Maree and Shieldaig, at Rannoch in Perthshire and in a south-western group near Tyndrum. Most occur on strongly acid, coarse-textured soils, most of which are strongly podzolised. They grade into acid alderwood, 7A, in peaty hollows, and into birchwood, 12A, and acid oakwoods (6Ab and 6Bb) on acid brown earths. Most examples fall within type 27 of the Merlewood classification.

The field layer, which is generally composed mainly of dwarf shrubs and bryophytes, has been classified by Steven and Carlisle (1959) and Bunce (Bunce and Jeffers, 1977) into numerous types. McVean and Ratcliffe (1962) recognised two types which broadly corresponded to open and dense stands. The frequent species in my 16 sample stands were: *Anemone nemorosa*, *Blechnum spicant*, *Calluna vulgaris**, *Deschampsia flexuosa**, *Galium saxatile*, *Luzula pilosa*, *Molinia caerulea*, *Oxalis acetosella*, *Potentilla erecta*, *Pteridium aquilinum**, *Vaccinium myrtillus**, *V. vitis-idaea** and *Viola riviniana*, but many of the species found in acid oakwoods occur here sparsely. Several rather infrequent herbs are characteristic of the pinewoods (White, 1898): *Corallorhiza trifida*, *Goodyera repens*, *Linnaea borealis*, *Listera cordata*, *Moneses uniflora*, *Pyrola media*, *P. minor*, *P. rotundifolia*, *Orthilia secunda* and *Trientalis europaea*.

8.12.2 Stand type 11B. Acid oak–pinewoods

Pine woodland containing *Quercus petraea* and *Q. robur*. Mixtures of oak and pine were considered by McVean and Ratcliffe (1962) to be the present–natural woodland type of the lowland belt of northeast Scotland lying to the north and east of the Cairngorms, and indeed fragments of such communities survive in Darnaway Forest (Moray), Cawdor (Nairn) and Torr Alvie on Speyside. Both *Q. petraea* and *Q. robur* occur sparingly in some pinewoods, such as Shieldaig in the west and Glentanar in the east. They seem to be associated with soils which lack a substantial mor humus horizon, for example on rock outcrops and steep slopes. They may represent vestiges of a type which linked 11A and 6Ab or 6Bb in the range of variation. Only three samples conforming to this type have been recorded, all on very dry,

strongly acid, coarse-textured soils. Only *Deschampsia flexuosa*, *Pteridium aquilinum* and *Vaccinium myrtillus* were present in them all, and *Calluna vulgaris* and *Holcus mollis* were present in two. Secondary stands of this type are common on heathland in southern England.

8.12.3 Stand type 11C. Calcareous pinewoods

Although 'native' pinewoods no longer occur on calcareous soils, in Britain, they are a frequent feature of continental woods, and there is some evidence that they once occurred in northern England. Professor F. Oldfield (personal communication) has found high pine pollen counts and pine stumps in lake sediments around Malham in the Craven Pennines on limestone.

8.13 Birch woodland (Group 12)

Stands containing *Betula*, but not the other species and genera characteristic of groups 1–11. Birches are found throughout Britain and on most types of soil, though they are rare on moist, base-rich soils and most abundant on freely-drained sands or peat. *B. pubescens* tends to grow on moist, peaty soils and *B. pendula* is more likely to prevail on dry, mineral soils, but the two overlap greatly and their relative proportions in the majority of woods are difficult to predict. Birch woodland occurs in two main circumstances. In northern Scotland birch grows beyond the range of oak, pine, ash, etc., and some of the northern-most ancient semi-natural woods consist of birch woodland. In most of Britain within the range of the oaks, birch woodland only forms temporarily as an early seral type on heathland, grassland and dried out bogs and in disturbed woods. The status of individual birch woodlands in central and northern Scotland is difficult to determine, but many are certainly secondary, and some of the birch woodland on ancient woodland sites may have been formed after felling out pine or oak. Because of these difficulties it has been impossible so far to construct distribution maps showing where birch woodland is a stable component of ancient, semi-natural woodland. In Scotland birch woodland occurs mainly on acid brown earths leaving the more podzolised soils to pine.

McVean and Ratcliffe (1962) recognise two types, the *Vaccinium*-rich birchwoods, which tend to occur on black, mildly acid, crumbly humus, and the herb-rich birchwoods, which grow mainly on mineral soil with a mull humus. The nearest approach to this in terms of tree and shrub constituents is to divide birch woodland into two stand types, based on the absence and presence respectively of *Corylus* (Table 8.10).

8.13.1 Stand type 12A. (Rowan-) birchwoods

Birch woodland without *Corylus*. Either or both the main birch species may be dominant, and indeed there is some uncertainty about the taxonomic status of many populations. *Sorbus aucuparia* is the most consistent associate, but it only occasionally dominates a stand, perhaps those which have been undisturbed long enough for regeneration to occur beneath the shade of a pre-existing stand. *Juniperus* is frequent and often abundant in the inland districts of north-east Scotland where it forms an underwood beneath the birch, and appears to replace *Ulex europaeus* in the continental climate. Clones of *Populus tremula* are scattered throughout, and not just on the moist soils which are associated with this species in the lowlands. The soils are mostly mildly acid, sand or loam, with little or no mor humus accumulation. Some appear to be flushed, but most are freely drained. Good examples occur in Deeside (e.g. Morrone, near Braemar), Speyside (e.g. Craigellachie) and on the northern shores of Loch Rannoch. Some may have been derived from pine woodland following felling and others may be secondary, e.g. the birch woodland which fringes the Letterewe oakwood on the north side of Loch Maree. This stand type is closely related to types 6Ab, 6Bb and 11A, but the exact relationships between them are complex (Steven and Carlisle, 1959). Most examples fall with the birch–pinewoods in type 27 of the Merlewood classification.

(Rowan-) birchwoods also occur sparingly throughout Britain on light, acid soils, but in ancient woods they are confined to the fringes of bracken glades and develop temporarily on disturbed birch–oakwood, acid beechwood and oak–limewood stands.

The field layer is normally dominated by grasses or dwarf shrub species. Frequent species are *Agrostis tenuis**, *Anemone nemorosa**, *Anthoxanthum odoratum*, *Blechnum spicant*, *Calluna vulgaris*, *Campanula rotundifolia*, *Carex pilulifera*, *Deschampsia flexuosa**, *Festuca ovina* agg*, *Galium saxatile*, *Luzula campestris*, *L. pilosa**, *Oxalis acetosella*, *Potentilla erecta*, *Pteridium aquilinum*, *Succisa pratensis*, *Vaccinium myrtillus**, *V. vitis-idaea*, *Veronica officinalis* and *Viola riviniana*. Other species present include *Carex binervis*, *Conopodium majus*, *Lysimachia nemorum* and *Melampyrum pratense*. Many stands have light, open canopies, so grassland and bog species are commonly present. The only stand recorded in the south of Britain (Newball Wood, Lincs) had a field layer of *Pteridium aquilinum* and *Holcus mollis*.

8.13.2 Stand type 12B. Hazel–birchwoods

Birch woodland with *Corylus*. *Sorbus aucuparia*, *Crataegus monogyna* and *Prunus padus* are commonly present in northern examples. *Salix caprea* is sometimes important but not in the samples recorded. The soils beneath the recorded stands were somewhat heavier and less acid (pH 4.1–5.8) than those associated with 12A. Hazel and birch are the most abundant species, both of which are commonly coppiced. In most of Britain this type is undoubtedly a transitory phase following felling of oak, ash or lime from a stand in group 3, 4, 5 or 6. The same probably applies to many stands in the north, for example, at Craig Nordie and the lower end of Crathie Wood in Deeside. The few hazel–birchwoods in the extreme north of Scotland may be beyond the range even of oak and ash, and thus a permanent form of ancient semi-natural woodland. A fine example occurs at Strathbeag, near Loch Eribol, and there is evidently another example at Berriedale on the Orkneys. The recorded stands are randomly scattered through the Merlewood types.

The field layer is generally very rich. The following species occurred in two or more of the five stands which were recorded in detail: *Ajuga reptans*, *Anemone nemorosa*, *Anthoxanthum odoratum*, *Campanula rotundifolia*, *Cardamine flexuosa*, *Cirsium vulgare*, *Conopodium majus**, *Digitalis purpurea*, *Dryopteris dilatata*, *D. filix-mas*, *Endymion non-scriptus*, *Galeobdolon luteum*, *Geranium robertianum*, *Heracleum sphondylium*, *Holcus lanatus*, *H. mollis*, *Lapsana communis*, *Luzula pilosa*,

Lysimachia nemorum, Mercurialis perennis*, Oxalis acetosella, Potentilla sterilis*, Primula vulgaris, Pteridium aquilinum*, Ranunculus acris, R. ficaria, R. repens, Rubus fruticosus, Rumex acetosella, Senecio jacobea, Stachys sylvatica, Taraxacum officinale, Teucrium scorodonia, Urtica dioica, Veronica chamaedrys, V. officinalis, Viola riviniana*.

8.14 Conspectus of stand types in British ancient, semi-natural woodland

1 ASH–WYCH ELM WOODLAND
    1A. Calcareous ash–wych elm woods
        a. Southern variant
        b. Northern variant
    1B. Wet ash–wych elm woods
        a. Heavy soil variant
        b. Light soil variant
    1C. Calcareous ash–wych elm woods on dry and/or heavy soils
        a. Eastern variant
        b. Sessile oak variant
    1D. Western valley ash–wych elm woods
2 ASH–MAPLE WOODLAND
    2A. Wet ash–maple woods
        a. Typical wet ash–maple woods
        b. Wet maple woods
    2B. Ash–maple woods on light soils
        a. Variant on poorly-drained soils
        b. Variant on freely-drained soils
    2C. Dry ash–maple woods
3 HAZEL–ASH WOODLAND
    3A. Acid pedunculate oak–hazel–ash woods
        a. Heavy soil form
        b. Light soil form
    3B. Southern calcareous hazel–ash woods
    3C. Northern calcareous hazel–ash woods
    3D. Acid sessile oak–hazel–ash woods
4 ASH–LIME WOODLAND
    4A. Acid birch–ash–lime woods
    4B. Maple–ash–lime woods
        a. Lowland variant
        b. Western variant
    4C. Sessile oak–ash–lime woods
5 OAK–LIME WOODLAND
    5A. Acid pedunculate oak–lime woods
    5B. Acid sessile oak–lime woods

6 BIRCH–OAK WOODLAND
    6A. Upland sessile oakwoods
        b. Upland birch–sessile oakwoods
        c. Upland hazel–sessile oakwoods
    6B. Upland pedunculate oakwoods
        b. Upland birch–pedunculate oakwoods
        c. Upland hazel pedunculate oakwoods
    6C. Lowland sessile oakwoods
        b. Lowland birch–sessile oakwoods
        c. Lowland hazel–sessile oakwoods
    6D. Lowland pedunculate oakwoods
        b. Lowland birch–pedunculate oakwoods
        c. Lowland hazel–pedunculate oakwoods
7 ALDER WOODLAND
    7A. Valley alderwoods on mineral soils
        a. Acid valley alderwoods
        b. Valley alderwoods on neutral–alkaline soils
    7B. Wet valley alderwoods
        a. Sump alderwoods
        b. Base-rich springline alderwoods
        c. Base-poor springline alderwoods
    7C. Plateau alderwoods
    7D. Slope alderwoods
    7E. Bird cherry–alderwoods
        a. Lowland variant
        b. Upland variant
8 BEECH WOODLAND
    8A. Acid sessile oak–beechwoods
    8B. Acid pedunculate oak–beechwoods
    8C. Calcareous pedunculate oak–ash–beechwoods
        a. Dry lime–wych elm variant
        b. Moist wych elm variant
        c. Maple variant
    8D. Acid pedunculate oak–ash–beechwoods
    8E. Sessile oak–ash–beechwoods
        a. Acid variant
        b. Calcareous variant
9 HORNBEAM WOODLAND
    9A. Pedunculate oak–hornbeam woods
        a. Birch–hazel variant
        b. Ash–maple variant
    9B. Sessile oak–hornbeam woods
        a. Acid sessile oak–hornbeam woods
        b. Calcareous sessile oak–hornbeam woods

10 SUCKERING ELM WOODLAND
  10A. Invasive elm woods
  10B. Valley elm woods
11 PINE WOODLAND
  11A. Acid birch–pinewoods
11B. Acid oak–pinewoods
11C. Calcareous pinewoods
12 BIRCH WOODLAND
  12A. Rowan–birch woods
  12B. Hazel–birch woods

Chapter 9

Management variants of stand types

The foregoing stand types have been defined on the basis of samples recorded in semi-natural stands growing on ancient woodland sites which have been managed (or treated) as coppice. Together they span the site dimension of variation, whilst differences due to management (and time; Chapter 10) have been minimised. Man has, of course, had a considerable effect on their composition (Section 3.5), but this is believed to be a relatively insignificant determinant of the differences between stand types.

The management dimension of variation can be introduced by considering the stand types described in Chapter 8 as 'coppice types', each of which has, or could have, a counterpart 'high forest type' and 'wood pasture type'. These three silvicultural treatment variants cover the range of conditions found in ancient, semi-natural stands. Thus, for instance, acid sessile oak–beech woodland (stand type 8A) occurs as coppice, high forest and wood pasture, and there may be consistent differences in composition between the variants. Furthermore, the composition of any stand is affected to a greater or lesser degree by man, whilst remaining semi-natural, in such a way that 'improved' and 'disturbed' forms of each management variant can be recognised. Thus an 'improved' form of acid sessile oak–beech high forest may have a high proportion of beech due to planting, whereas a 'disturbed' form may have a high proportion of birch.

9.1 High forest variants

Consistent differences between coppice and high forest variants of one stand type are difficult to find. Coppice and high forest have tended to merge with the retreat of coppicing, so many quasi-high forest stands have had to be included in the descriptions of stand types. Some stand types occur only or mainly as high forest [8C in south-east England, 8D (Fig. 4.5), 11A, 12A and 12B], whereas others occur only as coppice and derivatives from coppice. Where a direct comparison between coppice and high forest stands of one stand type is possible, as in the case of types 2A, 3A and 6Dc, which occur fairly commonly in both forms in the English lowlands, the high forest stands usually have a higher cover of oak and a lower cover of shrub and other tree species, e.g. Alice Holt Forest (Hants) and Salcey Forest (Bucks and Northants). Likewise, in the Cotswold beechwoods, where the same comparison can be made, beech is dominant in the high forest canopy but in the coppice it is merely one species in the mixture. However, even this comparison is invalid, because the high forest stands are strictly 'improved' forms (Section 9.4) created usually by planting oak or beech. Thus the differences in composition between coppice and high forest are obscure and are confounded with different levels of improvement, and there seems to be no justification for recognising distinct high forest variants of each stand (coppice) type.

9.2 Wood pasture variants

Most wood pasture stands can be referred to stand types 6Bb, 6Cb, 6Db, 8A, 8B and 9A. The New Forest unenclosed woods for example are mainly beech–oak stands of types 6Cb, 6Db, 8A and 8B (Fig. 2.1) and their secondary variants (Chapter 10) with coppiced alderwoods of types 7A and 7B along the valleys. Similar stands are found in the old, unenclosed stands of Dean, Windsor and Ashdown Forests, whilst many parts of Epping Forest have beech–oak–hornbeam woods of types 8B, 9A and intermediates. Outside southern England many wood pasture stands are birch–oakwoods of types 6Cb (e.g. Sherwood Forest, Notts; Sutton Park, West Midlands) or 6Db (e.g. Staverton Park, Suffolk). Some parklands contain a few ash, maple, wych elm and lime amongst the prevailing oaks, for example at Moccas Park (Hereford) which can partly be described as the wood pasture variant of types 1A and 1D. Many wood pasture stands contain a generation of trees which developed after a reduction in grazing pressure: these stands, in which birch and other pioneer species are well represented, can be regarded as a form of wood pasture in which the equivalent scrub type (Chapter 10) is also present.

Whilst it would be possible to have wood pasture variants of each stand type (with the exception of types 11A and 12A, which are already a form of wood pasture), it is convenient to group them into three categories. Wood pasture is too infrequent to justify a large number of types, but more importantly the wood pasture variants of many types have converged to mixtures of oak and exotic species. Other native species, which may have been present, have been reduced or eliminated, whilst exotic species have been repeatedly planted. Thus, the wood pasture variants of stand types in groups 1–6, all of which contain oak, have been 'improved' into a common form of oak parkland with exotics.

Accordingly, three main wood pasture types seem worth recognising:

(1) Mixed oak wood pasture. Variants of stand groups 1–7. Beech and hornbeam are rare or absent, and oaks are generally abundant. Apart from planted exotics, some stands contain ash, maple, wych elm, lime or alder, showing that they are wood pasture variants of stand groups 1–5 and 7, but other stands lack these species and may be wood pasture variants of stand group 6 (Fig. 18.5). Widespread in Britain.

(2) Beech–oak wood pasture. Variant of stand group 8. Usually takes the form of beech-dominated mature stands on acid soils. Confined to southern England, where good examples occur in several old forest areas (Figs. 2.1 and 13.3).

(3) Hornbeam–oak wood pasture. Variant of stand group 9. Contains hornbeam and oak with occasional ash. Found mainly in East Anglia and south-east England where good examples exist at Hatfield Forest (Essex), Epping Forest (Essex) and Mersham Hatch Park (Kent).

Direct comparison of coppice and wood pasture variants of one stand type is sometimes possible. Lowland birch–pedunculate oakwoods (6Db) have more browsing-resistant holly in wood pasture stands, more or less birch (depending on whether natural regeneration in open conditions has been possible recently) and less rowan and other minor species. These are, however, only part of the differences. Evidence from pollen profiles and historical documents shows that some stands of type 6Dc in the New Forest have been changed under wood pasture management to type 6Db by the elimination of hazel, and thence to type 8B by the invasion of beech. Thus, the true wood pasture variant of 6D may be 8B.

9.3 Disturbed stands

Any stand may be disturbed by extensive felling, other management activities or natural calamities, but not completely destroyed. The stand which subsequently regenerates usually contains all the former constituents, and thus remains unchanged in respect of stand types, but shrub species and rapid colonists within the stand are greatly increased. For instance, acid pedunculate oak–ash–limewoods of stand type 4A are generally dominated by oak, ash, lime and hazel in their undisturbed state, but extensive felling followed by neglect causes hawthorn and birch to increase. In effect the disturbance form of this stand type is intermediate between the undisturbed stand and the equivalent scrub type (Chapter 10).

Disturbance forms of some stand types actually lack

Fig. 9.1 Chestnut coppice in mid-rotation, Maplehurst Wood, Sussex. Most chestnut coppices are plantations made in the 18th and 19th Centuries at the expense of more natural mixtures. They are still commercially cut in Kent and elsewhere for fence stakes and hop poles, thus maintaining the alternation of light and shade phases which was once characteristic of most British woodland management.

the definitive species and therefore fall within another stand type. For example, birch stands containing few oaks may spring up after a mature birch–oak stand has been felled, e.g. Sheephouse Wood (Bucks). Ash groups develop when mature beeches fall in an ash–beechwood. Thus, stand group 12 can be a disturbance form of stand group 6, and patches of types 2C and 3B develop after natural disturbance in an 8C stand. Indeed, certain stand types exist in ancient woods mainly as disturbance (or seral) forms in some regions (e.g. 12A over most of Britain; 2C and 3B in the beech regions) whilst forming stable (or 'climax') stand types elsewhere.

9.4 Improved stands

Stands can be 'improved' by the promotion of the most useful species, the result being an almost pure stand. This is particularly true of birch–oakwoods in western Britain and beechwoods on the southern chalk. Many hazel coppices in the south are improved forms of types 2A, 2C and 3A. Improved birch–oakwoods and beechwoods usually take the form of high forest stands, many of which were planted.

A more obvious improved type is chestnut coppice (Fig. 9.1). Chestnut was commonly planted in south-east England, Essex and parts of south-west England, mainly on light, acid soils where stand types 3A, 6C, 6D and 8A grow. The original stand type can still be recognised in most woods, which can therefore be described as, for example, chestnut forms of 6C, 8A and 9Bb (as in the chestnut coppices around Faversham and Canterbury, Kent). Chestnut, however, is a long-established species, though doubtfully native, and some stands have contained the species for centuries, e.g. the chestnuts on the eastern side of the

177

Forest of Dean (Glos) contained *Castanea* in the 13th century (Hart, 1966) and still contains it. The chestnuts in Chalkney Wood (Essex), which are mixed with lime, may also be a long-established population. Arguably, chestnut woodland should be recognised as a separate stand group comparable with suckering elm woods.

Sycamore has commonly been planted as well, but it has also spread into disturbed stands. Thus a sycamore-rich stand may be either an improved or disturbed form of a particular stand type. In Bedford Purlieus, for example, this species spread from the southern parts of the wood during a prolonged period of felling and replanting the old coppice with oak (Peterken and Welch, 1975).

9.5 Plantations of native species

Plantations of native species can just be tolerated within the definition of semi-natural woodland provided that the planted species is native to the site, i.e. to that soil type in that region. Such stands are an extreme form of improvement, but they can usually be assigned to an appropriate stand type. When native species are planted on ground which they would not naturally have occupied it is no longer possible to regard them as even tenuously semi-natural. Thus, the plantations of beech on Midland clays and further north are not beechwoods within the terms of stand group 8, but beech versions of other stand types, or simply beech plantations. Likewise, the Scots pine plantations which are now widespread cannot legitimately be regarded as pinewoods of stand group 11, except in the Highlands.

Naturalness is a continuous variable (Section 3.1). Stands can be arranged in a hierarchy of decreasing naturalness:

(a) Natural woodland, insignificantly affected by man.
(b) Mixed coppice; unenclosed forest stands; woods in inaccessible sites; native pinewoods and birch-woods.
(c) Improved coppice; high forest of species native to the site developed from coppice; medieval park woodland; disturbed versions of (b).
(d) Plantations of species native to the site.
(e) Plantations of species not native to the site, but with native physiognomy – mostly deciduous broadleaved.
(f) Plantations of evergreen broadleaves and conifers [Coniferous plantations on base-poor soils in the Highlands are included in category (e)].

Only stands in categories (a)–(d) can be assigned to a stand type, but those in categories (e) and (f) can be described as, for example, a Norway spruce plantation on a site which might naturally support stand type 8A.

Chapter 10

Succession and stand types

10.1 Secondary succession

The classification of stand types in Chapter 8 strictly applies only to semi-natural stands on ancient woodland sites, the composition of which is presumed to be fairly stable, at least in relation to the species and structural forms upon which the stand types are defined. Secondary semi-natural stands are not stable; their composition changes as succession proceeds. Nevertheless, they can be accommodated within the classification if successional variants of each stand type are recognised.

Each stand type has a seral counterpart, in other words the scrub and emergent woodland community which develops on the same site type. For instance, juniper scrub is a scrub counterpart of the maple variant of calcareous pedunculate oak–ash–beech woodland [i.e. S(8Cc)]. In so far as the composition of seral scrub is determined by site factors there is a direct relationship between stand types and scrub types. However, other factors, which seem to be unimportant in ancient woods, are highly significant in early seral communities, notably the proximity of seed sources, the circumstances in which the succession was initiated, and chance variation in colonisation. Thus, although scrub types can be defined on the basis of soil characteristics, the relationship is somewhat ill-defined. Heavy clays may, for example, be colonised by thickets of hawthorn, ash or suckering elm, depending on which species are available in the vicinity, and

conversely birch scrub can form on rendzinas, though it is far more likely to occur on acid sands.

The random element in scrub composition is not the only complication in the relationship between stand types and scrub types. Another is the inability of certain species to colonise, at least in the early stages of succession (Section 10.2). A third is the existence of certain scrub types which never develop into woodland, such as montane *Salix lapponum* and *S. myrsinites* scrub above the tree line, hazel scrub on exposed north-western limestones and sallow scrub on peat and acid flushes. A fourth complication is that some stand types are extinct or nearly so (10B, 11C; Section 11.2), yet scrub can still develop on the sites they would have occupied. Thus, each stand type has a range of possible equivalent scrub types within which there is one scrub type more likely to develop than any other. Most scrub types have a range of stand types to which they are equivalent and into which they may develop. But a few scrub types have no equivalent amongst the stand types.

Scrub communities are mostly the earliest stage of a natural succession leading to a stand which approximates to one of the stand types. Juniper scrub, for instance, is the scrub type of the 'juniper sere' described by Watt (1934), which passes through a sequence of secondary woodland forms equivalent to type 8Cc until a stand is formed containing beech, ash, pedunculate oak and maple. Succession does not necessarily restore the stand type which formerly occupied the site, and even if it does the process may

Table 10.1 Composition of sample stands in Cambridgeshire woodlands on calcareous boulder clay. The woods are arranged according to their age and origin. The tree and shrub species are arranged according to their general incidence in relation to woodland age and origin; this is not invariably reflected in the few samples recorded.

| | PRIMARY WOODS | | | SECONDARY WOODS | | |
| | Ancient woods | | | | Recent woods | |
	Undisturbed Hayley	Disturbed Knapwell	Pre-1279 Papworth	15th century Overhall	17–18th centuries	19–20th centuries
Slow colonist from ancient woodland						
Corylus avellana	5 6	4 4	5			
Moderate colonists from ancient woodland						
Acer campestre	7 6	2	1 5	3 1	3 3 4 2	4 2
Crataegus oxyacanthoides	5 5	6	H		2 2 6	6
Euonymus europaeus		2	2		2	
Lonicera periclymenum	4		2	2	3	
Malus sylvestris	4					
Populus tremula	5			3		
Quercus robur	8 6	5	5 3	2	+	
Rosa arvensis	6					
Thelycrania sanguinea		4	4 4		2	2
Fast colonists from ancient woodland						
Crataegus monogyna	6 H	4 4	5 3 5	4 4	6 7 6 H	7 8 6
Fraxinus excelsior	8 9	8 7	9	4	9 8 9	3 9
Prunus spinosa	4 D	5	4		2 4	4
Fast colonists which enter disturbed ancient woodland						
Sambucus nigra		4 5	5 4	4 5 2	2 5	4
Ulmus carpinifolia		5 7	9 10	10 10 9	4	9
Species which are more characteristic of secondary woodland						
Acer pseudoplatanus			1 1			1
Hedera helix			2 4			
Ligustrum vulgare				2	4 6	
Malus sylvestris ssp. mitis					5 4	6
Rosa canina					2	2
Viburnum lantana					2 2	2
Undefined species						
Betula pubescens	3					
Ilex aquifolium			1			
Ribes sylvestre			2			1
SOIL pH at 10 cm	6.1 6.5 6.3 6.5	7.7 7.2	4.3 7.5 7.9	5.4 7.6 5.1	7.8 8.3 7.9 7.7	8.0 8.0 7.4
Profile drainage	VP VP VP VP	P P	P P P	VP VP VP	P P P P	P P P

H = Hybrid hawthorn.

take hundreds of years. Comparison of stands of different origins on the calcareous, poorly-drained boulder clay of Cambridgeshire, where stand type 2A typifies the ancient semi-natural woods, illustrates this point (Table 10.1).

Hayley Wood represents the ancient (probably primary) woods (Rackham, 1975). Knapwell Wood is also primary, but has been disturbed by attempted reforestation and use as a farm dump. The other stands are in secondary woods which range from the early medieval Papworth Wood to scrub of recent origin. The undisturbed soils of the primary wood are mildly acid at the surface, but the secondary woods are more calcareous, better drained and, as the presence of *Galium aparine*, *Heracleum sphondylium*, etc., suggests, richer in phosphates and nitrates. Thus, the soils beneath the secondary woods are significantly different from those beneath the primary wood from which they were presumably derived, and it is not certain that they will ever return to the original condition even if they remain undisturbed for many hundreds of years. Nevertheless, the secondary woods are recognisably related to stand type 2A. They consist initially of a mixture of fast colonists from the type (*Fraxinus*, *Crataegus monogyna*), scrub species found less often in the type (*Rosa canina*, *Prunus spinosa*, *Ligustrum*, etc.), species found only in the early stage of succession (*Rhamnus*, *Viburnum lantana*), and species responding to the soil changes caused by cultivation, etc. (*Sambucus*). Other species in the type enter the succession at a later stage, mostly when the secondary woods are well established (*Acer campestre*, *Crataegus oxyacanthoides*, *Quercus robur*). Only *Corylus* of the species present in undisturbed primary woodland appears to be very slow to enter the succession. (The place of invasive elm in this scheme is discussed below under long-term succession.) Thus, in this instance, succession does restore a stand of type 2A, but many centuries are required and the result still differs from the 2A stands which occupy primary woodland sites, because of the enduring effects of cultivation on the soils.

10.2 Scrub types

The scrub types, i.e. the early seral stages which correspond to each stand type, may be inferred from the composition of the stand type itself by assuming

Table 10.2 Occurrence of tree and shrub species in the early stages of natural succession to woodland on various soils. A, abundant; ++, common; +, frequent; rare, not shown.

Soil type*	A	AB	B	BC	C	D
Clematis vitalba	A	+				
Thelycrania sanguinea	++	+				
Corylus avellana	++	+				
Euonymus europaeus	+	+				
Ligustrum vulgare	A	+	+			
Rhamnus catharticus	+	+	+			
Viburnum lantana	++	+				
Juniperus communis	++				+	
Sorbus aria	++				+	
Taxus baccata	++				+	
Acer pseudoplatanus	+	++		+		+
Crataegus monogyna	+	A	A	A		+
Fraxinus excelsior	A	A	A	++		+
Prunus spinosa	+	+	A	++		+
Rosa canina	+	+	A	+		
Salix caprea			+			
Sambucus nigra	+	++	A	+		+
Ulmus carpinifolia		+	++			
Quercus robur		+	+	+	+	+
Betula pubescens			+	+	A	A
Betula pendula	+		+	+	A	+
Ilex aquifolium	+			+	A	
Pinus sylvestris	+			+	A	+
Quercus petraea	+			++		
Sorbus aucuparia	+			+	++	+
Ulex europaeus	+			+	A	
Frangula alnus					+	+
Populus tremula				+		
Viburnum opulus				+		+
Alnus glutinosa						A
Myrica gale						+
Salix cinerea						++

* Soil types: A, dry, calcareous; AB, moist, calcareous; B, heavy, poorly drained; BC, light, poorly drained; C, light, dry, acid; D, wet.

that they consist of species which are known to be fast colonists throughout the range of the type and then adding scrub species which appear only in the early stages of succession. The results confirm that the scrub equivalents of many stand types are virtually indistinguishable, and that they may be grouped into four main scrub types and two intermediate forms (Table 10.2). These are similar to the scrub types of Ward (1974), and are listed here:

A. Scrub of dry, freely-drained calcareous soils
Scrub often dominated by *Taxus*, *Juniperus*, *Corylus* or a mixture of calcicole shrubs. This often includes species which are more frequently associated with dry, acid soils, e.g. *Betula pendula*, *Ilex*, *Sorbus aucuparia*. The main tree colonist is *Fraxinus*. Notable examples of secondary semi-natural woodland occur in the Peak district (Merton, 1970). Seral equivalent of stand types 1C, 2C, 3C, 4C, 8Ca, 8Cc, 8Eb and perhaps 11C. Falls within the *Prunetalia—Berberidion* of the class *Querco—Fagetea*.

AB. Scrub of moist, freely-drained calcareous soils
Compared with A, this scrub type has fewer calcicole shrubs and dry soil species, and more *Crataegus monogyna* and other species of heavy soils. Seral equivalent of stand types 1A, 3B, 4Bb, 8Cb, 9Bb and 10B.

B. Scrub of heavy, poorly-drained soils
Scrub usually dominated by spinose shrubs, e.g. *Crataegus monogyna*, *Prunus spinosa*, *Rosa canina*. Calcicole shrubs (e.g. *Ligustrum vulgare*, *Thelycrania sanguinea*) and wet soil species (e.g. *Salix cinerea*) are usually present. *Sambucus*, *Ulmus procera* and *U. carpinifolia* are frequent on eutrophic soils. The main tree colonist is *Fraxinus*. Seral equivalent of stand types 1B, 2A, 3A, 4Ba, 7C, 9A and 10A. Within the *Prunetalia* of class *Querco—Fagetea*.

BC. Scrub of medium and light-textured, poorly-drained or permanently moist soils
Compared with B, this has many species of dry, acid soils, and the tree colonists tend to be *Betula pubescens*, *Fraxinus* and *Quercus robur*. Seral equivalent of stand types 2B, 4A, 5A, 7D, 8D and 9Ba.

C. Scrub of freely-drained, acid (usually light) soils
Scrub dominated by *Betula* (Fig. 5.2), *Ilex*, *Pinus sylvestris*, *Sorbus aucuparia* and *Ulex europeaus*. A few species often regarded as calcicole, e.g. *Juniperus*, *Taxus*, *Sorbus aria*, occur rarely. The main tree colonists are *Betula*, *Pinus*, *Quercus*. Seral equivalent of stand types 5B, 6A, 6B, 6C, 6D, 8A, 8B, 11A, 12A and 12B. Said by Ward (1974) to fall within classes *Nardo—Callunetea* and *Erico—Pinetea*, but probably also falls within the *Quercetea robori-petraeae*.

D. Scrub of wet soils, often permanently water-logged and peaty
Scrub dominated by *Salix* spp., *Alnus* and restricted species, such as *Myrica gale*. Many species from scrub type B occur on the drier sites, including *Viburnum opulus* and *Rhamnus catharticus*. *Sambucus* is frequent on base-rich sites, and *Frangula* on acid soils. Seral equivalent of stand types 7A, 7B, 7D, 7E. Said to fall within classes *Salicetea purpurea* and *Alnetea glutinosae*.

Three stands types have seral equivalents which fall between scrub types A, B and C, namely 1D, 3D and 8Ea.

The difference between any given stand type and its corresponding scrub type varies considerably, depending on how many species in the stand type are slow colonists. Stand and scrub types in groups 11 and 12 probably differ least because all the constituent species of the stand type appear to be fast colonists. At the other extreme, stand types which include *Tilia cordata* differ most from their seral equivalents, because this species appears to be a slow colonist throughout its British range. In general the stand types of heavy soils appear to differ more from their seral equivalents than those of light soils.

Recent, semi-natural, secondary woodland which is incorrectly thought to be ancient will be inadvertently misclassified within the stand types rather than the scrub types. Scrub of types A, AB, B and BC will mostly be misclassified as hazel—ash woodland (group 3); scrub of type C will be classified as birch—oak, pine or birch woodland (groups 6, 11 or 12); and scrub of type D may be classified as hazel—ash, alder or birch woodland (groups 3, 7 or 12).

10.3 Cyclic changes

A birch—oak stand in group 6 will become a birchwood in group 12 if all the oaks are felled, but the change is reversed when a birchwood is colonised by oak. Natural fluctuations in the maple population of a natural ash—maple stand of type 2C, where maple is often sparse, may transform parts of the stand into type 3B until the maple increases again. Dutch elm disease opens up the possibility that stands in groups 1 and 10 will change to other stand groups, although in fact the elms are rarely quite exterminated, and even if they are, it is possible that they might recolonise when the disease abates. These instances show that any given patch of semi-natural woodland may oscillate between two or more stand types as its composition fluctuates in response to natural events, such as death and regeneration (Watt, 1947), and human intervention in the form of selective felling and planting. Beech is especially affected by these considerations. Many stands of types 1A, 2C, 6A, 6B, etc. beyond the native range of beech, have technically been transformed to beech woodland by sporadic planting of this species. Other stands of these types, where beech was once planted but has since been felled leaving few progeny, were once technically beechwoods but are no longer. Many ash—maple and hazel—ash stands of types 2A and 3A have lately been converted to beech plantations (which carries them beyond the scope of the semi-natural stand types), but they might revert to these types if the beech is felled.

The natural rotation of tree species in virgin forest and naturally regenerated managed stands appears to be widespread in continental woods (Köstler, 1956). Tree species tend to regenerate on ground occupied by other species, as in the selection stands (Section 4.3) of silver fir, spruce and beech, where silver fir regenerates under spruce and spruce under beech. It is possible that the same phenomenon occurs in British woods, even though it has not been convincingly demonstrated. Failure of western oakwoods to regenerate has led to speculation that they only survive by alternating through a phase of birch dominance. Beech rapidly colonises many acid oakwoods, but in the New Forest, where beech has been established for many centuries, there are signs of oak resurgence in some beechwoods (Flower, 1977). Sycamore is invading many woods, but sycamore saplings are seldom abundant in mature sycamore stands, and may be common in adjacent stands of other species (e.g. Dovedale, Peak District).

10.4 Long-term successions

Irreversible natural changes also seem to be taking place in the composition of ancient, semi-natural woods. Suckering elms have been invading many woods (Rackham, 1971; 1975) to form stand type

10A where type 2A formerly grew (Section 8.10). This succession was admittedly mostly made possible by planting, but once established, the elms appear to be capable of spreading and holding the captured ground, Dutch elm disease notwithstanding. Likewise, sycamore is an introduced species, which owes its presence in most woods to planting, but it has been able to spread into many stand types throughout Britain. Like the suckering elms it can grow on a wide range of soils and seems to be capable of invading all stand types, but – again, like the elms – it is concentrating its invasion into part of the range, in this instance the ash—wych elm woods (Group 1) and the moist wych elm variant of calcareous pedunculate oak—ash—beechwoods (8Cb). These, significantly, are the woodland types within which sycamore is naturally most frequent on the Continent.

The same pattern of invasion can also be discerned in the earlier invasion of beech and hornbeam into the 'mixed oak' forests: colonisation was assisted by man and proceeded across a wide edaphic front whilst concentrating upon certain soil types. Beech was widespread but rare in southern England during the Atlantic period, but it expanded later in association with human disturbance of natural forests (Godwin, 1975). Natural changes – possibly in the soil – may also have assisted this late expansion, for a similar increase in beech took place at the end of previous interglacials at a time when man's influence was hardly significant. Beech has mainly colonised acid sands and thin calcareous soils, partly by invading directly into existing woods and partly by entering into secondary succession on cleared ground. On acid sands beech presumably invaded mainly birch—oakwoods, acid oak—limewoods and unwooded ground which had been occupied by such stands, to form stands of types 8A and 8B. This process continued in recent centuries in the New Forest (Dimbleby and Gill, 1955; Barber, 1975) and continues today. On southern chalk slopes beechwoods of type 8C must have replaced stands of types 2C, 3B and perhaps 4B. Here replacement was probably indirect, for surviving examples of these stand types on the chalk (e.g. Alkham Valley, Kent, Gopher Wood, Wilts, Clouts Wood, Wilts) are not being invaded by beech today. These examples have relatively deep soils, suggesting that beech has invaded on rendzinas created after clearance and erosion of the scarp soils. Further west, towards the edge of the beech range, many stands of types 1A, 1C, 2C, 3B, 3C and 4C must have been changed to types 8C and 8Eb, whilst on acid soils stand types 5B, 6A and 6B were being changed to 8A, 8B and 8Ea.

Hornbeam, like beech, appears to have expanded in response to natural soil changes and disturbance to natural forests, colonising a wide range of soil types, but concentrating its advance on heavy, poorly-drained soils. Hornbeam invasion is rare in contemporary woods, but in the past it must have mainly colonised stand types 2A, 3A and 6Dc to form type 9A and type 6Cc to form 9Ba. Type 9Bb may have been formed from a virtually extinct group of type 4C stands, which survive as vestiges in Farningham and Covert Woods (Kent).

The Flandrian succession of woodland types can be described in terms of the existing stand types. By the early Boreal pine and birch woods of groups 11 and 12 were widely established, though the stands may not have been as pure as their modern successors. They were displaced over most of Britain by the Atlantic 'mixed oak' forests, which broadly correspond to stand groups 1 and 3–7. Apart from the wetter alderwoods, which are beyond the edaphic tolerance of most species, and planted stands, the modern successors of these forests tend to be mixed in composition with no species becoming absolutely dominant. Field maple was evidently not present at this time, and must later have displaced some stands in groups 3 and 4 to form ash–maple woods of group 2, whose modern successors are also rarely dominated by one species. Subsequently, during the Sub-Atlantic, beech and hornbeam expanded in the south at the expense largely of stands in groups 1–6. Both species could colonise a wide range of soils, but tended to concentrate on complementary soils, with beech on freely drained, and hornbeam on poorly drained, soils. Still later, suckering elms and then sycamore have been expanding in much the same fashion, mainly at the expense of groups 2 and 3 and group 1 stands respectively. Apart from maple, the post-Atlantic colonisers have tended to dominate stands far more than the species of Atlantic forests. Whilst beech and hornbeam were advancing, wych elm and lime have declined, so stands in groups 1 and 4 have changed to groups 2 and 3, and stands in group 5 have changed to group 6.

Chapter 11

British woodland types in a European context

British woods have never been thoroughly classified in terms of the main groupings of European vegetation, although a number of partial attempts have been made and a few associations have been described. This deficiency should be remedied by the forthcoming National Vegetation Classification, but meanwhile it seems worth considering where the stand types described in Chapter 8 fall within the main woodland classes, orders and alliances of Continental Classifications. Unfortunately, this is far from straightforward. Not only do the major groupings and the relationships between them vary from one author to another, but some British woods of more oceanic districts should probably be regarded as distinct Alliances and Sub-Alliances, which have not yet been properly described. The main groupings of Oberdorfer *et al.* (1967) for West Germany, to which the work of most other authors having a bearing on British woodlands can be related, seem to provide a reasonable path through the maze, and have largely been accepted here. Professor Frank Klötzli's (1970) little-known but valuable account of British mixed deciduous woods beyond the beech and hornbeam zones has been the major source of guidance. No attempt has been made at a comprehensive review of continental woodland classifications, but the works of Oberdorfer (1957; 1970), Westhof and den Held (1969) and Ellenberg and Klötzli (1972) have been consulted.

11.1 Class Betulo–Adenostyletea

Class   **BETULO – ADENOSTYLETEA** Br.–Bl.48

Order   **Adenostyletalia** Br.–Bl.31
'Tall-herb' communities of sub-alpine zone

Alliance   DRYOPTERO – CALAMAGROSTIDION Nord. 43

Birks (1973) describes a *Luzula sylvatica–Vaccinium myrtillus* association on Skye, growing on ungrazed and unburned ledges on strongly acid, deep raw humus. Trees are normally absent, but occasionally birches are present, placing the association within stand type 12A. Similar communities occur sparingly in western and northern Britain, Ireland, western Norway and the Faeroes.

Alliance   MULGEDION ALPINI Nord. 43

Birks (1973) describes three associations on Skye within this alliance, of which one, the *Betula pubescens–Cirsium heterophyllum* association, commonly occurs within woodland on freely-drained, neutral, brown earths with a high base status. The rich field layer is invariably ungrazed and includes *Cirsium heterophyllum*, *Filipendula ulmaria*, *Geum rivale* and *Trollius europeaus*. This association corresponds with stand types 12A and 12B, for on Skye the associated trees are birch (*B. pubescens*), hazel and rowan, whilst the Morrone birchwood on Deeside (which is a good example of 12A) is also placed by Birks within this association. The association is found south to the Pennines. It is related to sub-alpine birch scrub of Scandinavia, Greenland and Iceland. McVean and Ratcliffe (1962) describe a *Betula*–herb nodum, which appears also to fall (at least partly) within both stand type 12A and this alliance.

11.2 Class Salicetea Purpureae

Class   **SALICETEA PURPUREAE**
Moor 58
Willow and poplar woods of river banks and wide valleys
Order   **Salicetalia purpureae** Moor 58
Alliance SALICION ALBAE (Soó 36) Tx. 55

Woodlands of poplar, willow, elm, alder, etc. on river flood plains are inconsistently treated in continental phytosociological literature, and it is not at all clear how they relate to British vegetation. The broader valleys of the English lowlands contain no ancient woodland now, but hedgerow trees and patches of scrub and secondary woodland contain what appear to be fragments of woodlands once dominated by *Alnus glutinosa*, *Fraxinus excelsior*, *Populus canescens*, *P. nigra*, *Quercus robur*, *Salix alba*, *S. fragilis*, *S. triandra* and *S. viminalis*. (See also *Ulmion carpinifoliae* below.)

11.3 Class Alnetea Glutinosae

Class   **ALNETEA GLUTINOSAE**
Br.–Bl. et Tx. 43
Order   **Alnetalia glutinosae** Tx. 37
Alliance ALNION GLUTINOSAE (Malc. 29)
Meijer.–Dr. 36
Communities of alderwoods on damp soil

Klötzli (1970) divides the valley alderwoods of Britain into an *Osmundo–Alnetum* and *Pellio–Alnetum*. The *Osmundo–Alnetum* encompasses the alderwoods of fens and nutrient-rich swamps including the woods fringing the Norfolk Broads, Cothill Fen and the valley alderwoods of the New Forest. It is equivalent to types 3b and 3c of McVean (1956), i.e. the wetter woods of moderate or high base status, and is largely

co-incident with stand type 7B. Klötzli regards it as the oceanic equivalent of the *Carici elongatae–Alnetum*. This and related associations are widely distributed in Europe.

The *Pellio–Alnetum* of nutrient-deficient, acid and occasionally mineral soils is said to be equivalent to McVean's types 2b and 2c, although it also seems to extend into types 1a, 1b and 2a. Broadly it is equal to stand type 7A. Klötzli's examples are found in northern and western Britain, with no indication that the community can occur in the south and east. A number of related communities have been described from western Europe, but otherwise this type of alderwood seems to be absent from the continent. Klötzli regards the *Pellio–Alnetum* as a replacement for the *Alno–Padion* alliance in oceanic climates.

11.4 Class Vaccinio–Picetea

Class   **VACCINIO – PICETEA** Br.–Bl.39
Order   **Vaccinio–Picetalia** Br.–Bl.39
Spruce, pine and birch communities of northern, eastern and montane Europe on acid, podzolic soils.
Alliance VACCINIO – PICEION Br.–Bl.38

McVean and Ratcliffe (1962) placed four woodland and scrub associations of the Highlands within this alliance, namely the *Pinetum Vaccineto–Callunetum* of open pinewood and pine–birch mixtures, *Pinetum Hylocomieto–Vaccinietum* of moderately dense pinewoods, *Betuletum Oxaleto–Vaccinetum* of grazed birchwoods with rowan and juniper, and the *Juniperus–Thelypteris* nodum of the Cairngorm juniper scrub. Birse and Robertson (1976) place two groupings here, namely the *Erica cinerea–Pinus sylvestris* plantations and the *Betula pubescens* community. The last two communities, which are from lowland southern Scotland, are respectively plantations and secondary woodland, and thus not strictly within the scope of the stand types. Although these communities are clearly equivalent to stand types 11A and 12A, they are not coincident: type 12A also overlaps into the *Betulo–Adenostyletea*, and, since both hazel and oak may occur in the *Betuletum Oxaleto–Vaccinetum*, this association must extend into stand types 6A, 6B and 12B, where it shows some of the characteristics of the

Quercetea robori-petraeae (McVean, 1964). The *Betula pubescens* community is comparable to the *Vaccinio–Betuletum pubescentis* Tx. 37 of north-west Germany (Birse and Robertson, 1976) and both this and the *Erica cinerea–Pinus sylvestris* plantation appear to be similar to the secondary pine and birch woodlands on the heathlands of southern England. McVean (1964) equates the *Pinetum Hylocomieto–Vaccinietum* with the *Pineto–Vaccinetum myrtilli*, which is widely distributed in north-west Europe, and likens the *Pinetum Vaccineto–Callunetum* to the *Mastigobryeto–Piceetum* spruce forests of central Europe.

11.5 Class Quercetea Robori-Petraeae

Class   **QUERCETEA ROBORI-PETRAEAE** Br.–Bl, et Tx. 43

Order   **Quercetalia robori-petraeae** Tx. 31

Alliance   QUERCION ROBORI-PETRAEAE (Malc. 19) Br.–Bl.32

West European deciduous woodlands on acidic to mildly basic soils

These, the acid birch–oak woodlands, largely correspond to stand group 6. They have been divided by Doing (1975) into two sub-alliances. The *Vaccinio–Quercion* consists of the heathy oakwoods in strongly acid soils often with mor humus, which tend to occur in colder, wetter climates and have many species in common with natural conifer forests. They can be broadly equated with the birch sub-types of stand types 6A to 6D. *The Violo–Quercion* is a relatively species-rich form of oak woodland which develops on less acid and more fertile soils. It has a tendency to occur in warmer, drier climates and to have floristic links with the *Quercion pubescenti-petraeae* (see below). These are broadly equivalent to the hazel sub-types of stand types 6A to 6D. Doing notes that both sub-alliances may occur in fine grained mosaics within single woods, but that alternatively some natural regions may be dominated by associations from just one sub-alliance. In Britain the *Vaccinio–Quercion* tends to predominate in the west and north, the *Violo–Quercion* being better represented in the south and east.

Klötzli (1970) recognised two associations in Britain which appear to cut across Doing's sub-alliances. The *Blechno–Quercetum* broadly corresponds with the concept of 'western oakwoods', i.e. principally stand type 6A, but also 6B and perhaps some stands in 6C and 3D. The *Querco–Betuletum* comprises the lowland oak–birchwoods in a less oceanic climate, i.e. stand types 6C and 6D. Birse and Robertson (1976) described a *Galio saxatilis–Quercetum* association on dry, acid sands in lowland Scotland, which also overlaps Doing's sub-alliances. This appears to be closest to Klötzli's *Querco–Betuletum*, yet it falls within stands types 6A and 6B. It is closely related to McVean and Ratcliffe's (1962) *Betuletum Oxaleto–Vaccinetum*, which includes oakwoods in the southern and east central Highlands. Birks (1973) describes a *Betula pubescens–Vaccinium myrtillis* association, examples of which fall within stand types 6A and 12A, and a *Corylus avellana–Oxalis acetosella* association. The latter occurs as hazel scrub which is perhaps best regarded as an attenuated form of stand types 1D and even 3C, and comes close to the *Querco–Fagetea*.

The *Blechno–Quercetum* is abundant in Ireland (Kelley and Moore, 1975) from where it was first described by Braun-Blanquet and Tüxen (1952). Related woods are widespread in north and west France, Belgium, Holland, south Sweden and Germany, where many stands are perhaps derived from beech forest by coppice management, and also in central Europe. Klötzli sees many similarities between the *Blechno–Quercetum* and acid oakwoods in Mediterranean climates of Galicia and Cantabria, all of which link with, and are physiognomically related to, evergreen forests in the Mediterranean region and the northern coniferous zone.

The precise extent of the *Quercetea robori-petraeae* in relation to the stand types is uncertain. Doing places the *Luzulo–Fagion* (i.e. the acid beechwoods of stand types 8A and 8B) here, but others do not, and these woods are mentioned under the *Querco–Fagetea*. Some stand types seem to be transitional between the *Quercetea robori-petraeae* and *Querco–Fagetea*, notably types 1D, 3D, 5A and 5B. Oberdorfer (1957) places some hornbeam–sessile oak–birchwoods of southern Germany within the *Violo–Quercetum*, so perhaps stand type 9Ba fits here.

11.6 Class Querco–Fagetea

Class **QUERCO – FAGETEA** Br.–Bl, et Vlieg. 37
Central and west European deciduous woodlands and scrub on base-rich soils and their semi-natural derivations

Order **Prunetalia** Tx. 52

Most secondary scrub communities in Britain fall within this order

Order **Quercetalia pubescentis** Br.–Bl.31
Alliance QUERCION PUBESCENTI-PETRAEAE Libb. 33

The mixed oakwoods of warm climates and dry, freely-drained soils, which are found mainly in southern Europe where the temperate deciduous woodlands impinge on evergreen Mediterranean woodlands, are not represented in Britain, but the mixed woodlands on drier soils of south-western limestones in Britain are closely related. Klötzli (1970) christened these limestone woods *Hyperico–Fraxinetum*, an association which more or less matches stand types 1C, 3C and 4C. Woodlands on the northern limestones of the Pennines, Cumbria and north-west Scotland have been assigned to a *Fraxinus excelsior–Brachypodium sylvaticum* association (McVean and Ratcliffe, 1962; Birks, 1973) which overlaps the *Hyperico–Fraxinetum* and *Dryopterido–Fraxinetum* of Klötzli and the *Corylo–Fraxinetum* of limestone woods in Ireland (Braun-Blanquet and Tüxen, 1952). Shimwell (1971) considered that these associations should be a new alliance, the *Fraxino–Brachypodion*. This would cover stand types 1C, 3C, 4C and the northern extremites of 1A.

Order **Fagetalia silvaticae** Pawl. 28
Alliance ALNO – PADION Knapp. 42
Sub-alliance *Alnion glutinoso-incanae* Oberd. 53

These continental valley alderwoods are rare in Britain where they are largely replaced by the *Pellio–Alnetum* (see above) (Klötzli, 1970). The most likely examples in Britain are the drier-valley alderwoods of southern and eastern Britain especially the highly localised type 7Ea, but perhaps also the drier forms of both 7A and 7B. Even though it is found in an oceanic climate, it may be reasonable to include

stand type 7Eb here also, for some rare British species – *Circaea alpina*, *Chrysosplenium alternifolium*, *Impatiens noli-tangere* – are loosely associated with this type and characteristic of the *Alno–Padion* on the continent. Associations within this sub-alliance are widely distributed in continental Europe.

Sub-alliance *Ulmion carpinifoliae* Oberd. 53

Valley oak–elm woodlands are widespread in the major river valleys of the Continent, where they are accompanied by ash, poplars, maple, sycamore and many other tree species. Such woodland may once have been present in the larger river valleys of lowland England (Jones, 1959), but it has long since been cleared and *Populus nigra* is no longer regarded as a woodland species. Perhaps the sole surviving stand type within this sub-alliance is 10B, although many of the suckering elm woods within type 10A seem to be closely related.

Alliance CARPINION BETULI (Issl. 31) Oberd. 53

The mixed broadleaf woodlands on mainly poorly-drained soils in Britain have not apparently been properly related to continental types, but two of the types described by Klötzli (1970) fall within this alliance. One, the *Hyperico–Fraxinetum* has been discussed above, and the other is the *Querco–Fraxinetum*. Klötzli's treatment of these woods and the examples upon which he draws show that the *Querco–Fraxinetum* corresponds to the sum of stand types 1B, 2A, 2B, 3A and 7C, the southern and eastern examples of 2C, and may extend to the drier forms of 7B and some examples of 6Dc. Type 10A may also be best placed here (but see above). Klötzli does not mention lime, but even so types 4A and 4Ba appear to fall within this association. Essentially, these are the lowland mixed deciduous woods beyond the main ranges of beech and hornbeam, but which may include a limited amount of hornbeam. Similar woods are common in the coppices of northern France, western France and Belgium, and related types evidently occur in central European coppices. However, hornbeam is much more abundant on the Continent, and the relationship is therefore closer with stand group 9 (see below). Birse and Robertson (1976) place their *Aceri–Ulmetum glabrae* association in the *Carpinion*; this corresponds to stand types 1Ab and 1D.

The *Querco–Fraxinetum* can perhaps be regarded as an attenuated form of the mixed hornbeam woods which are so widespread and abundant on the Continent, whereas the hornbeam stand types 9A and 9B are geographically and floristically closer to the main body of these woods. Apart from sub-type 9Ba, which may fit better in the *Quercion robori-petraeae* (see above), the British hornbeam woods can all be placed within the main types of continental *Carpinetum*. The *Stellario–Carpinetum* contains a wide range of types with pedunculate oak, which correspond to most of stand type 9A. Most of the stands within this group appear to be species-poor, growing on poorly drained light to medium-textured, acid soils. The *Querco–Fraxinetum* can be regarded as an extension into an oceanic climate of a group which as a whole appears to be more characteristic of a sub-atlantic climate (Oberdorfer, 1957). The *Galio–Carpinetum* of heavy, base-rich loams consists mainly of the sessile oak–hornbeam woods of central Europe, which are represented in Britain by the rare type 9Bb.

Alliance   FAGION SILVATICAE Tx. et Diem. 36

Beechwoods of various kinds are the major alliance of central and southern European mixed deciduous woodlands. They certainly encompass the British beechwoods of stand group 8, but there seems to be uncertainty about how far they extend beyond them. For example, Birks (1973) places the *Fraxinus excelsior–Brachypodium sylvaticum* association of north-western limestones within the *Fagion*, but notes Shimwell's view that this should be placed into a separate, new alliance. Likewise there is some doubt about whether the acid, species-poor beechwoods should be placed here or in the *Quercetea robori-petraeae*.

Sub-alliance *Luzulo–Fagion* Lohm. et Tx. 54

Klötzli (1970) and others agree that the species-poor beechwoods on acid soils in southern England (stand types 8A and 8B) are closely related to the widespread acid beech–oakwoods of north-western Europe and grade into the montane acid beech–fir woods of central Europe. The *Ilici–Fagetum* and *Fago–Quercetum* associations correspond closely to British communities. These are described in Gehu (1975).

Sub-alliance *Asperulo–Fagion* Tx. 55 em Th. Hull 66

Beechwoods of moist, moderately base-rich soils are uncommon in Britain, although they are abundant and widespread in central Europe. In Britain they occur mainly on the margins of acid beechwoods, and include Watt's (1934) types A_0 and A of the Chiltern plateau, in an area where, significantly, *Cardamine bulbifera* is present. Thus, type 8D, and perhaps sub-type 8Ea can be regarded as the extremity of this sub-alliance. Klötzli (1970) notes that an *Endymio–Fagetum* association occurs on brown earth in Britain.

Sub-alliance *Daphno–Fagion* Th. Hull 66

This group of calcareous beechwoods seems hard to distinguish from the next sub-alliance, and indeed it partly corresponds to *Cephalanthero–Fagion* of Oberdorfer (1957). It appears to correspond best to the ash–beechwoods on deeper soils especially type 8Cb. Klötzli (1970) notes that British beechwoods on limestone correspond to a *Daphno–Fagetum* on the Continent.

Sub-alliance *Cephalantherio–Fagion* Tx. 55

These beechwoods of dry, thin, calcareous soils of central European limestones include most of stand types 8Ca, 8Cc and 8Eb in Britain, especially the steep, scarp woods of the chalk and oolite and the mixed beechwoods on carboniferous limestone in the Wye Valley area.

Alliance   TILIO – ACERION Klika 55

Klötzli (1970) described a *Dryopterido–Fraxinetum* of base-rich, freely-drained and generally alkaline soils, which occurs mainly in western Britain. This corresponds to stand types 1A, 4Bb, 7D and parts of 2C, and may extend on to a few stands within 1D, 3D, 6Ac and 6Bc. Klötzli regards these as *Tilio–Acerion* woodlands with a distinctive dominance of evergreen ferns which is characteristic of the oceanic climate in which they occur. Similar woods are common in Ireland, where they form part of the *Corylo–Fraxinetum* of Braun-Blanquet and Tüxen

(1952), and in Brittany. Associations within the *Tilio–Acerion* are frequent on base-rich, moist, freely-drained soils in central Europe, and seem to be especially characteristic of ravines in montane districts. Given the importance of sycamore to this alliance, it is no surprise to find this species so frequently in the related British stands.

PART THREE

Woodland nature conservation

Chapter 12

Objectives and priorities of nature conservation in British woodlands

12.1 Exploitation, management and conservation

Nature conservation has been defined variously as an attitude of mind and a form of land management, and of course it is both. It is an expression of concern that man should have a sensitive relationship with his environment which can be sustained indefinitely, but it must be expressed practically in the way in which land (i.e. soils and vegetation) and populations of individual species are utilised. Other attitudes to and forms of utilisation are encountered which range from the short-term materialism of exploitation to the long-term view of conservation, which tempers materialism with scientific, aesthetic and spiritual considerations.

12.1.1 Man in relation to renewable resources

Before considering nature conservation in relation to forestry, it is necessary to consider some general features of man's relationship with renewable resources. Simplifying enormously, man's initial impact on a renewable resource has usually been at low intensity, but without heed for the long-term consequences. Later, as the intensity of use has increased, man has tended to exploit the resource, i.e. to utilise it at a rate which cannot be sustained or whose adverse effects could become irreversible. Eventually, after perhaps a crisis has shifted the balance from the exploiters to those with a longer view, exploitation is replaced by sustainable management. The timing of such changes has varied from one resource to another. For example, deer populations in Britain were managed sustainably more than one thousand years ago, whereas utilisation of much of the world's rain forest will pass from natural conditions through exploitation to sustained management within the present century. Some resources have been so badly exploited that they are no longer available to manage, e.g. the northern hemisphere whales.

Exploitation implies high-intensity utilisation, but sustained management can be conducted at both high and low intensities. For example, traditional agriculture in Britain was essentially a form of low-intensity sustained management (though it contained some elements of exploitation), whereas modern agriculture is high-intensity management whose capacity for material yields is believed to be indefinitely sustainable. In a material sense, therefore, neither traditional nor modern agriculture can be classified as exploitation, but the difference in intensity of management is significant. Under traditional agriculture, semi-natural grasslands, wetlands and other uncultivated ground survived within the system, but under modern agriculture the grasslands are converted to leys, the wetlands are drained and man exerts a tight control. The natural elements in the landscape diminish as the intensity of land use increases.

Woodland Conservation and Management

Man's relationships with species vary in sustainability and intensity, just as they do with resources in general. The character of the relationship varies from pest control through sustained yield harvesting to preservation from extinction, and the intensity varies from nil to, say, harvesting intensively from specially bred stock. Native species are harvested according to their utility and either preserved or controlled according to their tendency to decrease or increase in the face of man's activities and the extent to which they compete with man. Changes occur in relationships. Rabbits, originally introduced and carefully farmed as a crop, became a pest in need of control. On the other hand some former crop varieties have had to be preserved like wild species, e.g. by old breed societies.

12.1.2 Conservation and nature conservation

The term 'conservation' refers to part of man's relationship with renewable resources, but the exact limits of usage are vague. Moreover, 'nature conservation' is used by some in a narrower or different sense and by others as interchangeable with 'conservation'. These uncertainties cause significant misunderstandings in the relationship between forestry and nature conservation.

'Conservation' is taken to include all forms of sustainable management, irrespective of the intensity of use. One can conserve the capacity of soils or fish populations to yield food, but where this is done at a high intensity of use the relationship cannot be described as a form of nature conservation. 'Nature conservation' relates to low-intensity utilisation, i.e. that which permits a significant number of natural features to coexist with the features created by man. Thus, traditional agriculture was reasonably good for nature conservation (as well as being good conservation), but modern agriculture is bad for nature conservation (though, hopefully, is good conservation). Since 'nature conservation' relates to intensity rather than sustainability, one can include within its meaning non-sustainable, low-intensity relationships, such as the classic retrogressive successions under primitive exploitation. These relationships are summarised in Fig. 12.1.

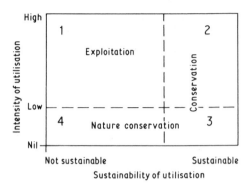

Fig. 12.1 A diagram representing the difference between 'conservation' and 'nature conservation' as applied to woodlands and their management. Examples of activities in each box of the diagram are:
(1) Clear felling without regeneration due to fire, grazing, erosion, etc.
(2) Intensive plantation forestry in which the main characteristics of the stand and site are closely controlled by foresters.
(3) Low intensity treatment yielding some timber and other products in which some natural features are allowed to survive, e.g. traditional coppicing.
(4) Retrogressive succession from woodland to a less productive, non-woodland ecosystem, e.g. by use of woods as pasture.

The link between 'nature conservation' and low-intensity human impact is reasonably clear as far as habitats are concerned, but less so with species. Here the different kinds of relationship with man are significant, for 'nature conservation' tends to be applied to the preservation of species against extinction and not to pest control. Sometimes the term extends to the harvesting of native species, e.g. the Nature Conservancy's harvesting of Red Deer on the island of Rhum, or of native pines from the extraction zones of native pinewoods (Section 18.4).

Nature conservation is therefore not just the antithesis of exploitation, but also both a special form of and separate entity from broad resource conservation. In practice it includes:

(a) Low-intensity ecosystem management which enables some natural features to survive in a managed environment;
(b) Retention of some areas in a natural state without any significant impact by man;

194

(c) Preserving threatened species against extinction, and perhaps cropping native species. [Maintaining a rich ambient wildlife of native species forms part of item (a).]

12.1.3 The benefits of nature conservation

The benefits to man of nature conservation cannot easily be expressed as a list, because the items are clearly inter-related. Material benefits come directly from, for example, the venison of deer, the fur of seals and the fruit of *Rubus fruticosus*. Some wild species are important sources of drugs and other useful chemicals. Less directly the value of wild genotypes in breeding particular strains of crop plants and animals is sufficiently important to justify, for example, the conservation of wild wheats and the creation of gene banks. The more spectacular natural environments and aggregations of species are considerable tourist attractions, which provide pleasure for tourists and a stimulus to local economies. Likewise, those wild animals which are hunted provide sport for some and money for others. Wildlife and natural communities are a focus of interest and provide an absorbing hobby for naturalists across a wide range of knowledge and intensity of interest. They are also the subject of scientific research, at once a means to creative pleasure for those involved and those who appreciate the results, and a source of new opportunities for more material benefits.

Another benefit of nature conservation which is not immediately obvious is the preservation of controls against which the impact of man on the environment can be measured. For example, the total effects of man on soils since they were cultivated can be measured if undisturbed, natural soils survive for comparison with cultivated soils on the same kind of site. A comparison between the structure and function of invertebrate populations in natural woodland and monoculture plantations might, for example, provide clues for the biological control of Pine Beauty moth in Scottish pine plantations, far healthier for the environment than the current blanket insecticide spraying, which from North American experience has to be maintained indefinitely once it has begun. Controls are an essential feature of the scientific method: hence the frequent use in nature conservation of the phrase 'scientific interest'.

12.2 Wilderness and nature conservation

To the more obvious and direct benefits of nature conservation we must add another which is more fundamental and perhaps more difficult to appreciate: satisfying the human need for wilderness. Wilderness is a particularly difficult condition to define, but essentially it refers to uncultivated and undeveloped land where man exercises no control. The word originated in northern Europe, where it referred specifically to primaeval forests, but translators of the Bible applied it also to the deserts beyond the cultivated lands of the Middle East. Expanded in this way, wilderness represented for centuries a threat to civilisation, something over which man had no control and which had to be removed or subjugated. Pioneers who tamed the wilderness were regarded as heroes. Those who made the desert bloom and the waste ground fruitful were unquestionably doing good. However, an alternative view of wilderness as a beautiful, sublime condition in which man could find solitude, live close to nature and witness the work of the Christian God developed by the 18th century and led to modern civilisation's dual and ambivalent attitude towards wilderness. Now, whilst there are many who espouse what is essentially the original view that uncultivated land should be cultivated and who even regard the absence of cultivation as morally wrong (literally 'waste'), others regard wilderness as an essential safeguard against over-civilisation, where they can 'get away from it all', or know that they can do so if the need arises. This does not just take the form of a need for large tracts of wilderness to be preserved against management and cultivation; any wild creature going about its business can give the human observer a balancing perspective on his own existence.

In America colonists from Europe encountered a huge wilderness. Pioneers colonised and cleared assiduously until by the late 19th century all but a few patches had been civilised and cultivated. The pioneer spirit had, however, given Americans a distinctive culture which in turn generated public support for preserving the remaining wilderness areas. Despite

Fig. 12.2 Lodore Woods, Borrowdale, Cumbria, an English wilderness with trees. Sessile oakwoods clothe the rugged slopes on the margin of Derwentwater. Above and beyond the open hills stretch out. This is by no means a wilderness in the strict sense, for the hills are grazed by sheep and the woods have been intensively coppiced. Nevertheless, it is relative wilderness where man's influence is unobtrusive enough to allow visitors to feel and be refreshed by a sense of wilderness. Forestry in only a small part of the view could easily damage the essential illusion without greatly changing the wildlife of the valley. (Photo. D. A. Ratcliffe)

enormous counter-pressures, the Yosemite Valley was reserved in 1864, Yellowstone National Park was designated in 1872, and now the United States has a series of reservations which are huge by British standards. The significance of wilderness in general and to Americans in particular has been reviewed by Nash (1973).

Britain does not have wilderness in the strict sense, only what Derek Ratcliffe (personal communication) has called 'relative wilderness', where the hand of man is not absent but is unobtrusive (Fig. 12.2). The most extensive areas are the upland moors and mountains where man's influence has been profound, but where it has been so even that it is not immediately apparent. Here the advent of forestry in the form of straight-sided plantations of one species growing in straight lines is a strong civilising influence which destroys wilderness. In the lowlands the relative wildernesses are small, but no less significant, as the following story shows.

Overhall Grove (Cambs) is a mature, 20 ha suckering elm woodland which has not been managed for 60 years and is now a nature reserve, famous for its spring displays of oxlips and bluebells. At weekends during April and May it is thronged with visitors, some of whom come from tens of miles away, and as an honorary warden I had many opportunities to find out what attracted them. Most, of course, spoke of the flowers, but many clearly thought there was something more, the freedom to walk anywhere beneath the massive elms, oaks and ashes in a wood where there were no obvious signs of management and the stream still meanders in an apparently natural course. As one visitor said: 'I come here every year. It makes me expand! I always go away feeling happy.' Not everyone is impressed with the clinging clay of the tracks, but for many visitors even a small encounter with wilderness is important.

The need for wilderness originated in cities as a counterpoint to the artificial contrivances of modern living and a release from the claustrophobia of urban crowds. This need is particularly acute in Britain, where the population on holiday heads for the beaches, the mountains and indeed any available patch of attractive (usually less-intensively cultivated) countryside. And yet, with the intensification of agriculture and forestry, the available relative wilderness diminishes annually. Though it is rarely cited as an objective of woodland nature conservation, I believe that maintaining an element of wilderness for the psychological health of the nation is important, perhaps more so than anyone can prove.

The need for wilderness can also be expressed in more practical terms. Large wilderness areas provide some safeguard against periods of political weakness or loss of political interest in the natural environment, when exploitation may be resumed, for they are ecosystems which (internally at least) do not depend on man for survival, and, being large, they would not quickly be destroyed. The species which is living wild is ultimately safer than the species which survives only in zoos and botanical gardens.

12.3 Forestry and nature conservation in Britain

12.3.1 Historical review

Woodland in Britain has long been exploited. The consequences were initially slight, but eventually vast expanses of woodland were cleared, some woodland types were completely lost, and many species became extinct. Admittedly this benefited man, who gained land for cultivation as well as freedom from attacks by wolves. The crisis and transition to management took place in stages and at various times in different places. The first change from clearance and wood-pasture exploitation to coppice management (a low-intensity system) took place in or before the Dark Ages in much of England, but as late as the 18th century in parts of Scotland. Simultaneously, some wildernesses were set aside as Crown Forests and chases, which were effectively a form of habitat-based nature conservation focused on certain beasts of the chase. Political stress, social instability and external attack allowed exploitation to return at intervals, such as the Dissolution, the Commonwealth and the World Wars of the present century. Inefficient management led to chronic semi-exploitation in monastic estates and Crown woodlands, which eventually generated another crisis in the 17th century, when usable timber was scarce in the vicinity of ship-building yards. Once again exploitation gave way to management and increased Government interest in forestry, but forestry was now more concerned with the production of timber by more intensive management, and less with its medieval preoccupation of game conservation. During the 17th, 18th and 19th centuries, traditional, low-intensity coppice management continued beside more modern, high-intensity plantation forestry, but then, in the late 19th and early 20th centuries, the effects of exploitation again increased, because exploitation in the Empire had spoiled the economic foundation of management at home. This and World War I generated a mixture of neglect and exploitation at home followed by transition under crisis to management from which the modern forestry policy emerged.

12.3.2 Foresters as conservationists and nature conservationists

The modern forester claims to be a conservationist, and in the sense defined above this claim is substantially acceptable. He regards afforestation of upland moorland and the conversion of neglected coppice woods into plantations as good management which not only maintains the productive potential of the land but also increases it. The consequences of exploitation in the past – clearance, podzolisation and blanket peat formation, erosion, diminution in the genetic quality of the tree stock, destruction of the reserve of standing timber – are being reversed by afforestation involving peatland drainage and fertilisers, and tree breeding programmes, the introduction of new species and the accumulation of a timber reserve. Moreover, the modern forester sometimes sees himself as the lineal descendant of the medieval foresters, who practiced conservation long before conservation was separately recognised. So much is true if one accepts the foresters' belief that the practices and level of production of modern forestry are indefinitely sustainable, but this is not the point for nature conservation. The essential characteristic of modern forestry is its pursuit of high production by means of intensive management, and this brings it into conflict with nature conservation.

Modern forestry, which in its most intensive form is aptly described by foresters themselves as 'tree farming silviculture', conflicts with nature conservation in several ways. Forestry is extending beyond the point at which the deleterious consequences of past exploitation are being reversed: species are being introduced which have never grown here in the past; peatlands which never bore woodland even in the Atlantic forest maximum are being planted; and stand structures which were not part of natural woodland are being created. High-intensity management generates a stand structure and composition which is more akin to an industrial conveyor belt than either the natural woodlands destroyed by exploitation or the managed woodlands formed by traditional, low-intensity systems. Modern plantation management attempts to direct all the productive resources of the land into timber production, leaving as little as possible over for nature: in a densely stocked, fast-growing plantation, for example, most of the light is intercepted by the trees and no shrub or herb layer can develop. Many natural and semi-natural habitats have been destroyed with the drainage of bogs, afforestation of sand dunes and the conversion of old coppices into plantations. Modern forestry also destroys the wilderness aspect of many parts of Britain: this is most obvious in the uplands where seemingly wild and untamed land is covered by manifestly organised plantations with their network of roads, and generates the familiar reaction against 'regiments of conifers marching across the hillsides', but it happens too in small lowland woods which provide perhaps the only opportunity to experience a touch of wilderness within easy reach of most urban centres.

Foresters are admittedly fairly sensitive to such matters nowadays, but they have become so reluctantly – at least at the policy level – and only because outside pressures have persuaded them of the need. Many individual foresters are sympathetic towards nature conservation and very knowledgeable about wildlife, but they must act within the forestry policy, which is to grow utilisable timber as economically as possible (Forestry Commission, 1974) and cover as much land as possible with trees (Forestry Commission, 1977), and thus inevitably their contribution is limited. Modern forestry may appear to be seeking a socially acceptable balance between the conflicting needs of intensive timber production and nature conservation (and amenity, etc.), but to many people including myself the balance is not right, the benefits of nature conservation are undervalued, and a separate effort is justified outside forestry in order to ensure that the basic requirements of nature conservation are met.

Foresters will justifiably point out in response that woodlands which are managed for timber production are in fact fairly rich in wildlife, and certainly far richer than intensively cultivated ground. Several factors enable this to be so. Many plantations are still in the first 10 years of the first rotation, when wildlife is enriched. Despite the creation of huge plantations in some districts, much land remains unplanted outside the forests, providing food for vertebrates which find abundant shelter but insufficient sustenance within the plantations. Both of these advantages are ephemeral. Some land is too wet or rocky to be planted, so some semi-natural habitat remains. Likewise rides, roads and firebreaks are required, providing incidental benefit for wildlife. Trees are a long-rotation crop compared with barley, so some ecological stability can develop both in the stands and the rides. Forest management is not as intensive as it might be, so that some structural variety remains in the form of marginal belts of mature trees and wind-thrown plantations. Management in many woods is modified in the interests of amenity and game by retaining some broadleaves and moderating the pace of change. Timber trees are planted for a future market whose needs cannot be predicted, so there is always a case for diversity in planting, including broadleaf species. And, of course, some patches are managed as small nature reserves, sheltering rare species or natural features, and some larger woods are managed by the Forestry Commission (or by others with their agreement) as nature reserves (e.g. Black Wood of Rannoch, Perthshire). Whilst those who are concerned for nature conservation must welcome these facts and measures, forestry remains primarily concerned to grow timber and regards other objectives of management as constraints.

Interestingly enough, the modern debate in Britain about forestry in relation to landscape and nature conservation repeats a dispute which took place in the 1890s about the purpose of forest reserves in America and which is described by Nash (1973). John Muir argued in favour of wilderness, but was initially prepared to accept some selective cutting of mature trees as a compromise. Foresters, led by Gifford Pinchot, believed that forests should be used, albeit wisely, to produce the maximum sustained yield, and should be regarded as an economic rather than an aesthetic resource. However, the existence of wilderness was not compatible with productive forest management: under foresters' control the woods would be preserved against unregulated lumbering but the wilderness quality would be destroyed.

Modern agriculture and forestry have both been blamed for causing considerable damage to nature conservation interests (e.g. Nature Conservancy Council, 1976), but this is less a question of exploitation versus conservation (provided always that modern agricultural and silvicultural methods are indeed indefinitely sustainable), and more a matter of nature conservation versus high-intensity management. Nature conservation seeks to maintain as much as possible of the diversity, energy flows, community structures and natural features characteristic of natural ecosystems by means of low-intensity management methods and the preservation of at least some tracts of absolute or relative wilderness. Man retains a light, remote, but firm control. High-intensity management seeks to maximise the crop and

channel the energy flow through relatively few species, but in order to achieve this the ecosystem must be closely controlled, extracted nutrients have to be replaced and as a result the natural elements in the system are minimised. We see in this competition man's ambivalent attitude to wilderness, which on the one hand is something to be cherished, enjoyed and preserved, and on the other, is land ripe for reclamation, cultivation and economic development. Society, taking a long-term view and remembering that some elements of the wilderness cannot be recreated if they are ever destroyed, must weigh the benefits of nature conservation against the material benefits of high-intensity management.

12.4 Objectives of nature conservation in woodlands

The aims of nature conservation in British woods are:

(1) To maintain naturally self-perpetuating populations of all native plant and animal species throughout their range.

(2) To maintain adequate examples of all semi-natural woodland communities, including communities of trees and shrubs, field layer, epiphytes, animals and the soils and other physical features upon which they depend.

(3) To maintain other features of interest.

(4) To contribute to maintaining an element of wilderness in the British landscape.

These aims are discussed in subsequent sections. They can be achieved by the following general approaches:

(1) Maintaining or restoring traditional low-intensity forms of management and the semi-natural features they incorporate.

(2) Maintaining or restoring the few woods whose characteristics are almost natural.

(3) Directly protecting threatened species.

(4) Persuading foresters to limit the intensity of management for timber production.

(5) Maintaining a mature structure in existing woodlands and zoning the distribution of future afforestation in order to preserve an element of wilderness.

199

12.5 Priorities in species conservation

Some writers have suggested that particular species are intrinsically more valuable than most because of their interest or material value to man. Moore (1969) emphasised that the most valuable species are those of known economic value; species which can throw light on the behaviour of man; 'living fossils' which are important in the study of evolution; species which give aesthetic pleasure; those which are especially suitable for teaching; others which have been studied intensively; and those which are evolving or spreading rapidly. The number of species included by all these criteria taken together is relatively small. Helliwell (1973) has proposed three other factors which increase the value of a species: (i) species which support or influence a large number of other species, such as oak and fox, (ii) emotive species, such as national emblems or those which contribute to the character of a region, such as red deer or Scots Pine, and (iii) taxonomically isolated species, i.e. in British woods *Adoxa moschatellina* would be considered more valuable than any one of the *Viola* species which occur in woods.

Without disagreeing in any way with these generalisations, my own approach to priorities in species conservation tries to be more practical. I accept that some species may be intrinsically more valuable, but I do not depend on this point being accepted. I take as my objective the maintenance of self-perpetuating populations (i.e. in semi-natural habitats, not zoos and botanical gardens, where species depend on the unbroken continuity of concern by man) of *all* native species, from which it follows that greatest efforts and resources will have to be spent on the most threatened species. Accepting that, for example, we wish to maintain both *Cypripedium calceolus* and *Anthriscus sylvestris* as members of the British flora, we must obviously expend more effort on the former, whether or not it is intrinsically more valuable. And in this sense, any site containing *C. calceolus* will be more valuable in furthering the objectives of nature conservation than one containing *A. sylvestris*.

Although rare species must be a focus of effort and interest, it seems more useful to define priorities in terms of extinction-prone species (Section 6.2). If

these include, as Terborgh (1974) stated, most of the species of original–natural conditions and these species are now found mainly in stable habitats, then woodlands are more likely to contain extinction-prone species than any other habitat in Britain. Furthermore these species are most likely to occur in the most stable woodland type, i.e. ancient woodland. Rather than expend our efforts on rare species alone, all extinction-prone species should be given high priority for conservation resources.

Much remains to be done before this concept from Island Theory (Section 6.2) can confidently be applied to woodland nature conservation in Britain, but we can nevertheless tentatively identify the following overlapping groups of extinction-prone species in British woodlands:

(1) Rare species. With the exception of a few species, such as the endemic whitebeams (e.g. *Sorbus subcuneata*, found in the birch–oakwoods of Somerset and Devon), most rare woodland species are fairly common and widespread in other countries. *Maianthemum bifolium*, for example, is very rare in Britain, but is common in continental woods and extends to the Far East. This and other rare British woodland species, such as the New Forest cicada, are thus largely species on the edge of their international range where their populations are mostly small and isolated. The concept of rarity also includes species which are common in parts of Britain but present only as few, small isolated populations elsewhere, e.g. *Paris quadrifolia* in Scotland. Britain contains many bryophytes, lichens and ferns (e.g. *Dryopteris aemula*, *Hymenophyllum tunbridgense*) which thrive only in the moist, equable Atlantic climate. These are not all rare in Britain, but are rare or restricted on a global scale.

(2) Species with a slow dispersal/colonising ability. These are the ancient woodland species discussed in Sections 3.3, 3.6, 3.7. Many are widespread and may have large populations where they occur, e.g. *Primula elatior*; but others are rare. A species may come within this category in only part of its British range.

(3) Widespread species which tend to occur as small, isolated populations, e.g. *Cephalanthera longifolia*, *Chrysosplenium alternifolium*, the Purple Emperor butterfly. Now that semi-natural habitats are mostly

reduced to small, isolated patches in the lowlands, the species which depend on them and cannot survive on cultivated land are reduced to this condition also.

(4) Woodland species with no alternative habitat. For example, in Central Lincolnshire *Convallaria majalis*, which is confined to woodland, is more extinction-prone than *Teucrium scorodonia*, which occurs also on heathlands.

(5) Species which depend on a declining habitat. In this sense, *Endymion non-scriptus* is extinction-prone because it depends on semi-natural habitats, even though it is still common, widespread, grows in a wide range of habitats, colonises fairly readily, and can be very abundant where it occurs.

(6) Species which are subject to wide fluctuations in number. At the time-scale which practical conservation has to consider, these are mainly species with annual cycles. Many woodland butterflies are included here, and *Impatiens noli-tangere* is a good example amongst woodland herbs.

(7) Species on the top trophic rung. The top carnivores have been exterminated, but one might include the kite, buzzard, honey buzzard, and other raptors which nest in woodland.

(8) Colonial nesters. Whilst the wood ant must be included here, a better example is the Black Hairstreak butterfly, which concentrates its populations into small portions of large woods. *Primula elatior* would be included here even if it did not also qualify under (2), (4) and (5) above: although there must be hundreds of millions of oxlips in Britain, they occur as concentrated populations, and the clearance of an oxlip wood would destroy many millions of plants.

It is sometimes claimed that evolution is a natural process and therefore that there is no need for species conservation, that natural fluctuations in numbers, evolution and competition between species, and natural responses to climatic shifts will ensure that species continue to become extinct at a local, regional or even absolute scale. This is true, but it is not extinction as such that concerns nature conservationists but the accelerated rate of extinction due to human activities. In much the same way, conservationists are not concerned about erosion as a natural process, but accelerated erosion due to mismanagement.

The practical objective of species conservation is to maintain viable populations of all native species throughout their geographical and edaphic ranges. A species which is spread over a wide area is less likely to be damaged by local deleterious changes. On the edge of its ecological range a species provides opportunities to study the limits of tolerance of populations to environmental features. At the centre of its ecological range a species is likely to be most abundant, vigorous and resilient in the face of habitat change, and is most likely to be a functionally important component of the ecosystem. Populations of species should be kept as large as possible, within the limits set by interspecific competition, in order to give them the best chance of long-term survival. Some species require no assistance because they are at home in a modern urban, agricultural or plantation environment. Others are not so fortunate and need help in proportion to the size of the threat they have to meet.

12.6 Priorities in woodland habitat conservation

The second practical objective of woodland nature conservation is to maintain adequate examples of all semi-natural woodland communities. Are some types of semi-natural woodland more important than others?

Existing British semi-natural woods exhibit past- and future-naturalness. Primary woodland is the main repository of past-naturalness, whereas secondary woods have future-naturalness. It seems to me (Peterken, 1977a) that past-natural conditions and therefore primary woodlands are more important than future-natural conditions, because they can provide first-hand insights into the nature of primaeval, original-natural woodland in which man had only insignificant influence, and they are the nearest approach we have to control ecosystems (in the scientific sense) through which we can measure and appreciate the extent and nature of man's long-term impact on the environment. In addition to these intrinsic values there is the practical fact that primary woodland cannot be re-created once it has been destroyed (Peterken, 1974b): if it has, or is likely to have, any value, then we have to keep the examples we have now. Future-natural conditions embodied in secondary woods also have some value: they allow natural succession to be studied, an important ecological process in which instability is replaced by natural stability, and demonstrate the degree and speed with which past human influence on the environment can be eliminated. These features may be as valuable as those of primary woods, but in practical terms there is less concern to protect them, for they can be re-created if necessary. The same logic can be extended to plantations. Their value as habitat types is rather less than the value of semi-natural woods (they hardly satisfy the need for wilderness, for even the location of individual trees in the stand is determined by man), and they too are re-creatable.

An example from the Cambridgeshire boulder clay woodlands may clarify and amplify these points (Peterken, 1977a; Section 10.1). An ash–hawthorn woodland, developing by natural succession on abandoned fields, is developing towards a future-natural state (Section 3.1), which may take many centuries to stabilise. It matters little if one such stand is cleared away, for if in the future we need such a stand we can always allow the succession to be repeated: the only disadvantage is that we would have to wait a long time. The clearance of, for example, Hayley Wood (Rackham, 1975), on the other hand, could not be so lightly regarded. It is an ancient, probably primary woodland whose soil and plant communities have never been destroyed by clearance but have been present on the site continuously since the area was occupied by natural woodland. It exhibits past-natural features in, for example, its soil profiles, the distribution of tree and shrub communities, and the composition of the field layer. If these were destroyed by clearance and subsequently the need for these features were to be increased to the extent that our successors wished to re-create Hayley Wood as it is now, we might assume that in time this would be possible. Such an attempt has, in effect, been made at Papworth Wood and Overhall Grove, two secondary woods in the same area which have been in existence for respectively 800 and 400 years, managed under the same coppice system as was operated at Hayley Wood. Despite this long period, the isolation of these woods has proved a complete barrier to colonisation by many woodland species to one or both of these sites (e.g. *Mercurialis perennis*,

Primula elatior, *Orchis mascula*); others are apparently recent arrivals which are still spreading (e.g. *Corylus avellana*, *Anemone nemorosa*); and the effect of past cultivation and habitation on the soil is still obvious in the great abundance of, for example, *Urtica dioica*, *Poa trivialis*, *Galium aparine*, which are rare in Hayley Wood. Complete re-creation of the present Hayley Wood would clearly take at least 800 years, and possibly much longer. This alone makes Hayley Wood non-re-creatable in practice, for no society would embark on and sustain an objective which takes more than 800 years. But it is also non-re-creatable in another sense. How would we know when restoration is complete if we do not have available, as a control, another wood of the Hayley type, which has not passed through the sequence of clearance, cultivation of the cleared ground and restoration? Even if an exact facsimile of past-natural conditions can be made in reasonable time, we still need examples which have not had to be re-created in order to prove that restoration has been achieved.

Although primary woods are particularly important, two types of secondary woodland must be carefully considered. Many examples of natural succession exist which have enjoyed less than 100 years of development, but few can be identified which have progressed much further because fewer successions were initiated before 1870, and most of those which dated from that period have subsequently been modified by management. The few stands which have developed their structure and composition naturally for more than 150 years are especially valuable in studies of natural succession. Furthermore, they are the least capable of re-creation, for not only would restoration to the existing condition take 150 years, but the circumstances under which the succession proceeded cannot be repeated. The second type is ancient secondary woodland, which is important for much the same reasons. Several hundred years of ecosystem development have appreciated, and the initial conditions (i.e. relatively slight modification of the site and surroundings by early agriculture) cannot easily be repeated. Many ancient secondary woods have rich wildlife communities containing many extinction-prone species. Indeed, they are often very difficult to distinguish from primary woods on the ground.

With this admission of ancient secondary woodland we can now state that ancient semi-natural woods (i.e. primary *and* ancient secondary woods) and those recent secondary woods which have developed naturally for over 150 years are the most important for nature conservation. The former have mostly been managed by a traditional system which has largely determined which past-natural features have survived. Thus, the following five categories, collectively termed 'special woodland types', justify top priority for nature conservation:

(1) Ancient, semi-natural stands surviving from medieval wood-pasture management (Sections 2.2, 2.3).

(2) Ancient, semi-natural high forest stands, notably pine- and birch-woods in the Highlands (Section 2.6).

(3) Ancient, semi-natural coppices and high forest stands which were formerly coppiced (Sections 2.4, 2.5, 4.4).

(4) Ancient, semi-natural stands which have not been managed. These occur only in inaccessible sites such as deep, narrow gorges and cliff faces.

(5) Secondary stands originating before 1800 which have not been managed, and have therefore a natural structure.

These special woodland types cover an estimated 300 000–350 000 ha, or about one-fifth of all British woodland (Chapter 19). The great majority fall within the coppice category. Some are now managed as nature reserves and more have been scheduled as 'Sites of Special Scientific Interest' under Section 23 of the National Parks and Access to the Countryside Act of 1949, but most have no statutory recognition at all.

12.7 Other features of interest

Woodlands possess other features which are of value for nature conservation. These falls into five broad groups:

(1) Geological and physiographic features of natural origin. Many of these features, such as exposures of particular strata on a cliff face or an old quarry within a wood, are essentially independent of the presence of woodland – they would remain if the woodland were cleared – but other features owe their survival to continuity of tree cover, e.g. minor surface irregularities left by retreating glaciers in boulder clay, and the meandering courses of headwater streams in intensively cultivated areas where most have been straightened into ditches.

(2) Other habitats which occur within woods in small amounts, e.g. ponds, streams, rock faces, grassland on rides, etc. Although these are not wooded, they are part of a woodland for management purposes and the boundary between them and wooded ground is often rich in species.

(3) Physical artefacts, such as earthworks.

(4) Qualitative habitat characteristics which determine whether a stand is a good or bad example of a particular type, e.g. freedom from invasion by exotic species, a diverse structure, continuity of coppicing.

(5) Features not manifested on the ground, notably a good historical record and use for important research.

These points are considered further in Chapters 13 and 14. They do not usually determine general nature conservation priorities, but can be important factors in comparing examples of the special woodland types.

The value of relatively undisturbed soils in ancient woodlands has been summarised by Ball and Stevens (1981). Such soils can be used in the following ways:

(1) As controls for comparison with cultivated soils on the same parent material, enabling the effects of cultivation to be measured. Where the woodland remains semi-natural, the effects of planting alien species can be measured.

(2) For characterising the natural phase of a soil in support of mapping and classification.

(3) For studying the natural processes in soil formation and displaying these for educational purposes.

(4) As a source of information on the local history of vegetation and land use. Protecting evidence of geomorphic and geological history.

(5) To monitor chemical changes due to airborne pollutants.

12.8 Integration of objectives and priorities

The four objectives of woodland nature conservation

are interdependent, the distinct strands of a single rope, which has been teased apart in the preceding sections in order to show its structure. Both species conservation and habitat conservation are focused on relatively natural habitats and the species they contain rather than artificial habitats and the camp-followers of man. Extinction-prone species are found mainly in ancient woods, not as a convenient chance, but because both are functions of habitat stability and are destroyed or diminished by the instability induced by man. Thus, a programme of action for woodland nature conservation can place higher priority on maintaining a selection of the remaining special woodland types, secure in the knowledge that these will contain most of the extinction-prone species. Only rarely will sites be selected purely because they contain a rare species, e.g. Glen Diomham on Arran, which contains the endemic whitebeams *Sorbus pseudofennica* and *S. arranensis*.

Separate consideration of species and habitat conservation leads to two important practical results. First, in selecting sites to represent the range of variation, one can justifiably select more examples of species-rich types than species-poor types. Second, certain individual woods may be significant for species conservation rather than habitat conservation, and vice versa: recognition of this may clarify the type of management to be undertaken. Species in those woods, which were selected mainly as representatives of a type, will mainly be expected to look after themselves.

Conflict can arise between the wilderness aspect and the need for management. The latter by its very nature tends to diminish the sense of wilderness, yet it is often necessary to maintain particular species and communities, the general diversity of habitat or evict invading aliens. Apart from this the various objectives and the derived priorities are mutually supportive.

Chapter 13

Observation and recording in woodlands

13.1 Objectives

Woodland recording for nature conservation is essentially straightforward in intention. The recorded observations are designed to show which species are present, where they occur, how large their populations are, which communities they form together and what factors are affecting the site. Such observations relate to the moment when they are made, but they are often designed also as a baseline for measuring how fast and in what direction the observed features are changing. Survey and monitoring can also form part of a more detailed investigation within a research programme, and in these circumstances both the objectives and the methods will generally be more precise. Unfortunately, some recorders, whose objectives are concerned with nature conservation, have tried to inject a level of complexity which is more appropriate to a research project, and have thereby endowed 'recording' with an unnecessary mystique. The intention here is to discuss only those observations and records which can usefully be made for nature conservation purposes, i.e. site assessment and management. Specialised matters, such as soil survey and the monitoring of populations for demographic studies, are beyond the scope of this work.

13.2 Sources for woodland history

Past events are so important in understanding the structure and composition of existing woods that one must establish the origin and past management of any site under observation. Is a wood primary or secondary? If it is secondary, when did it originate? This is not just a matter of sources of evidence, their availability, strengths and limitation, but also a question of logic and levels of proof.

Knapwell Wood, a coppice on the Cambridgeshire boulder clay (Rackham, 1969), provides a simple example (Fig. 13.1). As a small (4.5 ha) wood abutting Knapwell Wood Farm one might think it had recently been planted for shelter, yet it is present on all editions of the Ordnance Survey back to 1824. The inclosure map of 1775 (Fig. 2.9) shows the same wood as part of a group of old inclosures in a parish which was otherwise occupied by open fields outside the village core: these inclosures formed a rounded shape which fitted unconformably into the pattern of arable strips. The only records of a wood in Knapwell before 1775 are in Domesday Book (1086), which mentions a small wood for fence material; a 12th century claim to the grove of Knapwell and the arable land which was inside the ditch surrounding the wood; and a 1279 record of an 8 acre (3.2 ha) wood, the property of Ramsey Abbey. The 12th century description 'implies a wood with adjoining fields surrounded by a ditch, which describes the present wood with its closes admirably'.

Rackham argued that the early medieval wood and the modern wood are one and the same: 'Any other

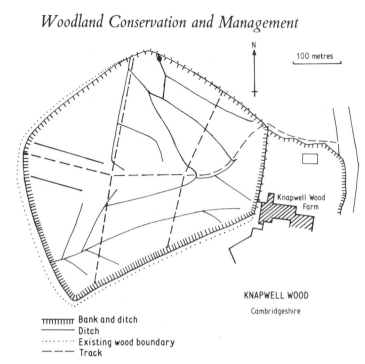

KNAPWELL WOOD
Cambridgeshire

ттттттт Bank and ditch
———— Ditch
·········· Existing wood boundary
— — — Track

Fig. 13.1 Plan of earthworks in Knapwell Wood, Cambs (redrawn from Rackham, 1969). A large medieval boundary bank and external ditch encloses the woodland. Within the enclosed area the only signs of physical disturbance are drainage ditches and a small pond. Remains of the former wood boundary run away to the east (see Fig. 2.9). On the west and south margins the wood has expanded slightly on to adjacent fields.

explanation requires both that the medieval wood has disappeared (in which case its site would have remained as an "ancient enclosure") and that the modern wood originated at some time before 1775 (in which case it would have an angular shape from the former fields and strips). It is thus almost certain that the present wood has been part of the landscape for at least 850 years.' The wood is linked to the parish boundary by a field which is still known as 'Stocking Furlong', which implies relatively late clearance of ancient woodland. In fact, Knapwell Wood appears to be a relic of the natural woods which survived on the clay plateau until post-Roman times (Taylor, 1973). The evidence from maps, documents, parish topography, boundary characteristics, field names and the general local history of clearance and settlement combines to demonstrate beyond reasonable doubt that Knapwell Wood has not been cleared since the main phase of woodland fragmentation in the area, i.e. that it is primary woodland. Similar reasoning has been used to establish that other individual woods are almost certainly ancient, and probably primary, e.g. Hayley Wood, Cambs (Rackham, 1975), Monks Wood, Cambs (Hooper, 1973), Bedford Purlieus (Rixon, 1975) and Staverton Park (Peterken, 1969a).

Four important general points emerge. First, the evidence is entirely non-ecological; if it were not, then it could not be used to characterise the site for ecological purposes without the danger of circular reasoning. Second, many independent lines of evidence, which individually merely suggest a certain course of development, are combined in mutual support to produce a reasonably complete picture. The sources are almost inevitably very incomplete, but if they all point towards the same conclusion, then they gain strength from each other. Third, it is not possible to prove absolutely that a wood is primary or even ancient, for there are gaps in the record during which a wood could have been cleared and restored. During the 1279 to 1775 gap in the Knapwell record, the site could have been cleared and restored to woodland several times, and even if the gap had been 50, not 500, years, a break in woodland continuity would have been possible. However, such a course of events is illogical and vacillating; it would involve a considerable effort, which would not have been lightly undertaken; and in any case there was a continuing market for wood and timber throughout the period. It is thus likely, but not certain, that Knapwell Wood survived throughout the historical period and its status is thus established beyond reasonable doubt. One might add that neighbouring

(a)

Fig. 13.2 Old photographs can be a valuable record of former and possibly vanished conditions. The view (a) from the Linn of Dee road at Braemar, upper Deeside, north to the lower end of Glen Quoich was recovered from a postcard bought in a second-hand bookshop in Edinburgh, and originally posted from Aberdeen station on 14 July 1932. Photographed again in March 1979 (b) we see how much of the native pine woodland and adjacent plantations were felled in the Second World War and the extent to which the population of birches in the valley has deteriorated in nearly fifty years. (Photos J. B. White Ltd, and D. Morris)

(b)

Table 13.1 The accuracy of the Ordnance Survey first edition as a predictor of ancient woodland in the East Midlands (Peterken, 1976).

	Rockingham Forest	Central Lincs	West Cambs	Total
Woodland present on the first OS and still present in 1972/3 (ha) Of which:	3854	2780	834	7468
Ancient woodland	3558	2558	598	6714
Secondary woodland originating 1650–*c*1820	296	222	236	754
Proportion of existing woodland marked on the first OS which is ancient woodland (%)	93	92	72	90

woods of the same kind were recorded in greater detail (Rackham, 1975), and thus help to fill the gap by analogy. Fourth, the status of secondary woodland is fundamentally easier to prove, for one has to establish a positive event (clearance), and not the absence of one. Thus, one can establish that a wood is *definitely* secondary, or *probably* primary.

Rackham (1971, 1976) has summarised the sources available for studies of woodland history, particularly the study of individual woods. Pollen analysis and building timbers are both useful, but the provenence of the pollen and the timber may not easily be localised. Artefacts such as banks and ditches are relevant to the wood where they occur, and are discussed below. Documentary and cartographic sources can be immensely detailed and precisely dated. The study of old documents is rarely easy for the ecologist, who needs the specialised skills of reading old script, transcribing and translating what is read, and evaluating its significance, as well as patience and serendipity. The best places to start are the County Record Office and a reference library. The former will give valuable assistance, as well as access to local papers, usually indexed by parish. The latter will provide rare local books which are sources of both information and leads. Calendars of State archives can also be very useful. The County Record Office will also have a collection of local estate maps. Specialist knowledge and skills are certainly needed for the finer

points, but an immense amount of valuable information can be obtained by intelligent persistence and luck (Fig. 13.2).

Short cuts on woodland history are available, taking advantage of the fact that most secondary woods arose recently. The earliest Ordnance Survey maps, dating from the early 19th century, show woods clearly, and any wood which has arisen since then will be obvious. Most other secondary woods arose not long before, and their origin is often obvious in their shape, position or name. Thus, the assumption that any existing wood which is present on the earliest OS map is ancient will usually be correct. Refine this by eliminating those woods on the first OS which are clearly recent in origin, and the accuracy of the assumption improves. In three study areas in Eastern England (Peterken, 1976) where the ancient woods had been identified by detailed research, the short cut to identification was highly effective (Table 13.1). In these areas at least, informed examination of the early OS map is almost as good for identifying ancient woods as a detailed historical study (see also Section 2.7).

13.3 Bank and ditch maps

Maps of artefacts – often known as 'bank and ditch maps' – are a valuable source of evidence on the history of a wood. Not only can one detect primary, ancient and secondary woodland of various ages, but one can also locate the exact boundary of such woodland on the ground (Fig. 13.3). Furthermore, if the map is extended on to the surrounding land and combined with historical and cartographic sources, the history of the existing wood and its former extent can be thoroughly established. Bank and ditch mapping is a specialised form of field archaeology which demands experience in seeing the artefacts, as well as in mapping, presentation and interpretation. The result is a record of the nature and location of all ground features which have been made or modified by man. The leading exponent is Dr O. Rackham, and his and other published examples may be found in Rackham (1971, 1975, 1976, 1980), Peterken and Welch (1975), Steele and Welch (1973) and Peterken (1976). The earliest example appears to be the 18th century map of Drayton Park (Northants) (RCHM, 1975).

Fig. 13.3 Historical evidence on the ground. Ridley Wood in the New Forest, Hampshire, was a coppice until the early 17th century (Sumner, 1926), since when it has been unenclosed. Almost four centuries later the medieval coppice boundary bank survives with a vestige of an external ditch: in the upper picture the coppice lay to the right of the bank. During the 17th and 18th centuries the wood expanded slightly (the spreading beech in the upper picture stands outside the wood bank) and, under continued grazing, developed an open structure: trees from this period can be seen in both pictures. Then a generation of narrow-crowned trees grew up in the mid–late 19th century, mainly beech and oak in the core of the wood (lower picture) and holly round the margin. In terms of the classification this is the wood pasture form of acid pedunculate oak–beech woodland, type 8B.

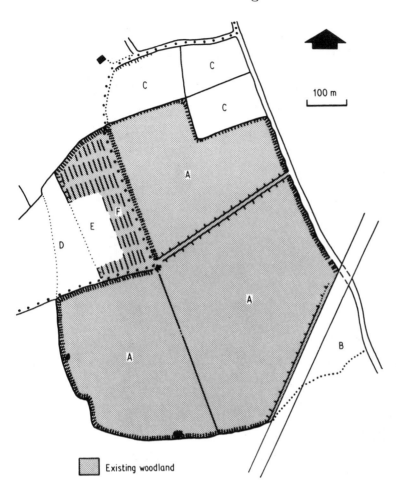

Existing woodland

Fig. 13.4 Plan of earthworks in and around Wickenby Wood, Lincs. Area A is mostly bounded by a large bank and external ditch which appears once to have extended round areas B an C. Within A the only artefacts are ponds, rideside ditches and drainage ditches on a dendroid pattern (not shown). Area F is underlain by ridge and furrow and bounded on its north side by another large bank and external ditch. The western margins of areas D and part of E have been ploughed out. Areas A and F constitute the modern Wickenby Wood, but areas B, C, D and E are arable. For interpretation see text below.

Surveyors must walk the length of all linear features, armed with a 1 : 10 000 or 1 : 25 000 map, a compass and a sketch pad. Distances should be measured by pacing. The features are mainly interpreted on the ground, and only their broad topographical relationships are inferred from the completed map: thus, one must distinguish on the site between a boundary bank, a disused sunken track and drainage ditches. A fair copy of the map should be drawn in the laboratory from the field sketch maps.

Wickenby Wood, Lincs, is presented as a worked example (Fig. 13.4). Area A is presumed to be the surviving ancient woodland which, apart from the rides and ponds, includes no signs of soil disturbance. Area B was similar until it was cut off and cleared to arable when the railway was built in the 19th century. Likewise, Area C was also once part of the medieval

Wickenby Wood contained within wood banks, but this was cleared before the 19th century. Area D + E + F also seems to have been wooded within historical times, for there is a surviving portion of wood bank, and the outline of Wickenby Wood and fields in Friesthorpe parish would be difficult to explain otherwise. Presumably D + E + F is the site of the former Friesthorpe Wood, which was cleared before the 19th century when the parish fields were enlarged. Later E + F reverted to woodland, and finally E was once again cleared in recent times.

Combination of bank and ditch maps with the historical record is the main line of evidence on woodland history which is independent of ecological features. The two sources complement each other, because the maps are precise about the location and sequence of events, whilst the historical record can

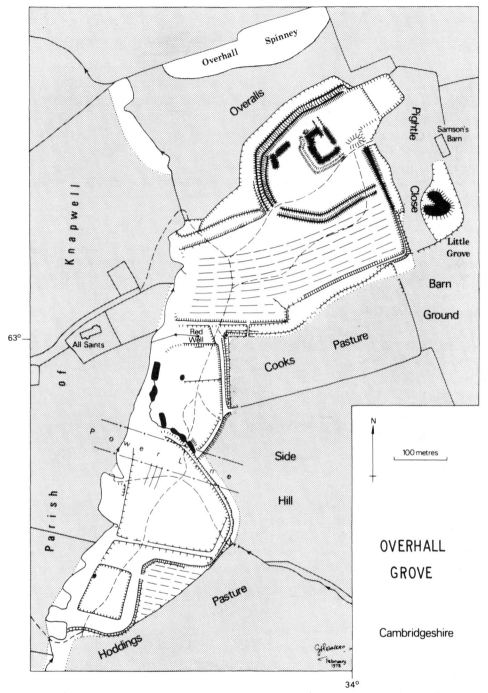

Overhall Spinney

Overalls

Knapwell

Pightle Close

Samson's Barn

Little Grove

Barn

Ground

63°

All Saints

Red Well

Cooks

Pasture

Parish

of

Power line

Side

Hill

N

100 metres

OVERHALL

GROVE

Cambridgeshire

Pasture

Hoddings

G. H. Petken
February
1973

34°

Fig. 13.5 Plan of earthworks in Overhall Grove, Cambs. Beneath the wood lie the remains of a moated manor house site within a multiple-banked enclosure, fishponds, field boundary banks and ridge-and-furrow cultivation remains, but the woodland edge is not defined by large wood bank and external ditch. An estate map shows that the wood was well established by 1650. Documentary and archaeological sources indicate that the manor house was occupied in the 13th and 14th centuries. The wood has therefore existed for about 400 years.

provide a precise date. Knapwell Wood (Fig. 13.1) is surrounded by a substantial medieval bank and external ditch within which there is no sign of past cultivation or habitation, and only some drainage ditches: this not only reinforces the conclusions from historical and topographical sources, but shows exactly where, along one margin, the wood has slightly invaded an adjacent field. Ancient, secondary woods can usually be identified by a historical record demonstrating continuity since the Middle Ages, combined with artefacts beneath the wood proving that the site must have been cleared at some earlier time. Prehistoric earthworks underlie many ancient woods on the southern chalk and oolite, e.g. Doles Wood (Hants), which appears to have originated in the Dark Ages (Dewar, 1926), and some woods in Rockingham Forest (Peterken, 1976). A fine example of both the potential complexity of a bank and ditch map and an ancient, secondary wood is Overhall Grove (Cambs), which certainly existed before 1650, but which overlies the site, fields and ponds of a medieval manor house (Fig. 13.5).

13.4 Species lists

An obvious, basic need is to know which species of animals and plants are present in a wood. However, the number, even in a small wood, is dauntingly large, and it is doubtful if we can ever obtain a complete list for all groups. Even after many years of specialist attention to Monks Wood (Cambs), some of the contributors to Steele and Welch (1973) confessed that the lists were certainly incomplete. Nevertheless, some 3682 species had been recorded, including 2842 species of animals (Table 13.2), and this was still far short of the 5000 species of animals which Elton (1968) considered were present in Wytham Wood (Oxon).

Some groups, notably birds, butterflies and vascular plants, are relatively easy to record, because they are either stationary or conspicuous, many of the individual species can be recognised at a glance, and the total number of species is not large. For the same reasons many people are capable of identifying them. Most groups are more difficult because they are inconspicuous, secretive and can only be readily identified by a few specialists. Vascular plants are perhaps

Table 13.2 Numbers of species of various groups recorded in Monks Wood, Cambs (Steele and Welch, 1973).

Flowering plants and ferns	372
Mosses and liverworts	97
Lichens	34
Fungi	337
Crustacea (woodlice, etc.)	21
Collembola	48
Insecta	
Odonata (dragonflies)	6
Orthoptera	10
Psocoptera	13
Hemiptera (plant bugs, froghoppers, etc.)	209
Neuroptera (lacewing flies, etc.)	21
Trichoptera (caddis flies)	11
Lepidoptera (butterflies and moths)	459
Diptera (flies)	452
Siphonaptera (fleas)	17
Hymenoptera (bees, wasps, ants, etc.)	219
Coleoptera (beetles)	1017
Arachnida	
Araneae (spiders)	122
Chelonethi (pseudoscorpions)	3
Opiliones (harvestmen)	16
Acari (mites and ticks)	147
Myriapoda (millipedes and centipedes)	22
Mollusca (slugs and snails)	42
Annelida-Lumbricidea (worms)	12
Chordata	
Amphibia and Reptilia	10
Aves (birds)	115
Mammalia	24

the easiest and therefore the most suitable as a basic record for any woodland. Furthermore they form the habitat and provide the food for other groups, as well as being familiar to many people and worth recording as species in their own right. Birds and butterflies may be just as easy to record, and there may be many competent recorders available, but they depend on vascular plants for food and shelter, and generally occupy a less fundamental position in the ecosystem. Vascular plant recording is therefore considered in more detail below.

The recording process has been facilitated by the printed 8″ × 5″ (20.3 × 12.7 cm) cards produced by the Biological Records Centre at Monks Wood, mostly in connection with distribution mapping schemes. These cards have spaces to record site name, grid reference, habitat type, etc., but otherwise bear lists of species which can be marked whilst recording in the field or transcribing from pre-existing sources of records. They can conveniently be stored as a simple card index and, since the species all have a code number, the records can readily be fed into an automatic data processing system, from which maps, list of species in sites and lists of sites for species can be produced.

Recording is subject to a law of diminishing returns. The vascular plant list of any wood rises rapidly at first, but the rate of accumulation of new species falls steadily as recording continues until a point of diminishing returns is reached when new species are found only at long intervals. This is predictable from species–area relationships (Section 6.2): the area searched is roughly proportional to the time taken; the number of species seen is proportional to the logarithm of the area searched; so the number of species recorded is proportional to the logarithm of the time spent recording. The species–area–time relationship is often distorted by surges of additional species which are encountered when the recorder reaches a new habitat for the first time (e.g. a stream or a geological boundary). Surges also occur on a longer time-scale when a recorder returns to a wood at a different season and when another recorder first visits a wood. This is partly explained by the behaviour of the species themselves, some of which are conspicuous and identifiable only in a certain season, but clearly the recorder's ability, concentration, enthusiasm and selectivity also influence the outcome.

Over much longer periods colonisation by additional species becomes an additional factor: for example, Hayley Wood (Cambs) has probably been colonised by 25–30 species since recording started in 1860 (Rackham, 1975). This, and the logarithmic species–time relationship, implies that the list for the larger woods is never complete, and experience confirms that this is so. Despite regular visits by botanists from Cambridge, a recent addition to the Hayley Wood list was *Betula pubescens*, not as a sapling as one might expect, but as an adult tree. And two herbs which I introduced into Monks Wood were never detected, even though they survived for several years beside a ride along which biologists passed daily.

The initial plant list is perhaps best compiled by walking the wood margins and making at least two traverses across the site at right-angles to the main topographical features (usually from low ground to high ground, or from light to heavy soil zones as revealed by a walk around the boundaries). Where there are no clear features to follow or cross, the search should follow a figure of eight path. Large woods should be divided into separate search areas, for which separate lists can be made with advantage. Most species can be seen and recognised on an early summer day, say in May or June, but even so it is better to make at least one other visit later in the season. Season is important in recording, for not only are some species actually invisible by late summer (e.g. *Anemone nemorosa*, *Ranunculus ficaria*), but species in some critical groups can only be reliably identified at certain phases of growth (e.g. *Carex*, *Epipactis*, *Hieracium*). Eventually, it may be possible to quantify the various factors which influence the number of species, or rather the proportion of the total actually present, which are recorded in any visit, i.e. duration of recording, area of wood, woodland type(s) present, season and recorder's ability. Thus, when, say, 125 species are recorded in a 50 ha wood in 4 h in late May by a competent recorder, one could apply suitable multipliers and claim that, say, c.230 species are likely to inhabit the wood.

Table 13.3 Summary of the number of vascular plant species in Bedford Purlieus, Cambs (Rixon, 1975).

Total number of species recorded	462
Native and still present	424
Introduced species	8
Extinct and possibly extinct species	30
Woodland and wood-edge	194
Grassland and disturbed ground*	288
Aquatic	24

* The wood includes two modern quarries, one of which is active, the other levelled and afforested.

The richest wood according to available records appears to be Bedford Purlieus (Peterken and Welch, 1975), but the 462 vascular plant species on record include some which are probably now extinct, and, as in so many woods, the number is greatly inflated by grassland species on the rides (Table 13.3). The list for

Table 13.4 The number of vascular plant species found so far in a selection of well-recorded woods (Nature Conservancy Council Records, site files).

Site	Species	Hectares	Woodland type
Scotland			
Den of Airlie, Tayside	200	84	Gorge woodland. Predominantly ash–wych elm
Boturich Woodlands	195	37	Oak–birch grading to ash–alder
Inchcailloch (NNR)	252	55	Acid sessile oak; alder
Cragbank, Roxburgh	177	9	Grazed ash–wych elm with alder
St Boswells, Roxburgh	202	19	Ash–wych elm on flushed clays
Arriundle (NNR)	123	120	Oceanic, sessile oak
Wales and Western England			
Roudsea Wood (NNR), Cumbria	452	118	Ash–oak on limestone; acid sessile oak; mires with trees
Ling Gill (NNR), Yorkshire	301	5	Open Pennine ash–wych elm
Coed Tremadoc (NNR)	145	24	Sessile oak in oceanic climate
Rodney Stoke (NNR), Somerset	167*	26	Mendip ash wood
Wistmans Wood (FNR), Devon	41*	4	High altitude acid pedunculate oakwood
English Lowlands			
Bedford Purlieus, Cambs	462	185	Wide range of mixed broadleaf stands
Monks Wood (NNR), Cambs	372	157	Ash–maple on heavy soils
Hayley Wood, Cambs	215*	49	Ash–maple on clay
Bradfield Woods, Suffolk	370	131	Plateau alder and other stand types on wide soil range
Bovingdon Woods, Essex	112	64	Mainly acid oak–lime
Wychwood (NNR), Oxon	364	531	Mixed broadleaf on limestone and clay, wide rides
Savernake Forest, Wilts	252	918	Acid oak–beech woodland; now many plantations
Bix Bottom, Oxon	390	103	Mixed broadleaf on chalk; open areas
Alice Holt Forest, Hants	428	865	Oak forest on clays, now much replanted
Weald Edge Hangers, Hants	302	238	Acid and alkaline beechwoods on chalk scarp
Burnham Beeches, Berks	117	450	Strongly acid beech, oak, birch
Ham Street (NNR), Kent	261	97	Oak over chestnut, hazel, hornbeam
Blean Woods (NNR), Kent	122	67	Predominantly acid sessile oak, beech, hornbeam and chestnut

* Species found only on unwooded ground within these nature reserves have been excluded.

several other woods must be nearly complete, and the examples given in Table 13.4 show what can be found in examples of various woodland types.

13.5 Distribution of species within woods

Distribution maps of species within a particular wood are made for two reasons. Some species are sufficiently rare or localised to justify special management provisions and clearly it is essential to know where they grow. Other species, which indicate particular ecological conditions, can be mapped as a short-cut to a map of these conditions. Thus, a map of *Geum rivale* is an effective, rapid method of mapping moist, base-rich soils. No special methodology exists for the first group of species: one simply plots their location when they are discovered. Searches for such species become efficient once the searcher has an appreciation of their ecology within the site, but systematic searches of an entire wood should not be rejected, because they often show that a species which was thought to be localised is actually more widespread. The map of wood ant nests by Welch (Peterken and Welsh, 1975; see Fig. 13.6) was made by a combination of systematic survey of the whole wood, whilst paying special attention to the rides and clearings where the species was found to be concentrated.

Species which are abundant or widespread within a wood can be mapped completely or by compartments or by regularly-spaced samples. These three

Fig. 13.6 The distribution of wood ant (*Formica rufa* L.) nests occupied in 1974 in Bedford Purlieus (Cambs). Mapped by Dr R. C. Welch. Rides are shown by broken lines. Reproduced from Peterken and Welch (1975).

approaches each have their separate strengths, weaknesses and applications:

(1) Complete mapping. Maps which show the complete occurrence of an abundant species are generally more useful for ecological purposes than compartmental maps (b) and provide a more complete and interpretable pattern than sample maps (c). Large-scale maps of this kind, e.g. of all the individual trees

of a given species in their exact position, can hardly be bettered, but the effort required to compile them is immense and rarely justified. Alternatively, a sketch map can be prepared at a smaller scale in which the observer seeks to represent the main features of the real distribution, whilst simplifying some of the detail. A map made by one surveyor will differ in fine detail from one prepared by another surveyor, because no two people will simplify in the same way, but the

Fig. 13.7 Distribution of three generations of holly, *Ilex aquifolium*, in Staverton Park (Suffolk) (Peterken, 1969a). (a) Old generation trees originating before 1820, mostly pollarded trees well over 200 years old, (b) middle generation trees originating about 1820–1860, not pollarded, (c) saplings and a few trees originating after 1870. The ages of most trees were estimated from their size and shape, but ring counts of a few were taken to check the estimates.

gain in speed more than offsets this disadvantage in most circumstances. The distribution of all tree species in several age classes in Staverton Park were mapped simultaneously in three days by this method (Peterken, 1969a; Fig. 13.7). Likewise, the five ground flora dominants of Bedford Purlieus were mapped simultaneously (Peterken and Welch, 1975; Fig. 13.8), and the maps could be used individually and in combination to show the distribution of significant soil features.

(2) Compartmental mapping. If a wood can be divided into compartments along rides or other features which are long-lasting and easily recognisable on the ground, the distribution of a species can be recorded and shown by its presence and absence (or by estimates of abundance) in each compartment. Such maps are objective statements, simple to record and related to management in so far as this is planned on a compartmental basis. The scale, however, is far too coarse for any ecological interpretation: the maps are

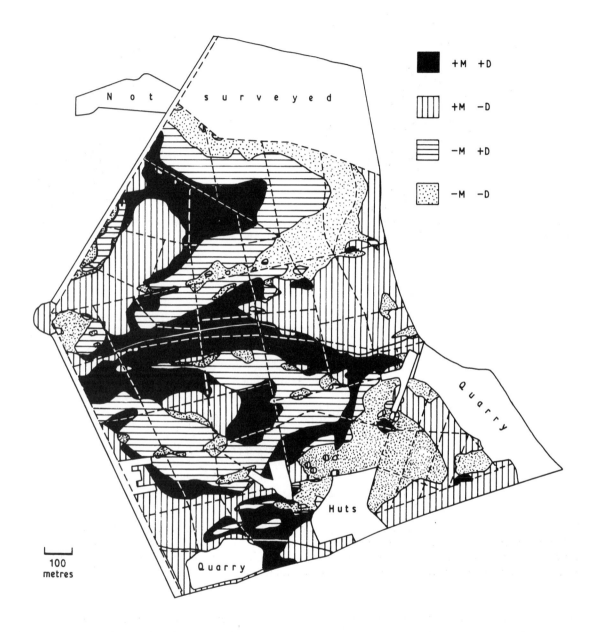

Legend:
- ■ +M +D
- ▦ +M −D
- ▤ −M +D
- ░ −M −D

Map labels: Not surveyed, Quarry, Huts, Quarry

100 metres

Fig. 13.8 Superimposed distributions of *Deschampsia caespitosa* and *Mercurialis perennis* in Bedford Purlieus (Cambs), showing the areas where these species formed a significant amount of the field layer. Redrawn from Peterken and Welch, 1975). *Mercurialis* picks out the soils with a high pH, whilst *Deschampsia* occupies the poorly drained and moist soils, irrespective of their pH. Combined, these species divide the wood into four broad site types:
+M +D, Base-rich brown earths and gleys on clay and clay loam.
+M −D, Rendzinas and calcareous brown earths, all freely drained.
−M +D, Acid gleys developed in heavy soils.
−M −D, Acid, medium-light soils, well-drained at least in the surface horizons.

equivalent to vice-county maps of national distributions, which are quite unsuitable for showing, say, the affinity of some species for limestone districts.

(3) Sample mapping. A grid of regularly-spaced samples will generate distribution maps whose accuracy is determined by the scale of the real distribution patterns and the density of sampling. Fine detail is invariably lost. (This point is discussed in Section 13.6 in relation to vegetation maps.) However, if the sample points are accurately located, the records will be objective enough to be used for monitoring, but only at the cost of a great increase in time. Sample mapping of species is thus mainly useful for monitoring, whereas 'complete' mapping is no more laborious and much more informative for general woodland management and ecological interpretation.

The distribution of bird territories within woods can be mapped by the techniques of the Common Bird Census. The method is fully described by Williamson (1964). It consists of plotting all instances of territorial behaviour on a large map of the study area. The records from repeated visits during the breeding season are placed on to separate maps for each species. These tend to form clusters, which are assumed to be separate territories, even if the nest is not found.

13.6 Vegetation maps

Vegetation maps can be also produced by any of the approaches mentioned in Section 13.5. The method used should be determined by the objectives, the accuracy required, the resources available and the uses which will be made of the result. A vegetation classification is needed, the choice of which can greatly influence the effort required to complete the map and its value. The classification can either exist before mapping commences (i.e. 'external' to the survey), or be generated from data collected during the survey (i.e. 'internal' to the survey). In the former case a field survey consists of identifying the defined and often familiar types, whereas a survey based on an internal system consists of systematically collecting data for subsequent classification.

Woodland vegetation maps have been made in terms of the stand types defined in Part Two. The example of Swithland Wood (Leics) (Fig. 13.9) was made by recognising stand types continuously whilst walking around the wood, and plotting the types and boundaries between them on to a sketch map. An Ordnance Survey map at scales of 1 : 25 000 or 1 : 10 000 would have been suitable for navigation if a special site map prepared from the OS had not been available. The walk is, in effect, a transect which crosses and recrosses itself. As it progresses the observations are transformed from a meandering, linear plot into a map as boundaries recognised at one point are linked to those recognised nearby. Mapping started at the north end where the types subsequently proved to be relatively complex, due to the presence of much invading sycamore and small fragments of both ash–wych elm (1B) and valley alder (7A) amongst the prevailing sessile oak–lime (5B). Nevertheless, all the main stand types present had been identified and the general pattern of ecological variation was apparent by the time that the surveyor first reached the centre of the wood. Initially the surveyor kept to paths and boundaries, which enabled him to move freely and to be sure where he was on the map, but the emerging pattern could only be completed by plunging into the thicket. The wood was already well divided by paths into several compartments, so each compartment was completed before moving on to the next. Swithland Wood proved to be unusually complex, both in the large number of stand types present and in the small-scale mosaic of their pattern. Both oaks were present and demanded much attention in order to identify possible patches of 6D amongst the 6C, or 5A amongst 5B. A confusingly large network of additional tracks had formed since the OS base map had been made, so the surveyor was occasionally unsure of his position. Sycamore had invaded the north-east quarter abundantly, but stand types could still be identified there albeit with some hesitation. The map (Fig. 13.9) was finished in nine hours, at roughly 7 ha per hour. Maps of ground vegetation can be made by the same approach (Fig. 13.10).

The relationship between this map and the reality on the ground is illustrated by a hypothetical example (Fig. 13.11). The distributions of maple and hornbeam overlap (Fig. 13.11a), but hornbeam takes

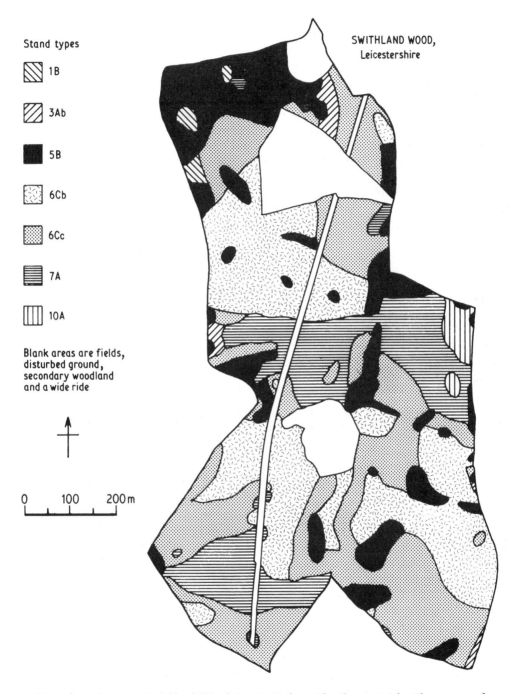

Fig. 13.9 Map of stand types in Swithland Wood (Leics). Light, acid soils associated with outcrops of pre-Cambrian rocks are occupied mainly by birch–sessile oak woodland. Alder woodland occupies the low-lying ground. Keuper Marl, which underlies most of the site, has given rise to generally acid, poorly drained soils occupied by hazel–sessile oak and sessile oak–lime woodland. The wood was once managed as coppice, but some parts have been uncut since 1850 and have now developed into a magnificent mature high forest stand.

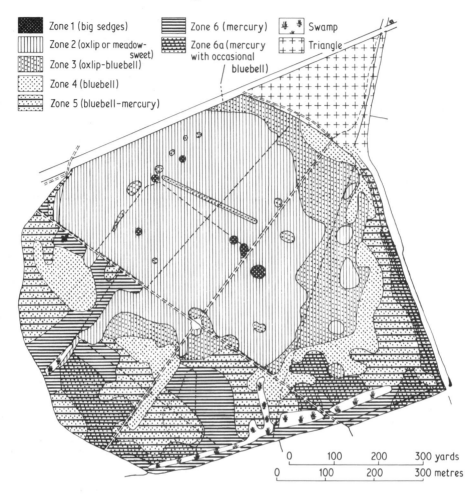

Zone 1 (big sedges)
Zone 2 (oxlip or meadow-sweet)
Zone 3 (oxlip-bluebell)
Zone 4 (bluebell)
Zone 5 (bluebell-mercury)
Zone 6 (mercury)
Zone 6a (mercury with occasional bluebell)
Swamp
Triangle

0   100   200   300 yards
0   100   200   300 metres

Fig. 13.10 Map of ground vegetation zones in Hayley Wood (Cambs) (Rackham, 1975). The wood lies on a very shallow dome of boulder clay so the wettest zones are in the centre. The swamp has developed where water is impounded behind the ancient wood bank. The triangular patch to the north is recent secondary woodland originating after 1920.

precedence in defining stand types. The stand types present are therefore 2A, 3A and 9A. The map (Fig. 13.11c) reflects the actual distribution fairly accurately, but the scatter of maple has been simplified, and the boundaries between 2A and 3A – which are gradations on the ground – are represented by sharp boundaries on the map. The surveyor has perceived the patterns of the locally significant species, i.e. maple and hornbeam in the example, and represented them as accurately as possible.

The relationship between 2A and 3A in Fig. 13.11 exemplifies a persistent mapping problem: is it right to map large tracts as 3A when maple is at least thinly scattered through most of the stand? More generally, how much maple must be present amongst the ash, hazel and pedunculate oak to assign a stand to 2A rather than 3A? Stands containing a thin scatter of maple can occur on acid clays at the edge of maple's edaphic range, and, in theory, where minute patches of calcareous clay occur in an otherwise acid site. This problem can be faced by notionally dividing the stand into 30 × 30 m squares (the basis of the classification), referring each square to a stand type, and producing the map shown in Fig. 13.11b. Clumps of 2A squares

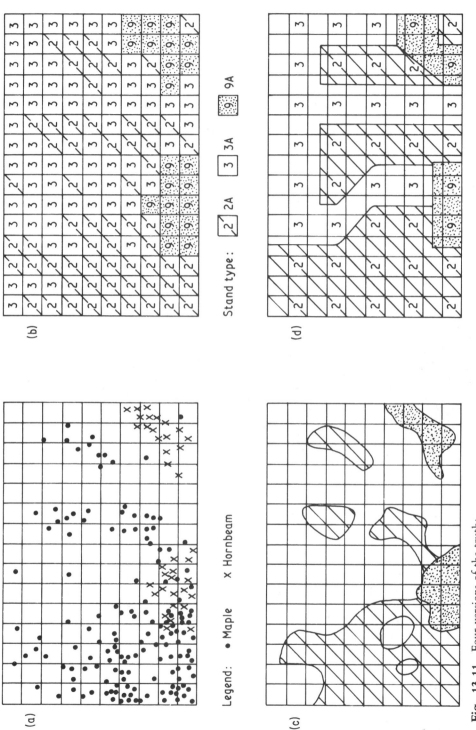

Legend: • Maple   × Hornbeam

Stand type: | 2 | 2A | 3 | 3A | 9 | 9A |

Fig. 13.11 Four versions of the truth:

(a) The actual distribution of adult maple and hornbeam in a stand otherwise composed of ash, hazel and pedunculate oak. The stand measures 300 m × 450 m.

(b) The same stand divided into 10 × 15 squares of 30 m × 30 m, showing the stand type to which each square belongs.

(c) A map of stand types as it might be prepared by a surveyor mapping in the style of the map of Swithland Wood (Fig. 13.9).

(d) A map of stand types prepared by sampling 25% of 30 m × 30 m squares, then interpolating stand type boundaries.

The stand types are 9Ab where the hornbeam is present, 2A where it is absent but maple is present, and 3A where neither hornbeam nor maple is present.

have to be mapped as such, but isolated 2A squares are ignored. One could arbitrarily define the transition between the two types as a stand where one square in four is in 2A, but Fig. 13.11b shows clearly how difficult it is in practice to apply such a criterion. We must accept that the composition of semi-natural communities varies continuously; that boundaries between types are defined arbitrarily; that we can choose how much of a wood is mapped as intermediate between two or more types and how much is placed firmly within a type; and that any map of the kind shown in Figs. 13.9 and 13.11c is a subjective statement. The surveyor readily perceives that much of the stand in Fig. 13.11a is transitional between the cores of types 2A and 3A, but the map would be very confusing if large areas of transitional forms were shown. Map B in Fig. 13.11 is probably the most acceptable in theory, but it would be immensely laborious to produce it in practice. Map C is an alternative representation of the truth which looks as accurate and is much quicker to make. No doubt two surveyors mapping in the style of Map C would differ in fine detail, but, in cost–benefit terms, it is probably better to accept the speed and subjectivity of Map C than seek the laborious but objective Map B. These arguments apply where two stand types differ by the presence or absence of one species, i.e. maple in the example. When a stand contains any two of the species which define stand groups 7–11, it must be mapped as a mixture of two stand types.

Similar difficulties are faced by soil surveyors, who must map soils in terms of classes defined from a number of profiles. Their solution, which has been described by Avery (1973), is as follows: 'Whereas a profile class groups profiles according to their similarity, a soil map must group them according to their contiguity. Profile classification serves to guide the creation of map units and provides a uniform basis for describing and identifying them. Accordingly a map unit is usually identified by the name of a profile class, implying that most of the soil in each delineation conforms to that class, and that inclusions of other kinds of soil conform to one or more closely related classes or occupy an insignificant proportionate area. More heterogeneous units (complexes, associations, undifferentiated groups) are similarly identified by the names of two or more classes.'

Substituting 'stand type' for 'profile class', etc., this exactly describes the relationship between the classification of stand types and any map of them.

Swithland Wood could alternatively have been mapped in terms of the Merlewood classification. Spot identifications, made during the course of mapping, showed that stand types 1B, 3A, 7A and the stands of 5B at the north end fell within Merlewood types 9–12, whilst stand types 6Cb, 6Cc and the rest of 5B were in types 23 and 24. No attempt was made to complete the map of the Merlewood types, so the following assessment is tentative. The Merlewood type at any given point took 3–5 min to identify, for the presence or absence of up to 50 species had to be ascertained. Some identifications were uncertain, presumably because the stand was actually intermediate between two or more types. Boundaries between types could not be recognised easily within groups of related types, i.e. 9–12 and 23–24, because they were determined by the presence or absence of many species. The species which determine the types include inconspicuous bryophytes and herbs: one could not assume that they were absent if they were not seen at a glance. The survey was undertaken in late summer when most herbs were still visible, but the Merlewood system could not have been reliably used in winter. Thus, it seems that the Merlewood classification is impractical as a basis for mapping, although it reflected the ecological variation within Swithland Wood at least as well as the stand types, and was superior in so far as it separated the disparate forms of 5B.

The strength of the stand types described in Chapter 8 lies in their practical advantages, relevance to management and ecological basis. They can be recognised at a glance, especially when the range of types present has been established. Only a few species need be known: the map of Swithland Wood is determined by eight species. Maps can be made even in winter, provided that one is prepared to map oaks from fallen leaves and ascertain soil characteristics (when necessary) from direct observation or by implication from the attenuated field layer. Boundaries are defined by the presence or absence of one species, i.e. maple or hornbeam in Fig. 13.11. The map of stand types is directly relevant to management, and it adequately reflects ecological variation within the

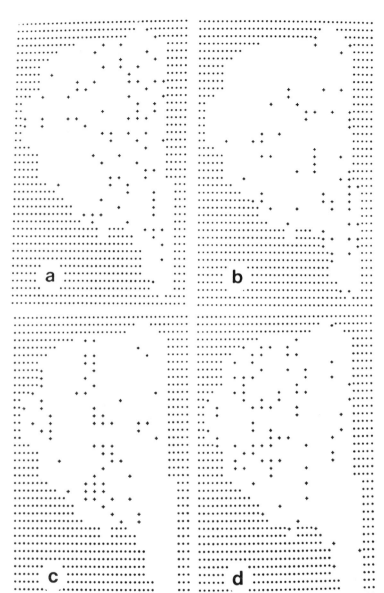

Fig. 13.12 The distribution of four of the ten vegetation types recognised by Indicator Species Analysis on Inchcailloch, a wooded island in Loch Lomond (Horril *et al.*, 1975). The physiognomic dominants are: (a) type 1, *Pteridium aquilinum*, *Luzula sylvatica*, (b) type 2, *Luzula sylvatica*, (c) type 3, *Rubus fruticosus* and (d) type 4, *Rubus fruticosus*.

wood. Neither variation in dominance nor the intensity of invasion by exotics affects the definition of stand types, but these features can be shown as overlays. Where a stand is not semi-natural, then it can be mapped in the style of a forestry stock map.

Map B in Fig. 13.11 is based on a sampling system, but in this instance the sample takes in 100% of the real ground. This approach would be brought within practicable bounds if a useful map were made at a much lower sampling density. A grid of sample points accurately located within the wood can be recorded in terms of an existing classification, or some systematic observations can be made from which a classification is later generated. Either way, each sample point is assigned to a type. Fine maps of this kind were produced by Horrill *et al.* (1975) for the field layer of Inchcailloch, a wooded island in Loch Lomond, where species lists were made in 4 m² samples spaced at 35 m intervals (i.e. a 0.33% sample), and classified 'internally' by Indicator Species Analysis. The example maps in Fig. 13.12 show the broad pattern of vegetation and, provided that sample points can be accurately re-located, the record would serve as an objective basis for monitoring any changes. Maps in the style of Fig. 13.10 might be drawn by interpretation, but the result is likely to be a gross over-simplification of the real pattern, except where the pattern is unusually simple: Map D in Fig. 13.11 shows how distorted such a product would be, even with a sampling density as high as 25%. One could, of course, re-examine the ground between sampling points in order to improve the definition of boundaries, even though the vegetation types generated by the Indicator Species Analysis are not instantly recognisable.

13.7 Stand structure

The considerable effort of forestry enumeration and inventory techniques can be justified if detailed and accurate knowledge of stand structure is required, as it is in the case of woodland nature reserve management, but not for general site survey, for which quicker and cheaper methods are needed. In such circumstances it seems more useful to know how to recognise the effects of past stand management, to estimate the age of trees, and to record the structure rapidly.

Managed coppices, plantations, naturally regenerated high forest and wood pasture can normally be readily recognised. Mature, high forest plantations may be difficult to distinguish from semi-natural stands, partly because plantations develop some natural features in the intervals between management operations, and the visible effects of management – cut stumps, for example – are obscured. Mature plantations usually retain some trace of the straight planting lines, straight margins and even-agedness imposed with the original planting. Their composition is often limited to one or two species grown in pure stands. Exotics are frequently found, and the boundary of a species within a stand is sharp and not usually associated with a soil boundary. Caution is, however, necessary, for many exotics are able to regenerate naturally; a pure stand can be formed by

Table 13.5 A shorthand notation for recording the natural and managed growth forms of woodland trees and shrubs.

TREES
Single-stemmed individuals in the mature canopy and tall subordinate strata.

Ts Standard trees in a coppice-with-standards system.
Tn Self-sown trees. Reserved in a coppice for trees which are contemporary with existing coppice growth.
Tc Trees which have been promoted from coppice by singling or neglect. This can include mature trees with two trunks.
Tp Planted trees.
Tx Other trees, generally those of uncertain origin.

Note that combinations are possible. Thus Tsp is a standard which has been planted.

COPPICE
Trees and shrubs which have been cut and allowed to grow again from the stump.

Cs Small coppice, cut close to the ground.
Cl Large coppice, cut well above the ground.
Ct Coppice grown from the cut stump of a tree.

POLLARDS
Trees cut more than one metre above ground level and allowed to grow again from the stump.

Pc Low pollards, or very high-cut coppice, say 1.0–2.5 m above ground.
Px Pollards cut more than 2.5 m above the ground.

SHRUBS
Ss Small shrub.
Sl Large shrub, more than 5.0 m high with a broad crown.
Sc Climber.
St Young tree usually above 3.0 m high (post-dating last coppicing).

JUVENILES
Js Seedlings, germinated in last two years.
Jp Saplings.
Jv Suckers.
Jl Layers.

cutting out all other species; and a sharp boundary can be formed by planting on one side, not necessarily on both. Former wood pasture can be recognised by the presence of large, wide-spreading trees (often formerly pollarded) which are distinctly older than the younger, straighter individuals surrounding them. Coppice management and individual trees which were formerly coppiced can be recognised from

Table 13.6 An example record of a semi-natural woodland stand. For explanation of shorthand notation see Table 13.5.

Species	Cover	Trees	Coppice	Pollards	Shrubs	Juveniles
Acer campestre	4		Cl			
Acer pseudoplatanus	4	Tn				Jp
Betula pendula	5	Tn	Cl			
Clematis vitalba	2				Sc	
Corylus avellana	7		Cs			
Crataegus monogyna	4				Ss	Jp
Daphne laureola	2				Ss	
Euonymus europaeus	1					Jp
Fagus sylvatica	4		Cs			
Fraxinus excelsior	1		(Cl)			Jp Js
Ilex aquifolium	1					Jp
Prunus avium	+					(Jp)
Quercus robur	5	(Tn)	Cl			
Salix caprea	3		Cl			
Taxus baccata	1				(Sl)	Jp
Thelycrania sanguinea	+				(Ss)	
Tilia cordata	6	Tc	Cl Cs	Pc		
Total cover	100%					

Coppice last cut about 1920. Some very large, high-cut stools of lime nearby. Note invasion of various species now that coppice has passed vigorous phase. No standards or stumps of standards – therefore coppice originally simple.

the coppice stools and the multi-trunked structure of many trees. Coppice which has been promoted to high forest by singling can usually be recognised by surviving swellings at the base of the trunk.

The age of individual trees can be estimated with sufficient accuracy for most general survey purposes, provided one is 'calibrated' from ring counts of trees in similar sites. Surveyors should therefore make a habit of counting the rings of any cut stumps they find, and recording age and diameter over bark. Growth rates vary enormously: trees in good sites with room to spread their branches grow much faster than those on poor soils and in shaded situations, so allowance must be made for this. Surveying in Welsh oakwoods I have obtained a correlation coefficient of +0.95 between estimated age and age subsequently counted from growth rings. Experience must be gained in the right area: it is no good extrapolating age/diameter relationships from Welsh oakwoods to oak in south-east England. The absolute chronology of the generation structure of the New Forest Woods (Peterken and Tubbs, 1965) was established from

opportunist counts of growth rings combined with the broad appreciation of growth rates which these counts provided.

Recorders can thus describe a stand as, say, 'an even-aged plantation of about 70–80 years', or 'former coppice-with-standards, last cut in the 1920s, from which most of the standards were removed at the last coppicing', and this will suffice for rapid survey. When more detailed records are needed, a representative sample of a stand can be described using the shorthand shown in Table 13.5. When this is combined with a short description and estimates of age a full, valuable and relatively quick record is made of the existing structure and the processes which led up to it.

An example of a simple record of a semi-natural stand is given in Table 13.6. This table records the trees and shrubs in a 30 m × 30 m sample of Coombe Woods, Gwent (Grid reference 31.459930). The sample was chosen to represent a particular part of the wood, in this instance the mixed broadleaf woodland without clear dominants growing on freely-drained, calcareous clay over carboniferous limestone. With eight coppice species this is one of the richest mixed coppices in Britain. The sample falls within stand type 8Ca, but since beech is rare it comes close to 4Bb. The record consists of a list of species, an estimate of cover in terms of a modified Domin scale (Table 8.1), and the growth forms of each species (Table 13.5). Brackets indicate growth forms found near, but not within the sample. Species may occur as several forms, but if one predominates this is underlined: thus most of the limes were large coppice. Together with supplementary notes on the height and cover of tree and shrub layers, estimated age of main components and any notable features, the record takes no more than ten minutes to make. It saves one from copious notes, which tend not be be used, and the records can be processed automatically. The supplementary notes actually made in this instance are shown. The course of development can be inferred. This stand had last been coppiced 50–60 years ago. Natural regeneration had been excluded by the dense canopy for many years, but the dominant trees have lately become less vigorous, and ash and other species have begun to regenerate. However, the light-demanding birch and oak, which regenerated from seed after the last

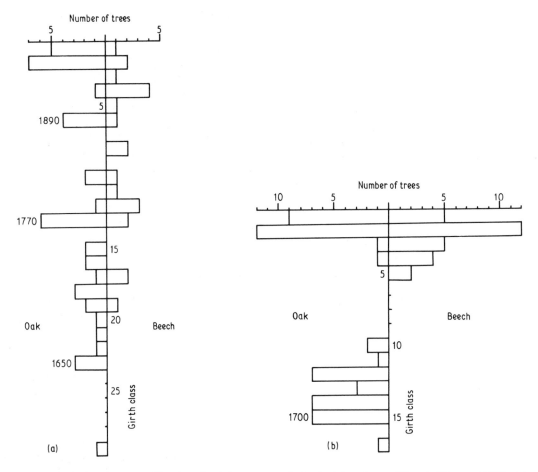

Fig. 13.13 Girth distributions of beech and oak in two woods in the New Forest (Hants) (Flower, 1977). (a) Whitley Wood, SU 297059, (b) South Bentley Inclosure, SU 234129. The diagram shows the number of oak (to the left) and beech (to the right) trees recorded in a random sample of 80 trees obtained by pacing on random compass bearings through the wood, stopping after a random number of paces and measuring the nearest tree in each quadrant about the north point at each stop. Girths shown in 0.2 m classes. Since size and age are roughly related approximate dates can be assigned to prominent girth classes.

Whitley Wood, one of the least disturbed stands in the New Forest, contains trees from all the recognised generations (Peterken and Tubbs, 1965, Flower, 1977), including one huge pre-A generation oak. Oak is presently regenerating faster than beech. South Bentley Inclosure was an Elizabethan coppice, but it was planted to high forest with oak in 1700 and 1775. Selective felling in the last 30 years has encouraged prolific beech and oak regeneration.

coppicing, have not been able to regenerate lately beneath the closed canopy.

When the desire for greater detail can justify the greatly increased costs of recording, two techniques are available. The girth or diameter at 'breast height' of all trees and shrubs above a convenient height or girth can be measured. This, together with measure-ments of height, enables volumes of standing timber to be calculated. Histograms of girth classes display the generation structure of each species (Fig. 13.13). Alternatively, pictorial representation of 'slices' of a stand cutting through a representative area display the vertical and horizontal structure admirably (Fig. 13.14).

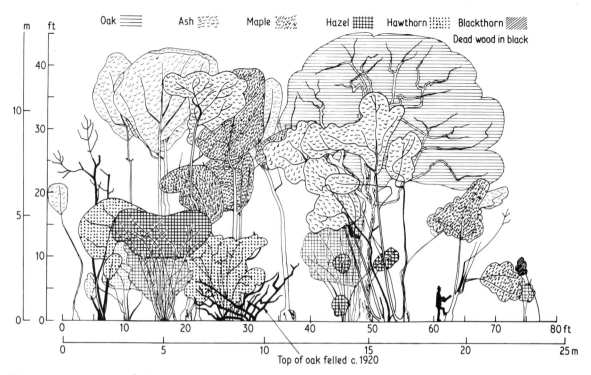

Oak ≡≡≡   Ash   Maple   Hazel ▦   Hawthorn   Blackthorn
Dead wood in black

Top of oak felled c. 1920

Fig. 13.14 Structure of ash, maple, hazel coppice with oak standards (stand type 2A) in Hayley Wood (Cambs) (Rackham, 1975). The diagram shows a measured profile of trees and shrubs over 2 m high in a 4.5 m × 24.5 m plot.

13.8 Monitoring

Woodland managers need to be sure that management (or a policy of non-intervention) is having the desired or desirable effects on communities and species. The rate and direction of change can be detected and measured from the difference between observations repeated at intervals.

One might undertake a survey with the intention of measuring subsequent changes. For example, the distribution of a species might be mapped because it is thought to be spreading or declining; or stand structure might be recorded in order to monitor subsequent natural changes under a non-intervention management policy. Such monitoring can only be successful if the initial record (the 'baseline') is accurate enough for the anticipated changes to be detected. The method by which the record was made must also be accurately recorded. The observations and the record of the method must be safely stored (and not lost in a 'dead' file), so that it survives the often long interval between records. The exact location on the ground of the features observed must be marked on maps by reference to features which are permanent and independent of the features being monitored or by physical marks (e.g. posts) on the ground. The method of relocation must be durable, but not interfere with the population or community which is under study. Maps leave no mark on the ground, but the rides, boundary banks, identifiable trees or other features used as datum-points may not prove as unchanging as the baseline recorder had expected. Posts, pegs and buried metal markers (located with metal detectors) are unequivocal and precise markers but they are liable to be moved by curious people, destroyed by vandals, or just deteriorate naturally. Populations of holly seedlings in the New Forest were monitored using quadrats marked by short, almost-buried wooden pegs which were located by maps and conspicuous trees nearby

227

(Peterken, 1966). Suitably treated against decay some still survived 16 years later, but others were dug up within weeks of being set out.

Monitoring by repeating the kinds of survey described above in Sections 13.3–13.6 is not the only possible approach. An alternative is to repeat exactly a process of observation, the results of which have comparative but not absolute significance. Butterfly populations can be monitored by counting the numbers of each species seen along a standard transect (Pollard, 1977) which is divided into sections according to the habitats through which it passes. Seasonal changes become apparent, and many useful observations can be made on the biology and ecology of individual species, but the main object is to detect year-to-year changes. At no stage is it necessary to know the total population of each species in the sites being monitored.

The main problem with monitoring is to determine at the outset what is worth monitoring, for a poor decision may be expensive in wasted time and materials. Perhaps the most obvious features are plant populations, especially small populations of rare species, the individuals of which have to be accurately mapped and relocated by permanent markers (Tamm, 1956). Natural regeneration should be monitored, especially after a management change (e.g. fencing to exclude sheep), or a decision to allow a wood to develop naturally. Transects were established by Dr E. W. Jones in Lady Park Wood (Gwent) in about 1944, when the Forestry Commission set the wood aside as a natural reserve, and later repeats of the record are providing valuable insights into long-term changes. Another record made by Dr Jones has formed the baseline for measuring long-term floristic changes due to afforestation (Hill and Jones, 1978). A remarkable record of plots recorded in 1847 (Řehák, 1968) in the Boubin Virgin Forest (Šumava Hills, Czechoslovakia) by Joseph John (Fig. 13.15) has enabled the rate of decay of fallen trees to be measured, and the long-term trend from fir dominance towards beech and spruce to be quantified (Section 1.1). The maps and the plot markers were lost for decades and were found again by a lucky chance.

Monitoring need not involve elaborate and time-consuming recording. Photographs, taken from fixed, relocatable points, are quick and cheap. They have the additional advantage that the record is not edited (unlike baselines which are made according to a careful recording system and which are therefore limited by the features and level of accuracy recorded) and can be useful for unforeseen changes, as well as providing a more immediate, subjective impression of change. For example, photographs within a woodland originally taken as a quick, cheap method of recording the growth of ground vegetation, saplings and shrubs following exclosure of deer, might 30 years later be more valuable as a source of information on the health or rate of degeneration of the canopy trees. Photographs are therefore a suitable method of establishing a baseline when the purpose is uncertain, the priority low or resources are limited. They should, of course, also be taken as a supplement to more detailed records.

If monitoring is simply repeated survey, and if survey can range from a single photograph to an elaborate record, then some estimate of change can be made today from baselines which were fortuitously established in the past. For example, numerous old photographs of the famous Wistman's Wood (Devon) in old travel books and articles have been relocated on the ground by Mr G. M. Spooner (Proctor *et al.*, 1980) using as markers the background view and distinctive boulders. Not only does the resulting series of prints form a fine basis for studying the fluctuating fortunes of this remarkable stand, but past and present photographs – held side by side – bring home as nothing else can, just how the trees have changed. Photographs of an archaeological excavation in 1912 provided confirmation that the old oaks and hawthorns of Staverton Park (Suffolk) were changing only slowly (Peterken, 1969a). Prints of Winston Churchill inspecting a range on Dungeness in World War II showed, in the background, hollies which had since changed (Peterken and Hubbard, 1972). (Both these cases show how important the incidental and unforeseen benefits of photographs can be for monitoring.) Historical records, including such obvious sources as Ordnance Survey maps back to about 1800, have proved immensely useful in monitoring some aspects of long-term changes, e.g. Hayley Wood (Rackham, 1975) and Bedford Purlieus (Rixon, 1975).

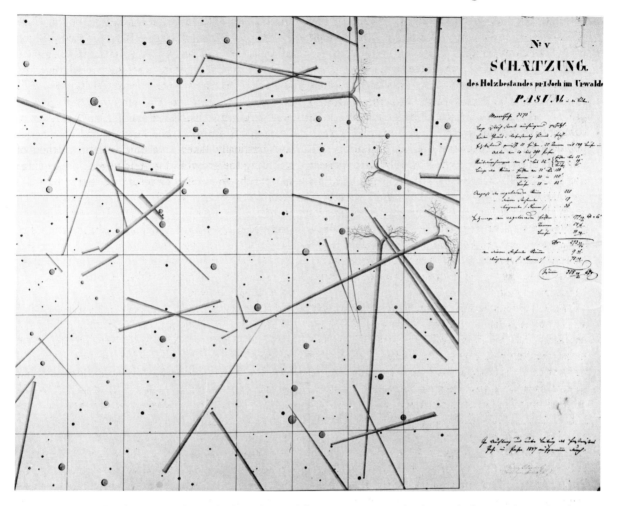

Fig. 13.15 Joseph John's original record of standing and fallen trees in a sample plot marked out in the Boubinsky Prales, a virgin forest in the Šumava Hills of southern Bohemia, Czechoslovakia. Recorded first in 1847 and subsequently (Řehák, 1968) from 1954 onwards, John's foresight has enabled the changes in composition and the rate of turnover of fallen trees to be measured over a very long period. The oldest comparable record in an unmanaged British stand appears to be that made by Dr E. W. Jones in Lady Park Wood, Gwent. Reproduced from the original in the State Regional Archives, Trebon.

13.9 Management records

Anyone who is concerned with the management of woodland should keep records of actions actually taken on the ground (Chapter 17, Principle 12). These need not be elaborate. Indeed, they may be no more than the notes, made in preparation for management, which are not thrown away, and which will thus be available in the future when there is any need to assess the impact of management on the wildlife (or indeed, rates of timber growth).

Systematic recording, storing and indexing is more worthwhile on nature reserves, where the records are likely to be frequently consulted. An Event Record system (Perring *et al.*, 1973) was devised for this purpose, among others. Record sheets bearing a simple map of the reserve are filled in whenever an event occurs. The event, which can be a natural happening or a site inspection, etc., as well as a particular act in a programme of habitat management, is described by recording what happened, the immediate consequence and its exact location (Fig. 13.16). Several hundred sheets may eventually be filled for the larger and more active reserves, so indexes are needed in the form of a classification of events and compartments within the site. Thus one can discover from the record what has happened in any compartment, but on looking up the record one can see exactly where the event took place, and then return to exactly that point on the ground to see what the longer-term consequences have been. Or, say, one can look up all instances of coppice management, wherever they occurred, prior to reviewing the effectiveness of this form of management. The Event Records for National Nature Reserves are processed automatically and can be used at a regional or national level. For example, the machinery could list the reserves in which fires occurred during the 1976 drought, or all reserves in which an attempt has been made to control the *Rhododendron* in the last 10 years. Features cards can be used instead when an index is required to the events on a single site (Peterken, 1969b). Whilst an Event Record may seem either superficial or over-elaborate at the start, its value increases as events, which were very obvious and memorable at the time they occurred, gradually fade from the individual and corporate memory.

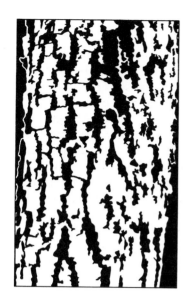

Chapter 14

Assessment of woodlands for nature conservation

14.1 General review

14.1.1 Introduction

How valuable is a particular wood for nature conservation? Which site best represents a certain type of woodland? Are some parts of a wood more valuable than others? These and similar questions are repeatedly faced by those concerned with nature conservation, and the answers form the basis of practical policies and decisions on the future of individual woods. Good answers, however, are difficult to find.

Site assessment is basically concerned to establish the extent to which a site contributes to the objectives and priorities of nature conservation. This is not a matter for the ecologist/scientist alone, even though such a person can provide a basic evaluation and judgement which constitutes an assessment of the intrinsic nature conservation value, if there is such a thing. Decisions on action, however, are also influenced by more practical matters. When an ecologist is asked by a Naturalists' Trust if a particular wood is worth buying as a nature reserve, he can place it in its ecological context, identify any special features and (if the information is available) compare it with other woods in the same region and of the same type. The decision about purchase, however, has to be taken by the Trust, taking into account not just the ecological evaluation but many other factors such as: (a) the price in relation to ecological value and financial resources, (b) whether the purchase price can be raised by appeal, (c) the cost in time and money of management, (d) competing demands for purchase of other land for nature reserves, (e) the likely demand for use of the site for education and/or research, (f) whether the Trust membership is more interested in conserving a bluebell display or a population of a rare but inconspicuous species, etc. The ecologist must be a party to such a decision but so must the treasurer. That said, the rest of this chapter will concentrate on ecological factors in evaluation.

Assessments have traditionally been based on the judgement of ecologists, eminent and otherwise, and many decisions in nature conservation continue to be based on subjective experience. The assessor would, whilst examining a site, react subjectively to what he found, and then after a short reflection, pronounce the place to be valuable, worthless or somewhere in between. Reasons given to support this judgement vary in detail and accuracy according to the assessor's character, experience and powers of observation. The experience of most people is, however, limited. The local naturalist is liable to regard the wood down the road as easily the most important wood in Britain, unless he has travelled to a number of woods elsewhere. Most professional ecologists who manage to fit in some fieldwork are restricted in their work to one county or a group of counties, where they are not well placed to judge the national significance of a particular wood. Likewise most research and teaching ecologists seem to know only a few areas well, though the areas may be widely scattered about Britain. Despite these limitations, ecologists have generally agreed on the value of sites, and the measure of their agreement has seemed to increase in proportion to their ecological experience. Perhaps, too, most assessments are made by ecologists brought up with a broad Tansleyan tradition, who tend to take account of many viewpoints in forming a balanced judgement.

14.2 Woodland attributes: their value for nature conservation

It is important to specify as exactly as possible the features of value in a particular wood in order to compare any two sites precisely, give an overall judgement an informed and credible basis, and plan management at a detailed level. Accordingly, there follows a summary of the several but inter-related attributes of woods which are significant for nature conservation. The points made in relation to invertebrates are based on the advice of Mr Alan Stubbs.

14.2.1 Species

Species lists should be compiled for as many groups as possible. A reasonable vascular plant list can usually be drawn up from a single visit, but other groups demand more time and specialist knowledge. In general, a wood is more valuable if it contains a large number of species, provided they are natural components of the woodland types present. Rare, local and other extinction-prone species are more valuable than common and widespread species (Section 12.5), though one rare species in isolation rarely justifies a high rating for the whole site. Almost every wood will contain a few rare invertebrates but the best sites are usually found to have as many as 3–5% scarce and rare species in several groups. Groups of species should be considered, e.g. the large number of butterfly species in Shabbington Wood (Oxon) makes this site important (Fig. 14.1), and the unusually wide range of calcifuge plants in Swanton Novers Woods (Norfolk) is just one reason for rating this site highly. Populations of species on the edge of their range have

potential for research into limiting conditions. Special genetical situations can also be of scientific interest, e.g. the hybrid swarm between *Geum rivale* and *G. urbanum* in Ditton Park Wood (Cambs). Populations of colonial species such as Herons and Black Hairstreak butterfly are important. Spectacular floral displays, e.g. of bluebells, are appreciated by all. On the other hand alien and introduced species can be detrimental if they compete vigorously with native species: many woods on light soils and in oceanic climates have been severely damaged by rampant *Rhododendron*.

14.2.2 Woodland types

Ancient, semi-natural woods are the most valuable (Section 12.6). Their value extends to all components – tree and shrub layers, ground flora, undisturbed soil profiles – individually and in combination. Within this broad group it is important to consider woodland types by reference to at least one classification. A wood comprising an extensive example of a single type might be a valuable representative of part of the range of variation. It would be even more valuable if it also contained other types and transition zones where the ecological range of species and communities can be observed and which are often rich in species: in that sense Gamlingay Wood (Adamson, 1912) would be more valuable than Hayley Wood (Rackham, 1975) as an example of stand type 2A in Cambridgeshire. Sites containing numerous types may be too diverse to produce a really good, i.e. typical, example of any one type but the range of transitional conditions available is enormous: such is the case with the stands in Lady Park Wood (Gwent). It is important to consider 'types' against several classifications if they are available, especially classifications covering stand, ground layer and epiphytes (Section 7.2). Since any classification is a more or less arbitrary division of a continuum, and any one wood may fit neatly into a type in one classification but fall confusingly between types in another, it is even desirable to apply two 'competing' classifications. To some extent types can be treated like species, with greater value being placed on rare and local types, such as 7E (Section 8.8) and 9Bb (Section 8.10), and a stand of a type on the edge of its range, e.g. Pencelly

Fig. 14.1 A ride in Shabbington Wood, Oxon, an ancient wood now almost wholly stocked with plantations. Open rides are a rich wildlife habitat, the bare ground and short turf in the centre, the verges of tall herbs and the marginal scrub each providing distinctive opportunities for a range of species. The exceptionally rich butterfly fauna of this wood has so far survived the conversion of semi-natural woodland to plantations – indeed, the active management has probably been beneficial – but it is likely to decline if the maturing conifers are allowed to shade out the marginal scrub. (Photo. P. Wakely, Nature Conservancy Council)

Forest (Pembs) containing an outlying example of stand type 3Aa (Section 8.4). Patches of secondary woodland are a valuable diversifying feature of a predominantly ancient wood, especially where, as at Hintlesham Woods (Suffolk), they cover a range of ages and thereby demonstrate the course of woodland development. Special communities such as the moss carpets of western woods are valuable if well developed. Extreme conditions are of scientific interest, such as the depauperate oakwoods of Dartmoor (Devon) and the sea cliffs of Millook (Cornwall), the alder carr of Barton Broad (Norfolk) and the tree-line pinewoods on Creag Fhiaclach on the edge of the Cairngorms.

14.2.3 Stand structure and management

Nature conservation has an ambivalent attitude to silvicultural management and stand structure. On the one hand management creates structural diversity, more niches and thus provides for more species, but it may diminish the mature timber habitat and eliminate natural structural characteristics and any semblance of wilderness. Lack of management generally leads to a uniform structure in the short and medium term though in the long term the natural structure may become more diverse. Thus, a stand can fall between two stools and it seems best for evaluation to decide whether a stand is or should be managed before proceeding to judgement.

Unmanaged stands become more valuable as the effects of past management attenuate with the passage of time. Staverton Thicks (Suffolk) (Peterken, 1969a) has not been managed for 150 years or so and is well on the way to becoming a natural stand. On the other hand, Lady Park Wood (Gwent), reserved as a natural stand, is only now beginning to have value in this respect because it was extensively cut over in about 1940.

Many of the valuable features of managed stands are stated or implied in Chapter 17, but some particular points should be noted:

(1) *Vertical structure.* Stands are usually more valuable if the foliage is widely dispersed vertically. An even canopy height with small depth of foliage, no understorey and sparse ground flora presents only a limited range of habitats for invertebrates. In upland woods, which are mostly grazed, even a sparse understorey is valuable.

(2) *Active coppicing.* The structure of an actively managed coppice-with-standards stand, though artificial, provides an excellent combination of vertical and horizontal diversity, and moreover it is the system to which the plant and animal communities have become adapted.

(3) *Mixed age structure.* A range of age classes provides structural variety now and a good prospect that this will be maintained in the future. If the mixture is achieved as a mixed-aged stand, rather than as patches of even-aged stands of different ages, then the resulting open structure benefits the invertebrates and epiphytes.

(4) *Mature trees.* Trees which a forester might describe as overmature, but which are regarded by biologists as merely mature, provide habitats which are not available in young stands. Hollows in live trees and sap runs are highly valued for invertebrates.

(5) *Dead wood.* A highly valued habitat for woodland fauna (Stubbs, 1972). This is important in all woodlands, but especially so in wood-pasture woods, where the timber-utilising invertebrate fauna is rich, and continuity of the dead wood habitat is necessary.

14.2.4 Subsidiary habitats within woods

(a) *Natural glades.* Natural glades in coppice woodlands appear to take three forms: bracken areas, bogs and mires and rock outcrops. These may not be completely treeless, and some are only temporary. Bracken glades are fairly common and widespread in lowland areas where the soil affords free drainage in surface horizons. Some such 'glades' are merely gaps in the coppice, with no gap in the standards. Wetland glades are much less common, but the wettest parts of cut-over and neglected coppice often retain an open canopy long after the canopy has closed elsewhere: the woody growth on very wet ground is generally poorly stocked and depauperate, with a correspondingly vigorous marsh-type ground flora. Large rock outcrops are restricted to areas of high relief, and may occur as inland cliffs. The best series are probably in the gorges of the Lower Wye and Middle Clyde valleys, where a thin covering of stunted trees is maintained. All these glades are valuable as distinct habitats which add diversity, and as subjects for studying, say, the manner in which shade-intolerant species survive in natural woodland.

(b) *Rides and other grassland within woodland.* A modern paradox is that the richest 'unimproved' grassland is often to be found in woodlands. This is especially true of the lowlands, where most of the surviving grasslands are temporary leys, and roadside verges are frequently disturbed, but it may be true anywhere, for woodland rides are usually grazed less heavily than grassland *per se*.

Rides (Fig. 14.1) are valuable for several reasons. They may be examples of unimproved grassland, sometimes richer than other grasslands in the district, and occasionally they are the only grasslands left which are not on a roadside verge. On wet soils rides may contain fine marsh communities, which are absent elsewhere. Ride margins provide an interface between two major habitats which is especially attractive to the woodland fauna. Certain plant species appear to thrive best in semi-woodland conditions, e.g. *Lathyrus sylvestris*, *Melampyrum cristatum*, *Pimpinella major*, *Vicia sylvatica*: these occur best in woodland margins and rides, but are also characteristic of hedge-bottoms. Ruts and ride-side ditches provide a standing water habitat, which may be absent elsewhere in the wood. Open rides and glades which receive plenty of sun are generally best for invertebrates. Shaded rides have a flora which is normally more closely related to the ground flora below coppice than to (unshaded) grassland. Nevertheless they are generally different from the adjacent ground flora because the soil structure and drainage has been changed by compaction and disturbance. Both ride edges and wood margins are better for invertebrates if they are irregular and diffuse rather than sharp and hard: some of the best margins occur where an old, hedged lane runs beside a wood on the south side. Grassland glades are fairly unusual in lowland woods, and many have recently been created within nature reserves. An ancient deer lawn, rich in local grassland species, survived as grassland grazed by deer in Weldon Park Wood, Rockingham Forest, until very recently. A small field within Swithland Wood (Leics) (Fig. 13.9) is still rich in grassland species. Glades are much more common in the north and west, where woodland boundaries are less sharply defined and rock outcrops break up the canopy.

(c) *Watercourses and pools.* Streams (Fig. 14.2), pools and seepage zones are very valuable. They form distinct habitats and provide on their margins conditions in which many woodland species thrive. Very often a river or stream forms part of a woodland boundary, whilst in the north, west and parts of the Weald, many woods are centred on ravines containing fast flowing streams and waterfalls, which are exceptionally important for bryophytes and ferns. Small streams originating within a wood are perhaps more valuable than those originating above it, because they should be uncontaminated by run-off from agriculture and approximate to natural conditions: some cold springs in the Weald, for example, contain relict populations of northern 'glacial' species. Those which have been canalised are less valuable than streams in their natural course, because natural streams have greater habitat variety and are worth studying as sources of information on processes in land formation. Some small woodland streams found in hornbeam woods meander spectacularly (e.g. Wormley Wood, Herts).

Woodland pools are not necessarily man made. Some appear to be natural depressions, perhaps deepened into ponds. Many are artificially created ponds (the spoil from which is usually found on site) or the by-product of shallow quarries or marl digging (from which the spoil will have gone). Whatever their origin, permanently unshaded pools provide the best conditions for aquatic plants and fauna. Those on a wood margin are more vulnerable to damage by agriculture, but they are more likely to remain permanently unshaded.

(d) *Rock faces and outcrops.* Shaded rock outcrops within woodland are potentially very valuable. Natural outcrops, which often occur close to streams, are amongst the most sheltered habitats in Britain. They are often excellent habitats for bryophytes and invertebrates, e.g. Edlington Wood (Yorks), and notable for the accumulation of rare species. Fine examples exist in the gill woodlands of the Weald, on outcrops of Tunbridge Wells sandstone, and in many western oakwoods (Ratcliffe, 1968).

(e) *Artefacts.* Banks, ditches, ponds and other artefacts provide some habitat diversity and evidence on the management history of the wood. Banks, for example, provide drier and often more alkaline soils than undisturbed soils nearby. Their historical significance is discussed below under 'historical record'.

14.2.5 Size

Large woods are generally more valuable than small woods because they are likely to contain more species, larger populations of species, more woodland types and a greater range of structure and subsidiary habitats. However, the relationship is not always close and size as such is an unimportant factor where other attributes are separately observed, though a few species (e.g. Purple Emperor butterfly) appear to require large woods.

Nature conservation value per unit area decreases with size. For example, doubling the size of the wood doubles the cost of acquisition as a nature reserve but less than doubles the number of species present, though the chance of each species surviving is improved. This law of diminishing returns implies that there may be an optimum size for nature reserves when costs have to be considered.

14.2.6 The Fourth Dimension – Time

(a) *Habitat continuity.* The fundamental value of

habitat continuity in time has been discussed in Sections 12.5 and 17.2. Its significance for invertebrates is difficult to assess, but it appears to be one of the most important factors which affect the general variety of species and the presence of rare species. Thus open rides, sunny glades, a coppice-with-standards structure, and diffuse wood margins are valuable wherever they occur, but are very much more valuable if the desirable condition has been present for a long time. This is a direct reflection of the ecological isolation factor acting in space and time: continuity of desirable habitat conditions is most important in geographically isolated woods.

(b) *Historical record.* Whilst a full historical record of management is especially valuable it is difficult to lay down rules about what constitutes a good record. Two important and relatively accessible elements are records of wood sales and other aspects of stand management since 1800 and detailed estate maps showing the wood at any date between 1580 and 1850. The 'record' extends beyond documentation of past management and condition to include artefacts (banks, ditches, etc.) visible in the wood as well as records from early ecological studies. A few examples may help. Llanmelin Wood (Gwent) is important because it contains a hill fort of the *Silures* which was abandoned when the Romans settled nearby at Caerwent: the mixed native woodland of this demonstrably ancient, but secondary woodland is very similar to the probably primary woodland nearby, and demonstrates the recovery power of natural woodland. Chalkney Wood (Essex) seems more valuable because the boundary banks, which were mapped in detail in 1595, all survive.

(c) *Potential.* Particular woods are often said to have potential, implying that they can be improved for wildlife. Often this is because there has been reversible deterioration in the recent past, e.g. increasing shade

of rides, regeneration restricted by grazing, or simply rubbish dumping in a neglected wood. Potential is, however, a dangerous factor to consider in site assessment because any wood can be improved for nature conservation, and in some circumstances potential can be less than present value, e.g. a wood recently isolated by nearby clearance may lose species (Section 6.2): the succession of mature oaks may not be sustainable in parklands which lack maturing trees.

14.2.7 The wood in relation to its surroundings

(a) *Adjacent habitats.* Adjacent, semi-natural habitats add to the value of a wood in several ways. The other habitat may be valuable in its own right, and if so it could form a single management unit with the wood, e.g. as a diverse nature reserve. The other habitat is liable to support at least some of the woodland species, and thus effectively reduce the isolation of the populations within the wood itself. Adjacent meadows can be of considerable value for insects if a good variety of flowers is maintained, and will supplement the rides. Marshes, disused railway lines, old drove roads and river banks are also valuable adjacent features. The wood margin, which is an important habitat, especially for invertebrates, is likely to be richer than the normally sharp and disturbed boundary between woodland and arable. Wood margins with a broad scrub transition zone are possibly the richest.

(b) *Shelter.* Woods in sheltered situations or on a south-facing slope are likely to be richest for invertebrates, whilst an exposed wood on the top of a hill or by the coast is likely to be poor. Large woods can produce their own shelter internally if a diverse structure is maintained. Long, straight rides are less sheltered than rides with bends or T-junctions.

Fig. 14.2 Ceunant Cynfal, Maentwrog, Merioneth, one of the finest river gorges in Britain, formed by the Afon Cynfal flowing across Cambrian slates. Sheltered by the sessile oak woodland on acidic sites and ash–wych elm woodland on flushed, base-rich soils, a rich community of bryophytes, ferns and epiphytic lichens thrives, especially in the vicinity of the stream. This and similar woods in the oceanic climate of western Britain harbour many species of 'Atlantic' bryophytes, for which Ireland and western Britain are the European strongholds. (Photo. D. A. Ratcliffe)

(c) *Isolation.* Which is more valuable, a 10 ha ancient, semi-natural wood situated many miles from another wood, or an otherwise identical 10 ha of woodland which forms part of a large block of woodland? The former will be the only indication of woodland characteristics in its area and a valuable diversifying habitat, but vulnerable to the impoverishing effects of isolation. The latter is likely to be richer, but more expendable in the sense that more similar woodland remains nearby. My answer is that both are valuable, but for different reasons. Isolated ancient woods, such as those in the Borders region of Scotland, have rarity value as the last examples of a part of the natural range of variation, whereas portions of large woods help to support species such as the Purple Emperor butterfly, which has died out from isolated woods in much of the English Midlands.

(d) *Landscape.* Woods are undoubtedly important elements in the landscape, but many would doubt that this is a valid factor in their value for nature conservation. I include wilderness conservation with nature conservation and hence, for example, regard the Borrowdale (Cumb) oakwoods (Fig. 12.2) as that much more valuable because they are a prominent feature of a relatively natural landscape.

14.2.8 Research needs

Whilst any wood could be the focus of a great range of research, it can be quite difficult to find exactly the right circumstances for a particular project. Provision for future research is inevitably imprecise, for such needs are unpredictable: the only practicable approach is to maintain as many sites and as wide a range of conditions as other factors will allow. The main exceptions are the so-called 'classic sites', whose value is increased by virtue of past research or recording, provided the records of past work are retained. Examples of sites whose value has been increased in this way include Gamlingay Wood, Cambs (Adamson, 1912), Hayley Wood, Cambs (Rackham, 1975), woods near Faversham, Kent (Wilson, 1911), Abernethy Forest (Birks, 1970; O'Sullivan, 1973a, b), Bookham Common (various reports of the London Natural History Society) and Caeo Forest, Dyfedd (Hill and Jones, 1978).

14.2.9 Level of disturbance

Disturbance in various forms can reduce the value of a wood. If large numbers of people visit a wood, such as the beechwoods and box scrub of Box Hill (Surrey) soil is compacted and eroded, birds are disturbed in the breeding season and wildflowers are picked. Visitors have indirect effects: for example, they do not actually fell mature trees or lay tarmac paths themselves, but officials do so on their behalf. Rubbish dumped into a wood will eutrophicate the soil and possibly pollute any streams. Civil works, such as the huge pipe laid along the bottom of part of Castle Eden Dene (Durham), may have little effect on the ecosytem but damage the relatively natural appearance of a wood.

14.3 Site assessment: the example of Foxley Wood

Foxley Wood provides a convenient example of site assessment aimed at detailing both the importance of the wood as a whole and defining those parts which are of greatest interest. It was the largest wood included in the Norfolk survey (Section 14.5). Although it is privately owned, it is mostly managed by the Forestry Commission.

The itemised statement of importance runs as follows:

(1) It is an ancient, semi-natural woodland, one of the largest in Norfolk.

(2) Stand types, which are infrequent and especially characteristic of eastern England, are present, notably 4B, 7C and a form of 2A which is on medium-textured soils and thus close to 2B in the range of variation (Fig. 14.3).

(3) It contains more vascular plant species than any other ancient woodland in the county.

(4) Apart from numerous locally rare species, the flora contains an unusually wide ecological range from calcifuge species (e.g. *Hypericum humifusum, Sphagnum squarrosum*) to calcicoles (e.g. *Orchis mascula, Viola reichenbachiana*).

(5) Bracken glades and marshes occur in small patches which are naturally lightly wooded. Rides remain open and well stocked with grassland species which have largely been exterminated on the surrounding arable farms.

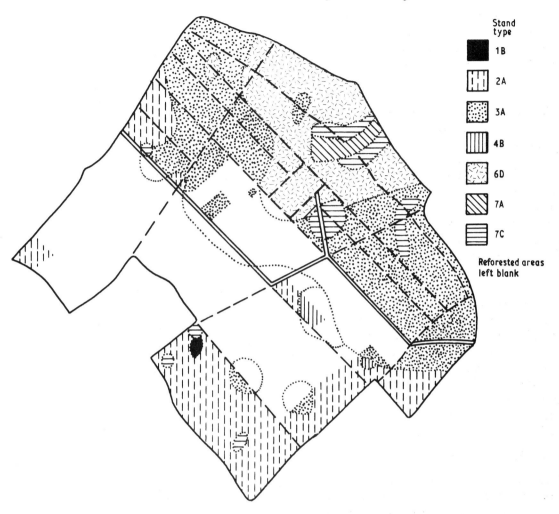

Stand type

1B

2A

3A

4B

6D

7A

7C

Reforested areas left blank

Fig. 14.3 Map of stand types in Foxley Wood, Norfolk, surveyed by S. Goodfellow and G. F. Peterken. Large parts of the wood have been replanted with conifers and it is no longer possible to recover accurately the former distribution of stand types within these areas, though an attempt has been made to show approximate boundaries. Mapping was also very difficult in the semi-natural woodland, which has been considerably disturbed by exploitive felling of saleable timber and indiscriminate application of herbicides.

(6) Coppicing has continued until recently in parts of the wood.

(7) A stream, which arises in the wood, is unaffected by surrounding agriculture, and still flows through the valley alder woodland (7B) which would have surrounded it under natural conditions.

(8) Since the wood is divided between two parishes, it may be possible to detect any differences in management and their effects on composition. Historical records of management have not been investigated.

This wood has been partially replanted with pure conifers. The remaining semi-natural woodland was cleared of saleable trees before the Forestry Commission took responsibility for management, but later, parts of the semi-natural woodland were indiscriminately treated with herbicides and there was no subsequent planting. Forest roads and deep ditches have been driven through. Ecologically, therefore, Foxley Wood has been badly damaged by modern forestry, yet its hydrology remains apparently intact.

The main pattern of semi-natural woodland types can still be seen, even though little remains of the 2A–3A transition zone, and type 4B has been much reduced. The north-eastern half of the wood was undoubtedly more important than the south-western half, and to that extent replanting has not been as damaging as it might have been, but further in-roads into the semi-natural woodland could hardly fail to cause a significant decline in the value of this wood for nature conservation.

14.4 Comparisons between woods

14.4.1 General considerations

The basic, general purpose requirement in site assessment is, in my opinion, a ranking system which will place any number of woods into an order of value for nature conservation. Admittedly one often wants to value a single wood, but the judgements about one site cannot be made in isolation from others of a similar kind or in the same area.

Comparisons between sites often raise difficulties arising from the relationships between attributes. Whilst some woods are valuable on several counts – such as a large, rich wood containing many rare and local species, covering a wide range of soils and possessed of a mature structure – many are good in some respects, but poor in others. For example, the famous Wytham Woods near Oxford are large, have a mixture of ancient, semi-natural and secondary woodland and a long history of biological research and survey, but they have a very limited range of stand types, a poor woodland flora in relation to their size, and sycamore is abundantly colonising some parts of the mature woodland. Thus, woodland attributes can vary independently – a wood can be outstanding for one feature and ordinary for all others – and site evaluation cannot proceed until the relative priority and importance of each attribute is determined. Nevertheless, the incidence of one attribute is often positively correlated with that of another. A rich wood is more likely to contain rare species than a poor wood, and a wood with a wide range of subsidiary habitats is generally richer than a wood without them. The difficulties of relating attributes and producing a harmonious assessment can easily be exaggerated.

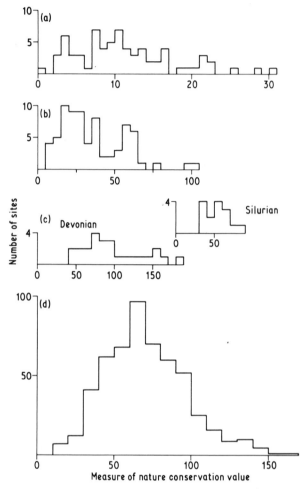

Fig. 14.4 Examples of the distribution of sites in relation to a measure of their nature conservation value. (a) Ancient woods in central Lincolnshire, measured by the number of ancient woodland indicator species (Peterken, 1974a). (b) Ancient woods in Norfolk, measured by the number of selected woodland plants (Goodfellow and Peterken, in press) (c) Semi-natural woods in Powys, measured by a floristic score derived from the total list of species. Data by kind permission of Dr R. G. Woods and Mr M. Massey. (d) Limestone pavements in Britain, measured according to a floristic index by Ward and Evans (1976). Data from all sites included by kind permission of Dr S. D. Ward.

The spread of site values often takes a characteristic skewed form, i.e. the mode lies below the mean and the spread of exceptionally high values is wider than the spread of exceptionally low values. Numerous quantitative measurements of site value have been attempted during routine conservation work (Fig. 14.4) and most confirm that, in any population of woods within a limited area or representing a particular woodland type, (a) conservation value is a continuous variable within which there is no sharp division between valuable and worthless; (b) a few woods are of outstanding value; (c) no woods are completely worthless; and (d) most woods are a little above or a little below average. Certainly this distribution corresponds well to the common experience that a few woods are easily recognised as and agreed by all to be outstanding, whatever evaluation system is adopted, whilst the bulk of woods are moderately valuable. The woods in this middle range are difficult to rank; their position in the league table depends substantially on the relative importance accorded to different attributes; but within this range it scarcely matters if individual sites are somewhat misplaced.

14.4.2 A Nature Conservation Review

The Nature Conservancy Council has had to evaluate a large number of individual woods in order to determine which should be scheduled as Sites of Special Scientific Interest (SSSI). The most important of these have recently been listed in *A Nature Conservation Review* (Ratcliffe, 1977) (NCR). These evaluations have been based on the approach adopted in Cmd 7122 (1947), and the attributes and priorities are described in Sections 12.5, 12.6 and 14.1 above. They were elaborated in the NCR as ten criteria, which can be summarised thus in respect of woodlands:

(1) Size (extent). Where woods are isolated habitat islands, their importance generally increases with size of area. Many woods are undesirably small in their total area. Woods with large trees are better than those with only small trees.

(2) Diversity. Variety is better than uniformity. Sites with more species are better than those with few, provided that similar types of woodlands are being compared. Sites with a range of woodland types are better than those with only one.

(3) Naturalness. Woodlands which have been least modified in structure and composition by man are the most valuable.

(4) Rarity. Woods are more valuable if rare species or communities are present.

(5) Fragility. Fragile ecosystems and species have high value. Fragility is both intrinsic (e.g. from successional change or vulnerability of small populations) and extrinsic (i.e. probability of deleterious change by human actions). It can also be a combination, e.g. a population fragmented into small groups by human action. Ancient woods are more valuable than recent woods.

(6) Typicalness. Given that one objective is to maintain examples of all woodland types, good examples of commonplace types are as important as examples of rare types.

(7) Recorded history. Woods which have been used for an important piece of research or which possess a good historical record are thereby made more important.

(8) Position in an ecological/geographical unit. A wood which is contiguous with other semi-natural habitats is more valuable than one which is not.

(9) Potential value. Woods whose value has diminished due to past management, but not irreversibly, have a potential value greater than present value.

(10) Intrinsic appeal. This applies to species, not habitats. Species which appeal to the greater number of people are more important, i.e. birds and conspicuous flowers.

These criteria are not mutually exclusive, but overlap in several ways. Mostly they reinforce each other, as in the case of size and diversity. Some are complementary, notably rarity and typicalness. Some are potentially in conflict, notably diversity and naturalness. Pragmatic considerations loom large in some criteria, notably intrinsic appeal, whereas a more fundamental consideration underlies, for example, naturalness. In the case of fragility, these two aspects are equally balanced. The ten criteria do not contribute equally to each beneficiary of nature conservation: thus, typicalness and recorded history are especially important for research needs, but intrinsic appeal avowedly caters for popular interest. 'Permanence', which is given as a significant attribute of woodlands, is included within the naturalness and fragility criteria.

14.4.3 Quantitative assessments

Most site assessments are qualitative. Their reliability depends on the experience and lack of bias of the assessor. When, as is usually the case, the information available is not in a quantitative form, or when some woods are recorded far more thoroughly than others, qualitative assessment seems unavoidable. Fortunately the results are generally reliable in the sense that different ecologists agree with them, and they are available quickly, thus allowing subsequent action to be determined (which may be urgent in the face of a threat). Even so, qualitative, so-called subjective assessments have disadvantages. First, because the decisions which flow from an assessment may affect someone adversely, the assessment may be disputed with just as much confidence and in much the same terms as it is given: the criticism will not necessarily be constructive, for in public inquiries it may be sufficient just to cast doubt on the judgement. Second, any assessor, however honest and experienced, may have personal interests and prejudices which influence his judgement. Personal interest and knowledge were not least amongst the factors which determined the list of sites recommended as nature reserves by the Wild Life Conservation Special Committee (Cmd 7122, 1947).

Ecologists have therefore had to consider criteria more quantitatively, both to convince sceptics and for their own peace of mind. Quantitative assessments have often been considered 'objective', but this is incorrect: nature conservation is concerned with value judgements placed by society on certain features in the natural environment, so some degree of subjectivity is unavoidable. Rather, quantitative assessments should, and usually do, have three characteristics which make the results more secure in the face of criticism: (i) judgements are confined to the definition of criteria and the relationships between them, and removed from the assessment of particular features in particular sites, which should be recorded as objectively and quantitatively as possible; (ii) the criteria are explicit and the weightings given to each feature are precise and open to debate; (iii) the overall assessment of a site is the sum or product of separate assessments of several features, and it is not therefore usually vitiated by successful criticism of any one

judgement. Assessments made in this way have three further advantages. Inexperienced people, by applying a pre-determined system, can evaluate woods almost as reliably as experienced people. Woods can be compared on the basis of fine detail. Finally, the assessor's special interests and prejudices have to be declared at the outset and justified if they are to influence the final verdict.

Several authors have evolved assessment systems which attempt to be precise and quantifiable, but so far none has been widely accepted, nor successfully applied at a national scale. Helliwell (1973) has tried to value woods in financial terms, which, if successful, would have the benefit of a common language with valuation for commercial purposes. This method depends on introducing a financial value which then becomes a constraint into calculations, such as valuing a pair of Blue Tits at £1, or valuing the educational capacity/potential of a wood by a *scarcity index × number of obvious species × durability index × £D*, where $D = 1/5000$. The indices are admitted to be somewhat arbitrary at present, and no doubt D will change with inflation and any change made by society in the priority of education. Even if financial formulae were developed and agreed, one might still prefer to relate nature conservation to commercial factors through 'opportunity loss', i.e. an assessment in cash of the loss which an individual or a community is prepared to sustain in the interests of nature conservation. Nevertheless, Helliwell's approach is a commendable attempt to develop the necessary precise, explicit criteria and quantitative measures. Wright (1977), on the other hand, tests a system which adds a management appraisal and an assessment of potential amenity and educational use to the scientific criteria described above, and then compares his system with some others. His system, however, like many others, involves a semi-quantitative scoring system based on quantity-adjectives rather than precise measures. Some precision can be introduced into a points-scoring system, e.g. the unpublished work of R. G. Woods and M. E. Massey (Table 14.1).

The quantitative approaches depend on quantifying all attributes and values which form the basis of criteria. This is easy enough in some instances, such as the number of species or the area of habitat, but

Table 14.1 Subjective scores used by R. G. Woods and M. E. Massey (personal communication) for comparing the value for nature conservation of semi-natural woodlands in Powys.

Score	Attribute
STAND STRUCTURE	
10	High forest
5	Coppice-with-standards
0	Simple coppice
5	Trees of uneven age/size
5	Shrub layer covering more than 50% of the ground
3	Shrub layer covering 25–50% of the ground
GROUND VEGETATION	
5	Herb layer well developed, i.e. ungrazed
5	Moss carpets (ground cover above 50%) on north and west facing slopes
3	Moss carpets on south and east facing slopes
AREA	
1	For every 4 ha (10 acres)
ADDITIONAL HABITATS	
1	Rock or scree
1	Stream
1	Pond
1	Wet flush
1	Glade
1	Dead wood
1	Adjacent semi-natural habitat

exceedingly difficult in the case of, say, fragility or recorded history. When attempts are made to quantify difficult attributes by complicated indices or points scored for quantity-adjectives (e.g. 1 for 'poor', 2 for 'medium', 3 for 'good') the system is at best debatable and only justified if the results are credible, which leaves one very much where one stood with subjective judgements. Quantitative approaches also seem to rely on the independence of attributes, which allows the various values to be summed or multiplied as appropriate, yet in practice the attributes which are valued by society are neither independent nor free of some conflict. Clearly, quantitative methodology is in its infancy and nice points can be argued *ad nauseam*. Meanwhile, in areas where few semi-natural woods remain, there is such a large measure of agreement about which sites are best, and such pressure to damage their conservation value, that practical conservation measures cannot wait on the fine-tuning of assessment methodology.

14.5 Site comparisons based on species counts

One of the most important lessons to be drawn from quantified assessments so far is that useful quantification has its limits, set by both practical restraints and fundamental difficulties. No amount of survey will provide all the potentially relevant details of all the woods to be compared, and, even if the data were available, questionable assumptions would have to be made in order to relate each attribute to every other. (For example, is a good historical record equal in value to one rare species, ten or how many?) Thus, quantification is desirable only up to the point beyond which the increased costs and delays of obtaining additional data add little to the precision and reliability of the final result. Assessment procedures must remain a blend of clearly stated judgements and precise, simple quantification.

The best unambiguous, quantifiable and directly relevant attribute of any wood seems to be the number of species it contains. A wood containing more wildlife species than another can usually be regarded (albeit tentatively) as more important for nature conservation. However, the comparison must be confined to groups which have been equally well recorded: impressive lists of numerous groups are available for Monks Wood (Steele and Welch, 1973), but they do not enable us to compare Monks Wood with neighbouring, poorly recorded woods. When large numbers of woods are involved the basis for comparison is most likely to be the list of vascular plants. This is a group which is easy to record over a long season, which contains enough species to generate a sensitive scale of comparison, which responds as a group to the small-scale variety characteristic of woods, and which supports the great majority of the fauna.

Some ecologists argue that assessments should be objective, implying that any subjectivity is an undesirable failing. Subjectivity, however, is unavoidable: judgements are being made and this must at some point result in value statements on behalf of individuals and society at large. Rather, the plea for objectivity seems to seek (a) some quantitative observations which are free of recorder bias, and (b) restriction of subjective judgements to those parts of the procedure not involved with field observations.

My approach to survey and assessment for nature conservation of any group of woods is based on four stages (Goodfellow and Peterken, in press):

Stage 1 Having accepted that ancient, semi-natural woods are the most important (Section 12.6), each wood is classified as (a) ancient or recent and (b) plantation or semi-natural, using mainly maps and air photographs. Only the woods with at least some ancient, semi-natural woodland need be considered further if survey resources are limited.

Stage 2 Field survey of each wood, recording chosen features as objectively as possible.

Stage 3 Ranking all surveyed woods into one or more league tables on the basis of species recorded.

Stage 4 Selecting a limited number of woods, being those which, for various reasons, are most important for nature conservation. This selection is based on the league tables from stage 3, but other factors are taken into account.

This approach was first applied in central Lincolnshire (Peterken, 1974a). Detailed historical research had identified 85 ancient woods (stage 1). Vascular plant lists were collected for each wood (stage 2). Extensive field survey had identified 50 species which were more or less confined to ancient woods, and the woods were ranked on the number of such ancient woodland indicator species (stage 3). The best limewoods were selected by taking those woods containing lime which appeared highest up the league table (stage 4). The result corresponded closely to the selection made in Ratcliffe (1977) before the floristic surveys had been undertaken.

The Lincolnshire assessment was possible because detailed historical and floristic surveys were available from other studies. Modifications were necessary if the system was to be practicable for woodland surveys in general: ancient woods had to be identified without resort to lengthy historical research, and the definition of ancient woodland indicators (if they were to be used at all) must not depend on extensive floristic surveys outside woods. Furthermore, the selection stage had to be capable of yielding a list of important woods in a group, not just the best of any particular type.

Suitably modified the system was applied to Norfolk woods (Goodfellow and Peterken, 1981).

Ancient woods were identified by use of old Ordnance Survey maps (Section 13.2). Recent air photographs showed which stands were certainly plantations, the remainder being accepted as provisionally semi-natural (stage 1). Woods containing at least some ancient, semi-natural woodland were visited. The woodland vascular plants were listed. Stand types (Chapter 8) were mapped, though in fact it was necessary only to list those present and estimate their extent in terms of four area categories (<0.5 ha, $0.5–2$ ha, $2–10$ ha, <10 ha). Other features were recorded non-quantitatively (stage 2). All the 75 woods surveyed were ranked on the basis of the number of woodland species, after excluding fast colonisers and introductions and allowing for differences in the intensity of survey (stage 3). Several selection procedures were then tested (stage 4), the most satisfactory being selection in groups (Table 14.2) as follows:

Group 1 Fourteen woods were selected in a sequence of moves designed to form the shortest list containing two examples of each stand type (where available) from the largest available size classes. Three examples of stand type 7E were selected because this is particularly significant in Norfolk (Fig. 8.27). Stand types 10A and 12B were excluded from the selection process because the latter in Norfolk is merely a disturbed version (Section 9.3) of type 6D, and the former is present only as tiny marginal stands originating from planted trees.

Group 2 Three woods were added because they contained an unusually large number of stand types. This allows for the possibility that a different vegetation classification would have led to a slightly different selection in group 1, and ensures that transition zones are well represented.

Group 3 The three richest woods not so far selected were added to bring the total to twenty (Section 14.6).

Group 4 A further nine woods were held in reserve. They were chosen for specific attributes, e.g. the largest wood not so far selected; the third largest example of stand type 7E in Norfolk; stand well stocked with mature oaks, etc.

Finally, the selection was itself judged. Did each wood contribute something to the overall selection

Table 14.2 A summary of the main characteristics of the 29 woods involved in the final stages of selecting a list of important woods for Norfolk. The 16 firmly selected woods are listed in upper case. The 10 woods forming the reserve selections are in lower case. The three woods completely rejected from the final selection are shown in italics. For further explanation see the text.

League position		Area (ha)	Flora score	1B	2A	2B	3A	4A	4B	5A	5B	6C	6D	7A	7B	7C	7E	8B	9A a	9A b	10A	12B	No. of stand types
Group 1																							
1	FOXLEY WOOD	127	100	d	*A*	*A*		c		*A*	*A*		*A*			B							7
2	SWANTON NOVERS WOODS	61	99		d		d	d	*B*	c	*A*	*A*	B	c		B	*A*		d		B	c	8
4	Hindolveston Wood	69	67				d	d					d	*A*			*A*						3
5	GAWDYHALL BIG WOOD	30	64	d	*A*		*A*			d			d			*A*			*A*		d		4
6	ASHWELLTHORPE LOWER WOOD	36	64	c	c	B		d					d	*A*			*A*		*B*		B		6
9	WAYLAND WOOD	32	62			B										*A*			*A*	c	c		5
10	SEXTON WOOD	39	60	d	c			*B*	c	*A*									*A*	c			4
17	HOCKERING WOOD	90	56			c						B	B		c		d		d		d		7
25	HERONS CARR	16	38				d					c	c	*A*			d		d		d		5
29	FELBRIGG GREAT WOOD	130	36												*A*		*A*						2
30	*Hollands Wood*	4	35			B														d			1
31	*Wootton Carr*	25	35									B	B	B		B			B				4
56	EDGEFIELD LITTLE WOOD	11	20																B		B		2
72	*Gurneys Wood*	2	9															c					1
Group 2																							
3	HETHEL WOOD	25	76	d		d		*A*				c	c			B	B		B	c	c		6
7	BARNEY WOOD	18	64	d		d		c				B	B	c		B	d		d		d		8
14	FUNDENHALL WOOD	18	56	d		B		*B*				c	c	B		B			B		B		6
Group 3																							
8	HONEYPOT WOOD	10	62	d		B		B												c			3
11	HORSE WOOD	14	58	B	c			d								B			c	d			5
12	HORNINGTOFT WOOD	9	58	B		B		d											d				4
Group 4																							
13	*Hedenham Wood*	23	56	B		B													A		B		3
18	*Tindall Wood*	42	53																A	c	c		3
20	*Winters Grove*	9	52										d						B		d		2
21	*Hempnall Little Wood*	7	49	d											d				B		B		2
22	*Hazel Hurn*	9	49	d		B									d	d				c			5
24	*Toft Monks Great Wood*	13	42																A	d			2
28	*Pulham Market Big Wood*	5	36	d								c	c			c			A	d			5
34	*Shotesham Little Wood*	18	33	d								c	c			c			A	c			3
36	*Guybons Wood*	9	32		c							c	c				B				c		3

League position: As determined by the flora score.

Hectares: The area of the whole wood, including any plantations.

Flora score: Based on the number of woodland vascular plant species recorded.

Stand types: The extent of each stand type in each wood is given in four size classes, A = above 10 ha, B = 2.1–10 ha, c = 0.5–2.0 ha, d = less than 0.5 ha. These areas refer only to the semi-natural parts of each wood.

The attribute for which each site was selected at the penultimate stage of the selection process is shown in italics.

which no other wood contributed? Did any wood have any deleterious features or valuable features which did not affect the species present or the occurrence of stand types, e.g. concrete rides or good historical record? Hollands Wood was excluded because it contributed only a 2.0 ha stand of type 2B. Wootton Carr was excluded because it was a very poor example of type 7A and better examples were known from neighbouring Lincolnshire. Hindolveston Wood was relegated to the reserve group because very little semi-natural woodland now remains there: it was not excluded altogether because a large stand of type 7E survives beneath a poplar plantation. Gurneys Wood was eliminated because it contained only a minute fragment of the beech woodland for which it had been selected. Thus, the final selection from the surveyed woods contained 16 firm selections and 10 reserve selections.

The Norfolk system has several advantages. Economy of effort is achieved by confining survey to important woods, accepting approximate identification of ancient semi-natural woodland in stage 1, and by quantifying only the minimum necessary for selection. Quantification is simple, precise and free from recorder bias. Flexibility is built in by the last act of selection, when the list generated by the quantitative procedure is itself judged against other features which cannot be economically quantified. All the important NCR criteria (Section 14.4) are taken into account. Size (1) and diversity (2) are major determinants of the number of species in each wood. Rarity (4) is brought in by excluding fast colonisers from the ranking stage and using rare species as one of the factors in the final step of selection. Naturalness (3) and fragility (5) underly the concentration on ancient semi-natural woods from stage 1 onwards. Typicalness (6) is the main factor considered in the first steps of selection. Other criteria are considered at the final step of selection. Attempts are made to consider the study of woods in a wider context by weighting selection for and against certain stand types (+7E, −7A, 10A, 12B).

Perhaps the most questionable aspect is the use of woodland species rather than total species as the basis for ranking. Woodland species are regarded as a fundamental attribute of a wood which is largely unaffected by changes in the extent of unwooded ground within the wood. (Numerous species are lost when rides revert to woodland. Conversely, a quick way to enlarge a list in a uniform wood is to cut rides and clearings.) However, rides are often important grassland habitats and components of the habitat diversity upon which much of the woodland fauna depends, so there is a good case for including all species. Alternatively, ride characteristics can be considered qualitatively in the final step of selection: this saves survey time.

The system applied in Lincolnshire and Norfolk should work elsewhere, though the bases for selecting species for ranking and ensuring proper representation of the range of variation must vary according to local circumstances. In the uplands especially it may be necessary to include all vascular plant species on the grounds that woodland species are not so clearly defined as in the lowlands. In western woods (Fig. 14.2) it may be desirable to base assessments on bryophytes and ferns and in wood-pasture stands the epiphytes would be more appropriate if lists can be obtained. Alternatively, only rare local and attractive species (of all groups, not just vascular plants) might be used in accordance with NCR criteria (4) and (10). Representation can be considered using other floristic classifications, or a field layer classification might be used instead of or in addition to the stand types. Alternatively, it might be more useful to classify sites, so that each wood falls into one category only, using perhaps geology or whole-site features.

Quantitative, systematic assessments of woods in areas larger than a county have not been successfully attempted, except in the case of the native pinewoods, a well-defined and limited type, where even the assessment by R. Goodier and R. G. H. Bunce (1977) was not primarily concerned with ranking sites. National assessments are likely to remain semi-quantitative at best, for every wood is unlikely ever to be recorded to a consistent standard. In any case, each region has distinctive features, such as the bryophyte communities of western woods or the widespread mature timber fauna of south-east England, which must be taken into account, and in any national selection a widespread type must be represented throughout its range. In ecological terms at least a national assessment is almost the sum of its parts.

14.6 How many sites are needed?

How many woods must be safeguarded for nature

conservation? In Norfolk we judged that about 20 woods were required, because:

(a) this number exceeded the minimum required to represent the range of types, i.e. the 13 woods in groups 1 and 2;
(b) they remained a small minority of the ancient, semi-natural Norfolk woods, thus retaining the credibility of the claim that they were special;
(c) their total area of about 800 ha (depending on exactly which 20 woods are finally selected) is still only 2.7% of all Norfolk woods;
(d) the conservation organisations involved could cope with the amount of work generated by scheduling these woods as Sites of Special Scientific Interest and, where possible, acquiring some as nature reserves.

This is a most complex issue, which carries us far beyond the brief of the ecologist alone and any pretence that an objective limit can be stated. We may perhaps agree that there is an absolute minimum requirement based on ecological attributes alone, but above this the limit is set by practical and political considerations. Much also depends on what exactly is meant by the phrase 'safeguard for nature conservation', for this does not necessarily exclude other uses. These issues are discussed further in Part Four.

14.7 Assessments of parts of contiguous woodland

14.7.1 Densely wooded regions

The woodlands of central Lincolnshire and Norfolk are mostly small and isolated. There is little doubt about the boundaries of individual sites and consequently no difficulty about assessing and comparing the sites by counting species. Difficulties arise, however, when site boundaries are less obvious and have to be defined arbitrarily, i.e. in densely wooded districts and in comparisons between parts of a single wood.

In densely wooded areas such as the Chilterns there are large woods which run continuously for up to five miles across many property boundaries. They are physically divided by roads and historically split into a number of different portions, each with a separate

name. If such huge sites are simply treated as single woods they will tend to come at or near the top of any league table based on a count of species. A virtually similar woodland which, however, is divided into two woods by, say, a narrow field, would (twice) come lower down, not because it is less rich in aggregate, but because the constituent parts are less rich individually.

Faced with such problems one could define sites by boundaries which have some meaning in management terms, such as minor roads and historical subdivisions of larger woods (often with separate names and divided by banks and ditches), which are often also property boundaries. Alternatively, or in addition, contiguous blocks of semi-natural woodland can be treated as separate sites, thereby treating adjacent plantations as if they were not there. If the smallest feasible subdivisions into 'sites' are recorded separately, adjacent stands can later be amalgamated for the ranking process if they form a management unit.

The subdivision of more-or-less contiguous woods for assessment purposes is ecologically arbitrary. The individual portions are not isolated like the separate woods in sparsely wooded Norfolk, and there is less need to base conservation policies on the assumption that all species must be maintained where they now live. This leads us from the site-based approach to nature conservation, for which site assessment is required, into the special area approach (Section 15.4). This admittedly still involves the definition of important sites, but mainly to include a rare species or special features: measures to ensure the conservation of the range of ecological variation and of all species is not a primary consideration, because so much semi-natural woodland still remains. If separate sites cannot be defined for assessment purposes then the site level of conservation is likely to be inappropriate.

14.7.2 Identifying important parts of individual woods

A common requirement is to identify those parts of a wood which are particularly important. This usually arises when management is being decided, especially when a hitherto completely semi-natural wood is to be partly converted to plantations leaving the rest as a natural reserve. In many instances where the features

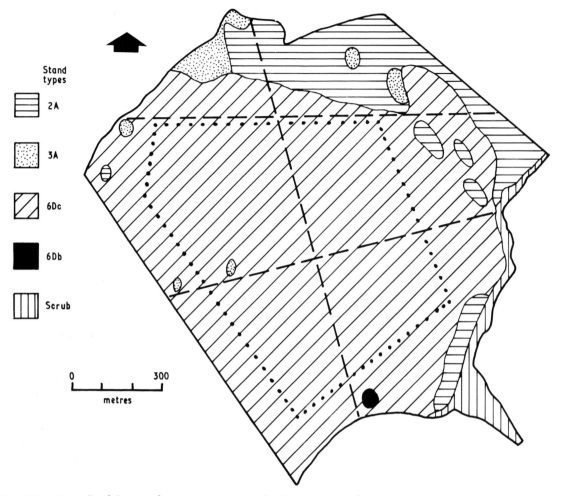

Stand types

2A

3A

6Dc

6Db

Scrub

0 ⊢ ⊢ ⊢ 300
metres

Fig. 14.5 Example of the use of a vegetation map in devising a strategy for nature conservation and intensive timber growing in an individual wood. This wood is diverse along the northern and eastern boundaries but uniform elsewhere. One form of compromise between nature conservation and forestry would be to convert the uniform core of the wood to plantations and leave the diverse zone under semi-natural mixed native broadleaf stands. If plantation forestry were kept within the dotted zone the requirements of landscape conservation would also be met.

of value have been clearly identified, there will be little room for doubt: in a wood notable for epiphytic lichens, one must choose the stands which are now the richest and have the best prospect (e.g. age structure, shelter, etc.) of remaining so.

Decisions are most likely to be difficult if a wood is simply a good example of a particular type or a rich wood without any other outstanding features. The object in selecting an area or areas within such a wood is to include (a) as many species as possible, (b) as much as possible of the range of habitat types and

their distributional pattern, (c) special features, e.g. gorges, natural glades, and (d) to achieve a shape which is either a coherent block or an interlinked network of small patches, and causes least inconvenience to other objectives of management. A map of woodland types is essential if the wood under consideration is diverse, because it shows where a loss of semi-natural woodland is least damaging to the range and pattern (Fig. 14.5). Where a wood is divided into compartments, the richest parts can be identified by compiling a list of woodland plants (all species or a

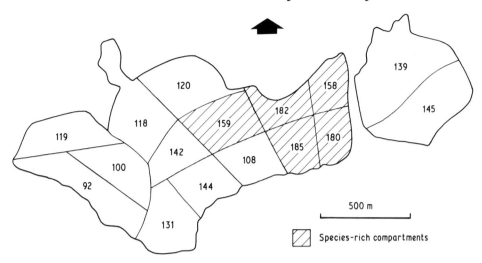

Fig. 14.6 Owston Wood, Notts, showing the arrangement of compartments and the number of vascular plant species in relation to area for each compartment. The figure for each compartment has only comparative value. It was calculated by dividing S, the number of shade-bearing vascular plant species recorded, by $A^{0.3}$, where A = area. A clearly defined block of relatively rich compartments is apparent in the central area, where small streams cut across the wood and a wide range of stand types is present. If there were ever any question of reserving part of this wood as a nature reserve then the five rich compartments are the obvious ones to select. (The plant lists were compiled for the Nature Conservancy Council by Mrs P. A. Evans.)

selection as in Section 14.4) and ranking compartments on this basis. If the compartments are of widely differing area one could allow for this by dividing the number of species, S by $A^{0.3}$, the area transformed, by assuming that $z = 0.3$ in the species/area relationship (Section 6.2) (Fig. 14.6). In the absence of any other guidance it is usually better to regard streamsides and wet areas, together with their immediate surroundings, as most important. These are often by far the richest part of a wood, especially in upland oakwoods and the acid oak–hornbeam woods of the lowlands, and they are least attractive to commercial forestry.

PART FOUR

Management for nature conservation

Chapter 15

Planning for nature conservation within forestry

15.1 The need for general principles

We stressed at the start of the previous part that nature conservation is both an attitude of mind and a form of land management. We then considered what nature conservation was trying to achieve in woodlands, which sites were most deserving of attention, and which features in them are most important. We discussed earlier the circumstances under which ancient, semi-natural woods had maintained some natural features through centuries of traditional management and exploitation, but we recognised too that these woods comprise no more than one-fifth of all woods, and that the remaining four-fifths which are plantations or recent, semi-natural woodland are also vast stores of wildlife which can hardly be ignored in any nature conservation policy. Now we turn to the practical aspects – the action necessary on the ground – which many would regard as the true test of an effective nature conservationist.

Nature conservation is only one demand made on woodlands. Indeed, the main objective of the Forestry Commission (1974) through which the State manages woods and channels financial assistance and advice to other woodland owners, is to produce utilisable timber as economically as possible. Woods are also prominent features in the landscape, so in many areas managers must ensure that their actions improve, or do not damage, visual amenity. Woodlands are also attractive places where people can wander freely in

relatively natural surroundings without damaging the woodland they have come to see, and without disturbing other people. Woods also provide the basis of more specialised recreation for the sportsman in pursuit of (or carefully positioned in wait for) fox and pheasant. Management of British woods as a whole must attempt to satisfy all these demands as well as satisfy the needs of nature conservation.

Unfortunately, the form of management which produces the most timber and the greatest profit, at least in the short-term, appears to be intensive monoculture of high-yielding conifers. This is the least attractive form for wildlife and habitat conservation, and it also seems relatively unattractive to those concerned with landscape, recreation and sport. Conversely, the economic justification for the traditional forms of management, which have left a legacy of semi-natural woods, has declined almost to vanishing point. One could arrange all management options for any given wood in order from best to worst on financial and timber-producing grounds, and generally find that this is the reverse of the order which would be produced on nature conservation grounds. This divergence between what is best in material economic terms and what is best for nature conservation, amenity, recreation and sport is a recent phenomenon, which has developed only since the advent of the modern forestry methods.

This basic and unhelpful relationship implies that, if we are to satisfy all the demands made upon woodlands, some timber-production must be fore-gone in the interest of other demands, but to what extent should we sacrifice timber production, profit margins in forestry and other economic benefits of forestry (e.g. employment in rural areas and process-ing industries) for the pleasure of sport and the non-material benefits of amenity, recreation and nature conservation? This, the central question of com-mercial forestry in relation to other woodland interests is surely a political question, and one which perhaps cannot have a final answer. The balance is achieved by discussion between the interested parties and competition between the interested organisa-tions. The political aspects take two forms, namely the climate of public opinion in which the discussions take place and the resources made available to the official bodies representing each interest, such as the Forestry Commission, Nature Conservancy Council and the Countryside Commissions.

One method of balancing conflicting claims has been to separate them on the ground. Thus, where nature reserves have been established access has been controlled and little timber has been produced for the market. Local Authorities, especially those in the vicinity of large towns and conurbations, have a long tradition of acquiring woods for public enjoyment (Eversley, 1910), but their attitude to management has been more preservationist than conservationist. Large numbers of the small and medium sized woods scattered through the agricultural districts have been used as game coverts, and they too have been largely neglected in silvicultural terms. On the other hand the Forestry Commission and other timber producers tended to grow timber with little heed to wildlife, sporting possibilities or effects on the landscape, occasionally perhaps 'setting aside' small areas for nature. The results were occasionally appalling: witness the stands of dead trees, killed by chemicals as part of a planting programme, which looked more like the aftermath of an atomic explosion than the considered result of British forestry.

More recently there has been a trend to greater integration of objectives, for several reasons. One, under the banner of 'multiple use', is part and parcel of the management philosophy which seeks high output from intensive management, in this instance by satisfying two demands in one wood. This can

work very well, notably in the widespread and successful attempts by the Forestry Commission to encourage recreation in their plantations, but even this generally admirable provision has limitations: if you establish a picnic site next to a colony of Martagon Lily, then sooner rather than later you have no lily. Second, it was becoming obvious that simple, single objectives were pointless, impossible or both. Amenity woods cannot be left unmanaged because moribund trees will eventually fall on picnickers: the trees must therefore be felled and replaced, so why not sell the timber? Such woods do in fact harbour wildlife – this is one of the amenities – and may be important in habitat conservation, so why not try to ensure that the necessary management is good for nature conservation? Nature reserves are often open to the public, so wherein lies the distinction between them and an amenity wood managed with nature conservation in mind? Gamekeepers acquired a reputation for antagonism to wildlife, but game preserves are often amongst the most important woods for habitat conservation, and game preser-vation must be thanked for the survival of so many ancient semi-natural, but derelict coppices. Since both nature reserves and game preserves can benefit from silvicultural management, especially of a traditional kind, why do it for nothing if you can sell the produce? Integration between timber production and game preservation has proceeded apace as the sporting potential of plantations has begun to be developed and plantations have been judiciously established in parts of game preserves. The third force towards inte-gration of objectives has been the recognition of possible ecological and economic limits to intensive, monoculture timber production. This takes several forms, e.g. (a) rising labour costs of planting provides a slightly greater incentive to try for natural regeneration as on the Continent; (b) fungal in-fections can build up in conifer monocultures to such an extent that an intervening broadleaved or mixed stand becomes necessary; (c) pure conifer stands on naturally broadleaf sites may cause soil changes leading to reduced productivity.

The best traditions of conservation surely demand that we seek maximum integration of objectives on a long-term, sustainable basis. There must be limits to integration, and it will no doubt be necessary to determine a primary objective for each wood or part thereof, but we need no longer apply apartheid in British woods. We ought in theory to produce a national woodland management plan in which the objectives for each wood are specified, but this is very difficult in practice: not only are there limits to planning of private property, but our requirements and our knowledge of sites and ecological and economic factors will change.

These planning limits were faced by those who formulated the forestry section of *A Plan for the Chilterns* (Chilterns Standing Conference, 1970). Their reply was a set of general principles which governed the rate and scale of felling and the composition of the nurse and final crops of the next rotation. Briefly these sought small-scale fellings spread over 30 years in any one wood and a final broadleaved crop, thereby maintaining the present appearance of the area, with its extensive stands of mature beech. This approach seemed reasonable to all concerned, though its effectiveness can only be measured after the actual results on the ground become clear.

With this seemingly successful precedent in mind, I formulated fifteen general principles for the integration of nature conservation with other objectives of management in British woods (Peterken, 1977b). These principles could be used (a) as a basis for discussion about the management of individual woods as well as the national forestry policy, (b) as a yardstick against which past management and present proposals can be judged, and (c) as a substitute for on-the-spot ecological advice when this is not otherwise obtainable. The eleven examples of their application to particular types of woodland which were given in the original publication are not repeated here, but the principles themselves form the basis for Chapters 15–19.

15.2 Strategic principles

Principle 1

Site grading   Distinguish between (a) individual woods of high conservation value, (b) woodland areas of high conservation value and (c) other woodlands.

Principle 2

Management priorities   Afford special treatment to special sites and special areas.

15.3 Scales of planning for nature conservation

Planning is a rather emotive term when used in connection with forestry, especially if it implies a formal bureaucratic procedure, but here it is used in a more general sense as an approach to reconciling the multifarious and often conflicting demands made on woodlands. We recognise that certain woods are more important than others for nature conservation (Chapters 12, 14) in much the same way that we recognise other woods to be especially prominent in the landscape or suitable for the growth of particular timber trees, and then grade them appropriately for each of these separate interests. Grading provides a practical means of integrating objectives: other things being equal, management priorities should be related to the importance of each site for each different demand. Thus, the management of a conspicuous wood in a National Park should pay particular regard to the effects on the landscape, even if it is capable of being managed intensively for timber production. A large, accessible wood with good communications is well suited to timber production, which might therefore take precedence over other factors. And those woods which are judged to be of special importance for nature conservation should, in the national interest, be managed primarily for nature conservation.

Nature conservation's contribution to this graded response can be stated at various scales, which can be broadly classified into four groups:
   (1) National and Regional Scale
   (2) Natural Area Scale
   (3) Site Scale
   (4) Locality Scale

15.3.1 Sites

Nature conservation priorities have generally been stated at the Site Scale, i.e. as a list of important individual woods. Each wood on such lists is a historical and management unit within which the range and distribution of woodland types is largely determined by edaphic factors. This scale is especially appropriate for considering the basic need to protect adequate examples of the semi-natural range of variation and transitions between types (e.g. Section 14.2, Norfolk Woods), thereby protecting viable populations of most native species. Nature conservation at the scale of individual woods has practical advantages. Most woods have just one owner, and few have more than two or three separate owners, except in those rare cases where a wood has been divided into a multiplicity of small lots (e.g. Kings Wood, Beds); agreement over management for nature conservation or outright purchase as nature reserves is therefore relatively easy. Individual woods vary enormously in size, but those of significance generally cover two orders of magnitude within the limits of 2–200 ha.

15.3.2 Localities

A locality in this context is a small patch of woodland containing, say, a population of some rare species, a pond or streamside, a herb-rich length of ride, or even an individual tree supporting a rich assemblage of epiphytic lichens. Most such localities will cover less than 2 ha. This scale is most appropriate for the conservation of particular species and localised features within otherwise ordinary woodland, notably within woodlands managed predominantly for timber production. A locality might also be a fragment of semi-natural woodland in an otherwise intensively managed plantation, a last resort in the conservation of representative stands of semi-natural woodland in the face of intensive forestry. Most foresters are interested and willing to protect localities within plantations, provided that they are accurately located, their significant features are defined and their management needs are stated. Of course, important sites also contain localised special features, but these are irrelevant in the planning process and become significant only when details of management are under discussion.

15.3.3 Natural areas

There are some well-wooded areas such as the Chilterns beechwoods, where it is arbitrary, difficult and often impractical to identify important sites as representative good examples of the woodland types in the area, even though the woods as a whole are undeniably important for nature conservation. Nature conservation needs in such areas are best stated as general policies combined with identification of special features on the locality scale, and important sites are only identified if they exhibit some extensive special feature. These special areas cover roughly 200–20 000 ha of woodland. They are discussed in detail in Section 15.4.

15.3.4 National and regional scale

Nature conservation needs at a national scale consist of intensive and extensive elements (Steele, 1975). The former embraces the proper management of special sites and areas, giving nature conservation high priority in the hierarchy of management objectives; this is discussed further in the following sections and Chapter 18. The extensive element must be an attempt on the part of managers of other woods to protect localised features of interest and maintain reasonably good habitats for wildlife generally: this is discussed further in Chapter 17. Both elements are needed. Intensive without extensive conservation would eventually isolate the best sites as islands of high value in a sea of mediocrity. Extensive without intensive conservation would maintain a moderately rich ambient wildlife, but the best sites would be damaged by intensive forestry.

15.4 Natural areas considered as management units

15.4.1 Conservation areas

'Conservation Areas', defined as a 'tract of countryside, the existing character of which it was desirable to preserve so far as may be possible', were proposed in Cmd 7122 (1947), in order to preserve their visual amenity and conserve nature. Certain natural features or mosaics of habitats in these areas were valuable in their entirety (not just their separate parts), but they covered such large areas that it was unreasonable to expect that schemes of major economic importance would be excluded from them. These Conservation Areas were to have been a planning measure, which ensured that nature conservation was properly considered whenever proposals for development or economic exploitation were made. Nature Conservation would not necessarily take precedence over other needs, except in certain parts which had been designated as 'scientific areas' or nature reserves. The 52 proposed Conservation Areas were separate from and supplementary to the 12 proposed National Parks, and both contained areas in which the Wild Life Conservation Special Committee considered that nature conservation should be the primary objective. Many contained well-wooded districts, e.g. Lake District, North Yorks Moors, Peak District, North Wales, Forest of Dean and Wye Valley, Cotswolds, Mendips, Dartmoor, New Forest, South Downs, Hindhead, Forest Ridges (Central Weald), Chilterns and Clipsham-Holywell. Many of these areas have since become National Parks and Areas of Outstanding Natural Beauty.

These proposals recognised that certain natural features, which are worthy of conservation, occur on so large a scale that alternative demands on the land cannot be excluded in a country as densely populated as Britain. They imply, first, that in the interests of nature conservation, amenity and recreation there should be some restraint on commercial use and development over the whole area, and, second, that the site-based approach to nature conservation must to some extent be replaced by a more general approach. This is essentially the same as my proposal in the first principle that nature conservation in certain groups of woods should be planned at an area rather than a site scale.

15.4.2 The Chiltern beechwoods as a natural area

The Chiltern Hills are a fine example of a Conservation Area and Scientific Area within the terms of Cmd 7122 and a Woodland Special Area as defined in principle 1. The beechwoods show within a densely wooded district (Fig. 15.1) a wide range of types varying in relation to soil (Watt, 1934), upon which minor variations in structure and composition due to differences in past management are superimposed. They are certainly an important group of predominantly ancient, semi-natural woods, but conservation through the identification and special management of individual important woods is inappropriate for several reasons:

(1) There are many very large woods which are divided into individual, named woods whose boundaries are ecologically arbitrary, even though they are long-established.

(2) Each wood is equally part of the ecological continuum. No individual woods (in so far as these can be meaningfully identified) stand out above the majority as especially important representative examples.

(3) Despite some conifer planting, there is still a very large area of beech woodland, and moreover beech is still being planted.

(4) As a densely wooded area, the ecological isolation of any wood is low and the capacity to recover from damaging change is correspondingly high.

The problems generated by a site-based approach to nature conservation in such an area can be illustrated by the hypothetical example of a plateau beechwood which has been identified as a good example of its type, probably because it is large, lacks alien trees and contains a rare plant. The owner may wish to fell one part and plant a conifer/beech mixture with the intention of growing eventually another mature stand of beech. Felling and planting will reduce the natural appearance of the stand, but ground vegetation will increase. As the new plantation passes through the pole stage wildlife will be reduced and the rare plant may be jeopardised, but eventually the conifers will be removed and the wildlife of beech high forest will move in again from surrounding woods. Provided that special care is taken of the immediate locality of the rare plant, and that restocking is not so intensive that all the mature beech is felled at once, it is unlikely that wildlife will suffer any long-term damage and in any case many examples of mature, plateau beechwood will remain elsewhere. The main danger is that, through some change of policy, the beech will be removed and the conifers allowed to form the final crop, much as oak was sprayed out of oak–conifer mixtures in eastern England in the 1960s.

In the Chilterns it seems better therefore to seek nature conservation within a planning framework by agreeing management policies which were broadly acceptable. The 'Plan for the Chilterns' formulated

+Fingest

1 km

Fig. 15.1 The pattern of woodland in the Chiltern Hills, Bucks and Oxon. The area shown measures 10 km × 10 km and lies to the north of Henley. Woodland, most of which is ancient, covers about 38% of the land.

under the Chilterns Standing Conference provides for the woodlands in the Area of Outstanding Natural Beauty to be retained; for the felling and restocking of individual woods to take place in fairly small coupes over a period of 30 years, and for the final crop for new plantings to be mainly beech. This should (grey squirrels permitting) maintain both the natural range of beechwood types, a substantial area of mature beechwood, and the well-wooded landscape. Special measures will be needed at the locality scale to protect some of the rare species of the region – *Dentaria bulbifera*, *Orchis militaris*, *Cephalanthera rubra*, *Epipogium aphyllum* and *Cephalanthera longifolia*. A few sites with special features may need protection, such as the scrub-coppice at Maidensgrove and samples of the woods studied by Watt (1934) in his classic work. Monitoring will be needed to check that action on the ground actually adheres to the general policies.

15.4.3 Natural woodland areas in general

The necessary features of woodland special areas appear to be (a) that they are well wooded; (b) much

of the woodland is ancient, semi-natural; (c) the individual woods are of approximately equal value for nature conservation; (d) the nature conservation value consists in part of the relationships between the individual sites over the whole area; and (e) there is some unifying factor of ownership or planning which enables general policies to be formulated and implemented. These conditions are more or less met in other areas, such as the New Forest, the Weald, parts of the Southern Chalklands, Cotswold Scarp, the southern Welsh borderland, Wyre Forest, parts of the Lake District, the Middle Clyde Valley and around the Cairngorms in Speyside and Deeside. Each special area has distinctive ecological characteristics and an individual permutation of demands and opportunities: generalisations about management policies are therefore inappropriate, and the policies for each area should be formulated only after careful survey. It would, for example, be entirely inappropriate to translate the Chilterns policies to the New Forest Ancient and Ornamental Woods, even though beech is a dominant tree in both areas, because of differences

258

in their structure, biological characteristics, surrounding land use and the recreational demands made upon them. In north-east Scotland both Darnaway Forest on the Findhorn and Abernethy Forest on the Spey are important woodland areas. The former is an irregular expanse of intensively managed woodlands in a lowland agricultural setting where retention of existing semi-natural oak, pine and beech woods as a hardwood working circle on a long-rotation will be satisfactory. Abernethy Forest is a large, native pinewood in an upland setting which should be retained in an unmanaged, natural state as far as possible.

Nature conservation at the area scale can be extended to a more dispersed group of woodlands, provided that they are under unified management. For example, many of the limewoods of central Lincolnshire are managed by the Forestry Commission, whose officers have for many years co-operated with the Nature Conservancy Council in order to devise appropriate management policies. Utilisable timber is being grown throughout these woods, but the semi-natural woodlands are being promoted to mixed high forest, whilst utilising the thinnings for fencing and other markets.

The 'special area' approach enables a reasonable balance to be struck between timber production, recreation, landscape protection and nature conservation without the artificialities and restrictions of the alternative strict segregation of objectives into separate sites. It is ecologically possible in regions with a high density of semi-natural woodland, where there is as yet less need to preserve representative examples and the low level of ecological isolation permits wildlife to respond more readily to changes. Special sites are still defined and protected, but these concentrate upon rare species and special features, not representative examples. If the special area approach fails because the policies are not generally satisfactory for nature conservation, one can revert to the special site approach. In the 30 years since Cmd 7122 was published the woodlands in the North Yorks Moors and the Clipsham–Holywell proposed Conservation Areas have been so intensively replanted that the priority for nature conservation is now to establish nature reserves in the remaining semi-natural woods.

15.5 Sites as units for nature conservation

The *Nature Conservation Review* (Ratcliffe, 1977) lists 234 woods covering 67 000 ha which were regarded as nationally important for nature conservation. In terms of the scales adopted in Section 15.3, the NCR woodland selection is mainly site-based. At one point it descends to the locality scale by including the Geary Ravine (Skye, 2 ha), but more importantly it includes many 'areas' as if they were 'sites'. Half the total area lies within only twelve listed sites greater than 1000 ha, some of which are large, coherent blocks of woodland (e.g. Epping Forest, Essex), whereas others are scattered over separate tracts of woodland within similar terrain (e.g. Speyside–Deeside pinewoods in north-east Scotland). In addition, the list includes woodlands intermediate in scale between sites and areas (e.g. Savernake Forest, Wilts, 930 ha) and constellations of similar but separate sites in one district (e.g. Bardney Forest, Lincs). This recognition of areas and sites in one list is not entirely consistent, for some natural woodland areas have been treated as sites, either as a single site (e.g. Blean Woods, Kent, only a sample of which is listed) or as a constellation of sites (e.g. Borrowdale Woods, Cumb).

The *Nature Conservation Review* is only the latest in a series of attempts (see Sheail, 1976) to list the nationally important sites. The earliest, the Rothschild list of 1915, contained 251 areas and sites of which approximately 30 contained significant woods, including Kingley Vale (Sussex), The New Forest and Rannoch (Perthshire). A further list of areas where reserves were urgently needed was drawn up in 1929: it named many other woods, such as Monks Park Wood (Suffolk), Symonds Yat (Glos) and Rothiemuchus in Speyside, as well as many well-wooded areas such as the Forest of Dean, Chilterns and Charnwood Forest. In 1943 the British Ecological Society proposed a list of National Habitat Reserves which included amongst those of outstanding importance Roudsea Wood (Lancs), Naddle Low Forest (Westm), Herons Carr (Norfolk), Kingley Vale and scrub at Bricket Wood (Herts). A Nature Reserves Investigation Committee considered this and lists drawn up by other groups and proposed in 1945 a list of national nature reserves which included Monks Wood (Hunts) and the beechwoods at Cranham (Glos) which had long been regarded as important, but the wood which received the highest vote (Bedford Purlieus, Cambs) was omitted, presumably because it had recently been felled and replanted by the Forestry Commission. Then, in 1947–49, the Huxley (England and Wales) and Ritchie (Scotland) committees produced a new national list of conservation areas, scientific areas and proposed national nature reserves.

It is instructive to see how the 25 woods in the 1947–49 list fared in the *Nature Conservation Review*, drawn up mainly around 1970, over twenty years later. Twenty-one sites appear in the NCR, 15 at grade 1. Two of the four woods omitted were damaged in the intervening years, Holywell–Pickworth (Lincs) by forestry and Cheddar Wood (Somerset) by quarrying: Lodge Wood (Bucks) remains unchanged and, like Bricket Wood Scrubs – a tract of recent secondary woodland – is not now considered to be important enough for inclusion. The two woods rejected late on the 1947 road to selection, Bedford Purlieus and Staverton Park (Suffolk), are both now grade 1 sites. Several other well-wooded areas were listed as conservation areas or scientific areas and, though direct comparisons are difficult, these lists have stood the tests of time equally well. Several lessons can be learned from this history. Successive lists differ because woods are damaged in the intervening years; new good sites are discovered in the course of further survey; some woods, whilst themselves not changing, become more important as other similar woods are damaged; and opinions change on how the selection is to be made. Even so, most sites continue to be regarded as important once they have been listed, provided that they are not damaged by treatment.

Perhaps the most significant trend has been the increase in the number of woods regarded as nationally important, which shows both in a comparison between the NCR and earlier lists, and in the increasing number of woods scheduled as Sites of Special Scientific Interest. Superficially this might suggest that the prospects for nature conservation have improved by leaps and bounds, but in fact the opposite is true. In 1915, when the Rothschild list was compiled, most coppice woods in the lowlands were still being cut on rotation and the semi-natural woods in the uplands, though not managed, at least remained unchanged. This prevailing low rate of woodland management change was very significant, because woodland wildlife was not threatened, any natural features were likely to survive, the extent of semi-natural woodland was not declining, and in general there was no need to protect sites just because they were good examples of a particular type. Indeed, it was not until the lists of the 1940s were prepared that it was felt necessary to include sites merely as good examples of a widespread type. Now, with the intensification of agriculture and forestry, it is becoming clear that an adequate quantity and range of semi-natural woodlands will not survive unless specific action is taken to conserve the best examples, and it has become increasingly necessary to define the needs of nature conservation in terms of numerous sites, so chosen that the full range of ecological variation can be perpetuated. In some districts it is still possible to state nature conservation needs at a more generalised level and concentrate special attention on special features, i.e. to define needs at the Area Scale, but in most districts the attrition of semi-natural woodland has gone so far that needs must now be defined at the Site Scale. One hopes that, in the future, semi-natural woods will not be reduced to the present parlous state of herb-rich meadows, of which there are so few surviving examples that all must be protected.

Chapter 16

Pattern and redistribution of woodland

16.1 Principles

Principle 3
Clearance

Minimise clearance. Necessary clearance should avoid sites and areas of high conservation value.

Principle 4
Afforestation

Accept afforestation, except on sites of high nature conservation value, but not so much that non-woodland habitats are reduced to small islands.

Principle 5
Woodland patterns

Develop (or retain) large blocks of connected woodland, whilst maintaining a scatter of small woods between large blocks.

16.2 Distribution and clearance of ancient woodland

16.2.1 Effects of woodland clearance

Woodland clearance (Figs. 16.1 and 16.2), i.e. the replacement of woodland by arable cultivation, ley grassland, quarries, housing estates and any other form of land use, has been widespread in the past and continues today in many areas. Whilst the total area of woodland has increased considerably in Britain as a whole, woodland clearance has been widespread in

Table 16.1 Woodland clearance in three East Midland study areas between 1946 and 1972 (Peterken and Harding, 1975).

	Rockingham Forest	Central Lincolnshire	West Cambridgeshire
WOODLAND IN 1946			
Total area (ha)	5494	3669	1476
Ancient	3997	2614	675
Recent	1497	1055	801
% total land area	13.2	7.4	3.4
CLEARANCE 1946–1972 (as % of 1946 woodland)			
Ancient	11.3*	2.2	11.4
Recent	2.4	5.7	21.8
Total	8.6*	3.2	17.1
Clearance due to agriculture (ha)	132	100	239

* Two thirds of the clearance in Rockingham Forest was due to open-caste mining and the associated expansion of Corby, which particularly affected several large ancient woods.

the intensively cultivated and increasingly urbanised lowlands. In three study areas in the East Midlands, Peterken and Harding (1975) found that between 3.2% and 17.1% of all the woodland present in 1946 had been cleared by 1972, mainly coppice woods which were replaced by arable fields (Table 16.1). The rate of clearance increased towards the end of the study period and continues unabated today. Although

Fig. 16.1 Bradfield Woods, Suffolk, comprising Felsham Hall Wood (top = north) and Monks Park Wood, are joined along a short boundary marked by a long pond. Half of Monks Park Wood has recently been cleared for arable cultivation, and clearance continues at the southern end where the torn-out coppice is bulldozed into windrows. Scattered oaks, formerly standards within the coppice, survive in the new wheatfield and a strip of woodland (which would be known as a shaw in the Weald) is retained. Surviving woodland is still managed as coppice-with-standards, the standards showing as large-crowned trees above the densely crowded underwood. *(cont'd opposite)*

Fig. 16.2 Clearance of derelict coppice at Chatcombe Wood, Glos. Bulldozed piles of old stools and coppice growth litter the newly ploughed slope, whilst in the background some mixed coppice of ash, hazel, maple, wych elm, beech and pedunculate oak survives. This land may have been wooded for thousands of years until clearance in 1978.

the intensively cultivated districts with least woodland were affected more than the well-wooded districts, at least in relative terms, even densely wooded regions such as the Weald are now seriously affected by clearance to arable and urban use.

Woodland clearance is almost always an unmitigated disaster for nature conservation. The immediate and most obvious effect is the complete loss of woodland habitat, species and any natural features they exhibited. If the woodland was, as it often has been, an ancient semi-natural wood the communities lost are irreplaceable and many populations of extinction-prone species will have been wiped out (Chapter 12). They are replaced by communities of arable crops

---

(Fig. 16.1 cont'd)
The irregular outline of the coppiced patches shows in Monks Park Wood. Soil marks to the north-west and north-east of Monks Park Wood show where this wood has been reduced by earlier clearance. Although the surviving woodland is still one of the richest in lowland Britain and contains extensive plateau alder woodland (stand type 7C), the clearance of half of Monks Park Wood ranks as a major modern disaster for woodland nature conservation, which carried away a woodland *Eriophorum* bog, acid oak–lime woodland (stand type 5A), the only patch of hornbeam coppice, hitherto undamaged earthworks of part of the medieval deer park and a substantial tract of traditionally worked coppice woodland.

with adventive weed species which are not only far more abundant than woodland communities, but are eminently replaceable and may well be far poorer in species. The loss is not so great if the cleared woodland was secondary and recent in origin, but it may still be significant. Furthermore, if the wood margin is allowed to survive as a hedge a few woodland species will remain, notably the plants which can grow in dry, calcareous, eutrophicated soils and the wood-edge birds and invertebrates, but this is only a small fraction of the original woodland community. The loss to nature conservation of woodland clearance is mitigated if the substitute land use permits long-term development of another semi-natural community, e.g. some quarries.

One might expect that the losses would be small and insignificant if some woodland remains, i.e. that the surviving half of a wood will contain everything that was present in the whole wood. This is not so. Some species and natural features will have been confined to the half which was cleared, and these will have been lost completely and immediately. Furthermore, the remaining half will be more isolated ecologically, the populations within it will be smaller and over an extended period some species which survived the original clearance are likely to become extinct as the communities within the reduced wood relax (Section 6.2) to the species complement appropriate to its increased isolation.

Clearance may also adversely affect amenity, timber production etc., but the damage is usually either less significant or more easily mitigated. Since small woods are already uneconomic for timber production, their clearance is not a significant loss. Fragmentation of a larger wood into a scatter of smaller woods may actually improve the shooting. Amenity may not be adversely affected if a belt of woodland is retained, or if an equivalent area is afforested as a substitute: but from the nature conservation angle 5 ha of newly created secondary woodland is no compensation for 5 ha of older woodland cleared away.

The management implications are clear. Keep the woods which already exist. Do not clear here and afforest there in order to 'move' woods to more convenient parts of the estate. Clear only if a net reduction in the extent of woodland is necessary. If it is, then clear recent secondary woods rather than ancient, semi-natural woods, i.e. adopt the redundancy principle of 'last in first out'. Remember that small woods provide valuable cover for game and a source of timber for use on the estate.

16.2.2 Design principles

Diamond (1975) stated six design principles which would help to minimise the rates of extinction through relaxation of species in habitat islands formed by fragmentation of a once-continuous habitat. Assuming, as is reasonable, that these apply to British woodlands, especially to ancient woods, they can be stated in the following form:

(1) More woodland is better than less woodland.

(2) One large woodland is better than many smaller woodlands of the same total area.

(3) Separate woodlands should be clustered rather than evenly dispersed.

(4) Separate woodlands in a cluster should be equidistant from each other, i.e. the cluster should be more or less circular, not in a chain.

(5) Separate woods should be linked by strips of similar habitat, e.g. wooded belts or hedges.

(6) Individual woodlands should be as nearly circular as possible.

These principles assume that the habitat is homogeneous, but British woods are a mosaic of different woodland types, each of which has its peculiar combination of species. If all the surviving woodland were in one large block, or if the separate woods were all clustered together, there is a danger that only one woodland type will be included and that other types will be completely omitted. Therefore, some dispersal of surviving woodland is desirable where this clearly enables a greater range of woodland types to be represented.

Diamond's principles also assume that all species are desirable species. This is true enough of the native species of relatively stable habitats, but aggressive aliens are not at all welcome. An isolated ancient semi-natural wood may contain fewer native species than an identical wood within a cluster of similar woods, but if it is spared invasion by grey squirrels, *Rhododendron* or sycamore then the number of species is more likely to remain constant in the long-term, and management will be easier.

16.2.3 The pattern of ancient woodland

Clearance from prehistoric times to the present has unconsciously observed these design principles fairly closely. Ancient woods tend to be compact rather than linear, distributed fairly evenly over the landscape, but with distinct clusters and concentrations at both the regional and the parish levels (Section 2.7). Retention of wood-relic hedges in the vicinity of such woods has further reduced isolation. Virtually all site types still carry some ancient, semi-natural woodland somewhere, and very few are quite unrepresented: ancient woods are absent from valleys, acid sands and light calcareous soils in some areas. We have inherited a woodland distribution which is close to ideal (given that only a small amount of woodland remains) but further clearance increases the risk that some semi-natural types will be eliminated.

16.3 Afforestation and nature conservation

16.3.1 The extent and limits of afforestation

As recently as 1895 Britain had just 1.1 million ha of woodland (Table 5.1), but with government aid the strenuous efforts of foresters and landowners have almost doubled this to 2.0 million ha. Despite increasing difficulties in finding more land to plant, the Forestry Commission (1977) believes that a further 1.7 million ha can and should be afforested by 2025. Eventually a time will come when further afforestation is impossible, but the limit will be determined, not by technical feasibility – trees would grow successfully on most of the land surface of Britain – but by the relative values placed by society on timber production, food production, water resources, landscape conservation, game, recreation and nature conservation.

Most of the plantations of the 20th and 21st centuries will have been established on the upland moors, peatlands and acid grasslands of the north and west, displacing in the process millions of sheep. Pasture improvement will compensate to some extent for the loss of sheep pasture, but even so agriculture will continue to compete with forestry for land in the uplands, and the manner in which they adjust to each other will substantially determine the future pattern of upland land use. Plantations intercept and transpire

more rainfall than grasslands: afforestation reduces the yields of water from catchments by about 15% (Clarke and McCulloch, 1975). This too limits the acceptable extent of afforestation, for the demand for freshwater and hydro-electricity is growing. Deer forests and grouse moors cover about 2 million ha, and if these remain profitable and socially acceptable little of this land will be available for afforestation. The bare, unplanted upland landscape is greatly appreciated by British and overseas visitors alike, being one of few remaining tracts of apparently wild scenery in western Europe, where people can enjoy the freedom of roaming largely at will. The Nature Conservancy Council (Ratcliffe, 1977) has listed 207 upland and peatland sites covering 0.5 million ha which are of national importance for nature conservation in their existing, unplanted state, and there are many more which are scheduled as Sites of Special Scientific Interest. All these factors constrain afforestation and prevent some plantable land from being planted (though an extension of woodland composed of native trees in the uplands would be highly desirable for wildlife), even though Britain now imports 92% of her timber needs and cannot expect to reduce this proportion much in the future. These factors have recently been reviewed in detail by the Centre of Agricultural Strategy (1980).

Despite the limitations, it is certain that substantial new plantations will be added to those already created since 1919, that almost all will be in the uplands, and that four-fifths will be in Scotland. From the narrow point of view of woodland nature conservation this is all to the good, for it provides a greatly expanded habitat for at least some woodland species and takes the pressure for expanding timber production off the remaining semi-natural woodlands. However, the moorland communities which will be lost also have a value for nature conservation, and the balance of advantage for nature conservation from afforestation depends entirely on whether the gains are larger or smaller than the losses.

16.3.2 Gains and losses on planted ground

The changes which take place when moorland is afforested (Section 5.4) may be resolved into losses, temporary gains and permanent gains.

265

(a) *Losses.* Most of the species and communities of moorland are exterminated or greatly reduced by the pole stage of the first rotation.

(1) Semi-natural moorland habitats. The plant communities are lost and the soils are usually drained and fertilised. Some of these communities, such as *Nardus* grassland and *Calluna* moors, are poor in species and are very extensive, so the loss of some is barely significant nationally. Others, such as base-rich grasslands and flushed tracts, are much richer and the losses are correspondingly greater. Afforestation has destroyed some important examples of semi-natural ecosystems, e.g. the raised bog of Lochar Moss (Galloway), Culbin Sands (Moray) and part of the sand dune system of Newborough Warren (Anglesey).

(2) Although upland afforestation greatly reduces the grassland and heathland species, it is the plants of mires and base-rich flushes which are most likely to be completely exterminated. Many moorland species appear to survive as buried seed whilst the plantations are growing, springing up again when clear-felling occurs, e.g. *Calluna, Juncus* sp.

(3) Moorland birds are reduced or completely excluded, notably the waders, such as curlew and golden plover, and raven. These have large territories and cannot survive in small patches of moorland.

(4) The sense of wilderness is diminished by the characteristic straight lines of modern forestry and the network of roads.

(b) *Temporary gains.* The change from moorland to plantation takes perhaps 15 years whilst the young plantation grows to the pole stage. During this phase certain features develop which are not apparently repeated at the start of the second rotation.

(1) The grassland communities grow unfettered by grazing and become tussocky. Herbs flower more freely and some birch and rowan scrub often develops. This is equivalent to the last flowering of the spring flora when a coppice is replanted with conifers.

(2) Vole populations increase, attracting large numbers of predators, notably short-eared owls and kestrels. Bird communities in general are enriched, for some woodland species move in before all the moorland species have moved out.

(c) *Permanent and long-term gains*

(1) Song birds become established in large numbers. These include species from natural coniferous woods, such as song thrush, chaffinch, coal tit and goldcrest, and rarer birds, such as crossbill and siskin.

(2) Red, roe and sika deer become established within the shelter of plantations, though they may have to forage on surrounding moorland and farmland in order to survive. Pine martens, polecats and wildcats have spread within plantations.

(3) Woodland plants colonise the plantations as they pass through the pole stage of the first rotation. Initially the main colonists are bryophytes and fungi. There is little sign of a rich flora developing.

(4) Rides and tracks are colonised by species which were not present before afforestation: they are attracted by the base-rich or disturbed character of the forest roads.

(5) Invertebrates from native coniferous woods colonise, but, under the artificially uniform canopy, some of the foliage-feeders become pests, e.g. the Pine Beauty moth.

16.3.3 Factors affecting the impact of afforestation on features of value for nature conservation

A balance sheet for nature conservation in relation to afforestation must take the above gains and losses into account and give greater weight to the permanent changes, but it must also consider the size of each gain and each loss. Here generalisations are difficult, because the quantitative impact of afforestation has not been thoroughly investigated and in any case it depends on the particular circumstances found in each district. The six factors which appear to be significant are the extent, pattern and exact location of plantations, time, plantation management, and the effects of plantations beyond their borders.

(a) *Extent.* Before 1919 most upland catchments were largely treeless or sparsely wooded and the various types of moorland were common and widespread. In these circumstances afforestation on a small scale adds a habitat without significantly reducing others, and provides opportunities for woodland species and those which need a variety of habitats. Further afforestation in a partly afforested area is less of a gain, even though the resulting plantations may become extensive enough to support more woodland species, including some with large territories. No new habitat is added, and the reduction in the non-woodland habitat may now be significant. Continued afforestation until non-woodland habitats are eliminated is a net loss, for the woodland gains are insignificant and the losses of other habitats and habitat variety itself are total, at least locally. Thus afforestation is a net loss in a densely wooded area. Between these extremes is a threshold where the gains and losses balance each other, and this is the point beyond which afforestation should not proceed. This threshold may be expressible as a proportion of the total land surface (or of those soil types which are affected by afforestation) but we do not have enough information to put a figure to it, and in any case it is unlikely that a standard proportion can apply in all areas. Nevertheless, it is quite clear that blanket afforestation is bad for species conservation within the district immediately affected.

(b) *Pattern.* The need from the point of view of woodland species is to facilitate colonisation. In theory this can be achieved by establishing large plantations and placing them close to seed sources, but in practice the soils beneath plantations often differ from those beneath woods in the same area and the number of potential colonists is small. Many upland ancient woods occupy mineral soils on slopes, whereas neighbouring plantations have been established on higher, flatter ground covered in blanket peat, so it is unlikely that, say, the bluebells in the ancient wood could colonise much of the plantation. Likewise, the fauna on the native broadleaves of the ancient, semi-natural wood is mostly incapable of colonising the conifers of the plantations. Another difficulty is that afforestation reduces and fragments the moorland habitats (Fig. 5.3) and the design principles of Section 16.2 now apply to the pattern of surviving moorland. From this opposite standpoint the plantations should be as small and as isolated from each other as possible. Clearly a compromise is needed if the wildlife of plantations is to be enriched without unduly impoverishing the surviving moorland, which would probably take the form of (i) separate blocks of plantations in a matrix of moorland, and (ii) an uneven distribution of plantations

within a catchment which enables at least some substantial blocks of moorland to remain.

(c) *Effects outside plantations.* Afforestation can alter the pattern of grazing outside plantations by preventing access to hitherto grazed land. Ungrazed moorland becomes leggy and its composition may change. In ungrazed woodlands natural regeneration may be possible where previously it was not. Fertilisers and sprays applied within plantations may drift on to surrounding land. More significantly the disturbance, ploughing and fertilising undertaken during felling and establishment can increase run-off and siltation and eutrophicate the streams for a period. Later in the rotation the acidity generated in the soils of mature plantations can increase the acidity of streams. This oscillation over the life of a plantation apparently disturbs and alters the natural communities of freshwater streams and lochs.

(d) *Location.* Many areas of moorland and peatland in the north and west are already rich in wildlife and possess relatively undisturbed communities. Some are fairly small, but many of those listed by Ratcliffe (1977) are very extensive, e.g. both the Beinn Dearg and Fionaven sites in North Scotland cover over 13 000 ha. Extensive afforestation with alien conifers in such sites will invariably be very damaging and unlikely to be compensated by future gains.

(e) *Time.* The wildlife and habitat losses due to afforestation mostly occur within the first 20 years of the first rotation, but the permanent gains mostly develop after this, and may continue for hundreds of years before a stable state is reached. The ultimate size of the gains cannot be measured because they are incomplete, and they have to be taken on trust, but they are likely to depend on how plantations are managed.

(f) *Plantation Management.* See Section 17.9 for a discussion on Plantation Management.

16.4 Assessment of changes in woodland pattern

16.4.1 Early afforestation

To what degree did afforestation before 1919 help or

hinder nature conservation? In the absence of a proper study the answer is necessarily speculative, but it is important to learn what we can of the longer-term balance sheet of afforestation versus the retention of unplanted ground, because some at least of the lessons may eventually be seen to apply to the present controversy over modern afforestation in the uplands.

Afforestation between 1700 and 1900 tended to be in small patches spread fairly evenly over the landscape (Section 5.3). Some clumping was apparent on certain estates and numerous shelterbelts were planted (Fig. 5.4). The total area planted was not large but it was distributed in a manner which ought to have been fairly good for colonisation by woodland species from medieval woodland nuclei, i.e. a scatter of stepping stones and belts and some denser groups of woods in which isolation ought to have been small and species numbers ought to have built up well. In fact, these woods were mainly colonised by common and widespread species and many species in the ancient woods have taken little or no advantage of the apparent opportunities to spread (Section 5.5). Many ancient woods were expanded by planting on adjacent ground: subsequent colonisation by most of the ancient woodland species has presumably helped to reduce the rate of relaxation in the ancient woods.

The pattern established by earlier afforestation was therefore reasonably good. What of the habitats replaced? Many small woods were planted on former arable in the wake of enclosure and here undoubtedly any loss of habitat was negligible beside the gain of woodland. Others, however, were planted on unimproved grassland, heath and wetlands: here the semi-natural habitats were replaced by plantations and many species must have been lost. At the time such habitats were still common and the losses might have seemed insignificant. Later, with the continued reduction of semi-natural grasslands, etc. by arable cultivation, the earlier losses to forestry would have been mildly regretted by a mid-19th century nature conservationist. Now, however, with the almost complete loss of unimproved grassland, heaths and wetlands throughout the lowlands, the verdict on the earlier afforestation has changed again. Consider a patch of, say, *Serratula tinctoria* or *Conopodium majus* growing in dry, acid grassland in central Lincolnshire in 1780: was its chance of surviving until 1980 better

or worse if the site were afforested? Until recently the better chance was to retain the site as pasture, where these species continued to thrive, whereas in the plantations they were confined to rides and margins or even exterminated. Now, unimproved grasslands have mostly been ploughed and the saw-wort and earthnut have gone, but most of the 18th century shelterbelts and small woods survive and with them a vestige of the grassland communities that they almost replaced.

The long-term verdict on early afforestation *vis-à-vis* nature conservation is thus somewhat paradoxical. The value of early afforested areas as woodland habitats is somewhat limited, even though the pattern was good, the direct losses of other habitats were largely insignificant at the time, many bare landscapes were undoubtedly diversified, and the species planted included a high proportion of native broadleaves. On the other hand their ultimate value in helping a vestige of grassland, heathland and wetland communities to survive has now proved to have been considerable in the face of arable intensification, even though, at the time, a conservationist would have regretted the planting.

16.4.2 Present changes in the woodland pattern

At the present time clearance is mainly concentrated in the lowland agricultural and urban regions whilst afforestation is virtually confined to the uplands. Earlier this century, between about 1920 and 1950, the contrast was not so stark: afforestation occurred in both the uplands and the lowlands, and the rate of woodland clearance in the lowlands was slower than it has been recently. The effects on the woodland pattern can be seen at various scales. Nationally, upland counties such as Nairn, Moray, Argyll and Kirkcudbright now have at least as much woodland in proportion to their total areas as the traditionally well-wooded lowland counties of Kent, Sussex and Surrey. Regionally, some traditionally sparsely wooded districts such as Breckland and the Sandlings have suddenly become densely wooded, whilst neighbouring districts with a moderate scatter of mainly ancient woods have lost more woodland through clearance than they have gained through afforestation.

In the uplands afforestation has tended to concentrate in some areas, such as Kielder, Galloway and Argyll, whilst ignoring others, such as the southern Pennines. In sparsely wooded lowland districts, forestry has not been worthwhile and agriculture has been paramount: consequently many woods have been cleared and afforestation has been negligible. In well-wooded lowland districts, by contrast, both state and private forestry has been worthwhile, agriculture has not been quite so dominant and so the rate of clearance has been slow. There was even a period around 1935–1950 when afforestation exceeded clearance. Thus, whilst woodland has become more evenly distributed nationally, there has been a distinct trend towards polarisation into well-wooded and sparsely wooded zones at a district scale. In the lowlands this polarisation has been intensified by the more thorough removal of hedges from intensively arable districts. Interestingly enough, the recent polarising trend in the lowlands exactly reverses the trend of the previous three centuries, when the combined effects of clearance and afforestation tended to produce an even scatter of small woods through most districts (Peterken and Harding, 1975).

What are the good and bad points for nature conservation in these changes? Woodland clearance is bad, but its effects are minimised if ancient woods are retained and recent woods are cleared (Principle 3), as was the case in two of Peterken and Harding's (1975) three East Midland study areas (Table 16.1). Most afforestation in the lowlands has been on heaths and old grassland where there has certainly been a great loss of semi-natural habitats as a result, but the alternative as we now see was often arable intensification which is far more damaging. Furthermore, some ancient woods have been expanded by afforestation on adjacent land, and this should have helped them to retain their existing wildlife. In the uplands, the principal defects in the afforestation pattern have been first the destruction of some of the finest examples of semi-natural habitats, and second the 'blanket' afforestation in some areas which has almost eliminated alternative habitats. The general verdict again depends on events outside the woodlands. The

new plantations are unlikely to develop rich wildlife communities but in the face of agricultural intensification they may be the only chance, albeit a poor one, of retaining some of the grassland and heathland species.

Looking to the future, nature conservation interests must seek to:

(1) Minimise clearance (Principle 3).
(2) Save the best moors, bogs, heaths and unimproved pastures from both afforestation and agricultural intensification.
(3) Minimise blanket afforestation.
(4) Ensure that more land is spared within plantations in order to retain a variety of habitats, notably small patches of scrub, streamsides and wetlands.

16.5 Afforestation by natural succession

The scrub woodlands which have developed mainly on old commons, downland pasture and heathlands throughout Britain (Section 5.2) present nature conservation with a dilemma. They are valuable as examples of natural succession and can in time develop into fairly rich communities, but the grasslands and heaths that they replace are lost. In most of the lowlands these unwooded, uncultivated habitats are now scarce, decreasing and reduced to isolated island habitats (Moore, 1962; Blackwood and Tubbs, 1970). Furthermore, it is doubtful whether they can be recreated if they are lost in the course of scrub development, because the soil is altered and there are now so few refuges from which the species can recolonise. In the uplands, however, heathland is still extensive and some patches of scrub woodland are a welcome diversification. In general, therefore, scrub development should be resisted in the lowlands if there is a reasonable long-term prospect of maintaining the grassland and heathland by grazing or burning, but in the uplands it should be allowed to develop. If the scrub has already grown so far that the canopy has closed and the pre-existing vegetation has been eliminated it is generally better to allow the succession to continue.

Chapter 17

Nature conservation aspects of woodland management

17.1 Principles

Principle 6
Change — Minimise rates of change within woods.

Principle 7
Stand maturity — Encourage maturity by maintaining long rotations. If this is not possible, retain a scatter of old trees after restocking.

Principle 8
Native species — Encourage native tree species and use non-native species only where necessary.

Principle 9
Diversity — Encourage diversity of (a) structure, (b) tree and shrub species and (c) habitat in so far as this is compatible with other principles.

Principle 10
Regeneration — Encourage restocking by natural regeneration or coppice growth.

Principle 11
Rare species — Take special measures where they are necessary to maintain populations of rare and local species.

Principle 12
Records — Retain records of management.

17.2 Change and the response of wildlife

When a change takes place (e.g. the change from broadleaves to conifer on a particular site, or vice versa), plant communities and their dependent fauna have to adjust to the new conditions. If the change is small (e.g. promotion of coppice to high forest), the adjustment may only be a quantitative change in the relative amounts of individual species. More substantial changes, such as the creation of a ride by clearing a strip of woodland, may cause qualitative adjustments, i.e. a complete change in the structure of the community accompanied by loss of some species and colonisation by others. Qualitative (and perhaps also quantitative) adjustments favour adventive and other fast-colonising species, because they colonise the newly suitable ground rapidly, achieve dominance, and maintain this advantage for some time as a result of the capacity of established communities to resist invasion by other species (inertia). If, however, the change is followed by a period of stability, the slow colonists may gradually infiltrate the communities of fast colonisers. Nevertheless, the sum effect of disturbance and increasing rates of habitat change within a wood is likely to be an increase in the populations of adventives and fast colonisers, and a decrease in the slow colonists: conversely, stability will favour the latter. Minimising the rates of habitat change therefore tends to maintain the importance of extinction-prone species in any wood (Section 6.2). In addition it minimises the rate at which irreplaceable features are lost, notably the semi(past)-natural stand types, which are characteristic of primary woodland.

Principle 6 is a plea for conservatism in woodland management, just as Principles 3 and 4 together are a plea for conservatism in the distribution and rate of change of land uses. Change is possible under this principle, but its rate and nature should be moderate, so that the species which are slow to adapt can keep up. Perhaps the most significant feature of modern forestry is that it has violated this principle almost everywhere in Britain, unlike traditional woodland management in Britain and much modern forestry in continental Europe.

Principle 6 can be applied only to those features which are under the control of management, such as the choice of species, the silvicultural system, rotation lengths and the distribution of broadleaves and conifers, and for practical reasons it can only be applied at a fairly coarse level. It is the main principle to consider when deciding on a programme of felling and restocking. It does not in any way prevent timber production, but it advocates a dynamic steady state within which timber can be produced whilst the characteristic plant and animal communities of woodland are maintained. It is especially important in the management of long-established woods where timber production is the main objective, and will become important in recent secondary woods (including most upland plantations) as they mature. The principle underlies nature reserve management, too, but there it is applied at a more detailed level (see Principle 14). It is an important element in the management of special areas, where the high density of semi-natural woodland allows faster recovery from reversible changes and therefore permits a greater rate of change in management.

This principle appears to be fundamental to maintaining visual amenity. People object both to felling of mature woodland and afforestation of bare hillsides, an apparent paradox which is really a resistance to change. Change is accepted, however, provided it is at a measured pace: indeed, if the pace is slow enough, change will not be noticed. Since special areas for nature conservation (Principle 1) are almost always important amenity areas (mostly Areas of Outstanding Natural Beauty), the case for strictly

applying this principle and the derivative management recommendations is reinforced.

17.3 Composition: choice and distribution of species

Trees play three roles in relation to nature conservation. Firstly they are, or may be, components of semi-natural communities – the stand – which are as worthy of conservation as the associated communities of herbs and fauna. Second, they are the essential components of woodland which provide the environment in which woodland species can survive; e.g. *Primula elatior*, the oxlip, would scarcely exist in Britain but for the presence of trees. Third, trees are food sources for phytophagous invertebrates, a substrate for epiphytic species and an essential resource for organisms which depend on wood, dead or alive.

Native species are the most valuable group in all three roles. Here 'native' is used in the strict sense of native to the site, i.e. a species is native only if it is likely to be a past-natural component of woodland on the site under consideration. Thus, *Pinus sylvestris* is certainly native only on certain sites in Scotland, although it may be native as a minor component of woodland on light soils further south. *Fagus sylvatica* conversely is only native in southern England and even there cannot be regarded as native on heavy, poorly drained soils. (In Britain it has been possible to regard species as native or introduced irrespective of the site or region under discussion because the country is a relatively small island. This makes little sense on the European continent, and not surprisingly their usage is closer to that defined above.) Only native species can legitimately form part of semi-natural stands, especially the more important semi-(past) natural stands. British wildlife is adapted to the particular conditions provided by native tree species, mostly the broadleaved, deciduous canopy and mull humus provided by birch, oak, ash, beech, lime, etc. Native species also provide the richest sources for more directly dependent groups such as phytophagous invertebrates (Southwood, 1961). Although all native species are equally valuable as members of the appropriate semi-natural stands, they vary in their

Table 17.1 The number of insect species in the groups Heteroptera, Homoptera (part), Lepidoptera and Coleoptera associated with various British trees (Southwood, 1961).

Native species and genera	Species	Introduced species and genera
Quercus robur and *Q. petraea*	284	
Salix spp.	266	
Betula spp.	229	
Crataegus spp.	149	
Prunus spinosa	109	
Populus spp.	97	
Malus spp.	93	
Pinus sylvestris	91	
Alnus glutinosa	90	
Ulmus spp.	82	
Corylus avellana	73	
Fagus sylvatica	64	
Fraxinus excelsior	41	
	37	*Picea abies*
Tilia spp.	31	
Carpinus betulus	28	
Sorbus aucuparia	28	
Acer campestre	26	
Juniperus communis	20	
	17	*Larix decidua*
	16	*Abies* spp.
	15	*Acer pseudoplatanus*
Ilex aquifolium	7	
	5	*Castanea sativa*
	4	*Aesculus hippocastanum*
	3	*Juglans regia*
	2	*Quercus ilex*
Taxus baccata	1	
	1	*Robinia pseudoacacia*
	0	*Platanus orientalis*

value as substrates and sources for dependent wildlife (Table 17.1).

Introduced species vary both in their overall value for nature conservation and in their contribution to the three roles of trees in relation to nature conservation. This is summarised in Table 17.2, which separates the Scottish Highlands (where pine is native) from the rest of Britain. The assessments are inevitably over-simple and to some extent they are uncertain: not only is *Nothofagus*, for example, so recently arrived that some aspects of its value have not been studied, but the richness of the dependent fauna is likely to increase in the very long-term, as wildlife species spread and adapt. The assessments are based on the assumption that, in the absence of native species, a native physiognomy is more valuable than an alien physiognomy, and that introduced species became more valuable with time. Despite its limitations the table indicates that, while native species are always the most valuable, many other groups of species contribute significantly to certain aspects of nature conservation.

Foresters are well aware of the value of native species for nature conservation, and markets do exist for the timber of native species (Garthwaite, 1977). The practical objections are equally familiar. Alien species exist which can grow faster than native species under British conditions, and in much of the uplands afforestation is economic only if coniferous, mostly alien, species are planted. Inevitably, therefore, conifers have predominated over broadleaves in recent planting and some compromise must be sought between mainly native broadleaf deciduous trees and alien, evergreen conifers. This can take three forms:

(1) Planting (deciduous) larch, (native) Scots Pine and the faster-growing introduced broadleaves, such as *Nothofagus*. The so-called southern beeches are seen by many timber-growers as a satisfactory solution to maintaining the traditional broadleaved, deciduous landscape of much of Britain whilst growing timber at a fast rate, but their value for wildlife is uncertain.

(2) Planting broadleaf/conifer mixtures, e.g. oak and Norway spruce, with the intention of harvesting the conifers as thinnings and growing the broadleaves as a final crop. Such mixtures seem to eliminate the shrub layer almost as much as pure conifers and the field layer is considerably reduced, but the resulting mature broadleaf stand is generally rich in wildlife. Unfortunately, many mixed plantings, especially in eastern England, have been thinned into pure conifer stands – alternatively, the broadleaves were eliminated with herbicides – so this form of compromise is viewed with misgivings.

(3) In accordance with Principle 6, conifers should be planted only where conifers grew before, or on hitherto unwooded ground, and existing broadleaved stands should be followed by more broadleaves. The pattern of broadleaves and conifers in established woods should change as little as possible from one rotation to another.

Table 17.2 Summary of the contribution of different trees to various aspects of nature conservation.

	NATIVE		INTRODUCED			
	Broadleaved, deciduous	*Conifer, evergreen*	*Long-established broadleaved*	*Broadleaved, deciduous*	*Conifer, deciduous*	*Conifer, evergreen*
	Oak, ash, beech, lime, birch	*Scots pine*	*Chestnut, sycamore*	*Nothofagus, red oak*	*Larch*	*Spruce fir, hemlock, other pines*
England, Wales and lowland Scotland						
(a) Component of semi-natural communities	+++	+	+	o	o	o
(b) Environment for wildlife	+++	+	+++	++	++	+
(c) Substrate for wildlife	+++	+	++	+	+	+
Scottish Highlands						
(a) Component of semi-natural communities	+++	+++	+	o	o	o
(b) Environment for wildlife	+++	+++	+++	++	++	++
(c) Substrate for wildlife	+++	+++	++	+	+	+

+++, very good; ++, moderate; +, low value; o, harmful.

17.3.1 Diversity of species

Greater diversity of habitat (Principle 9) enables more species to exist in a given area and assists more species to colonise. Understandably, therefore, diversity is regarded by nature conservationists as an all-embracing good whilst uniformity is correspondingly bad. By this reckoning any action which increases diversity will be good and any which decreases it must be bad.

There is obviously much to be said for this attitude, but it has drawbacks (see also Section 18.4). Diversity may be directly encouraged by planting additional species into a wood until it resembles a botanical garden. Such places obviously have a role in nature conservation, but unrestrained diversity of this kind is a disadvantage to habitat conservation, for the habitat is less natural. Natural woodland was possibly less diverse than managed woodland, so, at least at a local level, there may be some conflict between the aspirations of species and habitat conservation. Diversity is certainly good, but it must be encouraged with discretion.

17.4 Structure and management systems

Structure includes not just the physical distribution of trees and shrubs in a stand, but also the age class distribution in space and time, and the silvicultural systems which have been designed to create and maintain certain conditions. There appears to be little research on the direct effects of different structures on wildlife, so the recommendation that the ideal system combines stability (Principle 6) with diversity (Principle 9) is based on casual observations and inferences from the known habitat requirements of certain species. The value of a natural structure and traditional management system is discussed in Chapter 18.

17.4.1 Stability

Structural stability is valuable because it tends to favour extinction-prone species amongst the woodland fauna and flora. It can be expressed in three ways:

(1) Stand maturity (Principle 7, which is a special case of Principle 6) favours those species which are especially well-represented in wood pasture, i.e. timber-utilising invertebrates (Figs. 3.16 and 3.17), hole-nesting birds and epiphytic lichens (Fig. 3.14) and bryophytes. These depend on mature, moribund or dead timber. Many are extinction-prone species and all are at risk in the face of the short-term

273

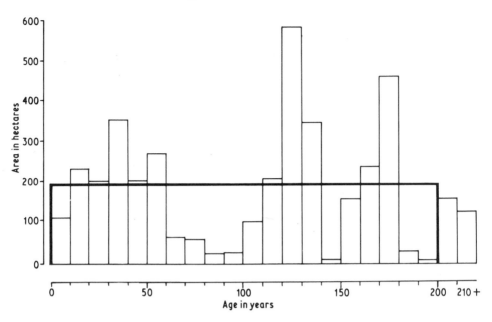

Fig. 17.1 Age structure of broadleaf stands within the New Forest Inclosures (data by kind permission of the Forestry Commission). The rate of broadleaf planting fluctuated wildly during the 18th and 19th centuries, but became steady from 1920 onwards. A planting rate of about 193 hectares per decade would have been required to produce an even spread of age classes over the existing area of broadleaf woodland on a 200 year rotation. If this rate were maintained from now on, a normal age distribution would not be approached until the 22nd century, but the overmature stands would be felled before the end of the present century.

rotations which are characteristic of modern forestry. Their needs are most likely to be satisfied if rotations are as long as possible; small numbers of mature trees are left when a stand is felled, and they are allowed to grow old and die where they stand (they can be poor timber trees); and dead wood is left lying in sheltered conditions on the ground (Stubbs, 1972).

(2) Normal forest. Long-term stability is best maintained by creating a forest in which all age classes are present in those proportions which, for the given area of the wood and chosen length of rotation, will maintain the existing age distribution of stands indefinitely. In this way any organism which depends on a particular age class will always have somewhere to go. Ideally, a normal age structure should be achieved in each wood, but failing this it should be created in each forest or private estate. Normal forests are rarely achieved because management plans change so rapidly. This is particularly true of British woodlands today, which at a local scale may be further from a normal age structure than ever before. In the New

Forest (Hants), for example, broadleaved planting was below the required rate from 1860–1920 and above from 1920–1970. (The requirement was determined after 1970 in this instance.) (Fig. 17.1.) Rapid afforestation of the uplands has created vast, even-aged stands. Neglect of management in long-established woods has allowed many to become uniformly over-mature. When management has been reinstated whole woods have often been felled and replanted at one time.

(3) Distribution of age classes. Each age class should be distributed through a wood or forest, so that any species which requires a particular age class can easily colonise from one part of the mosaic to another. For example, species which require the oldest age classes (many of which are poor colonisers) will periodically have to leap across expanses of immature stands in order to survive in a wood which has large, uniform, even-aged blocks. The desired small-scale patchwork of age classes can only be produced by small-scale working, but timber-production

274

economics push management towards large felling coupes and the uniformity of upland plantations virtually forces management into a large-scale mould. However, many woods now contain a spread of age classes mainly because of amenity and sporting needs, and the upland plantations will become more patchy in the future as windthrow, fire and deliberate forest planning expand the range of age classes in each forest.

17.4.2 Diversity

Structural diversity provides niches for more species than a uniform stand, but, as with the direct encouragement of diversity of tree species (Section 17.3), there may be a drawback. If the number of species in a wood is increased the population of each species will presumably be reduced and therefore more vulnerable to extinction. Thus, diversifying a site may increase the number of species, reduce the populations of pre-existing species, increase the turn-over of species and therefore tend to tip the scales against the extinction-prone species: the most vulnerable species may be made yet more vulnerable. This, however, is theoretical and on balance it is probably best to encourage as much structural diversity as possible.

Small-scale working provides a kaleidoscope of sheltered glades which encourages woodland insects and birds. Edge habitat is maximised to the advantage of those woodland species which are discouraged by deep shade. Large-scale working produces little edge habitat and the recently felled coupes are more like windswept heaths than glades. Selection systems are most diverse at the smallest scale, but they are uniform at an intermediate scale and tend to favour only shade-bearing trees. Absence of management encourages maturity, but minimises diversity: it creates short-term stability but eventually the whole stand must be felled and replaced in one operation (but see discussion of Natural Woodland, Section 18.2).

Where afforestation is taking place it is important to maintain existing diversity of species, habitats and woodland structure. Structural diversity is maintained by retaining small woods and patches of scrub (often along streams) which were present before their surroundings were afforested (Fig. 5.6).

17.4.3 Recommendations

In woods which are managed primarily for timber production, the interests of nature conservation can best be served by adopting systems which combine stability with diversity. This in practice means:

> long rotations
> small-scale group working
> creating and maintaining a normal age structure
> retaining existing diversity.

17.5 Restocking: natural regeneration and planting

Natural regeneration (Principle 10) has several advantages over planting for nature conservation:

(a) it favours native species (Principle 8);
(b) it favours the species already on the site (Principle 6);
(c) it tends to generate mixed stands (Principle 9);
(d) the stands produced tend to have a more irregular structure than plantations (Principle 9);
(e) natural genetic variety can be better maintained;
(f) the natural distribution of tree species in relation to soil types is favoured.

Natural regeneration is a disadvantage when aggressive introduced species are present which are threatening to dominate the wood. *Rhododendron*, which has spread into many woods in western districts and on light soils, impoverishes the wildlife and ought to be controlled: here planting will probably be necessary. Sycamore is a difficult case. Being an introduced species with a poor dependent fauna, it is not highly regarded for nature conservation, but it regenerates well and may eventually take up a fairly stable position in British woodland communities. In nature reserves and other special woodlands it should be controlled, but elsewhere it seems desirable to accept more sycamore if that is the ecological price of more n̲a̲t̲u̲r̲a̲l̲

Restocking by natura̲
enormous cost of planting,
free because fencing and
necessary. Despite this, a̲
planted. Obviously this is
moorland which is being af̲

276

Fig. 17.2 Underplanting of Norway Spruce beneath a shelterwood of mainly birch, the thinned residue of the former overgrown, semi-natural coppice mixture. Although underplanting with conifers is less immediately damaging than clear felling for both amenity and nature conservation, the end result when the overstorey is removed is much the same. Waterperry Forest Nature Reserve, Oxon.

equally inevitable when woodlands are being re-stocked. In Rockingham Forest, for example, all but one hectare of the 839 ha which were re-stocked between 1960 and 1972 were planted (Peterken and Harding, 1974). Several reasons can be suggested for this unfavourable state of affairs:

(1) Foresters generally intend to change the stand composition, usually from native broadleaves to introduced species of both broadleaves and conifers (Fig. 17.2).

(2) Even if the composition is not to be changed, growth of the available parent trees is generally [poo]d to be poor, so some measure of stock improve-[ment is] sought.

(3) Even if the species and growth qualities are acceptable, trees in Britain produce adequate quantities of mast irregularly and infrequently. For example, between 1921 and 1950 the beech mast in southern England was good or very good in 1922, 1924, 1929, 1934, 1944, 1948 and 1950 (Matthews, 1955).

(4) Even if good masts are frequent, the seedlings often encounter strong competition from ground vegetation and grazing animals; and in some woods they have to become established in mor humus or peat on impoverished soils.

Natural regeneration is therefore difficult to obtain even when it is wanted, and foresters understandably assume that planting is the only practicable approach. This assumption is officially reinforced by the Forestry Commission, whose financial support for restocking is a 'planting grant' which, although it can be paid for natural regeneration, seems to discriminate in favour of planting both in the conditions of the grant and the phraseology of the explanatory leaflet. Whilst planting is immediate and usually reliable, yet costly, there are nevertheless many instances of successful natural regeneration, such as that of the beech at Bradenham Coppice and Booker Common in the Chilterns, the oak at Salisbury Trench in the New Forest and in parts of Windsor Forest (where collected acorns were broadcast into prepared ground), the pine in parts of Glentanar Forest on Deeside and many southern sandy districts, and many instances involving pioneer and aggressive species such as birch, sycamore and ash.

Clearly conservationists cannot reasonably expect foresters in Britain to try for natural regeneration on a large scale or even to assign particular woods to a wholly natural system of restocking, but they can expect foresters to make more use of it when it occurs. Self-sown trees should be accepted into planted stands. Planting could be postponed in order to give natural regeneration a chance: if it fails, planting can follow. In southern latitudes on light and medium-textured soils it ought to be possible to manage some of the oak and beech woods on the Uniform System (Section 4.2) which is widely and commonly used just across the Channel. Natural regeneration might be more widely sought and accepted if deer were better controlled and if small-scale working (Section 17.4) were more widely practiced.

Restocking by coppice regrowth has all the advantages of natural regeneration for nature conservation and is generally more reliable. It is not, however, regarded as a valid method by the Foresty Commission, who do not give 'planting grants' for coppice, and thereby discourage coppice management and encourage high forest systems. This is understandable to some extent, but coppice is still a commercial proposition in parts of England and Wales, where chestnut fencing and other markets for coppice products are available. Coppice management for hardwood pulpwood can be economic even without state subsidy (Stern, 1973), but it depends on the personal circumstances of the owner and on guarantees from commercial companies. Since coppice management is especially appropriate for semi-natural woods, it is discussed further in Chapter 19.

17.6 Rides and other subsidiary habitats

Rides (Fig. 14.1), glades, ponds, etc. are important subsidiary habitats which increase the range of habitats within woods, add to structural variety and maintain a large number of species. Most are artificial in origin and survive only by active management. Rides in particular lose their distinctive communities if they become shaded, but can be rich habitats if they are managed on the 'Monks Wood' pattern in which a stepped margin of unmown grass, scrub and coppiced woodland is maintained by cutting on rotation (Steele and Welch, 1973). In the interests of ecological stability (Principle 6) the ride pattern must be maintained once it exists, because old rides are generally richer than new rides. This will apply as much to upland forests as to long-established lowland woods.

Hedges on ancient wood banks should be retained. They often consist of a mixture of shrub species which differs from the coppice mixture: for example, in acid clay woods field maple is often far more abundant in the hedge than in the body of the wood. In woods where the semi-natural coppice has been replaced by plantations, the hedge may be the last vestige of the former semi-natural stand. Since, furthermore, small populations of rare and local species are also often restricted to margins, hedge removal (Fig. 17.3) can be as damaging to a wood as it is to the wildlife of farmland.

Habitat diversity can be made where it does not exist, by opening rides and cutting glades, provided that the stands themselves are not unduly disturbed. The new upland forests will be better as wildlife habitats if patches of scrub woodland, streamside vegetation, rock outcrops and small bogs are left unplanted.

17.7 Care of small populations

One assumption behind these principles is that species will look after themselves if the habitat is maintained

Fig. 17.3 The margin of Waterperry Wood, Oxon, showing the shallow, spread wood bank, the external ditch (recently cleaned and deepened) and the sinuous line characteristic of ancient woodland boundaries. Hedgerow removal has facilitated the view, but destroyed a valuable historical and ecological feature for which the shorn margin of old coppice is no substitute.

in a suitable condition, and if the turnover and distribution of suitable habitat is not so high or dispersed that it exceeds the capacity of species to respond to change. This assumption is valid for abundant species which are well-dispersed, capable of colonising and not dependent on a declining habitat. Other species, however, are sufficiently rare or local to justify special conservation measures. The necessary measures have mostly been built into other principles: Principles 3 and 6 provide for poor colonisers and Principle 7 for the mature timber species. Furthermore, if the special woodland types are given special treatment (Principles 1 and 13–15) a high proportion of the population of extinction-prone species will be properly cared for. Nevertheless if Island Theory is correct in predicting some turnover of populations at the species level even in stable habitats, some populations of rare and local species will become extinct, even if all reasonable measures to protect them are taken. Under some circumstances therefore additional measures are necessary for the conservation of individual species. There are perhaps three general cases:

(1) In woods managed for timber or public recreation, where an abrupt change in management from favourable to unfavourable is likely, i.e. where some principles are not observed. In a timber-producing woodland replanting is usually with conifers, and small populations of rare and local plants usually have

to be protected simply by not reforesting their sites, e.g. the Forestry Commission have spared *Trientalis europea* and *Maianthemum bifolium* in the North Yorks Moors by leaving their immediate locality unplanted whilst the rest of the wood was replanted with pure conifers. The survival of Black Hairstreaks may be possible if a bank of old *Prunus spinosa* can be perpetuated in a corner of an otherwise reforested wood. The Forestry Commission have taken impressively thorough care of scrub, especially sallows, beside rides in the butterfly-rich parts of Alice Holt Forest (Surrey), where the Purple Emperor lives.

(2) In nature reserves and special woodland types managed according to Principles 13–15. Very small populations of very rare species are usually the only problem, and these may require careful habitat management and monitoring. In the Chiltern beechwoods for example, it may be better to maintain a localised Selection Forest over the immediate localities of the rare woodland herbs in the area, *Cephalanthera rubra*, *C. longifolia*, *Dentaria bulbifera*, *Epipogium aphyllum* and *Orchis militaris*

(3) In any area where the distribution of special treatments could be important. For example, if coppicing is to be maintained on a small scale, it may be best to select a site where the Dormouse survives. In any site where a minimal disturbance zone will be maintained, it may be best to select an area which already has mature trees (with their associated flora and fauna) or where sensitive species may abound, such as streamsides which tend to have rare bryophytes.

17.8 Management recording

Conservation is concerned with the relationship between man and the natural environment, and particularly with maintaining a balance which can be indefinitely sustained. This requires that we understand the impact which our actions have on the environment. One approach to understanding is to maintain control ecosystems which are absolutely or relatively free from human influence: this is one justification for habitat conservation. The other is to measure the changes which take place after management has been carried out, or to compare the conditions which are generated under different management regimes.

The second approach demands either that we set up trials whose results can be assessed in the future, or that we study the consequences of actions which were taken in the past: in both cases the people observing at the start are unlikely to be there at the end. Management records must be made which are quite independent of the management, for, if they are not, the study will either be impossible or endangered by circular reasoning. We may, for example, wish to compare the long-term effects on soil structure of different forms of site treatment before afforestation by examining the soils 50 years later at the end of the first rotation, but this is possible only if the records of the initial state were made and kept.

Many organisations are studying different aspects of forestry and the environment, and many long-term trials have been established. These are yielding useful information and will yield more. We should not, however, rely on them entirely (and not just because even official trials can be neglected, improperly recorded or simply forgotten by the successors of those scientists who established them) because they may not cover all the subjects which we may need to know in the future. We should, in fact, supplement such efforts by keeping records of woodland management, for some of this may chance to provide the answers to questions which only the managers of the future will think of asking (Principle 12).

It is recommended that:

(a) records of actions taken in each compartment of managed woodland should be made;
(b) surviving records of past management should be kept;
(c) both kinds of record should be properly archived;
(d) these records should not be confined to official trials.

The Event Record kept by the Nature Conservancy Council in many of its reserves is a start in this direction. So, too, are the private estate records which have been deposited with local Record Offices. On the other hand the extensive destruction of original records from the various Forestry Commission Censuses is undoubtedly a great loss. All original records from 1924 and all but a few samples from 1938 and 1942 have gone, but the records of private woods in 1947 fortunately survive. The Forestry

Commission does not retain an archive of old stock maps.

Curiously, Principle 12 is merely an echo of John Evelyn, who more than 300 years ago (1664) wrote 'I have often wish'd, that gentlemen were more curious of transmitting to posterity, such records, by noting the years when they begin any considerable plantation; that the ages to come may have both the satisfaction and encouragement by more accurate and certain calculations'.

17.9 Plantation management in the uplands

Plantations designed and managed for maximum timber production at minimum cost are generally least attractive for wildlife. All the land is planted, so habitat diversity is minimised. Monocultures of uniform age are characteristic. Many plantations remain unthinned, so the stand structure is uniform and no shrub layer can develop. Almost invariably the plantations consist of very few tree species: these are evergreen and coniferous, characteristics to which most native woodland wildlife species are ill-suited (Fig. 5.7). Nevertheless, diversifying measures are available to encourage wildlife (Steele, 1972a).

Foresters often suggest that the wildlife of upland plantations will be richer in the future, because (a) they will eventually be colonised by more woodland species, and (b) management will be more diverse. The first point is acceptable (Section 5.4), though on less fertile soils the population of buried seeds may fall with a consequent loss of floristic diversity (Hill, 1979). The second point, however, seems doubtful.

Future management possibilities can be considered through the contributions of E. J. M. Davies and D. C. Malcolm to a recent symposium (Ford *et al.*, 1979). Davies, a self-styled 'tree farmer', who believes that forestry has been slow to follow agricultural practice, advocates 'standard silviculture'. This involves ploughing before planting, heavy applications of fertilisers, line thinning and removal of well-grown trees in order to retain a uniform canopy, and finally clear felling at 35 years in large coupes of

10–100 ha. The object is to produce standard softwood sawlogs for industry by intensive, uniform and predictable methods. The skill of the forester 'can be measured in the way that he succeeds in dominating the environment'. Small variations would be allowed for wildlife and landscape, and stands liable to windthrow would remain unthinned. Malcolm sees future forest design as a choice between Davies' homogeneous standard and 'accepting that even with cultural treatments the environment is not homogeneous, particularly in respect of climate, that the existing heterogeneity can be used to provide variation without necessarily reduced production while meeting the other demands on the forest without strain'. Both writers believe that almost all plantations will remain even-aged conifer monocultures, but Davies' standard sulviculture assumes stability of both the environment and demand for a particular product, whereas Malcolm's second option, by allowing variation in plantation composition and the size and distribution of stands within the forest, confers flexibility on management and resilience in the face of change in, say, energy costs, market demands and social attitudes. Neither offers much hope of substantial improvements in upland plantations as wildlife habitats.

An alternative and more subtle plea is that even-aged conifer monocultures are, in fact, the (future–)natural condition of woodland on the nutrient-poor, peaty, windswept upland sites. This cannot be accepted completely, because differences between plantation management and hypothetical natural conifer forest would be significant. Plantation rotations would be much shorter than the cycle of catastrophes in natural forest. Nutrients lost through harvesting must be replaced by fertilising. Natural selection amongst tree species and genotypes depends on performance in competition, whereas in plantations it depends on productivity, wood quality and stem form (Malcolm, 1979). Even so, a valid point remains: the wildlife potential of upland forests is likely to be limited as much by natural factors as silvicultural treatments.

Chapter 18

Management of semi-natural woodland

18.1 Principles

Principle 13
Natural woodland

Manage a proportion of woods on non-intervention lines in order to restore natural woodland in so far as this is possible.

Principle 14
Traditional management

Maintain or restore traditional management where this is possible and appropriate.

Principle 15
'Modern' management

Where traditional management is not possible or appropriate, introduce alternative systems of management which retain or enhance the conservation value of special sites and areas.

These principles apply to these woodlands which qualify as special sites and areas (Chapter 15). These are mostly ancient, semi-natural woodlands, and amount to about 20% of existing woodland.

18.2 Natural woodland as a management objective

No woodland in Britain is believed to have escaped entirely the influence of man and thus remained in a completely natural state. Nevertheless, natural woodland is a commonly cited objective of nature reserve management, which, once achieved, would have several benefits for nature conservation (see Streeter, 1974). Natural woodlands are interesting subjects for study in their own right and as the environment within which much of our native wildlife and natural communities originated. They provide the best form of control conditions against which the effects of human influence on the environment can be appreciated and studied.

Management can be directed towards different kinds of naturalness. In woodlands where some elements of past-naturalness survive it is reasonable to try to restore a greater degree of such naturalness. In such cases one starts with a semi-(past) natural woodland and first ensures that the existing natural features are not further eroded, before attempting to eradicate the past influence of man by suitable measures. In other woodlands, where no past-natural features survive, one can simply allow natural succession to take its course towards future–natural woodland, or manage towards a simulation of original–natural woodland. In most cases the object is simple and clear, but the problems which bar the way are almost insurmountable.

The native pinewoods (Fig. 2.8) are believed to be closer to original–natural conditions than any other woodland type in Britain. Exploitation for timber and pasture has evidently caused degradation of the soil, much reduced the hazel, oak and birch which were formerly much more abundant within the pine-woods, and created stands which, if not even-aged, are probably more limited in age class distribution than completely natural pinewoods. Nevertheless, they retain the following past-natural features (Bunce and Jeffers, 1977):

(1) The site as continuously existing pine wood-land (confirmed by pollen analysis at Abernethy, Beinn Eighe and Rannoch).

(2) The soil profile, which in general has not been disturbed physically. Its present state results from millenia of natural development, accelerated latterly by human action.

(3) Genetic complement. The existing variety of phenotypes is believed to be mainly the residue of natural variation.

(4) Stand structure. The existing pattern of even-aged stands, even-aged components of mixed-age stands, and local occurrence of trees at the end of the natural life-span of pine is believed to be within the range of natural structural conditions.

(5) The associated plant and animal communities which include species that are not only characteristic of the Highland pinewoods but also apparently slow to colonise newly available pine woodland. These are believed to be survivors of original–natural pinewood communities.

If it is agreed that the ideal nature conservation objective in such woods is the restoration of natural woodland, then the first objective should be to maintain these five features. The long-term objective should be to maintain the woods in their present sites by natural regeneration, a course which in theory allows a form of natural pinewood to develop. Several problems are encountered:

(1) The past influence of man on soils and genetic complement can probably never be entirely removed.

(2) Many of the native pinewood species occur only in small, isolated populations, which are thus classic extinction-prone species. Past fragmentation of native pine woodland may be reflected in future local extinctions.

(3) The woods are heavily grazed by deer, which are tending to prevent natural regeneration.

Thorough and sustained exclosure of deer will be necessary to allow natural regeneration.

(4) Soil degradation due to natural maturation and the exploitation of recent times has encouraged the growth of mor-humus, blanket bog and dwarf-shrub heath beneath existing pine stands. This appears to be a factor in the slow rate of natural regeneration. Exposure of mineral soil in the absence of severe grazing permits copious regeneration, but only at the cost of disturbing the hitherto undisturbed soil profile. Nature conservationists may face a choice between losing the woods altogether by retrogressive succession, or restoring actively regenerating pine-wood by soil disturbance.

(5) Fire was a natural factor which tended to create conditions in which natural regeneration could flourish. If natural conditions are to be fully restored, fires must again be allowed to sweep unchecked through the pinewoods. However, most woods are too small to risk this, and in any case the modern incidence of fires is determined mainly by man.

(6) The scale of the mosaic of age classes in natural pine woodland was probably large, and the minimum area in which the net change in the mosaic was nil was also large. Most woods are likely to be below this size, and thus too small to function naturally. The mosaic appears to have included ground which temporarily lost its wooded cover.

(7) Most woods are in private ownership or held by the Forestry Commission. Both are more inter-ested in the considerable timber assets of native pine-woods. Reserving stands as natural woodland is expensive because existing timber on the ground cannot be harvested and future production is foregone.

(8) Pine plantations in the vicinity of the native pinewoods are likely to influence the genetic consti-tution of the native stands.

The unenclosed beech–oakwoods of the New Forest (Figs. 2.1 and 13.3) are similar in many ways to native pinewoods, and just as suitable for manage-ment as natural woodland. Superficially they are very different, but both comprise a mixture of primary and secondary woodland; have a near-natural age range including individuals in the oldest possible age classes; include many rare woodland species which are

Fig. 18.1 Derelict coppice at Kings Wood, Bedfordshire, last cut several decades ago, since when the rides have become shaded and partly blocked and the stand has developed towards a high forest structure with low timber value. Such stands pose considerable management problems for nature conservationists. Even as a nature reserve the structure would take many centuries to revert to a natural state (Fig. 1.1), and the present even-aged condition is a poor starting point for such an objective. Normally, therefore, one favours active management, not only because it provides opportunities to open up the rides and restore them as rich wildlife habitats, but also because utilisable timber is produced. However, coppice management cannot easily be restored because the stools are moribund, many have died and markets are poor or uncertain. Treatment as high forest of ash, oak, hornbeam, lime and birch would be satisfactory, because it enables woodland of species native to this site to be maintained, but it is unlikely to be adopted in the present forestry climate. If such stands are treated at all they are most likely to be converted to plantations consisting mainly of conifers and other species not native to this site, and understandably the nature conservationist often prefers continued neglect. (Photo. P. Wakely, Nature Conservancy Council)

evidently poor colonists; have a history of over-grazing and survival mainly through natural regeneration; have been raided periodically for useful timber; and have lost some species (e.g. hazel) as a result of grazing and soil changes. The New Forest presents the additional problem that beech has tended to replace oak. As some of the most natural woods in Britain, these are woods in which a natural woodland management objective can be most easily justified. Most of the problems in the pinewoods are repeated in the New Forest, but with two additions. One is the intensive recreational use, affecting the whole area,

283

but particularly those woods near roads, villages and camp sites. The other concerns the beech–oak problem. Is the succession a natural phenomenon accelerated by selective felling in the past? If so, do we aim for future–natural woodland by allowing it to continue? Or should we try to restore the beech–oak balance to what we think it would have been if man had not interfered? If the latter, then should we reintroduce hazel, small-leaved lime, etc., which have been virtually eliminated by man and pasture animals?

Certain points seem clear from the discussion so far. First, the number of sites in which a natural woodland objective can be pursued will be small, mainly because the cost is high. Second, natural woodland is a realistic objective only in large woods, and even there the influence of events beyond their borders will never be entirely absent. Third, since it may take several centuries to eliminate the effects of past human influence, such woods must be managed by an organisation which can sustain the objective indefinitely: even if it is, the duration of human institutions is shorter than the time needed, and we may ultimately be unsuccessful. The prospects are therefore poor, but in the case of the native pinewoods it seems worth trying. In the New Forest it is probably best to let beech and oak compete without silvicultural intervention in most woods, but to simulate original natural woodland on a small scale by controlling beech and alien species, re-introducing lime and hazel, and excluding commonable stock and deer.

The pursuit of natural woodland may start from ancient, semi-natural coppice woodland (Fig. 18.1), which may be allowed to develop without intervention, save for the control of excessive grazing and alien species. Here the prospects for restoring original–natural woodland are better in the long-term, for the tree, shrub and field layer communities and the soils are less modified than those of wood pasture. The structure imposed by coppice management will grow out in a few hundred years, and the development of a natural structure may even be accelerated by singling coppice growth. Although the basic need is to impose rigid non-intervention, the condition in which a stand is 'abandoned to nature' need not be even-aged. Indeed, if a wood is carefully managed into a mixed-age structure and then abandoned, a naturally functioning stand may form earlier than if an even-aged stand were abandoned.

The main strategic problem about restoring an ancient, semi-natural woodland to a natural state is to know whether it is valid or possible to restore past-natural conditions. This difficulty does not arise when secondary semi-natural scrub is allowed to develop naturally or when a plantation is abandoned in order to allow a natural structure and composition to develop, for in both instances the resulting state is unambiguously future–natural. This has important implications. Provided man's influence has been removed from within the site, whatever happens is right, even if, say, turkey oak were to spread rapidly and assume dominance, because the past effects of man (such as the introduction of species) are a valid determinant of the characteristics of future–natural conditions.

18.3 Traditional management: value and prospects

The existing nature conservation value of special sites and areas has developed under or survived through various forms of traditional management (in the broad sense, to include all forms of man's impact). This management has been remarkably stable, remaining fundamentally unchanged in some cases for a millenium. Under such conditions one assumes that the ecosystem has achieved stability, or at least that any changes now are oscillations about a stable mean which, as in natural woodland, produce no net change over a large area of woodland. If so, then the most certain way of maintaining the existing nature conservation value of such woods would seem to be to continue traditional management, for at least there should be no further changes. Furthermore, historical studies may be advanced by first-hand experience of land-management methods which were characteristic of former social and economic circumstances, and students of conservation in the broadest sense may wish to study a system which has demonstrably sustained its yield for hundreds of years.

This basic logic is partially undermined by three factors. First, traditional management has in fact changed and developed, and existing woods may not therefore be ecologically stable (see Part One).

Second, it is possible that, even under stable management, long-term changes may have been taking place over hundreds of years: we cannot assume that changes were complete as soon as natural woodland had been transformed into managed woodland. Degrading soil changes may still be taking place, generating beliefs that, for example, the soil of the Wyre Forest is 'oak sick'. Third, Principle 14 is a safety play, based on the assumption that no better alternative exists. In practice it is clear that some aspects of traditional management [e.g. continuous heavy grazing in upland woods (Fig. 8.18)] can be improved upon in purely nature conservation terms. Principle 14 must therefore be applied with discretion.

The case for maintaining traditional management is strongest in those mixed coppices which have not been planted. It is less strong in the hazel, ash, oak and alder coppices where one species is completely dominant as a result of planting, selection, or both. Mixed coppices occur mainly in the English lowlands and neighbouring districts in south-west England, the Welsh borderland and the Lake District, but some also occur in central and western Scotland. Three approaches to maintaining or restoring coppice management are possible: (i) maintaining the traditional markets for coppice wood; (ii) simulating traditional management in the absence of a traditional market for the benefit of nature conservation; and (iii) developing new markets for coppice wood. In each case we must face the fact that traditional management has already been abandoned in most sites, and that restoration of coppicing may be more difficult than maintaining it.

18.3.1 Traditional markets

Coppice management satisfied a mixture of generalised and specialised needs through the woodmen and other craftsmen who tended, cut and processed the coppice wood. The markets and woodland crafts have all declined dramatically in recent decades, and only some have been partly transformed to meet modern needs. Fuelwood was a major generalised use to which we may yet return, but few people use faggots now. One exception was Ye Olde Faggote Oven Bakery at Long Melford. Brushwood bundles are no longer used for field drainage, but they are used by some Water Authorities and Inland Drainage Boards for river bank stabilisation, for example in East Anglia. Bat faggots have been used recently in Sussex for coastal defence, and fascines have been used in large numbers for harbour works. Brushwood has been replaced by tarmac for 'roading'. Some specialised uses are extinct (e.g. clog making from alder) or on the verge of extinction; oak spelk baskets are made, if at all, only by octogenarians. Others have stabilised at a low level. For example, there are still some tanneries which use oak bark (e.g. Croggan's at Grampound, Cornwall) and are thriving, and there are rustic furniture makers who use the peeled oak branches. In parts of the Midlands where hedges are cut and laid, there is some demand for stakes and binders. Gate hurdles are still made from ash (e.g. in Wiltshire), and a number of wattle hurdlemakers still work some of the southern hazel coppices. However, the total demands of all these traditional uses and their modern modifications require at most 1000–1500 ha of coppice, an insignificant amount nationally. Only the chestnut fencing markets are thriving, maintaining chestnut coppice in south-east England and the Herefordshire area, but this is not strictly a traditional market.

There are several curious features about the surviving coppice crafts. For example, the demand is now so low that few woodland owners grow material especially for them, and the surviving craftsmen cannot easily find suitable material. Demand for the product exceeds supply, but this does not evidently allow people to raise prices or expand output. Furthermore, the product has gone up-market. The shoes and surgical equipment made from oak-tanned leather is expensive and high quality, but formerly this was the only form of tanning available to all and sundry. Thatching, for which hazel spars are needed, is no longer everyman's roofing material, but the choice of the relatively wealthy. The few practising craftsmen live, not in rose-shrouded thatched cottages, but in council houses.

Any attempt to maintain traditional crafts and markets must accept that we no longer live in the social and economic climate in which they developed. One possible method has been tried at the Bradfield Woods (Suffolk) which is managed as a nature reserve

285

Fig. 18.2 Traditional management is not yet dead. In Cranborne Chase, Dorset and Wilts, hurdles are still made from hazel coppice grown with oak standards. Using a screen made of completed six foot hurdles and brushwood bundled for pea sticks, the craftsman is well on the way to completing a five foot hurdle which will probably be used as a garden screen. With some ash and birch in the mixture, this stand appears to conform to type 3A, acid pedunculate oak–hazel–ash woodland. (Photo. P. Wakely, Nature Conservancy Council)

in association with Whelnetham Woodwork Ltd, manufacturers of rakes, scythe snaiths, mallets, tent pegs, thatching spars etc. The wood supplies much of the factory's needs, but not all. Likewise, the factory does much of the coppicing, but not all: they cut what is useful and volunteers cut the rest and burn the rubbish. Demand for rakes, etc., exceeds the supply, but the factory remains small, with an uncertain future. Not only must the owner philanthropically accept that the business may be unprofitable, but it is hard to find men who can do the work. Nevertheless, this is a model which, if it succeeds, is an ideal solution to nature conservation by traditional management.

Another approach might work with the wattle hurdlemakers (Fig. 18.2), groups of whom work woods in Cranborne Chase (Wilts and Dorset) and Kings Somborne (Hants). The markets are good, though the hurdles are now used more as garden and motorway reservation screens than for folding sheep. Most craftsmen are now in their forties or older, apprenticing having effectively ceased in the 1950s, and the hazel coppices are becoming less easy to work. Other problems exist (Peterken, 1972) and, if nothing

Fig. 18.3 Resumption of traditional management in Monks Wood National Nature Reserve, Cambs. Jeremy Woodward, the warden, is pegging down a hazel shoot in the expectation that it will root and eventually form a new hazel stool to fill a blank space in the coppice underwood. Whilst there is often a strong case for coppicing parts of nature reserves, overgrown coppice (in the background) presents many problems. Not only are living stools moribund, but many are dead. Bramble often smothers new growth. (Photo. P. Wakely, Nature Conservancy Council)

is done to change the circumstances of hurdlemakers, their craft will continue to decline. However, an actively coppiced hazel underwood is good for pheasant-rearing and there are still estates which would employ a gamekeeper/hurdlemaker. Hazel underwood nature reserves would prosper if the warden could also make hurdles. And, if the hurdlemakers could somehow be re-organised from their traditional self-employment into groups, then the market might be good enough to sustain them commercially and give better prospects for newcomers.

18.3.2 Simulating traditional management: coppicing for nature conservation

Maintenance of traditional management by maintaining the markets and crafts is an intricate business, even though it is richly rewarding when it succeeds. An alternative, which does not involve the livelihoods of any individuals, is to maintain the management without the markets, i.e. to coppice the wood and not sell the product. Obviously this is possible only with volunteer labour or as a direct charge on a conservation organisation, and it can only be limited in application. In practice it will be limited to nature reserves, and even here only a small part of the total area is likely to be treated. In the relatively large reserves of Monks Wood (Fig. 18.3) and Hayley Wood, amongst the first reserves in which coppicing was restored for purely nature conservation reasons, only about 10–15% of the total area is coppiced.

Several problems have been encountered, but finding volunteers has not been one of them, thanks to the British Trust for Conservation Volunteers and its local equivalents. Most woods are derelict and must first be restored to a proper coppice rotation. Many stools will be either dead or moribund, and usually they bear much heavier growth than they did when cut in rotation. Volunteers are willing, but not generally skilled, so some stools are slightly damaged. Disposal of wood is no problem in the short term: if it cannot be given away as firewood or used on the reserve, it can be burned with the brushwood or stacked to augment the supply of dead wood. Burning and stacking however, will eventually eutrophicate the soil and encourage the growth of species which are more characteristic of hedgerows and wastelands: the sites of decayed wood stacks and long-forgotten bonfires are often marked by a patch of nettles.

There are now probably about fifty reserves in which 'conservation coppicing' is undertaken, and in more than half of these there is no significant market for the wood. Most have started so recently that the first cycle of cutting is not yet complete. The second cutting should be much easier than the first (or restoration) cutting, and it may also be possible to sell the product: once derelict hazel has been brought back into rotation it can be used for thatching spars (as in Monks Wood) or wattle hurdles.

Coppicing without sale of produce is an expedient rather than a solution. It must always be better to find a market, for this brings in money which can be used for other aspects of reserve management and enlarges the area which can be coppiced.

18.3.3 New markets for coppice wood

New uses for coppice wood offer the only chance that coppicing will remain a significant feature of woodland management in Britain, rather than an interesting relic confined to a tiny minority of British woodlands. Coppice-grown wood is used for pulpwood, chipboard, temporary fencing for pipelines under construction, pallets for mechanical handling of goods, mining chocks, and turnery products such as brush handles (Garthwaite, 1977). The scale of these uses can be illustrated by the demand for hardwood pulpwood, which in about 1972 was equivalent to 40 000 ha of coppice managed on a 20-year rotation. The mills at Sudbrooke (Gwent) and Sittingbourne (Kent) tended to rely on a non-renewable resource, namely coppice growth cut when woods are re-forested with conifers, and on branch wood from hedge and woodland timber trees. Both have instituted registration schemes designed to secure the long-term supply by guaranteeing a market and a reasonable price to the grower. They obtain most of their wood from within 100 miles of the mills, i.e. mainly from southern England and south-east Wales, so in northern East Anglia and from the Midlands northwards it is necessary to seek other markets for coppice wood.

The pulpwood market has the great advantage that mixed species can be accepted in dimensions normally found in derelict coppice. Indeed, only elm is quite unsuitable for pulping. The brochure for the Bowaters Hardwood Pulpwood Registration Scheme showed as an ideal crop a picture of a derelict stand of mixed coppice in the Bridge Woods at Rochester, which could easily have illustrated a woodland nature reserve leaflet. Likewise, a picture of the same stand after cutting could equally have illustrated either the pulpwood crop of the 1990s or successful nature reserve management. Not surprisingly, the coppice wood cut on some national nature reserves is sold for pulp, e.g. Ham Street Woods (Kent). The other markets may not always be so catholic; for example, the brush companies prefer birch, alder or sycamore.

Inevitably there are disadvantages. Small growth cannot be used, and the timber merchant may cut only the material he can sell so that the owner may have to tidy up afterwards. More importantly, demand for brown corrugated paper or cigarette packets may fluctuate, and so may the cost of alternative hardwood sources (branch-wood from timber merchants and imported pulp). The result is a disconcertingly fast-changing market for coppice logs, which tends to discourage owners from any long-term commitment to these outlets. This problem is accentuated by the Forestry Commission's general refusal to recognise coppicing as a valid means of restocking for grant-aid.

Broadleaf trees produce more cellulose than conifers in some climates, but a much higher proportion is in leaves and twigs. If all this growth could be harvested, the productivity of broadleaf woods could be greatly increased. Coppicing on a short rotation may be one method of whole-tree harvesting which can be practised in the future (Crowther, 1977).

The present high price of oil and general uncertainty about the future cost of any energy source has stimulated interest in wood-burning stoves, especially those which can also burn other fuels. Present demand is met without apparent difficulty from hedgerow trees, notably dead elms, but in future, suppliers may turn to the remaining overgrown coppice woods, reviving in the process a traditional market, but for a new technology and in new social and economic circumstances. It would be ideal if this and other modern markets could create a commercial justification for maintaining semi-natural woodlands as coppice on a wide scale.

18.4 Alternatives to traditional management

Even if coppicing is restored to and maintained in as many woods as the costs and markets can support, the majority of special woodland sites and areas cannot now be managed as coppice or as natural woodland. Alternatives are needed which conform to the basic objective of national forestry policy – timber production – whilst retaining the features which are essential for nature conservation.

Alternative management systems for former coppice woods should seek to retain the main feature which survived through traditional coppice management, i.e. stands composed of species native to the site growing in more or less natural mixtures. Two general possibilities are available, both of which result in high forest:

(a) to promote coppice directly to high forest, and
(b) to plant native species at wide spacing.

Coppice can be promoted directly to high forest by neglect or by necessary singling and thinning. Some coppices have indeed been neglected long enough to develop into a form of high forest, but the results are seldom of good form or valuable as timber. Other coppices have been stored or promoted to high forest stands of respectable form by cutting out the less valuable species and leaving only one stem on each stool. Most of the stands which have been treated in this way were oak or beech coppice, and the result of storing has been a fairly uniform stand in which the associated species are present but sparse. More recently, attempts have been made to promote mixed coppices to high forest by thinning and singling, leaving a mixed stand containing all the species which were present in the unthinned canopy and an underwood enriched by fresh coppice growth. The results in Lincolnshire (Fig. 18.4), where the Forestry Commission has thinned 30–40 years coppice growth of mainly ash–lime stand types 4A and 4Ba into apparently worthwhile stands of mixed native broadleaves, have been encouraging. The cost of the conversion was met by sale of thinnings for temporary fencing, pallets and mining chocks, and eventually it is hoped that the main canopy species of pedunculate oak, birch, ash and small-leaved lime will regenerate naturally. Similar treatment of mixed broadleaves on a private estate at Easton Hornstocks (Northants) is also developing well.

The second course would be to clear-cut the coppice, plant oak, ash or any other species which grows naturally on the site, and allow the former coppice to survive as an underwood beneath the plantation. The 'free-growth' of widely spread oak (Jobling and Pearce, 1977) presents interesting possibilities, not least because a rich underwood develops which could be cut as a pulpwood crop in favourable circumstances. Indeed, the free growth idea seems only a shade different from coppice-with-standards. At the end of the rotation a new crop can be planted if natural regeneration is insufficient.

Whichever method is used to change from coppice to high forest, it is preferable to restock the mature stands by natural regeneration (Principle 10). The high forest system adopted need not be even-aged, and indeed there are advantages in an irregular structure (Section 17.4).

Fig. 18.4 Great West Wood, Lincs, a fine stand of acid birch–ash–lime woodland (stand type 4A). Long neglected as coppice, this stand had already developed a quasi high-forest structure when, in 1976, the Forestry Commission started a programme of thinning. Under an agreement with the Nature Conservancy Council this wood will be treated as native broadleaf high forest, thus yielding useful timber whilst retaining the semi-natural composition. (Photo. R. V. Collier)

Some guidance is necessary concerning the type of conversion system to apply in different coppices. Planting is generally more appropriate to the least-natural coppices, and to stands where the stock of stems capable of growing into good timber trees is insufficient. Promotion is preferred in the most-natural stands. Thus, promotion should be the first choice in the mixed coppices, whereas planting will

Fig. 18.5 Wood pasture in Bradgate Park, Leicestershire. A survival of a medieval deer park, fallow (shown) and red deer still graze the rough pasture beneath pollarded oaks. Several hundred years old, their 'high tops' in Shakespeare's vivid phrase 'bald with dry antiquity', the oaks will nevertheless live a long while yet, but the complete absence of successors bodes ill for the long term survival of the mature oak timber habitat. The pine planted in a cage (to the left) represents a laudable attempt to establish a new generation of trees, but the choice of species could hardly have been less appropriate to the character of the site and the needs of nature conservation.

be more acceptable or necessary in the uniform hazel coppices of the south and the oak coppices of the north and west.

The likely results of large-scale conversion to native broadleaf high forest can be seen in the Chilterns, where the ancient coppices were converted 200 years ago. There, however, beech was evidently planted well beyond its natural edaphic range and favoured exclusively in silvicultural operations, so the woods are now more uniform than they might have been.

Efforts have recently been made to find a suitable management system for the native pinewoods which

takes into account their high value both as semi-natural woodland and as timber resources. In the larger woods at least a zonation system seems acceptable. The most natural parts would form the 'Strict Reserve Zone' where the stand is allowed to develop naturally. This would be closely linked to the 'Extraction Zone', from which timber extraction would be permitted, but where regeneration must be natural: this, in effect, is the recent traditional pattern. The 'Planting Zone', extending over the rest of the existing pinewood area, would be managed primarily for timber production, but only pine of

291

local stock would be planted. (The Extraction and Planting Zones broadly correspond to the proposals above for old coppice woods.) Allowance for the expansion of these woods has been made under Basis III Dedication Schemes by supplementary grants for planting local native pine stock up to 1 km beyond the existing woodland boundary.

Unenclosed forest stands are few and important enough for most to warrant management as natural woodland (Principle 13), e.g. the New Forest. The general priority in the parklands is to retain the existing trees (Fig. 18.5) – young, mature, moribund and dead – as long as possible and to start a new generation of oaks, beech or whatever species are present in the oldest generation.

18.5 Nature reserves in woodlands

Nature reserves are areas within which management for nature conservation takes top priority. Most woodland nature reserves are ancient, semi-natural woodland, and the majority are managed by organisations for whom nature conservation is the primary objective. Other objects of management can be pursued on nature reserves; indeed it is entirely possible to take a crop of timber from some.

Nature reserves, like nature conservation in general, fulfil several functions, though not necessarily all on the same ground. Some are used for research; others for education and public enjoyment; and many protect rare species, natural features or rich communities containing numerous species. Most combine these functions in various ways and few require such careful protection that access is forbidden. Many serve several functions beyond nature conservation which themselves require restrictions on access, notably nature reserves in which pheasants are 'preserved' (Fig. 18.6). Private woods which are being managed as nature reserves by agreement with, say, a County Naturalists' Trust are understandably open to the public only on special occasions, even though members of the Trust can visit them at any time.

The very existence of nature reserves implies that management for other objectives does not satisfy all the needs of nature conservation. This is inevitable to some extent, for nature reserves are sometimes

required to maintain natural conditions which, by definition, must not be subject to management. Under traditional management the gap between the needs of timber production and nature conservation was relatively small, but under modern 'tree farming silviculture' the gap is so large that the only possible response has been to establish nature reserves (and also amenity woods and game coverts) as fast as possible. Most have been established by County Naturalist Trusts, the Nature Conservancy Council, the Woodland Trust and private landowners acting with these organisations, but local authorities, National Parks and the National Trusts have all contributed, and the Forestry Commission itself has followed its own early example of an ecological reserve in Lady Park Wood (Gwent) by reserving many small patches of woodland and open ground within plantations primarily for nature. In 1972 the total area of woodland National Nature Reserves in the English lowlands was 1950 ha, or 0.3% of all woodland (Steele, 1972b).

Once nature reserves are established, decisions must be taken on their management. Some areas are left unmanaged as a positive decision to achieve natural woodland (Principle 13). Others will be restored to traditional management or maintained as such (Principle 14), whilst some are converted to other systems which retain the important features (Principle 15). Subsidiary habitats, such as rides, glades and ponds, will be maintained or created. Small populations of rare and local species will be located and monitored, and, if they seem threatened and the warden thinks he knows their requirements, the habitat may be manipulated for them alone. Large areas are often left unmanaged because the resources for management are not available or there is no urgency to decide on more active measures: this is at least consistent with Principles 6 and 7, but problems may accumulate if eventually the stand is to be actively managed.

Each nature reserve should have a management plan containing:

(1) Description. Lists, maps and descriptions on vegetation, flora, fauna and soils. An account of past management and other events on and near the site.
(2) Evaluation and objectives. What features are considered to be valuable? What, therefore, are the

Fig. 18.6 Lime coppice in Swanton Novers Great Wood, Norfolk, both a National Nature Reserve and a valuable pheasant preserve. The stools, which bear one season's growth, have been cut well above ground level on roughly a ten year rotation. Many stools consist of several fused coppice shoots, apparently promoted from low-cut to high-cut coppice in the 19th century. In the background the line of scrub on the wood bank also consists mainly of lime. (Photo. P. Wakely, Nature Conservancy Council)

objects of management? Review management options and resources available.

(3) Prescription. Statement of the programme of management. This will have to be subject to periodic revision as management experience accumulates, resources change and new knowledge of the site becomes available.

Nature reserves have come to be regarded as something special, set aside from other forms of management (even though many are both nature reserves and, for example, game coverts or local recreational amenities). This exclusivity seems undesirable, except where especially vulnerable species or features have to be protected, because it is unnecessary and may tend to diminish public support for nature conservation. In particular, nature reserves are usually written-off as sources of utilisable timber even though many would benefit from a limited and controlled programme of felling, and extraction and from the resulting income. At the National Nature Reserve of Ham Street Woods (Kent) the woodland is coppiced regularly and the poles are sold for pulp. The main need in nature reserves is to know exactly what to aim for; to apply the fifteen principles strictly; and to recognise legitimate opportunities for harvesting timber. If, correspondingly, more timber-producing woods were managed with greater regard for those principles, then the need for nature reserves would diminish and the integration of nature conservation and timber production would be more effective.

Chapter 19

Integration of nature conservation with other objectives of woodland management

19.1 Introduction

The principal objective of forestry is to grow timber as economically as possible. Foresters in general try to maximise productivity and seek further ways of increasing it, hoping – and no doubt believing – that their methods and level of production are indefinitely sustainable. In theory all woodland could be devoted to growing as much timber as possible, but, under these conditions, most British woodland would take the form of coniferous plantations, native broadleaves would hardly ever be used, and semi-natural woodland would cease to exist. Some hardwoods would still be planted, but only where there is a chance of growing top-quality timber. If the same 'growth philosophy' were applied to land use in general, the medium-sized and small woods, which are now part of our farming landscape, would rapidly be cleared for arable cultivation.

Changes in these directions have, of course, been taking place, but they have not been pushed to their extreme conclusion because the harmful consequences are widely recognised. Amenity organisations become restless at the great changes in the landscape. Sportsmen would object to the deterioration of their shoots. Minor wood products would be difficult to obtain. Forestry would become too mechanised and uniform and many foresters would regret this reduction of their craft. Nature conservationists would object, not just to the elimination of semi-natural woodland, but

to the great reduction in wildlife generally. Many people would sense and regret the taming of wild places and the loss of distinctive character in the British scene. Accordingly, and for a variety of reasons, it is accepted that a proportion of the potential timber production must be sacrificed in the interests of amenity, nature conservation and other interests. The problem is to decide how and where it is done, how much is sacrificed and who pays.

The question 'How?' has already been discussed in Chapters 15–18, in which a general strategy of integrating management objectives is advocated, but with some segregation of objectives on the ground. The difficult question 'Who pays?' is briefly considered at the end of this chapter. The main point now is to consider 'How much?'.

19.2 Census of woodlands

The Forestry Commission has completed censuses of woodlands for the years 1924, 1947/49 and 1965/67, and they will complete another in the early 1980s. Two further censuses, in 1938 and 1942, were completed for several counties, but never finished. Earlier, the Board of Agriculture received returns from landowners giving, among other details, the areas of coppice, young plantations and other woodlands, so somewhat rudimentary woodland censuses are available for 1895, 1905 and 1913/14. The total area of woodland in Britain at these dates is given in

Table 5.1, where sources are also quoted. Some of the original returns and ground survey records on which the published figures were based still survive in the Public Record Office at Kew (PRO F22, 23, 30), but many have been destroyed. All the original records from 1924 have gone. Of the 32 counties wholly or partly surveyed in 1938, only the original records from Lincolnshire and Montgomery have been retained as examples. Likewise, only the Dorset and Cardiganshire records survive from 1942. Records from that apogee of woodland censuses in Britain, the 1947–49 survey, are fortunately better preserved: all the records from private woods survive and became open on 1 January 1980. The Forestry Commission does not hold records of any of its woods at these dates (see Section 17.8).

Essential though these surveys are to any quantitative consideration of British woodlands, they were of course designed to produce information on the production and standing reserve of timber, and their value for considering the needs of nature conservation is somewhat limited. The latest census, is, inevitably, out of date, though the Forestry Commission has produced a few more recent figures. The returns to the Board of Agriculture were probably incomplete, and, on the evidence of the original returns for the Rockingham Forest portion of Northamptonshire (Peterken, 1976), over-simple and inaccurate. The greatest problem, however, is the form of the record, which was not made in terms which are readily translated into ecologically useful categories. The emphasis is on plantations, stocking, growth potential and main timber species. Some coppice is recorded as such, but much has been disguised as 'scrub' and not clearly distinguished from secondary scrub. Minor species are often ignored, if not in the record, then in the published figures. Understandably enough, no distinction is drawn between plantations and semi-natural woodland, nor between ancient and recent woodland.

19.3 The main categories of British woodland

Every patch of woodland falls into one of the four main categories of Fig. 3.1, and the first step towards quantifying the needs of nature conservation is to apportion the total woodland area between these categories. In 1965/67 there were 1743 kha of woodland, of which 656 kha were in Scotland and 1087 kha were in England and Wales. In 1895 the total area was 1104 kha, so at least 639 kha of existing woodland has originated roughly since the start of the 20th century. This is a minimum figure which assumes that all the 1895 woodland survived to 1965/67. The actual survival rate was calculated from three areas in the east Midlands, where it ranged from 84% in Rockingham Forest to 95% in Central Lincolnshire (Peterken, 1976). These figures were believed to be higher than the national average, and in any case allowance had to be made for the fact that they referred to the period 1886–1972, so the national survival rate may have been nearer 80%. Taking this figure, 883 kha of the 1895 woodland survived to 1965/67, leaving the area of recent, largely 20th century woodland as 860 kha.

The 1895 woodland was a mixture of ancient woods and secondary woods arising in the 17th–19th centuries. The proportions of these two categories are unknown and not easy to estimate. In the east Midlands study areas 81% of the woodland present about 1886 was ancient (ranging from 57% in west Cambridgeshire to 87% in Rockingham Forest), but this figure is likely to be well above the national average, if only because the study areas were centred on groups of ancient woods. Guessing that the national average was about 65% produces a total area of ancient woodland surviving in 1965/67 of 574 kha, leaving 309 kha of pre-20th century recent, secondary woodland. On this admittedly shaky basis, therefore, the 1743 kha of woodland in 1965/67 consisted of:

574 kha of ancient woodland;
309 kha of secondary woodland, 17th–19th century origin;
860 kha of secondary woodland, 20th century origin.

How much of this was semi-natural? In the east Midlands study areas 52% of ancient woodland still supported semi-natural stands in 1972/3 (ranging from 48% in central Lincolnshire to 85% in west Cambridgeshire), whilst the balance of 48% was under plantation, most of recent establishment. Applying these figures nationally, the area of ancient,

Table 19.1 The estimated extent of various woodland categories in about 1965/67.

Woodland category	Thousands of hectares	Proportion (%)
Plantations		
(a) Recent, secondary plantations on land which was unwooded in about 1900	817	47
(b) Recent, secondary plantations created in the 17th, 18th and 19th centuries	295	16.5
(c) Ancient woodland, now under plantations	275	16
Semi-natural woodland		
(d) Recent, semi-natural woodland, mostly arising in the last 100 years	59	3.5
(e) Ancient, semi-natural woodland, much of which was coppiced until the 19th and 20th centuries	299	17
Total woodland area of Britain	1743	100

Table 19.2 The estimated proportions of various woodland categories in Britain now and in the recent past and future.

Woodland category	1965/67	1978	2025
Plantations			
(a) Recent secondary plantations	47	55	76
(b) 17th–19th century plantations	16.5	14	7
(c) Replanted ancient woods	16	14	8
Semi-natural woodland			
(d) Recent, semi-natural woodland	3.5	3	2
(e) Ancient, semi-natural woodland	17	14	7

semi-natural woodland was 299 kha, leaving 275 kha of plantations on ancient woodland sites.

Estimates of the proportion of recent woodland which was semi-natural in 1965/67 cannot be made by extrapolation from the east Midlands, where such stands are unusually few. In any case the distinction between semi-natural stands and plantations is even less clear than in the ancient woods. A guess therefore has been made that 5% of both pre- and post-1895 recent woodland is semi-natural, i.e. 59 kha, leaving 295 kha of pre-20th century secondary plantations and 817 kha of 20th century secondary plantations.

These estimates imply that the total area of plantations in 1965/67 was 1387 kha, or 80% of all woods, a figure which can be roughly checked by other sources. The census showed that 73% of all woodland was classified as high forest. Ryle (1969) notes that in 1965 the Forestry Commission had 627 kha 'under forest crops' and that by then they had made grants to private owners to plant about 228 kha, a total area of 855 kha evidently planted since about 1920. Comparing this with the estimated 1092 kha of 20th century plantations (both afforestation and reforestation of ancient woods), and making allowances for slight

differences in the interpretation of terms and the facts that some ancient woodland was planted before 1920 and some post-1895 planting was not grant-aided, it seems that the summary in Table 19.1 does at least give reasonable estimates of the amounts of major woodland types in Britain about 1965.

Since 1965/67 afforestation and woodland clearance have continued. In 1978 (FC, 1978) Britain had 1998 kha of woodland, so a minimum of 255 kha more secondary plantations had been added. Allowing for a limited amount of clearance and conversion of semi-natural woodland to plantations, the present-day proportions of the main woodland types must be roughly as in Table 19.2.

19.4 Ancient, semi-natural woodland

This, the most important category for nature conservation, occupied around 299 kha in 1965/67. There were about 275 kha of ancient woodland under plantations, some of which were broadleaved and, like Bedford Purlieus, not by any means lacking in value. Can we identify and quantify priorities within these?

A Nature Conservation Review (Section 14.1; Ratcliffe, 1977) identifies 234 woodland sites of national importance, whose area is given as 67 kha. However, some of this is not wooded: 1.5 kha are parkland which would not have been registered as woodland in a forestry census, and about 10.2 kha are heathland mixed with native pinewoods (using the figures of Goodier and Bunce, Table 1, in Bunce and Jeffers, 1977). Furthermore, about 4.3 kha of coniferous plantations in the New Forest were included in order to retain a rational boundary to the site. Allowing for other small patches of plantation in many other sites, the area of ancient, semi-natural woodland in NCR sites is about 50 kha. Inevitably, more important sites have been identified since 1970, when the main NCR lists were compiled, because gaps in the representative coverage have come to light and new important sites have been found. Allowing for this a reasonable estimate of the area of nationally important woods must be around 60 kha. Some of these woods are nature reserves and the rest are Sites of Special Scientific Interest (SSSI).

The next tier of importance is represented by woodland SSSI not included in the NCR, which are judged to be important locally and regionally, but not nationally. (They may become important if the corresponding NCR site is damaged.) In 1974 about 124 kha of woodland were contained within statutory sites – National Nature Reserves, Forest Nature Reserves, Local Nature Reserves, SSSIs – but this included at least 13 kha of coniferous plantation on land which had originally been scheduled as upland moorland SSSIs. Allowing also for additional woods scheduled since 1974 and de-scheduling of a few woods which have been damaged, the area of semi-natural woodland in statutory sites must be around 120 kha. This includes the 60 kha of NCR grade sites. The total area of ancient, semi-natural woodland has probably fallen by 5% from its estimated extent in 1965 of 299 kha, so the extent of such woodland which enjoys no statutory recognition is about 164 kha.

19.5 Future developments

The Forestry Commission as Forestry Authority believes that much more land should be afforested and is advocating this point of view strongly, notably in the *Wood Production Outlook* (FC, 1977). They naturally have the full backing of private timber-growers and wood-based industries, as well as the Centre for Agricultural Strategy (1980). The Forestry Commission estimates that 1660 kha in Scotland (Locke, 1976) and 650 kha in England and Wales (FC, 1978) of mainly upland grassland, heath and peatlands are plantable, but they recognise that this land is currently used as sheep ranges, deer forests,

grouse moors and nature reserves, and therefore expect only about half to become available for planting (Stewart, 1978). Nevertheless, their assumption in the *Wood Production Outlook* is that 1700 kha will be afforested by 2025 AD, and that 1500 kha of this will be in Scotland, where the proportion of land under woodland will rise from 10% to 30%. Assuming (i) that afforestation does in fact take place on this scale, (ii) that no woodland elsewhere is cleared, and (iii) that only a small amount of semi-natural woodland is converted to plantations – the latter assumptions both being highly optimistic – then by 2025 AD about 76% of all British woodland will consist of plantations on land which was not wooded at the start of the 20th century (Table 19.2). In fact, the *Wood Production Outlook* assumes that 100 kha of semi-natural woodland will be converted to plantations by 2025 AD.

19.6 Sacrifice of timber production

The level of sacrifice of timber production required in the interests of nature conservation can now be crudely estimated. Starting with an extreme and improbable case in which (i) no timber is henceforward cut in semi-natural woods, (ii) all existing plantations are managed to produce as much timber as possible, and (iii) no further afforestation takes place, and assuming that all land is potentially equally productive of timber, the long-term sacrifice of timber production would be 17%. Alternatively, if afforestation continued as assumed in the *Wood Production Outlook*, but the scenario remains otherwise unchanged, the sacrifice would be 9%. The short-term sacrifice would be greater, due to the bias of younger age classes to the 20th century plantations.

These approaches are impractical, not least because they would create an imbalance by virtually extinguishing the supply of British hardwoods. Furthermore, it is clear from Chapters 15–18 that nature conservation and timber production need not be so starkly segregated on the ground. There is no need to sacrifice all timber production from all semi-natural woods. Timber can be grown in the majority without significant damage to nature conservation interests, provided that the species grown are native to the site, that natural regeneration and coppice growth are

Table 19.3 Estimates of the relative timber production possible from various categories of British woodland, taking account of the needs of nature conservation.

Woodland category	Area in 1978 (kha)	Relative productivity (%)	Production index	Loss index
Plantations				
(a) Recent secondary plantations	1101	99	1090	11
(b) 17th–19 century plantations	281	80	225	56
(c) Replanted ancient woodland	275	60	165	110
Semi-natural woodland				
(d) Recent semi-natural woodland	59	80	47	12
(e) Ancient semi-natural woodland				
Sites of national importance	60	10	6	54
Other SSSI	60	30	18	42
Other ancient semi-natural woods	164	50	82	82
Future afforestation				
(f) Plantations on moorland, 1978–2025	1700	99	1683	17

Total of production indices a–e = 1633, or 82% of the potential 1998.
Total of production indices a–f = 3316, or 90% of the potential 3698.

Note:
(a) Relative production of 100% implies that all stands in a category should be managed to produce as much timber as possible.
(b) Production index is the product of area and relative production.
(c) Loss index = area – production index.
For further explanation, see text.

preferred to planting wherever possible, that efforts are made to develop a mature, mixed-age structure in high forest stands, that special care is taken of rare species and special features, etc. (Chapters 17 and 18).

An alternative estimate, which recognises the possibilities for integrating nature conservation and forestry, can be made by taking the area of each woodland category, estimating the proportion of potential productivity possible if management adheres to the general principles of Chapters 15–18 and summing the results. The areas given in Table 19.3 for each category of woodland are based on foregoing calculations, but have been adjusted to bring them up-to-date by assuming (a) that since 1965/67 the areas of ancient, semi-natural woodland and 17th–19th century secondary plantations have both declined by 5% due to clearance and replanting; and (b) that any losses of plantations on ancient woodland sites and recent semi-natural woodland are balanced by gains elsewhere.

The estimates of relative productivity need some explanation. A category which is managed to produce as much timber as possible has a relative productivity of 100%. No category reaches this level, for even in the new plantations one asks for small patches to remain unplanted, some scrub to be allowed along ride sides, and for a small proportion of broadleaf species to be planted, but the total sacrifice sought is very small indeed, perhaps as low as 1%. The area of rides is not, of course, counted as a sacrifice for nature conservation, valuable though rides are, because they are necessary for access, timber extraction and fire precautions. Nor is the need to avoid afforesting moorland which is already rich in wildlife, for this affects the distribution of afforestation, not necessarily the amount. At the other end of the scale the ancient, semi-natural woods managed primarily for nature conservation cannot readily produce much timber, but some should be managed on traditional lines or derivatives from them, and these would yield some timber, say 10% of potential. The majority of ancient, semi-natural woods should be managed to produce timber from native species: the relative productivity is perhaps rather high at 50%, but this makes allowance for woods in special areas (Section 15.4) where timber production can be more intensive without damage to nature conservation interests. The ancient woods which have already been converted to coniferous plantations can remain coniferous without further damage, but stands still broadleaved should be kept as native broadleaves: hence they will produce perhaps only 60% of potential. Finally, the recent semi-natural stands and the older secondary plantations, being relatively unimportant for nature conservation, can be managed more intensively still for timber, say 80% of potential.

On this basis the proper integration of forestry and nature conservation would result in a sacrifice of 18% of potential timber production if no further afforestation takes place, or 10% if, as is more likely, afforestation goes ahead as indicated.

The strategy summarised in Table 19.3 makes an important assumption, which may not be accepted by all those with an interest in nature conservation, namely that the greatest sacrifices should be made in ancient (and especially ancient, semi-natural) woodland, whilst minimal sacrifices are sought in recent and future secondary plantations, i.e. the sacrifice should be very unevenly spread. An alternative view might be that the new plantations cover an enormous area and that even a slight improvement in their wildlife carrying capacity would have a big effect. A much higher broadleaf component might be sought and extensive areas might be left unplanted or allowed to develop into semi-natural scrub. This would, of course, be desirable for wildlife, but if the relative productivity of all new plantations were reduced only as far as 90%, the overall sacrifice would have to be increased well beyond 10% if the ancient and semi-natural woods were to be properly considered. Given that there must be a limit to the resources available to nature conservation, I would prefer them to be concentrated into these woods which are already rich in wildlife and possessed of irreplaceable natural features, rather than dissipated over commercial plantations.

Another important feature of this strategy is that it accepts that there will be considerable further losses of wildlife and natural features in the interests of increased timber production. Most will be in the uplands where large tracts of moorland will be replaced by plantations, but there will also be significant losses in woodlands as the less-important semi-natural woods are managed more intensively. This leads on to another important point, that the proposed sacrifice is not actual but potential. No wood should be managed less intensively than hitherto: some woods can be managed more intensively without damaging nature conservation interests, but other woods cannot.

This strategy depends on two developments. First, the ancient, semi-natural woods are becoming almost insignificant nationally as potential sources of timber, because continuing afforestation is reducing them to a very small proportion of all British woodlands. Admittedly, since afforestation involves the loss of peatland and moorland, this implies that successful woodland nature conservation is to some extent in indirect competition with nature conservation in the uplands. Second, interest in growing hardwoods, particularly oak, is increasing, and there seems to be more willingness on the part of foresters to retain and manage existing broadleaf stands, rather than clear fell and replace them with alternative crops. This renewed interest is apparently a response to dwindling stocks of tropical hardwoods and the British over-emphasis on conifer planting in recent decades, which both indicate a rapid rise in the future price for hardwoods. These two developments combine to make realistic a policy of conserving ancient, semi-natural woods by active treatment of native trees for timber, rather than by preservation and reservation.

19.7 Discussion

The economics of nature conservation within forestry, which must take into account increased costs, decreased receipts and delayed income, are far too complex to discuss here, especially as numerous debatable assumptions must be made. Lorrain-Smith (1973) provides a valuable view of nature conservation from an economist's standpoint. Rudimentary calculations show that the cost of adequately taking account of the needs of woodland nature conservation (and other needs, see below) at 10% sacrifice of potential timber production is relatively small, though the absolute cost seems large. Britain imports 92% of her timber consumption at a cost in 1977 of over £2300 million, and produces 8% at home, saving presumably £190 million in imports. The annual sacrifice of 10% of potential home-grown timber thus costs around £20 million at present levels of production, which is less than 1% of timber consumption. Britain is likely to produce more than 8% of her timber demand in the future, so a 10% sacrifice of potential production levels in 2025 AD has a proportionately greater effect. Nevertheless, a 10% sacrifice of potential British timber production remains almost insignificant in relation to consumption of timber and timber imports. It could be afforded by cutting demand through more efficient use of timber by 1–2% of forecast.

It is important to recognise that this cost is not for

Table 19.4 Estimates of the area of all woodland (most of which was ancient, semi-natural) within statutory sites* in 1974, divided between various categories of owner (Nature Conservancy Council Files).

Ownership	Area (kha)
Nature Conservancy Council	2.7
County Naturalists Trusts	1.9
Society for the Promotion of Nature Conservation	0.3
Royal Society for the Protection of Birds	0.4
Forestry Commission	37.0
National Trusts	6.9
Ministry of Defence	1.3
Local Authorities	3.8
Private	69.3
Private, subject to common rights	0.8
Total area	124.4

* NNR, SSSI, FNR, LNR.

woodland nature conservation alone. Many semi-natural woods are in National Parks and Areas of Outstanding Natural Beauty, whilst many others are prominent features in less well-wooded countryside or fall within 'landscape' zones of Structure Plans. Some are owned by local authorities or a comparable body as public amenities. Many semi-natural coppices are used as game preserves.

Furthermore, management to satisfy these amenity, recreational and sporting needs is similar to the management required for nature conservation. Pheasants prefer coppiced woodland to conifer plantations. Landscape is best conserved by placing an emphasis on broadleaved, deciduous trees, maintaining mature stands by managing on long rotations, small-scale felling and restocking, and generally by moderating the pace of change. Of course, some semi-natural woods are insignificant in the landscape, whilst some species-poor shelterbelts are prominent, but to a substantial degree the needs of nature conservation, amenity and, to a lesser extent, sport can be satisfied together. To take some examples, Swanton Novers Woods (Norfolk) are not only a National Nature Reserve but also a well-run private pheasant preserve. The landscapes of the lower Wye Valley near Chepstow, Borrowdale in the Lake District, Loch Lomondside and the Spey Valley near Aviemore are made not just by their spectacular land forms, but also

by their mature, semi-natural woodlands: all are included in *A Nature Conservation Review*.

A woodland management policy which involves substantial further afforestation (in the right places) and a 10% restraint on timber production (also in the right places) seems to me to be a reasonable balance between the nation's need to produce and create a reserve of timber, yet maintain amenity, wildlife and natural features. Indeed, I believe it to be generally acceptable. The major difficulty, of course, is 'Who pays?' because private individuals and organisations own the majority of semi-natural woodland and the great majority of the smaller woods (Table 19.4), and many will want compensation if they have to bear the unrewarding economics of growing native hardwoods for timber. Fortunately, many important semi-natural woods are owned by Local Authorities, the National Trusts, the Forestry Commission and other bodies, such as Development Corporations (e.g. Castle Eden Dene, Durham), Corporation of the City of London (e.g. Epping Forest, Essex) and the Crown Estate Commissioners (e.g. Windsor Forest, Berks), who can and generally do take a broad view of their public responsibilities. This is not the place to explore in detail the political, economic and social implications of various forestry strategies for the future, but it is worth noting some points which should be taken into account:

(1) Private landowners, who enjoy a disproportionately large share of national wealth, should be prepared to bear a disproportionately large share of the sacrifices needed to provide national needs for amenity and nature conservation. Indeed, the continued political acceptability of private landowning may depend on recognising these and other public responsibilities.

(2) Reasonable markets do exist for native hardwoods (Garthwaite, 1977). Some progress has been made with adapting silvicultural systems to modern needs, e.g. free-growth of oak (Jobling and Pearce, 1977). A market exists for mixed coppice, which can be used to make pulpwood (Stern, 1973). Judging by the increasing popularity of wood-burning stoves, there will soon be a good firewood market, which could be satisfied by managing small farm woods as coppice – not, incidentally, the newest idea under the sun!

(3) The Forestry Commission has provided a supplementary grant for planting hardwoods under the Basis III Dedication Scheme, and they have also introduced a Small Woods Scheme. These grants were strongly biased to planting: indeed they were couched in terms which pre-supposed even-age plantation forestry, rather than natural regeneration and coppice as a means of restocking, or mixed-age forestry. These schemes were a mixed blessing in other respects, too, for planting often resulted in sycamore, *Nothofagus* and other aliens replacing locally native species.

(4) The importance of mixed-age forestry in maintaining visual amenity was recognised by Troup (1928) and is still recognised by foresters whose concerns extend beyond timber production (Garfitt, 1977), yet a grant system based on planting and clear-felling provides no incentive. Furthermore, the grant is available for planting, thus providing an incentive to fell stands even if this is not otherwise desirable. A grant system which places greater emphasis on management – maintaining woods in a productive condition – irrespective of whether restocking is by planting, is desirable.

(5) Recent studies by the Countryside Commission have shown widespread apathy amongst farmers for small woods on their farms. Perhaps co-operative management and marketing would help to ensure their future.

(6) Tree Preservation Orders are a valuable, short-term restraint on rapid management changes and woodland clearance. They are made by Local Authorities, generally in the interest of landscape conservation. Criticism that they simply stop management and are self-defeating in the long-term has some justification, though not as much as some people make out: after all, woodlands can perpetuate themselves. TPO's need not be negative, however, as has been shown by some counties where they have been used almost as a consultative procedure directed towards agreement on a management plan.

In conclusion: adequate provision for nature conservation, landscape conservation and many other benefits can be made with a relatively small sacrifice in the intensity of timber production. This is possible only if the distribution of various forms of woodland management (and of new afforestation) is carefully considered at national and local levels: the sacrifices must be the right kind in the right places. Even so, this strategy involves considerable further losses in the amount and variety of wildlife and natural features. Nevertheless, these further losses, justified by the need to increase production of British timber, are acceptable if, at the same time, the remaining important woodlands and uplands are securely protected against intensified timber production. Some means exist for putting such a strategy into practice, but further measures are needed to ensure that the burden does not fall unequally on some private woodland owners.

Chapter 20

British woodland management in a European context

Asholt Wood, near Folkstone in Kent, is closer to the Forêt Dominiale de Guines, south of Calais in France, than to the woods in north-east Kent around Rochester and Dartford. Asholt is also ecologically closer to Guines, for both are broadly pedunculate oak–ash woodlands (Plaisance, 1963), whereas the Rochester Bridge Woods and Darenth Wood contain much hornbeam, chestnut and sessile oak. These two woods demonstrate and represent the natural continuum linking British and Continental woods, which is best appreciated through the floristic similarity between the woods of south-east England, Hampshire and southern East Anglia on one side and the woods of Brittany, Normandy and Picardy on the other. They also, however, represent important differences, for the fifty kilometres which separate them cross an important divide: France and some other continental countries have had a history of landscape development and forestry which has differed somewhat from that of Britain and left them with much more woodland, a far higher proportion of mature forest and a greater range of silvicultural systems.

The immense variety of woodland, forestry and nature conservation needs in continental countries makes any comparison with Britain extremely hazardous. Furthermore, since this comparison is based mainly on my own impressions of parts of Czechoslovakia, France and Switzerland, the conclusions are liable to be biased or worse. Nevertheless, the comparison is worth making, because significant differences of emphasis emerge which enable us to understand better the particular relationship which exists between forestry and nature conservation in Britain.

20.1 Land use and the pattern of woodland

Britain is much less wooded than most European countries (Table 20.1). British woods tend to be small, scattered and ecologically isolated, whereas most continental countries have densely wooded regions and large, individual woods. There are of course some small woods standing isolated within continental farmlands, but they form only a small proportion of the total woodland cover. This difference is a particular expression of a general feature, namely that the land use pattern in many parts of the Continent is a large-scale mosaic, whereas in Britain it is mostly a small-scale patchwork. For example, the agricultural districts of France tend to be more intensively farmed, with fewer small woods, hedges, scrub and grassland than their counterparts in Britain but some parts of the lowlands contain extensive tracts of meadow and pasture, and very large forests are found almost throughout the country.

Differences between Britain and the Continent are perhaps greatest in hill and mountain districts. The British uplands have until recently been largely tree-less expanses of grassland, heathland and peat whilst the few woodlands that remained were generally

303

Table 20.1 The extent of woodland in various European countries in about 1970.

	Forest area (kha)	Forest as % land area	Change 1950–70 (kha)				Change 1950–70 as % total area in 1950		Management type (% utilised forest)		Growing stock in managed forest by volume		
			Afforesta-tion	Clear-ance	Net change	Re-forestation	Afforesta-tion	Clear-ance	High forest	Coppice + coppice with standards	Conifer (million m³)	Broad-leaf	Broadleaf % by area in managed forest
Norway	8 900	29									425	88	
Sweden	26 400	64									1 966	322	
Denmark	470	11	40	—	+40	—	8.5	0	100	0	22	23	37
Ireland	275	4	?	?		+175			100	0	13	2	10
UNITED KINGDOM	1 840	8	640	50	+590	238	34.8	4.0	97	3	68	53	27
Netherlands	290	7	35	6	+29	25	12.1	2.3	88	12	16	4	29
Belgium	617	20	20	7	+13	133	3.2	1.2	65	35	40	31	53
Luxembourg	83	32	1	1	0	3	1.5	1.1	81	19	2	11	68
France	14 000	26	634	180	+454	724	4.5	1.3	51	49	481	826	71
West Germany	7 207	29	200	144	+56	77	2.8	2.0	94	6	722	300	31
East Germany	2 900	28									257	93	
Poland	8 500	28									857	192	
Czechoslovakia	4 600	36			+196						598	203	37
Switzerland	1 100	27									186	84	
Austria	3 700	45									580	101	
Hungary	1 500	16									14	160	
Italy	6 193	20	530	87	+453	227	8.6	1.5	41	59	114	172	79
Spain	14 100	28									259	177	
Portugal	3 000	34									84	82	

Sources:
Forestry Problems and their Implications for the Environment in the Member States of the European Community. I. Results and Recommendations. Information on Agriculture *25*. Brussels, Commission of the European Communities, 1976.
Forest Resources in the European Region. Food and Agriculture Organisation, Rome, 1976.

small and confined to well-drained, uncultivable land on the steeper slopes. On the Continent the major ranges of hills and mountains are especially well-wooded and the trees often cover the majority of the land surface (e.g. the Vosges). Furthermore, much of the unwooded hill land is managed as hay-meadows or seasonal pasture, neither of which is intensively grazed or 'improved'. These continental herb-rich upland grasslands are often well-stocked with woodland herbs, and contrast strongly with the prevalent acid grassland and heathland of western Britain.

Continental countries have remained well-wooded throughout historical times. Not only was natural, retrogressive succession less frequent but settlement and clearance came much later to much of central Europe. Ecological isolation in space and time of woodland communities has therefore not been as great as in Britain. Ancient woodland, which in Britain covers approximately 2% of the land surface,

may well cover 20% of France and 30% of Czechoslovakia. (Consequently, the concept of primary woodland has only limited application on the Continent.) Apart from a few major undertakings such as the afforestation of the Landes in south-west France, most secondary woodland on the Continent has developed near to extensive ancient woodland and on soils which have not been greatly altered by cultivation, i.e. upland grassland.

These differences have had several ecological effects on woodland communities. First, few woodland species are as rigidly confined to woods as in most of lowland Britain, but occur also in neighbouring grassland and may be more abundant on wood margins. Thus, *Primula elatior* may actually be more abundant in woodland clearings than in the woodland itself, as in the Beskydy Hills of northern Czechoslovakia. *Cephalanthera rubra* seems to be concentrated in woodland areas, but within such areas may be frequent in grassland as well as the woods, e.g. Savoy Alps. Second, the flora of secondary woodlands is not markedly different from that of ancient woods, possibly because (i) the secondary woods are less ecologically isolated from the latter than in Britain, (ii) soil changes during the period of clearance have probably been slight, (iii) the land use preceding secondary woodland more often allowed woodland plants to be present on the site, (iv) the population pressure of species in woodlands is greater. In some parts of the Continent, however, the flora of secondary woods does differ markedly from that of ancient woods, forming the exceptions which prove the rule. For example, small, isolated secondary woods in the agricultural districts of Picardy and Southern Slovakia have much the same field layer communities of shade-bearing, adventives and nitrophilous species as recent secondary woods in the east Midlands (Section 5.5). The maritime pine (*Pinus pinaster*) woods of the Landes are poor in woodland species, but one of the few ancient woods in the area, the Forêt Usagère de Biscarosse, has a conspicuously richer flora.

20.2 Woodland management

The history of man's influence on continental woods has followed the same broad course as in Britain. Natural woodland was progressively cleared or brought into some form of management based on native species. Woods yielded timber, fuelwood, pasturage and many other products to local communities, but many survived in a more natural state as hunting reserves. Traditional systems were modified in time, but they have long been in decline, and more intensive, 'modern' systems are replacing them.

This broad similarity, however, conceals some highly significant differences of detail. Natural woodland, which survived until the late Middle Ages in remote parts of the Scottish Highlands, survived much more extensively and much later in parts of Europe, forming the original 'wilderness' (Nash, 1973). In Czechoslovakia, for example, substantial tracts of forest in hilly districts remained unaltered, save for occasional hunting, until the 19th century. By that time, fortunately, foresters were aware of the value of retaining some examples of natural forest, so now a number of virgin forest stands are reserved. Virgin forest also survives in other countries, such as

Białowieża Forest, Poland. Each country has its own peculiar history of woodland ownership and administration. This has resulted in varying patterns of ownership. In Czechoslovakia most woodland is owned and managed by the state. In Switzerland, Belgium, West Germany, France and indeed most European countries, much woodland is owned by local communities, and managed for them by state or local authorities. Britain has very few common woods and, until the recent rise of the Forestry Commission, most woods were privately owned.

British woodland management differs from that of continental countries in several ways, notably in the relative significance of afforestation and the management of long-established stands. Britain has been preoccupied with increasing the area of woodland, whilst long-established woodlands have been neglected, cleared or converted to even-aged plantations which somewhat resemble the products of afforestation. Continental countries have also in some instances afforested substantial areas (Table 20.1), but the forestry emphasis has been on the management of existing, often broadleaved stands. Moreover, the silvicultural systems adopted have frequently been based on long rotations, native species, small-scale fellings and sometimes mixed-age structures, whereas British forestry is largely concerned with even-aged coniferous high forest. Some European countries maintain traditional forms of management, but the traditional systems in Britain are largely ignored. Even-aged plantations of exotic conifers are, of course, found on the Continent, and not only in afforestation schemes, but they seem to be far less prominent in both the European landscape and the minds of continental foresters. These differences seem to have arisen in the last three centuries as each country has reacted distinctively to changes in the demand and supply of land, timber and wood, increased knowledge of management, and changes in social, political and economic structures. Before then, during the Middle Ages, continental woodland management was a mixture of exploitive fellings, coppicing and wood-pasture systems, much as in Britain.

Several interacting reasons may be advanced to explain this state of affairs:

(1) Since there is more woodland on the Continent

there is less need to afforest, and less chance that the techniques of afforestation will dominate forestry training and planning.

(2) Continental foresters have more opportunities for basing their silviculture on native species. *Abies alba, Acer pseudoplatanus, Larix decidua, Picea abies, Pinus cembra, P. halepensis* and *P. nigra* are all native, timber trees on the Continent which do not occur naturally in Britain. In addition, *Carpinus betulus, Fagus sylvatica, Pinus sylvestris, Tilia cordata* and *T. platyphyllos* are widespread as native trees on the Continent, but very restricted as native species in Britain.

(3) The dominant broadleaved tree species of British woods tend to grow faster and taller on the continental lowlands, presumably because the growing season is longer and warmer and the light intensity is higher. One has only to compare the magnificent stands of oak and beech in France with their counterparts in Britain to realise that there is a trend across western Europe which reaches its extreme in the scrub oakwoods of western Britain and Ireland. Sessile oak, which can grow to 40 m with 27 m of clean trunk in the Paris basin, is reduced to depauperate scrub on the cliffs of West Cornwall.

(4) Whereas continental foresters have more opportunities for basing their silviculture on native species and greater rewards for doing so, British foresters have found that conifers imported from the Pacific coast of North America are especially suited to the oceanic climate of Britain. This direct effect of climate is reinforced by indirect effects on soils. The soils available for forestry in Britain are generally impoverished, peaty or covered by mor humus and for the most part quite unsuited to commercial growth of broadleaf species.

(5) Continental foresters make greater use of natural regeneration for restocking, because they are mainly dealing with established woodlands rather than afforestation, and, in Czechoslovakia and perhaps other countries, there is not enough labour available to plant all the felled woodland. More fundamentally, natural regeneration is more reliable, because with warmer summers the frequency of good mast years is higher than in Britain, the seedlings are more likely to be rooting into a fertile woodland soil, and competition from shrubs and ground vegetation

is somewhat less. (The thickets of *Rubus fruticosus*, which are so often found in British woods are seemingly rare elsewhere.) Furthermore, the native species which are thereby favoured are more acceptable [see (2) and (3) above]; and the greater control of grazing restricts the damage to tolerable levels [see (9), below].

(6) The more mountainous topography of parts of the Continent obliges foresters to protect against avalanches, soil erosion and deterioration of water supplies. This encourages the use of the selection system, especially in Switzerland (Fig. 4.4), and other systems which depend on small felling coupes and permit continuous cover of trees. Such systems are possible on the Continent because the woodland structure is more mature and apparently diverse [see (7) below]. In Britain, clear felling of large coupes is common, and perhaps inevitable in the upland conifer forests with their even-aged stands, short rotations and limited spread of ages within individual forests. Even allowing for this, however, British forestry seems much less concerned with the dangers of soil erosion and the need for catchment control.

(7) Although foresters have existed in Britain for many centuries, forestry has virtually had to be reborn in the present century. Before 1919, when the Forestry Commission came into existence, Crown forests were few and not efficiently run for timber production, whilst private woodland management was suffering from the decline of traditional coppice management and the general agricultural depression. Furthermore, numerous woodlands were gutted of useable timber in the two World Wars. Understandably, therefore, foresters have approached long-established woodlands with rehabilitation, rather than maintenance, on their minds, and, in the absence of both long-term management plans and inherited management traditions, they have generally devised fresh plans which provided for a major break with past management on the site.

On the Continent there has been much greater continuity of management, perhaps best exemplified by the oak and beech forests of France, where management plans devised 200–300 years ago are still operated with only small alterations (Reed, 1954). The Selection Forests of Switzerland are the product of a long and consistent tradition of management which demands high skills. The long rotations of the French forests further enhance the stability of the system. Presumably the greater management continuity on the Continent has been possible because: (i) some at least of traditional continental management systems are more appropriate to modern needs – notably the high forest systems which were largely ignored by traditional management in Britain, (ii) continental forestry suffered no break in tradition, and (iii) continental forests were not so badly damaged by agricultural depression and war.

(8) The continental response to the decline of coppicing presents another instance of greater management continuity. Coppice management has been officially discontinued in Czechoslovakia and is declining in France, but many of the former coppice woods are being promoted to high forest of native, broadleaf species, not clear-felled and planted with exotics, often conifers, as in Britain. This is possible partly for reasons already outlined above, but it also seems to be related to the greater use of wood as fuel. Wood stacks are a common feature outside rural houses and forest roads are often lined with log piles. This, it seems, provides a market for thinnings from broadleaf high forest and, incidentally, enables coppicing to continue in many woods. In Britain coal displaced wood as fuel and the tradition of wood-burning hearths and stoves was broken. Those who supply the current demand for fuelwood in Britain seem to obtain their wood more from hedgerow trees, while the coppices, which could supply the demand indefinitely as they did in the past, stand neglected.

(9) The effects of mammals on the continental woods seem less harmful and more beneficial. Deer stocks seem to be controlled at a lower density. Fewer woods seem to be used as supplementary pasture and shelter for stock, as in upland Britain, and effective measures are taken to eliminate these practices in the continental uplands. Boar are present, which is said to facilitate natural regeneration. Grey squirrels are absent, but in much of Britain they are a major threat to hardwood forestry. As a conservationist one is tempted to claim that Britain is suffering from the failure to maintain a more-or-less natural balance of mammalian influence in woods through both unwise introductions and ineffective deer control.

(10) Finally, we return to another aspect of foresters' attitudes. Afforestation, and analogous major changes in long-established woods, require site treatment, clear-felling, planting, cleaning and, in general, detailed control of site and growing stock through machinery and chemicals (both fertilisers and herbicides). This is characteristic of British forestry. Indeed, one Forestry Commission District Officer once said to me that 'if the ecology isn't right, we will make it right'. Continental foresters tend to work more *with* existing conditions. In the French oak forests, much time and effort is devoted to maintaining the best humus conditions, even to the extent of introducing beech under the oak canopy for this purpose alone. Herbaceous species are known by continental foresters and used to interpret site conditions. Indeed, in Czechoslovakia the forests are all mapped in phytosociological terms and management is directed towards maintaining the best combination (in terms of production and long-term site maintenance) of largely native species for the site. Continental foresters seem, from my own contacts, to be good botanists and practical ecologists, whereas in Britain it is rare to find a forester who knows the native woodland flora well, and uses this knowledge to plan the distribution of trees in new plantings. The general impression is admittedly subjective but it is strong: continental foresters seem to be more sympathetic to the ecosystems they are managing.

20.3 Forestry and nature conservation

A conference on Forests and Landscape Conservation, held in Zürich in October 1977 under the auspices of the Council of Europe, provided a valuable broad view of nature conservation in relation to forestry (Bosshard, 1978). The modern intensification of forestry and agriculture, which has caused concern in Britain, has likewise worried European ecologists, countryside users and foresters, so it is hardly surprising that the recommendations for reconciling timber production with nature and landscape conservation have a familiar ring. Briefly summarised, these included:

(1) No clearance. The total area of European forests, small woods, hedges and isolated trees in open country must not be further reduced.

(2) Management systems. Priority should be given to the potential natural structure of forests and woodlands whilst planning management, irrespective of the immediate profitability. The selection system was regarded by some delegates as best, but management based on traditional silviculture, the maintenance of stability and continuity in management, and the creation of structural diversity were all believed to be important.

(3) Efforts must be made to preserve semi-natural deciduous woodlands and natural forests.

(4) Integration of management objectives is preferable to zonation of woodland into commercial and recreational areas.

(5) Primaeval or virgin forest areas, meadow and marshland woodlands, relict pine forests, lowland moorland and wet peatland forests must be conserved and preserved, with their plant and animal species.

(6) Forest management should be carried out in accordance with the principles of plant sociology; it should be better adjusted to site conditions.

(7) The use of pesticides and fertilizers should be discontinued.

(8) Greatest care should be practised in planting non-indigenous tree species.

(9) Habitat diversity should be maintained by not straightening margins, not planting areas rich in plant species, dry grasslands, rough pasture, and neither draining nor planting marshy and moorland areas.

(10) Game populations and grazing levels must be controlled.

(11) Recreational use of forests must be planned and controlled.

All these recommendations apply in Britain, but with important differences in emphasis. Perhaps the most obvious concerns afforestation, a major element of forestry and a major source of concern to landscape and native conservation in Britain, which on the Continent has apparently caused little concern [but see item (9) above]. Draft proposals by the European Economic Community to expand forestry at the expense of marginal agriculture with the consequent loss of many upland farms and their magnificent floral meadows may increase concern.

Within existing British woodlands the emphasis of nature conservation is on moderating the rate and degree of management change, especially in long-established woods, and on preserving representative examples of ancient, semi-natural woodlands. Both elements appear in the European view, but the former seems more important: items 2, 4, 6, 7, 8 and 10 above all seek to moderate both the rate of change in management and minimise differences between managed and natural stands, whereas only items 3 and 5 are concerned essentially with woodland nature reserves. This emphasis is entirely understandable, for semi-natural woods are still very extensive on the Continent; native species and natural regeneration are widely used in re-stocking, so there is a good chance that semi-natural woods will remain abundant; mixed-age forestry and small-scale working are frequent; and the fauna and flora are more resilient in the face of change. Woodland nature reserves are created, but less to protect examples of a range of semi-natural types, and more to protect virgin forest (Fig. 1.1) and stands of silvicultural interest, such as the spectacular oak–beech stand on the Beaux Monts at Compiègne, France (Fig. 20.1). It is mainly in sparsely wooded regions, such as the Netherlands, where the attitudes and practices of British nature conservation are found. Essentially, the difference between Britain and most of the Continent is that nature conservation needs can be cited at the Area scale (Section 15.3) in most of Europe, but in Britain the emphasis has to be on the Site scale.

This comparison emphasises the degree to which the relationship between forestry and nature conservation in Britain has been determined by the British climate and its effects, the history of extensive woodland clearance and fragmentation, and the rapid and profound changes now being wrought by the present forestry policy. The difficulties in reconciling timber production and the conservation of landscape and nature are largely understandable, and can only be partly blamed on misunderstandings and unenlightened attitudes on both sides. The relatively harmonious continental relationship is unfortunately difficult to achieve in Britain but it can be approached where ancient semi-natural woodlands are still common and native species can provide a fair crop of utilisable timber, notably in the Weald, New Forest, Chilterns, the Cotswold Scarp, the lower Wye Valley and adjacent areas, the Dartmoor fringes, Speyside

(a) (b)

Fig. 20.1 Les Beaux Monts, Forêt Dominiale de Compiègne, France, a magnificent stand (a) of sessile oak and beech on light, acid soils. Sound to the core even though it was 440 years old, the sawn face (b) of the fallen oak trunk shows how slow and even the growth has been.

and Deeside. Furthermore, there are many more-isolated semi-natural woods in Britain which contain native species capable of yielding useful timber, such as oak, lime, beech, ash, alder and birch. Here, the prospects for a more harmonious continental-style relationship between forestry and nature conservation, based on timber production from native species, if not good, are at least reasonable, and provide some hope that timber production and nature conservation will be more successfully integrated.

Chapter 21

Woodland conservation and management 1981–1992

The vignette recalls the 1987 meeting of the Nature Conservancy Council's woodland habitat network at Grizedale, where John Voysey of the Forestry Commission addressed NCC staff on nature conservation in plantations. It symbolises the change which came over the relationship between forestry and nature conservation in the 1980s. Less than a decade after foresters dismissed it as an alternative to and constraint upon forestry, nature conservation became a component of British forestry policy and practice. This was not the only significant change; during the last decade the historical perspective which this book presents became embedded in forestry policy and British foresters looked with increasing enthusiasm at a wider range of silvicultural systems.

This chapter outlines these and other developments. It summarises the substantial advances made in woodland classification, the consolidation of knowledge on woodland history and its ecological effects, and the new priorities in woodland conservation which have developed with both the advance of ecological knowledge and changes in forestry policy. It also reflects the fact that upland plantation forests have been seen increasingly as opportunities for wildlife enhancement, whilst the threat they pose to moorland habitats has necessarily been recognised.

21.1 Historical ecology

21.1.1 Historical studies

The historical approach to woodland ecology began in its modern form with Steven and Carlisle's *Native Pinewoods of Scotland* (1959) and Tubbs' *The New Forest: an ecological history* (1968), and reached its apotheosis with Rackham's *Trees and Woodland in the British Landscape* (1975) and *Ancient Woodlands* (1980). By 1980 the historical perspective was firmly embedded in ecological and nature conservation thinking, and was subsequently reinforced by Rackham's studies on south-east Essex (1986a), Norfolk (1986b), Hatfield Forest (1989), his prizewinning *The History of the Countryside* (1986c), Linnard's (1982) *Welsh Woods and Forests* and Tubbs' *The New Forest* (1986). Other detailed accounts of individual woods have appeared, for example those of Woodstock Park (Bond, 1981), Charnwood Forest (Crocker, 1981), Swithland Wood (Woodward, 1992), Gamlingay Wood (Rackham, 1992), Waterperry (Thomas, 1988), Mugdock Wood (Stevenson, 1990), Methven Wood (Hobson, 1988) and Argyll (Rymer, 1980)

During the 1980s interest in historical studies advanced in both North America and Europe. In the USA William Cronon's *Changes in the Land* (1983) inspired other studies, such as *A New Face on the*

Countryside (Silver, 1990). Following the early lead of Paul Sears, several ecologists used historical records to reconstruct the pre-Settlement state of eastern deciduous forests (e.g. Siccama, 1971; Lorimer, 1977; Grimm, 1984). In Europe the Forest and Environment Research Network of the European Science Foundation supported a historical section led by Professor Pietro Piussi of Firenze, which brought together an eclectic array of studies from many countries (Salibitano, 1988). Historical studies had not been neglected in either continent before 1980, as the Forest History Society of America and the Forest History group of IUFRO bear witness, but they had been focussed more on economic and social, rather than ecological, aspects. Just as in Britain, an intellectual tradition of forest history had developed which had little contact with individual, ordinary woods.

New work not only improved our appreciation of the minutiae of woodland history, but also enlarged our broader understanding. The interaction – or co-evolution – of woodlands and local people has run for over 5000 years in Britain, not just in the vicinity of the Somerset Levels (Rackham, 1979) but also in other districts, for example Maxey and Flag Fen (Pryor *et al.*, 1985; Pryor, 1991). In the fire-prone forests of North America co-evolution has continued for tens of thousands of years (Pyne, 1982), which at least raises questions about the concept of natural woodland. In Michigan less than 200 years of logging, clearance and land management altered fire regimes and created new forms of native woodland (Whitney, 1987). In Harvard Forest, Foster *et al.* (1992) have shown that even the primary woodlands have passed through a sequence of historically unique forms in responding to the impacts of local people, and that the modern forest, though seemingly near-natural, differs considerably from the pre-Settlement forest on the same site no more than 200 years ago. Even in the California oakwoods, the age structure can be explained by reference to changes in human impacts (Mensing, 1988).

Although Europe has a great deal of history, it is becoming increasingly apparent that advances in our understanding of the impact of people on forests are more likely to emerge from North America. A continent which prides itself in possessing numerous spectacular and extensive examples of natural woodlands it also has a relatively simple history of European impacts on its forests. Even after due allowance is made for the earlier impacts of native Americans, these features permit a clearer understanding than is possible in Europe of the effects of felling, fragmentation, pasturage and other impacts.

21.1.2 Relating species distributions to forest history

For me, historical studies are all about explaining the present. One of the most exciting aspects of work in the 1970s was discovering the degree to which plant and animal distributions were controlled by history, and recognising the implications for woodland conservation. Our studies in Lincolnshire (Peterken and Game, 1984) resulted in a list of 62 woodland herbs which were strongly associated with ancient woodlands and must therefore be poor colonists and incapable of significant survival outside woodland. However, many species, such as *Mercurialis perennis* (Peterken and Game, 1981), *Pteridium aquilinum*, *Myosotis sylvatica* and *Campanula latifolia*, varied in their degree of attachment to ancient woods within the study area. In fact, each species seemed to have a unique response to historical changes in the woodland pattern, and even those with the slightest association with ancient woods could be found sparingly in other habitats (e.g. *Anemone nemorosa* and *Sorbus torminalis*, Peterken, 1981; 1983a).

Many more studies have found that particular woodland plants are strongly associated with ancient woodland. Rackham (1980) listed many for East Anglia. Spencer (1990) described the ecology of selected species. Rose (1992) gave an up-dated list of epiphytic lichens with a limited colonising ability, the presence of which he uses as an Index of Ecological Continuity. Even within blocks comprising a mosaic of ancient and recent woodland, many woodland herbs remained mainly in and around the ancient parts (Gibson, 1988). Popular interest in 'Ancient Woodland Indicators' generated many candidate indicators in other parts of Britain, but most were anecdotal, i.e. not backed by independent evidence of woodland history. Thus, for example, *Dryopteris aemula* seems to be a good indicator, but this has yet to be demonstrated.

Even so, the concept of Ancient Woodland Indicators took firm root in conservation circles, whilst at the same time it became clear that no species was likely to be faithfully attached to ancient woods throughout its range. The important point for nature conservation was that there were slow colonists in the woods of every part of Britain, though the list of slow colonists differed from one district to the next.

Similar associations between plant species and ancient woodlands have been observed in those parts of Europe where woods have been fragmented, for example in Belgium (Hermy and Stieperaere, 1985) and Poland (Dzwonko and Loster, 1988). Whitney and Foster (1988) found that several woodland plants were significantly associated with the primary woodland fragments embedded in the landscape of secondary woodland in central Massachusetts. The essence of these findings was observed much earlier by, for example, Nichols (1913) and Marks (1942). Interestingly, the investigations in continental Europe and North America have been scientifically more rigorous than those in Britain.

The mechanisms underlying the failure of some plants to colonise newly-available woodland have rarely been investigated. Pigott and Huntley (1978–1982) demonstrated that *Tilia cordata* was unable to colonise towards the northern margins of its British range, where the species flowers in July and August, but rarely effects fertilisation in the short, cool summers. Other studies of slow colonists suggest a variety of contributory causes, such as edaphic, reproductive characteristics and land-use history outside woods, have combined in ways which are unique to each species, e.g. *Mercurialis perennis* (Peterken and Game, 1981), *Oxalis acetosella* (Packham, 1978) and *Galeobdolon luteum* (Packham, 1983).

Although the early work was focussed on plants, there is now a substantial body of anecdotal and scientific evidence suggesting that many animal species are strongly associated with ancient woodlands. In addition to the molluscs, for which this association was already well established we now appreciate that many beetles and other invertebrates are largely restricted to ancient woods (Garland, 1983; Harding and Rose, 1986; Heliövaara and Väisänen, 1984). These form part of a 'saproxylic complex' (Speight, 1989) of invertebrates, fungi, epiphytic lichens and

hole-nesting birds, which require large, old trees and rotting wood, and which are strongly associated with woodlands, such as old parkland, where such habitats have been continuously present. The association between black hairstreak and ancient woods, which was known before 1980, is now seen as one instance of a widespread conservatism between butterfly populations and habitats, which, in the case of butterflies, is often linked to continuity of open-space habitats. For example, Warren (1987) demonstrated the limited ability of heath fritillary populations to move around the extensive woodlands of the Blean. The fauna of ancient woodlands was given prominence in Marren's (1990) general account of the natural history of ancient woods.

The ancient woodland indicators in Britain and elsewhere are species which have evidently survived the transition from original-natural woodland to managed woodland. Elsewhere studies have demonstrated that some species of the original forest can hardly survive this change. One such species, the northern spotted owl of the Pacific Northwest (Simberloff, 1987), has become an issue on which the future of the finest remaining virgin forests of North America and a major timber industry will be resolved. Indeed, in the recent Presidential election, the Democratic Party candidates were castigated for policies which would leave the USA shoulder-deep in owls. Some of the invertebrates listed by Heliövaara and Väisänen (1984) may also be failing to make this transition as Scandinavian old-growth is felled. Amongst plants, Gustafsson and Hällingbäck (1988) demonstrated that several hepatics were restricted to primary forest in Sweden, and Duffy and Meier (1992) confirmed that plant species diversity declines when broadleaved old-growth is clear-cut in the eastern USA. We must assume that equivalent changes in British woods took place in prehistory.

The ecological issues underlying the phenomenon have also seemed increasingly complex. The faunal work has demonstrated the importance of continuity, not just of woodland, but of particular components of woodland. Thus, butterflies demonstrate the importance of open-space continuity. The restriction of saproxylics of all kinds to certain types of ancient woodland implies the need for continuity of large, old trees and dead wood. Faunal studies have also

Fig. 21.1 Distribution of small-leaved lime in Swithland Wood, Leics., in relation to medieval cultivation remains. Redrawn from Woodward (1992). Although many limes grow on the ridge-and-furrow, their distribution is largely confined to ground within 20m of possible relict locations on outcrops, quarries and boundary banks. Lime is a good indicator of ancient woodland at the landscape scale, but it has a limited capacity to spread within woodlands. Comparison with Fig. 13.9 also demonstrates the character of the approximations involved in rapid ground survey (see Section 13.6).

313

demonstrated that isolation favours the survival of sedentary individuals, which makes populations of poor colonist species even more sedentary (den Boer, 1990). In effect, habitat fragmentation and isolation may be forcing species to 'choose' between sedentary and good-colonising strategies, thus turning a continuous variable – colonising ability – into a bimodal distribution, and perhaps further limiting the capacity of sedentary species to colonise any new woodland we choose to create in the future.

21.1.3 The status of ancient woodland

When we first started to interpret existing woods historically, we thought in terms of primary and secondary woodland. It was Oliver Rackham who first recognised the need to distinguish ancient woods from primary woods. By 1981 we understood that a proportion of ancient woods were secondary; that even the probably-primary ancient woods had secondary inclusions and marginal expansions, and that the significance of continuity dated only from the era of woodland fragmentation, which had never happened in a few favoured areas. We were also well aware that many coppices on ancient sites had been modified by planting or selective promotion of valuable species. Nevertheless, we still considered that most ancient woods were relatively unmodified remnants of the original 'wildwood', preserved as oases of stability in a changing landscape.

Three significant changes have overtaken these perceptions and assumptions. First, we now recognise that a higher proportion of ancient woods must be secondary than we originally thought, and that the distinction between primary and secondary is not clear. Taylor (1983) provocatively doubted that primary woodland could exist in Britain, given the intensity of all-pervasive landscape use and change over millennia. Several studies have shown that substantial ancient woods are largely or wholly secondary. For example, Woodward (1992) identified cultivation remains under much of Swithland Wood (Figure 21.1) which, to my shame, I had overlooked. Excavations along the line of the M3 motorway at Micheldever Wood by Fasham (1983) showed that the site was largely clear of woodland long before the Roman occupation, and that the existing wood formed on abandoned land in or about the sixth century. In both the extreme west (Hannon and Bradshaw, 1989) and the English lowlands (Day, 1991) pollen deposits in small hollows have shown that supposedly ancient and primary woods passed through open or sparsely wooded phases in prehistoric and historic times. Indeed, the combination of evidence from archaeological investigations, pollen deposits and early historical records suggests that most of the ancient woods on chalk and southern limestone are secondary.

Secondly, we have become more aware that the composition of semi-natural stands in ancient woods may differ substantially from that of their original-natural antecedents. Bennett (1989) presented a new map of the natural British forests of 5000 years ago, which confirmed that lime had been dominant in lowland England (Greig, 1982), hazel, oak and elm were dominant in the upland west and north, that pine and birch occupied parts of central Scotland, Northern England and Wales, and that alder was widespread in low-lying districts. We already knew that the Chiltern beechwoods, Epping Forest and Loch Lomondside oakwoods were highly modified, but studies by Edwards (1986) in north Wales, Kelly (1981) and Mitchell (1988) in western Ireland reinforced the view that western oakwoods have been drastically modified from mixed woodland by felling and retrogressive changes in the site. The changes already recognised in the Borrowdale oakwoods (3.5) were traced by James Marsden (personal communication) to planting in the northern part of the valley by Greenwich Hospital between 1748 and 1832. In the lowlands, Richens (1983) explained the distribution of elms in ancient woods and elsewhere as prehistoric transplants from continental Europe, though this was disputed by Rackham (1986c). In fact, most modern plant associations probably originated in the last 1500 years as a response to direct and indirect human actions (Birks, 1993). On the other hand, the composition of the part of Roudsea Wood growing on carboniferous limestone has changed little in the last 5200–6700 years (Birks, 1982).

North American research has helped us to interpret British ancient woods. Comparisons have been made between pre-Settlement forests and their successor second-growth woods in the eastern deciduous and mixed forests of, for example, Vermont (Siccama, 1971), Michigan (Whitney, 1987) and Massachusetts (Foster *et al.*, 1992) and between modern old-growth and second-growth on adjacent, similar sites (Muller, 1982). These woods are all primary in the British sense, but the various impacts of European-origin culture have greatly modified the local distribution and relative abundance of the constituent species. Extrapolating to British conditions, we can infer that the detailed distribution and abundance of trees and shrubs within even primary ancient woods preserves the patterns of the wildwood only in the most general fashion.

The third important development has been to recognise that natural – or at least semi-natural – changes are still taking place in ancient woodlands today, and would presumably have been taking place throughout the historical period. Elms have been much reduced in biomass, if not in numbers, by disease. Birch has increased and oak regeneration has failed within the last century (Rackham, 1989). Ash, a minor component of the underwood in Lady Park Wood, increased substantially early in the 20th century. On a longer timescale, forest fragmentation and coppicing evidently arrested the spread of beech during the last 2000 years (Birks, 1989); planting in the last 300 years has allowed this species to reach regions which it might have occupied naturally.

A simplistic view of ancient woodlands is thus being replaced by a long history of complex interaction and co-evolution of woodland and local human communities. This has been taking place over thousands of years, in which substantial natural changes have occurred in the distribution and amounts of the dominant tree species. The primary ancient woods were derived directly from the wildwood and the species present may be much the same as in the wildwood, but species patterns and interactions have undergone substantial modification. The most modified semi-natural woods are perhaps the oceanic oakwoods, the beechwoods and coppices of the south-east and southern chalklands and the woodlands which once covered floodplains and other low-lying, fertile ground. Ironically, it is the mixed broadleaved woodlands containing lime, which the founder ecologists homogenised as 'oak-ashwoods', that

probably comprise the least modified remnants of the original forests.

21.2 Other aspects of woodland ecology

21.2.1 Natural disturbance

Early recognition that natural forests were disturbed by wind, fire and other factors (Clements, 1910; Sernander, 1936), and that this should influence thinking about the structure and dynamics of natural woodland (Jones, 1945; Watt, 1947) had long failed to make much impact on vegetation theory and conservation rationale, but during the 1980s there was a veritable flood of papers on natural disturbance and its implications (White, 1979; Pickett and White, 1985). Disturbance influences diversity, composition, structure and dynamics of forests at all scales, from the microdistribution of bryophytes on logs to the pattern of forest on a continent. As disturbance has been increasingly recognised as an important ecological factor, so the notion of climax as a stable end-point of succession has been virtually discarded.

In Britain we too have lately become more aware that disturbance affects our native woodlands. At Lady Park Wood (Peterken and Jones, 1987; 1989) the heat and drought of 1976 were just the most influential of many disturbances which accelerated mortality and transformed the relationships between the various tree species. Over the whole of southern England the growth of beech was affected by drought (Lonsdale *et al.*, 1989). Other long-term studies at Rannoch (Peterken and Stace, 1987), Denny Wood (Manners and Edwards, 1986) and Clairinsh (Backmeroff and Peterken, 1989) indicated that disturbance is varied in time and space, but all-pervasive. It recurs at intervals of decades in woods whose dominant trees live for centuries, and is therefore a permanent factor in natural forest ecology.

The storms of October 1987 and February 1990 impressed this on the general public as never before (Ogley, 1987; Whitbread, 1991a,b) (Fig. 21.2). In the early hours of October the 16th, 1987, 15% by volume of all the timber in south-east England and East Anglia was blown down, and many more trees were damaged but left standing. Some woods, such as those at Toys

Hill, Kent and Slindon, Sussex, were almost completely levelled, but in most woods the storm caused a heavy and irregular thinning. The principal feature was the patchiness of the impact, at all scales. Some districts were severely damaged, whereas others escaped almost unscathed. Within individual woods some portions were flattened, whilst the majority remained totally untouched. This patchiness could sometimes be explained by topography, stand age and forest composition: mature stands of beech on high ground and slopes facing the wind, mature conifer plantations, any trees rooted on soft, alluvial soil, and trees just inside the windward edge of woods bordering fields proved to be especially vulnerable. However, many features of the damage pattern, from a regional to a micro scale, remained unexplained, and the overwhelming impression was of a random, unpredictable strike.

Some of the storm-damaged woods were left as they fell so that the long-term responses of native woodlands to natural events could be studied. They joined the small collection of woods which were already under long-term observation in their unmanaged state (Peterken and Backmeroff, 1988). Although some woods were approaching a state which we suppose to resemble 'virgin forest' or original-natural woodland, Britain still had no sites where truly natural old-growth could be observed. In partial compensation new work became available on the nearest near-natural example of old-growth to Britain, the acid beech-oak stand at Fontainebleau (Lemée, 1985; Koop, 1989) and the famous Bialowieza Forest (Falinski, 1986). Studies of processes and structures in the old-growth stands of the eastern deciduous forests of North America (for example Lorimer, 1980; Runkle; 1981; 1982) quantified gap formation, age structure, rates of change in composition and other features of natural temperate broadleaved woods.

21.2.2 Dead wood

Pioneering work by Elton (1966) and Stubbs (1972) brought the importance of dead wood to the attention of ecologists and conservationists, but it was only in the 1980s that 'coarse woody debris' (CWD) became widely appreciated. Much of the advance was stimulated by research in the Pacific Northwest, where large standing and fallen dead trees form an essential part of the old-growth Douglas-fir, western hemlock, Sitka

Fig. 21.2 Windthrown beech near Tring, Hertfordshire. The storms of October 1987 and February 1990 were the most dramatic ecological events to affect woodlands during the last decade. They particularly affected beech woods on chalk slopes, where they generated a bizarre, white pock-marked pattern as seen from the air. Equally strange was the survival of fringes of standing trees on the margins of woods whose cores were largely levelled. In those blowdowns which had not been salvaged, many of the fallen trees were still alive in 1992. (Photo. P. Wakely, Nature Conservancy Council)

spruce forests and the whole landscape (Franklin *et al.*, 1981; Maser *et al.*, 1988). Largely as a result of this research, the importance of CWD has been firmly established in ecosystem processes and in the conservation of biodiversity (Harmon *et al.*, 1986).

In Britain Kirby *et al.* (1991) made a preliminary comparison between semi-natural and natural woodlands, but otherwise there has been little research on dead wood. The interest in Britain and the rest of Europe has been more in the saproxylic complex of beetles, fungi, epiphytic lichens, hole-nesting birds and other groups which survive as small widely separated populations in ancient wood pastures, remaining as relics of formerly widespread populations in original-natural woodland (Buckland, 1979). These species are now under threat throughout Europe (for example Heliövaara and Väisänen 1984; Speight 1989) and the need to work out practicable means of conserving them has been recognised (Warren and Key, 1991; Kirby, 1992).

21.2.3 Birds in semi-natural woodlands

The populations of woodland birds have been fairly stable through the 1980s, with only minor fluctuations (Table 21.1). The enduring popularity of birds has continued to be translated into a substantial output of new research, which has included studies on the ecology of birds in semi-natural woodlands. Fuller (1992) reviewed work in coppice woods and found substantial variation between individual woods. Trends through the coppice cycle were strong: whereas whitethroat was strongly associated with 0–4 year coppice regrowth, a variety of warblers peaked at about the time of canopy closure, and robin, blackbird

Table 21.1 Recent trends in the populations of some British woodland bird species

Increasing	Grey Heron
	Goshawk
	Goldeneye
	Redstart
	Nuthatch
	Siskin
	Crossbill
Stable	Honey Buzzard
	Wood Warbler
	Pied Flycatcher
	Hawfinch
	Crested Tit
	Goldcrest
	Tree Creeper
Decreasing	Capercaillie
	Woodcock
	Green Woodpecker
	Great Spotted Woodpecker
	Lesser Spotted Woodpecker
	Tree Pipit

Sources: Stroud and Glue (1991), D. Stroud, personal communication.

and wren were amongst the species which increasingly dominated the avifauna as coppice approached the end of ordinary coppice rotations. At its peak in 4-year-old coppice in Bradfield Woods, song bird territories reached 150 per 10ha, well in excess of the 27–87 recorded for all species in the interior of Bialowieza Forest (Tomialojc, 1991). Derelict coppice has more hole-nesters and fewer warblers than coppice in rotation, a characteristic which develops further as the former coppice approaches a high forest structure. Elsewhere, Hill *et al.* (1991) demonstrated the adverse effects on most common woodland birds on grazing in woods.

Several other trends in research may be tentatively identified. The precise influences of components of the forest on individual species have been identified: for example, Hill *et al.* (1990) related the incidence of crested tit, goldcrest and coal tit individually to combinations of tree density, tree height, dead wood amount and location in native pinewoods. Amongst the immense amount of valuable research on birds to emerge from Fennoscandia in recent years, some studies have revealed the effects of felling and fragmentation of old-growth habitats at various scales (e.g. Väisänen *et al.*, 1986), which may have implications for the management of plantation forests in Britain. Attempts to understand bird populations at the landscape scale have been made in more complex and less natural environments, for example, the study of nuthatch metapopulations in the Netherlands (Verboom *et al.*, 1991).

21.2.4 Landscape ecology

Island biogeography and its implications for nature conservation was a major pre-occupation of the 1970s, but by 1980 it was clear that ancient and secondary woods could not be treated simply as land-bridge and oceanic islands respectively. Whilst it was attractive to use island theory to assert that larger woods should be chosen as nature reserves, the SLOSS debate generally (Spellerberg, 1991a) and specific studies in ancient woods (Game and Peterken, 1984) showed that the issues and answers were more complex.

During the 1980s landscape ecology absorbed and replaced island biogeography as a way of looking at habitat patches in relation to their surroundings and how species and processes function on a large scale (Forman and Godron, 1986). Detailed consideration was given to the ecology and conservation of remnant habitats (Harris, 1984; Saunders *et al.*, 1987) and the ecological effects of fragmentation of woodlands (Burgess and Sharpe, 1981). In Britain landscape-scale perspectives became commoner (e.g. Stroud *et al.*, 1987).

In the lowlands the increasing influence on ancient woods by factors emanating from the unwooded matrix made a landscape perspective ever more necessary. The spread of deer, which use both woodland and farmland, has increasingly influenced management (21.6) (Ratcliffe, 1992). Eutrophication of ancient woodland soils by nitrate rain and lateral drift of fertiliser may be affecting all woods, much as it does in continental Europe (Ellenberg, 1990; Falkengren-Grerup, 1986). Set-aside agricultural land raised issues about the distribution of new woodlands in relation to existing woods and the possible spread of more non–native tree species into ancient woods. The status of ancient woods as habitat islands was recognised as a product of both millennia of forest clearance and recent decades of agricultural intensification (Peterken, 1992). Until the ecological isolation imposed on ancient woods could be alleviated by habitat re-creation in the surrounding farmland, it became ever more necessary to manage woodland nature reserves as self-sufficient habitats (Pickett and Thompson, 1978).

21.2.5 Plant succession in secondary woodland

The new interest in farm forestry and in creating new woods on surplus farmland raised questions about how new woods would develop as habitats. Up to a point these can be answered immediately, for farmland has been set aside as secondary woodland for centuries, and we can look now at the fauna and flora of farm woods which have existed for 300 years or more. However, the new woods of the late 20th century will orginate in a poorer and more altered landscape than those of earlier centuries.

The classic study was by Woodroffe-Peacock (1918) of Poolthorn Covert in Lincolnshire. This wood has since been cleared, but our own studies nearby (Peterken and Game, 1984; Peterken, 1993a)

detected no significant differences between the woodland plant species *v.* area relationships of woods dating from 1600–1817, 1817–1885 and 1885–1947. This result, when combined with the recognition that many species are virtually restricted to ancient woods, suggested that woods rapidly acquire the catholic and gap-phase species from hedgerows, pastures and disturbed ground, but that other species take centuries to arrive, if they arrive at all. A few secondary woods had been colonised by slow-colonist plants, but these were woods which had developed on land adjacent to an ancient wood, or beside probable refuges of woodland plants along streamsides, wood-relic hedges and meadows. They were, in fact, the exceptions which proved the general rule, that isolation is a factor which inhibits the floristic development of most new woodland. Some of these findings were confirmed by Usher *et al.* (1992). Pigott (1977) had earlier described the spread of plants into Geescroft wilderness.

The capacity of plants to colonise new secondary woodland varies regionally with variation in the reproductive capacity of each species and the availability of source populations (Le Duc *et al.*, 1992). Even so, given the inability of secondary woodlands in many districts to develop full woodland floras even under the more favourable conditions of past centuries, attention has turned to methods of speeding up the process. Buckley and Knight (1989) stressed the value of transferring soils from existing woods: the inoculum of seed in suitable substrate can greatly accelerate succession. Direct sowing of seed and planting of mature plants has enabled some ancient woodland species to become established even in new roadside belts (Francis *et al.*, 1992). General guidance on plant introductions to woodlands is given by Packham and Cohn (1990), who emphasise the need to use local stock of native species and to record procedures.

21.3 Classification

21.3.1 Classification of stand types

In a review of the first edition of this book, Eustace Jones (1983) expressed the view that one of the strengths of British ecology had been its avoidance of classification, and I have some sympathy with this attitude. Classifications all too easily become mechanisms for rigidifying and bureaucratising thought and action, and lifelines for those who lack the confidence to use them with the flexibility of a language. My stand type classification was developed to describe semi-natural woodlands when no alternative was available. The classes were not intended to be final, nor were they meant to imply that every part of a semi-natural wood could be unambiguously assigned to one of the classes. Clearly they could be used for survey or to structure thinking about, for example, the completeness with which nature reserves represented the full range of semi-natural woodland types, but not to the exclusion of other classifications.

Rackham (1980) devised a similar classification to structure his historical treatise. Later he set out the correspondences between his classification and mine (Rackham, 1986c; 104–5). My stand types were used in woodland surveys in several regions (e.g. Colebourn, 1983) and were summarised in other publications (e.g. Evans 1984; Lane and Tait 1990).

This accumulation of information and experience, when combined with the further 200 samples I recorded, would have allowed me to produce a revised stand type classification. The principal changes would have been:

(1) Chestnut woods would have been included as a stand group, as in Rackham's classification.

(2) The oaks would not have been given such prominence. For example, in group 6, the birch–oakwoods, I would have raised the importance of site characteristics by distinguishing birch–oak from hazel–oak types before using oak species to separate sub-types. The revised types, each containing their sessile and pedunculate oak subtypes, would be:

6A Upland birch–oak woods
6B Upland hazel–oak woods
6C Lowland birch–oak woods
6D Lowland hazel–oak woods.

(3) Types within the alderwood group would have been redefined, based on survey experience in Hampshire and Herefordshire.

21.3.2 National vegetation classification

A revised stand type classification has not been made partly because other classifications appeared. The Merlewood classification was published (Bunce, 1982; 1989), but it has not been widely used. More significantly, the woodlands part of the National Vegetation Classification (NVC) (Rodwell, 1991) became available in draft from 1986 onwards. The classification itself has since been increasingly used and promoted for survey and assessment. John Rodwell's text forms an excellent general source on the ecology of British native woodlands.

The NVC is phytosociological, that is, the classes have been defined according to the presence and quantity of trees, shrubs, vascular herbs and ferns, bryophytes and lichens, omitting only epiphytes on trees. All plant species count equally, so the humblest bryophyte is no less important than the dominant oaks in forming the classes. Unlike the stand type classification the NVC excludes site, historical and geographical factors from the construction of the classes, though each class has a particular geographical and edaphic range, and often a particular historical status.

The NVC recognises 18 classes of woodland, five classes of scrub and two classes of edge habitats, each with one to eight sub-classes. The relationships between these and the units of other classifications is set out in detail by Rodwell (1991). The simplified presentation prepared by Keith Kirby (Fig. 21.3) shows clearly that many familiar combinations have emerged, notably the beechwood types recognised by A.S. Watt (Tansley, 1939). In common with other classifications of British native woodlands, the NVC recognises that the main directions of variation lie between (i) upland and lowland, i.e. degree of oceanicity, (ii) wet soils and dry, and (iii) acid, base-poor and alkaline, base-rich soils.

The NVC and the stand type classification recognise the same level of detail. Counting both woodland and scrub types, 65 units were recognised as stand types and sub-types and 74 units were recognised as NVC classes and sub-classes. However, the NVC identifies far more units in wetland woods and the woods of the upland north and west, whereas more stand types than NVC classes are recognised in the mesic lowland woods. These differences reflect the bases of the two classifications; in upland and wetland woods the diversity of trees and shrubs is very limited while the diversity of ground vegetation is often high, whereas in lowland woods the reverse is often the case. Direct comparison of the two classifications using survey data revealed no 1:1 correspondences between units (Cooke, 1992).

I still believe it would be better to make three separate classifications of (i) trees and shrubs, (ii) ground vegetation and (iii) epiphytic assemblages. Logically, stand types should be complemented by a ground vegetation classification, which I would expect to have about the same degree of correspondence with the stand types as can be seen between the stand types and the NVC

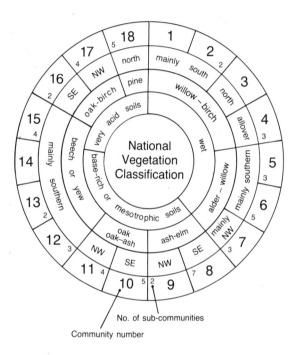

Community number

Scrub types and number of sub-communities

W19	northern juniper scrub	2
W20	montane dwarf willow scrub	–
W21	hawthorn scrub	4
W22	blackthorn scrub	3
W23	gorse scrub	3
W24	bramble underscrub	2
W25	bracken–bramble underscrub	2

Fig. 21.3 The NVC dartboard, a simplified key to the 18 woodland types in Great Britain recognised by the National Vegetation Classification, with a list of the scrub and undershrub types. This valuable diagram, which was devised by Keith Kirby to assist initial training in woodland classification, summarises the geographical bias, dominant tree species and associated soils of each type.

or the Merlewood classifications. The NVC agrees, in so far as it excludes epiphytes, but it treats stands and ground vegetation together. Unfortunately, the NVC types have been given 'stand type' labels, such as oak–birch woods and beechwoods, even though it is possible for oak–birch woods to lack oaks, non-beechwoods to contain over 90% cover of beech and beechwoods to contain less than 4% cover of beech. Given this kind of potential confusion, and the loss of familiar units from the NVC (such as limewood, hornbeamwood, ash–elm wood) there will be occasions when the stand type classification will be more useful. Certainly, general audiences will have difficulty in calling a birch wood a '*Quercus–Betula–Dicranum* woodland', and foresters conscious of the contrasting silvics of lime, hornbeam, ash, elm and oak will have to make distinctions within W8 and W10.

21.3.3 Relationship between British and continental European woods

The chapter relating stand types to continental European woodlands has been thoroughly superseded by several comprehensive volumes. The NVC (Rodwell, 1991) discussed the phytosociological affinities of each type within Britain and the rest of Europe. The English translation of Ellenberg's *Vegetation Ecology of Central Europe* (1988) gave detailed descriptions of native woodland types and their ecology. More recently G. Jahn, in the volume on *Temperate Deciduous Forests* in the *Ecosystems of the World* series (Röhrig and Ulrich, 1991) places British woods in a European context. This volume also includes a chapter by B.V. Barnes which allows European woods to be compared with the forests of eastern North America.

21.4 Survey, assessment and inventory

Whilst the mechanisms for recording and assessing individual woods scarcely changed over the last decade, fundamental shifts started in the way woods in general were perceived and evaluated by both ecologists and the general public.

21.4.1 Survey

Numerous surveys have been carried out for nature conservation. Unfortunately, ecologists have never followed the lead of the Royal Commission on Historical Monuments, which publishes *Inventories*; most ecological survey data remain unpublished, although some excellent summaries and digests are available (Colebourn, 1983; Barfield, 1984).

Kirby (1988) brought together guidance on various aspects of woodland survey. Recording methodology separates the broadscale mapping of major habitat types ('Phase 1') from the more detailed listing, mapping and description of individual sites ('Phase 2'). The former was pioneered by Smith (1900) and Moss (1907) and in modern times by the chalk grassland survey of Blackwood and Tubbs (1970). We pioneered the phased approach in woodlands with the survey of Norfolk woodlands (Goodfellow and Peterken, 1981) and this was subsequently developed as the Ancient Woodland Inventory (21.4.4). Phase 2 woodland surveys have generally involved listing vascular plant species, listing and mapping woodland types and recording the presence of habitat components, such as rides and pools. Kirby *et al.* (1986) tested seasonal variation in the efficiency of recording vascular plants and found, somewhat alarmingly, that on a three-hour walk in a 30ha wood the proportion of species recorded by the standard surveyor in early spring and late summer was only 46% of the total known to be there, rising to 61% in May. Sykes *et al.* (1983) found that individual observers were reasonably consistent when estimating the cover of plant species in quadrats. In recent years the principal development has been the increasing inclusion of vegetation mapping in Phase 2 surveys, initially in terms of stand types, latterly in terms of the NVC.

21.4.2 Assessment

The criteria expounded by Ratcliffe (1977) have remained the basis for assessment, though they have been subjected to critical analysis (Margules and Usher, 1981). The application of these criteria has been systematised and quantified wherever possible, both in woodlands (Kirby, 1986; 1988) and for SSSI selection in all habitats (NCC, 1989). Many woodland surveyors have found it useful to rank woods by counting the number of a select list of woodland plant species present in each to form a preliminary structure for assessing a group of woods (for example, the lists

compiled by R.J. Hornby and F. Rose in Marren (1990, 88–9). Assessment methodology has been extensively reviewed (Usher, 1986; Goldsmith, 1991).

The notion that assessment should be objective has been refined. In the Norfolk survey (Goodfellow and Peterken, 1981) we sought a degree of objectivity by (i) quantifying real features, not by subjective scores, (ii) confining judgements to pre-survey decisions on criteria and methodology and post-survey evaluation of results, and (iii) giving clear reasons for modifying the results from survey when identifying which woods are important for nature conservation. The various approaches to selecting important woods were tested on vascular plant lists from ancient woods in Lincolnshire (Game and Peterken, 1984). Only minor differences were found between various selection procedures, but selections based on a combination of representation and rare species were the most effective.

21.4.3 Value changes in woodland conservation

Much of the recording and assessment methodology is based on the site approach to conservation and the notion that nature reserves and other protected sites are separate from the rest. The principal statement is still *A Nature Conservation Review* (Ratcliffe, 1977) (NCR), combined with lists of SSSIs.

The agreed list of the most important woodlands for nature conservation has required constant revision as the focus has shifted to groups which had hitherto been largely ignored. In the 1960s and 1970s a growing interest in epiphytic lichens brought many ancient parklands into consideration and influenced the selection of sites for the NCR. Later the British Lichen Society carried out a carefully structured assessment of all important woodlands, which resulted in substantial revision of the relative importance of particular sites (Fletcher, 1982). In the 1980s the NCC's Invertebrate Site Register (Ball, 1989) revealed the importance of woods for invertebrate conservation, especially in the agricultural lowlands where few other semi-natural habitats remain. The important sites for invertebrates have not always been the good examples of semi-natural woodlands which dominated the NCR selection, but have tended to be both actively managed woods (including those with plantations) and woods with old trees. Thus a revision of the NCR is now highly desirable, following both new vegetation survey and the increasing recognition of the importance of the saproxylic assemblage (Harding and Rose, 1986; Speight, 1989; Warren and Key, 1991).

Over the same period the development of landscape ecology (Forman and Godron, 1986), the recognition of the impact of surroundings on individual woods (21.2.4) and the importance of wildlife and features outside special sites have combined to generate the feeling that nature reserves — indeed, any wood — cannot be understood or managed in isolation from its surroundings. This is not a new idea: the recognition of ancient woodland indicators and the fascination with island biogeography represented a landscape–scale perspective before British ecologists recognised landscape ecology as an intellectual entity. The historical interest in the whole landscape is clear (Rackham, 1986c). Grazing and browsing is an important ecological factor in upland woods (Mitchell and Kirby, 1990), but its impact and management depends on control throughout the landscape. Butterfly species function as metapopulations spread over large areas, which move with changing management throughout a large wooded area, ignoring completely the boundaries of nature reserves (Warren, 1987). Increasingly, ancient coppices have been seen as part of many interlocking networks of habitats (Peterken, 1992). Sites remain important, but the greater good flows from habitat retention and restoration in the whole landscape.

Another change has been towards a broader definition of which woods are important and a partial breakdown of the distinction between woodland nature reserves and other woods. Any list of important woods carries the implication that excluded woods are not important. Although nature reserves have positive roles in research and education, they are to some extent an admission of failure: they would not be necessary if management for other purposes satisfied nature conservation needs (Peterken, 1991). Accordingly, rather than segregate important woods from the rest, all woods should be managed in a wildlife-friendly fashion. In practice, the segregationist and all-embracing approaches have been combined in the widespread recognition in both political circles (e.g. by the House of Lords in 1982) and by conservation protagonists

(e.g. World Wide Fund for Nature in an *Observer* supplement, May 1992) that we should aim to conserve all ancient woods. Various measures have been introduced to conserve all ancient woods (21.6), so, whilst the number of woodland nature reserves has greatly increased in the last decade, the distinction between them and other woodlands has declined.

The general support for this broader approach was greatly stimulated by Oliver Rackham's books, particularly his *Trees and Woodlands in the British Landscape* (1975). This stimulus coincided with an increasing appreciation of the wider aesthetic, cultural and social values of woodlands (Thomas, 1983), which has influenced the selection of nature reserves (Goldsmith, 1991a). Nature conservation, which deals with the conservation of wildlife, natural features and preservation for scientific research, has been augmented in its aims by recognition of woods as historical monuments. They are important elements in the landscape, components of our sense of place, places for peaceful recreation and reflection, havens from urban environments and mechanised agriculture, and resources which have supported a particular economic and social relationship between people and the natural environment. Common Ground's *In a Nutshell* epitomised these wider values.

21.4.4 The Ancient Woodland Inventory

The NCC's main practical contribution to these developments was the Ancient Woodland Inventory (Fig. 21.4). Starting in 1981, Ian Bolt, Jonathan Spencer, Graham Walker, Alison Roberts and many others used old maps, air photographs, the accumulated results of ground surveys and many other sources to construct lists for each county and Scottish district of all the woods which originated before 1600. In addition, the area within each wood which could be classified as 'semi-natural' (as opposed to 'plantation') was determined. Differences in sources required different definitions and methods in England and Wales (Spencer and Kirby, 1992) and Scotland (Walker and Kirby, 1987).

The idea for the inventories was provoked by a remark from Roger Parker-Jervis in discussion at a conference at Monks Wood in 1972. As a woodland manager, he said that he and his contemporaries would

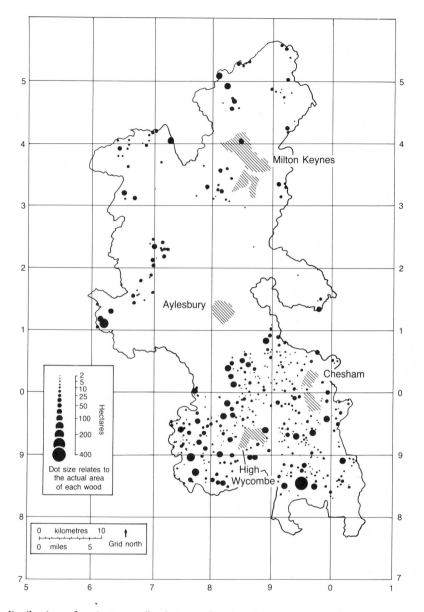

Fig. 21.4 The distribution of ancient woodlands in Buckinghamshire, as recorded in the Nature Conservancy Council's Ancient Woodlands Inventory. The well-wooded Chilterns in the south-east contain many ancient woods but the adjacent belts of Gault and Kimmeridge clays are virtually free of ancient woods. Elsewhere, as in much of lowland England, the ancient woods occur in clusters.

do their best to manage ancient woods sympathetically if only someone would tell him which woods were ancient. This idea developed into a major project with four main aims: (i) to provide a factual basis for site-based forestry policy for nature conservation; (ii) to improve the efficiency of ecological survey (iii) to provide a basis for selecting and justifying woodland SSSIs; and (iv) to form a basis for measuring past and

323

future change. By 1991 preliminary inventories were available for all counties and Scottish districts, each backed by a data sheet summarising information on most sites. The details were widely circulated amongst organisations, but wholesale publication was not practicable. Peter Marren (1992) wrote a popular summary.

21.4.5 Monitoring changes within woods

Monitoring remains a 'soap-in-the-bath' subject: something is there, but it is difficult to grasp. It is also a pantechnicon, a jumble of subjects thrown together under one label in varying combinations and degrees of disorder. It includes surveillance of change in the physical environment, detection of change in the vegetation or biota of reserves in response to management, and basic research conducted on a long-term basis. Goldsmith (1991b) and Spellerberg (1991b) recently brought together general reviews which may help to clarify objectives and methods.

Some of the difficulties may be fundamental to long-term studies (Taylor, 1988). The eventual significance of long-term observations is inevitably uncertain, however clear and precise the initial intentions may have been. A project which originated as a simple surveillance operation may reveal something of fundamental importance, whilst a carefully designed long-term research project may yield only banal results by the time it becomes long-term. Add to that the difficulties of sustaining individual and institutional interest and the flow of resources, and one can readily appreciate why monitoring in both its surveillance and research forms remains a hazardous, poorly-coordinated and uncertain enterprise.

The survival of long-term projects in British woodlands has been just as subject to the vagaries of nature and people as any other long-term work (Peterken and Backmeroff, 1988). Even so, when projects have survived for long enough they have usually realised their fundamental role of verifying ideas, generated by other approaches to the study of ecological change and their capacity for generating unexpected insights (Strayer *et al.*, 1986; Likens, 1989; Peterken, 1993b).

21.5 Changes in woodland pattern and extent

Several new sources have enabled us to form a more precise view of how British woodlands have changed and are changing. Additional information on hedges, non-woodland trees and other habitats has improved our understanding of change in the environment of woods.

21.5.1 Sources of information on woodland change

The Forestry Commission's 1979–82 census of woodlands and non-woodland trees was completed (Forestry Commission, 1984). Their *Forestry Facts and Figures* was updated annually and the changes agreed during the first three years of the Broadleaves Policy (21.6) were quantified. The Ministry of Agriculture published annual figures of farm woodlands, but the method of data collection was so flawed that the results were worthless (Peterken, 1983b; Essex, 1984). The Countryside Commission's Monitoring Landscape Change project (Hunting Surveys, 1986) and the ITE's Countryside Survey (Barr, 1990) generated national figures on woodland, hedgerow and other habitat changes. The NCC Inventory of Ancient Woodlands (21.4.4) was virtually completed.

Changes at a regional and county scale can be determined from some of these national sources. Additionally several independent studies of change over several decades were completed. Under the thirty-year rule, surviving records from the 1947 census of the Forestry Commission became available (Watkins, 1984; 1985). Studies of change in southern Scotland (Langdale-Brown, 1980), the Snowdonia and Brecon Beacons National Parks (Harkness, 1982) and Nottinghamshire (Watkins, 1983; Wheeler, 1984) were particularly thorough. The first of these was developed into the National Countryside Monitoring System by the NCC, which yielded figures for several counties (Budd, 1989). Several sub-county studies were completed, for example for west Berkshire (Young, 1981) and east Hampshire (Fleming, 1983). Information about changes in woodlands, hedges and non-woodland trees was summarised and reviewed by

Allison and Peterken (1985) and Peterken and Allison (1989).

21.5.2 National changes in forest pattern and composition

By March 1991 the total woodland area had increased to 2.34 mha. This was more than double the area recorded in 1895, the increase being the most sustained and widespread reversal of British deforestation since the post-Roman resurgence of woodland on the southern chalk and limestones. In the process, the focus of woodland moved steadily northwards from its traditional heartlands in southern England and the Welsh borderland. By about 1960 the area of coniferous woodland exceeded the broadleaved area for the first time since approximately 6000 BC. By 1991 at least 65% of all woods were coniferous and Sitka spruce was the commonest tree.

Change in the area of broadleaved woodland became of political importance, partly because of a tendentious editorial in a journal of the timber trade, which claimed that the broadleaved area had greatly increased and castigated the conservation lobby for insisting that it had decreased. Unfortunately, Forestry Commission census methodology had changed with each census so direct comparisons over time were inexact, but, in an attempt to resolve the issue, Mike Locke and myself tried to reduce the figures to a common base. The outcome (Table 21.2) showed that both sides were right: there had been a significant post-war decline, but only to levels which still exceeded those of the early 20th century.

Trends during the 1980s do not emerge easily from these sources, but the broad pattern is clear. Conifer plantations continued to expand in the uplands, but the broadleaved element within them started to increase. The rate of woodland clearance to agriculture in the lowlands almost certainly declined substantially. Threats to ancient woods were contested ever more strongly, as the imbroglio over a road building proposal through south-east London's Oxleas Wood amply demonstrated. The rate of replacement of broadleaved woods by conifers also decreased – instead the Broadleaves Policy is ensuring that broadleaves generally follow broadleaves, though usually at the cost of converting semi-natural woods to plantations.

Table 21.2 Changes in the extent of broadleaf woodland, 1924–1980 (kha)

Census date	England	Wales	Scotland	Great Britain
1924	520	70	160	750
1938	570	80	155	805
1947	585	80	170	835
1965	550	70	155	775
1980	560	70	140	770

Estimates were made to the nearest 5 kha from Forestry Commission census records by Mr G.M.L. Locke. The aim was to quantify broadleaved woodland as habitat, not as a timber resource, so the areas include not only broadleaf high forest, coppice, coppice-with-standards, broadleaf scrub and half the area of mixed high forest, but also a proportion of the felled, cleared, devastated and uneconomic woodlands. Minor adjustments were made to bring the figures from the various censuses to a common base of 0.25 ha, and to allow for incomplete data in 1938.

Changes in agricultural policy created an opportunity for more planting on hitherto cultivated land, and many new but usually small secondary woods were established.

21.5.3 Changes in ancient woodlands

The area of ancient woodland cannot increase, by definition. Since 1600 its area has decreased at varying rates and at different times in different districts. Some of the greatest losses were sustained after deforestation of medieval forests, such as Hainault and Rockingham. Clearance was particularly rapid in the mid-19th century, when agricultural land became so much more valuable than woodland, but it slowed considerably with the agricultural depression from the 1870s. By the start of the Second World War the ancient woodland area had been more or less stable for decades, but the coppice system, by which most ancient woods had long been managed, was obsolescent and coppicing had ceased altogether in many woods.

Extrapolation from several local studies suggested that only 10% of the ancient woods which existed in 1945 were still traditionally managed by 1980. The rest had either been destroyed by clearance (10%), converted to (often coniferous) plantations (30%) or

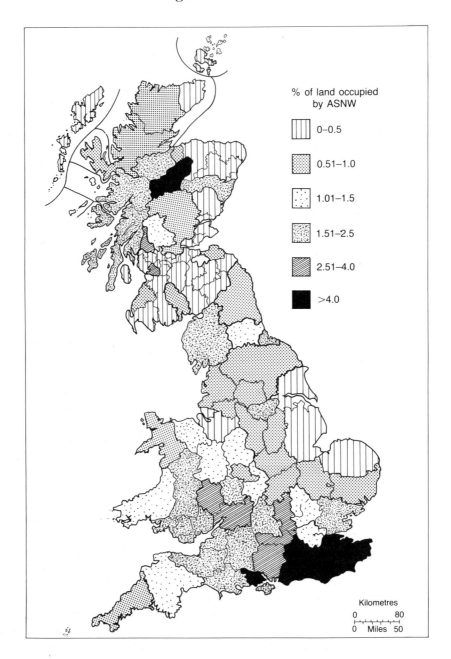

Fig. 21.5 The distribution of ancient, semi-natural woodlands in Britain, expressed as a proportion of the land area of counties and Scottish regions. Source: Nature Conservancy Council's Ancient Woodland Inventory (Spencer and Kirby, 1992; Roberts *et al.*, 1992).

left unmanaged (50%). These were rounded figures, of course, and assessments depended on what was accepted as traditional management or neglect. The Ancient Woodland Inventory eventually provided more precise figures, which showed that in England and Wales some 8% of all the ancient woods which existed in about 1920 had since been cleared, 37% were plantations and 55% remained as semi-natural woodlands (Spencer and Kirby, 1992). The figures for Scotland are not directly comparable, partly because the sources of information are different, but there too 41% of all the woods which had survived from at least the mid-18th century (i.e. ancient woods) are now stocked with plantations (Roberts *et al.*, 1992). By the late 1980s south-east England and the southern Welsh borderland contained the most ancient woodland (Fig. 21.5), as they had done for centuries.

The great majority of ancient woods are small: 83% of all individual woods in England and Wales are less than 20 ha and these amount to 45% of the total area. Only 501 of the 27 876 individual ancient woods exceed 100 ha and of these just 144 contain more than 100 ha of semi-natural woodland. In the English lowlands, complete extinction of individual woods was more or less restricted to woods below 20 ha (Fig. 21.6), but most of the larger woods were reduced (Fig. 21.7). The chance of a particular 1 ha patch of ancient woodland surviving from the 1920s to the 1980s was least in the 11–20 ha size class (Fig. 21.8), even though nature reserves and SSSIs were concentrated in medium-sized woods. Conversion from semi-natural woodland to plantation has been faster in the larger woods (Fig. 21.9), many of which are managed by the Forestry Commission or owned by estates which are large enough to employ foresters. Thus, only 17% of the ground in ancient woods of 2–5 ha is now occupied by plantations, whereas 59% of the ground in ancient woods over 100 ha has been restocked. Increasingly, therefore, the surviving ancient semi-natural woods are small, isolated woods or fragments of larger woods. Only 200 ancient woods greater than 50 ha (including 40 woods greater than 100 ha) remain completely semi-natural in the whole of England and Wales.

During the 1980s many ancient woods remained unmanaged. As the underwood grew taller, so the weaker stools died, low-growing components such as hazel diminished disproportionately, and the lower branches of remaining standard oaks were progressively killed by shade from overstood underwood. However, the management of ancient woods as mixtures of native trees increased markedly through the 1980s (21.7). Whereas in 1965 it was still commonplace to find whole ancient woods as unmanaged semi-natural woodland, by 1992 most woods contained at least small patches of recent felling, thinning or planting in addition to the usual rideside cutting for pheasant shooting.

21.5.4 Changes outside woodlands

By 1980 many ancient woods were more thoroughly isolated ecologically than they had ever been, due to hedgerow removal, the reduction of trees on farmland and the destruction of semi-natural grasslands and semi-woodlands (such as the shaded banks of unstraightened streams) (Peterken and Hughes, 1990). However, the devastation wrought by Dutch elm disease led to the formation of the Tree Council and to increasing rates of tree planting in and around villages, on roadsides and in corners within farmland. The work of Farming and Wildlife Advisory Groups stimulated an interest in habitat creation amongst some farmers. Nevertheless, Britain's position as the main European country for ancient trees (Rackham, 1986c) had been seriously eroded and old habitats were rapidly vanishing.

Although it is fashionable to believe that all the problems have been solved, the 1980s saw further erosion of our stock of old trees and habitats outside woodlands. The rate of destruction had probably declined, but this was inevitable, for there was by then so much less to destroy. The remaining hedgerow trees continued to age, whilst fears increased that their increasingly moribund state was due less to their age than to some widespread environmental change. Excessive water abstraction lowered water tables so far in lowland England that many streams and wetland habitats dried out. Eutrophication from farm fertilisers reduced remaining herb-rich semi-natural habitats towards the condition of rank herbage with a few common species. Tree planting and conservation remained popular, and habitat restoration

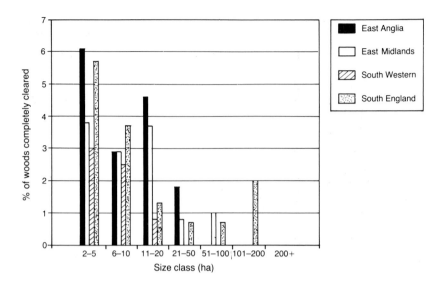

Fig. 21.6 The proportion of ancient woods present in the 1920s which was completely cleared by the 1980s, in relation to size in the 1920s. Source: NCC's Ancient Woodland Inventory, aggregating county figures for four of the 1990 NCC English regions. The selected regions cover most of lowland England, but exclude the south-east, where much of the woodland occurs as extensive tracts within which individual woods can be delimited only by arbitrary or legal lines.

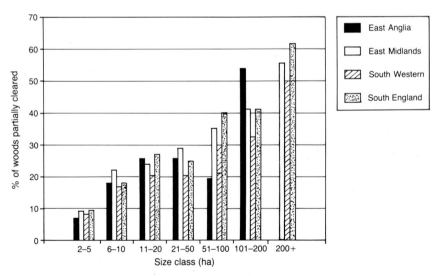

Fig. 21.7 The proportion of ancient woods present in the 1920s which was partially cleared by the 1980s. For other details, see Fig. 21.6.

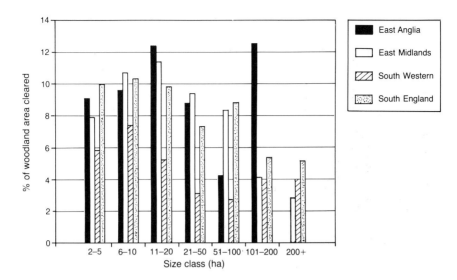

Fig. 21.8 The proportion of ancient woodland cleared between the 1920s and the 1980s. For other details, see Fig. 21.6.

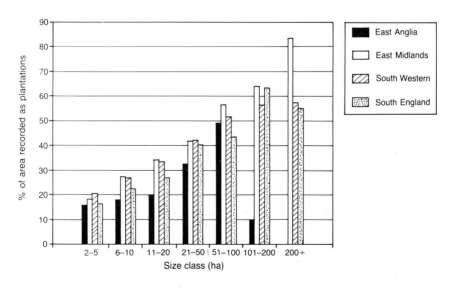

Fig. 21.9 The proportion of ancient woodland which was recorded as 'plantation' in the 1980s. Virtually all of this was converted from 'semi-natural' woodland between the 1920s and the 1980s. For other details, see Fig 21.6. The incursion of plantations into ancient woodland was generally greater in the larger woods, but in East Anglia the largest woods were spared by the provisions of the 1878 Epping Forest Act.

Table 21.3 The eight regional ecological groups of native woodland recognised by the Forestry Commission

Type	Zones	Main soil characteristics	NVC types	Stand types	Main historic management	Emphasis in future management
1 Lowland oak– beech woods	4	Acid	W15, W16	6C, 6D, 8A, 8B	C.WP	HF(WP)
2 Beech–ash woods	4	Alkaline – neutral	W12, W13, W14	[1A], [3C], 8C 8D, 8E	C.HF	HF
3 Lowland mixed broadleaf woods	(2), 3, 4	Various	W8(a–d), W10	1B, 2A, 2B, 2C 3A, 3B, 4A, 4B, 4C, 5A, 5B, 7C, 9A, 9B, 10A, 10B	C	C, HF
4 Upland mixed ash woods	2	Alkaline – neutral	W8 (e–g), W9	1A, 1C, 1D, 3C 3D, 7D, [8A–E]	C.HF	HF(C)
5 Upland oakwoods	2	Acid	W11, W17 (oak dominant)	6A, 6B, [8A, 8B]	C.HF grazed	HF grazed
6 Upland birch woods	1, 2	Acid	W11, W17 (birch dominant)	12A, 12B	HF grazed	HF grazed
7 Pine woods	1	Acid	W18, W19	11A, 11B, 11C	HF grazed	HF grazed
8 Alder–willow woods	1–4	Wet	W1, W2, W3, W4, W5, W6, W7	7A, 7B, 7C	C	Minimum intervention

The zones are those shown in Fig. 21.10. The NVC types are described in Rodwell (1991) and summarised in Fig. 21.3. The stand types are described in Chapter 8. The main forms of historic management are coppice (C), wood pasture (WP), high forest (HF) and grazed forms of each. The Codes of Practice published by the Forestry Commission recommend a general shift towards high forest management, but not to the exclusion of coppicing and forms of woodland pasturage.

schemes were initiated by the Countryside Commission. The Countryside Council for Wales started the Tir Cymen scheme, which is designed both to encourage retention of habitats on farms and to create more. However, such restoration schemes still have a long way to go: woods will remain isolated and in danger of impoverishment unless a great deal more woodland and semi-woodland habitat is created on farmland.

21.6 Policy developments

21.6.1 Policies for broadleaves and native woodlands

Developments in policy and advice relating to broadleaved or native woodlands were dramatic. The 1982 Loughborough conference on *Broadleaves in Britain* brought out widespread concern for the future

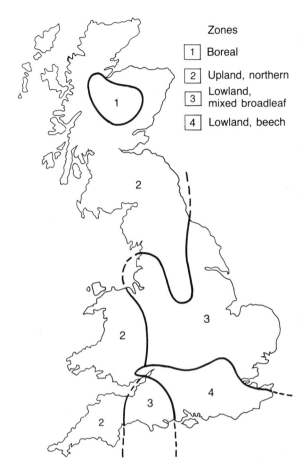

Zones

1 Boreal

2 Upland, northern

3 Lowland, mixed broadleaf

4 Lowland, beech

Fig. 21.10 Regional divisions of Britain based on the ecological characteristics of ancient, semi-natural woodlands. Region 1 circumscribes the main native pinewoods. Region 4 delimits the native range of beech, strictly defined. Regions 2 and 3 divide the rest of Britain into the north-western zone of upland, oceanic woods in a mainly pasture environment and the south-eastern zone of lowland woods in a mainly cultivated environment, respectively.

of broadleaved woodland (Malcolm *et al.*, 1982) and demonstrated that most of the positive thinking had taken place outside the Forestry Commission. The contribution by the Nature Conservancy Council (Steele and Peterken, 1982), which appears to have been the first quantitative expression of forestry strategy relating to existing woods, proposed four treatment classes. This approach was endorsed by the House of Lords Select Committee on Science and Technology in their supplementary report on *Scientific Aspects of Forestry* (4th report, 1982). After extensive consultations, the new Broadleaves Policy was announced in October 1985. It sought, through a mixture of controls, incentives and advice, to maintain existing broadleaved woodlands as broadleaved woodland, minimise further destruction by clearance, and add to the broadleaved area by new planting. More significantly for nature conservation, ancient woodlands were recognised as an important category requiring special consideration; they were to be identified through the NCC Ancient Woodland Inventory and their management was to be governed by guidelines.

The effectiveness of the Broadleaves Policy was reviewed by Miller *et al.* (1988). This survey concluded that most policy objectives were being met, but showed that ancient semi-natural woods were being converted from native mixtures of coppice origin to broadleaved plantations. The methods of restocking were: (by area) broadleaved planting 87%, natural regeneration of broadleaves 10%, conifer planting 2% and coppice 1%. Moreover, there was a tendency to homogenise the smallscale diversity of such woods: although 80% of the trees planted were 'regionally native' most were in fact oak, ash and cherry. Woodland clearance rates appeared to decrease during the later 1980s, but there were no figures with which to check.

Following this review management grants were reintroduced in April 1992 with a supplement for woods of high environmental value, which generally includes ancient, semi-natural woodlands. Codes of Practice were developed which clarified the suitable options for managing these woods (21.6.2).

Throughout the 1980s other agencies promoted and supported forms of management which attempted to strike a balance between timber production, nature conservation, landscape and public access. The Countryside Commission and Local Authorities supported schemes in various counties: those in Suffolk and Sussex, for example, were particularly effective in reviving management in small woods. This produced many environmental benefits, including some income and improved shooting for farmer-owners, some new rural employment and some production of fuel wood, quality timber and specialised products (e.g. thatching spars). Most Local Authorities now employ foresters and ecologists to develop and implement woodland conservation and management policies, e.g. Hampshire (Colebourn, 1983). In Wales, Coed Cymru was initiated by a consortium of conservation agencies led the Countryside Commission, to implement practical measures to restore and increase native woodlands. In the West country, Project Sylvanus was initiated by the Countryside Commission and other agencies with the Dartington Amenity Research Trust. In 1989 the National Small Woods Association was formed.

21.6.2 Guidelines and codes

Management guidelines for ancient and semi-natural woodland were formulated by the Forestry Commission as part of the Broadleaves Policy. Their broad aim was to encourage management based on the locally native tree species. However, the guidelines were meant to apply to all kinds of woodlands and necessarily took a rather generalised form. Birchwoods in the Highlands and, for example, beechwoods in the Chilterns have very different management needs and possibilities, and inevitably there was early pressure to produce 'regional' guidelines, which were more specific to each woodland type.

Following advice from a number of quarters, the Forestry Authority sought to fulfil this need by issuing in 1993 guides to the management of ancient and semi-natural woodlands. The guides were based on a classification (Table 21.3) devised by Richard Brittan of the Forestry Authority and myself, which grouped these woods into eight regional and ecological types. This classification recognised the main directions of variation which had emerged from the

classification of stand types and the NVC, but it aggregated stand types and NVC types into groups according to common management histories, existing management constraints and silvicultural needs and opportunities.

The basic division (Fig. 21.10; Table 21.3) recognised (i) the distinctiveness of the boreal birch and pine woods, (ii) the distinction between Highland and Lowland Britain, which is correlated with climate, land use, geological and geomorphological features, (iii) the distinctiveness of the region of beech and hornbeam woods in both biogeographical and silvicultural terms and (iv) the distinct ecological character of wetland woods, which occur in all four regions. Discrimination at this point would have given five types, but it was felt necessary also to (v) recognise birch woods as separate from pinewoods, both geographically and silviculturally, (vi) distinguish between oak–beech on acid soils and beech–ash on alkaline soils within the beech zone, and (vii) distinguish similarly between upland oakwoods on acid soils and upland ash woods on alkaline soils in the Highland zone. This left one large group, the lowland mixed broadleaved type, which was ecologically fairly homogeneous (NVC W8 and W10) but contained numerous stand types with a wide variety of silvicultural possibilities. Further subdivision would have generated many practical problems.

The guidance given in the guides accepts that owners will have a range of objectives from timber growing to nature and landscape conservation. Broadly, the aim is low-intensity management using the tree species native to the site which incorporates measures and restraints designed explicitly for wildlife. Non-intervention is accepted as a long-term aim in some nature reserves. Elsewhere, it is assumed that woodland should be managed and that the mature habitats (such as dead wood), which develop under non-intervention regimes, would be incorporated within ordinary treatments.

21.6.3 Agricultural change and set-aside

Changes in agricultural policy interacted with

a general concern about trees in the countryside (21.5.4) to generate new initiatives for tree planting outside woods and forests. Initially, under the stimulus of Dutch elm disease and the Tree Council, the focus was on planting individual trees or groups in hedges, field corners, along motorways and in towns, but during the 1980s the response to agricultural surpluses led to policies for planting more woodland on farmland. The scale of tree planting beside roads was such that in 1992 the Ministry of Transport could claim that it had planted more broadleaved trees each year than the Forestry Commission. On farmland, however, despite the introduction in 1988 of the Farm Woodland Scheme (Insley, 1988), farmers were rarely willing to make a long-term commitment of land to trees. In 1990 the Tree Council initiated a parish Tree Warden scheme with the aim of both looking after existing trees and planting more.

Much of this planting was on a small scale – blocks of less than 10ha – but in addition some far more ambitious schemes were mooted in the 1980s and are currently being implemented. The Countryside Commission's imaginative proposal for a new National Forest has resolved itself in a scheme to link the ancient Charnwood and Needwood Forests. This currently poorly-wooded region will not be planted with wall-to-wall trees, but will eventually develop into a tract of high density woodland (40% cover) in a mosaic of different habitats. Another Countryside Commission proposal has led to the joint Countryside Commission/Forestry Commission programme for Community Forests planted on the margins of large cities, including north-east London, Birmingham, south Yorkshire, Merseyside, Nottingham and Bristol. In Wales, Coed Cymru, whose main aim was to manage existing woods, also contributed marginally to the increase in native woodland. In Scotland a succession of measures, from the Native Pinewoods scheme and the Broadleaves Policy to the Central Scottish Woodlands Project, generated considerable interest for restoring the native woodland cover (Wightman, 1992). An organisation – 'Reforesting Scotland' – has been created to enlarge and direct resources into native woodland planting.

Planting so far has not made a great impact on the landscape, but even so the process of reversing the post-war trend towards isolating woods in a treeless landscape has started. This is important, because conservationists were increasingly being forced by the intensification of agriculture to adopt a siege mentality to woodland conservation. Each wood had to be ecologically self-sufficient, perpetuating habitats for the species it harboured on the assumption that any losses would never be made good by recolonisation from other woods or from relict populations in the farmed landscape. If the present trend towards more woodland and proper links between woodland continues, a more flexible approach to woodland conservation will be possible.

21.6.4 Woodland and forest nature reserves

The number of woodland nature reserves (in the broadest sense) increased substantially throughout the 1980s. The Woodland Trust continued its remarkable growth by acquiring about one million pounds' worth of woodland properties each year, many of which were ancient woods. By 1992 it owned 6700ha of woodland in 500 separate sites spread throughout Britain, including many outstanding sites, such as Wormley Wood (Herts) and woodland on the Doward (Hereford). The Royal Society for Nature Conservation, County Trusts and the Royal Society for the Protection of Birds acquired numerous woodland reserves, including such important places as Abernethy Forest. The National Trust employed advisers on nature conservation and forestry, and have treated an increasing proportion of their woodland estate as *de facto* nature reserves (Russell, 1992). More National Nature Reserves were established by the NCC, including important woodlands at Downton Gorge (Hereford) and Castle Eden Dene (Durham). In 1989 the Forestry Commission designated 46 sites as Forest Nature Reserves, many of which were semi-natural woodlands.

Following the 1981 Wildlife and Countryside Act, SSSIs were reviewed and re-notified. The SSSI system is designed to identify and protect a network of sites which collectively represent the national range of features of greatest value to wildlife, with the general aim of guaranteeing the survival of a necessary minimum of British wildlife and natural physical features (NCC, 1984). Management agreements became possible in order to compensate woodland owners for restrictions on management. By 1991, the last *Report of the Nature Conservancy Council* reported that 5671 SSSIs totalling 1 778 474ha had been scheduled and management agreements covered 98 545ha. Most of this was not woodland: the area of ancient semi-natural woodland in all kinds of conservation area was 44 667ha in England, 5116ha in Wales (Spencer and Kirby, 1992) and 51 106ha in Scotland (Roberts *et al.*, 1992). The provisions in the agreements varied according to the character of each site: whilst some were treated as minimum intervention reserves, many others were felled and restocked on an agreed programme with mainly locally native tree species. No compensation was paid to conservation organisations which owned SSSI woods because their objectives prevented them from adopting financially-attractive intensive forestry, but grants were available for specific measures.

The objectives of woodland National Nature Reserves (NNRs) had originally focussed on research, demonstration and education (Ovington, 1964; Peterken, 1991). Provided their wildlife and natural features remained safe, they were to be used primarily for research, so that techniques for conserving wildlife could be tested and demonstrated to other land managers. In the 1950s and 1960s many research projects and trials were initiated, some of a long-term character, and considerable success was achieved with demonstrating results to the public at reserves such as Beinn Eighe, Inchcailloch, Monks Wood and Ham Street. However, from about 1970 onwards the use of NNRs for research steadily declined and many potential long-term projects and trials were abandoned. Throughout the 1980s the philosophy of NNR acquisition and management remained unclear: whilst priorities for acquisition were defined by the NCR and treatments were designed mainly to diversify structure and wildlife within the limits of native woodland and natural responses, it was not clear what special

role, if any, the NNRs were intended to have (21.10).

21.7 Woodland conservation management

During the 1980s many more publications have considered the ecological effects of management (e.g. Mitchell and Kirby, 1989) and advised on management (e.g. Evans, 1984; Crowther and Evans, 1986; Lane and Tait, 1990; Watkins 1990). Buckley (1992) brought together new ideas and research on coppice ecology and management.

21.7.1 Coppice

Between 1965 and 1980 the coppice area increased from 28kha to 40kha. Although a little of this increase could be ascribed to commercial markets, such as hardwood pulpwood, much was due to conservationists. Coppicing not only continued the traditional form of management to which woodland wildlife had become adapted, but it was also well-designed to maintain habitat diversity within a limited area.

This revival of coppicing continued during the 1980s. Conservation coppicing became more technically competent with the work of the British Trust for Conservation Volunteers and work forces supported by unemployment schemes. The market for fuelwood allowed many overgrown coppices to be cut, while the traditional specialised coppice markets, such as hazel wattle hurdles, slightly increased as the more affluent members of the public sought more distinctive, less machine-made products. Direct assistance from county-based woodland management schemes, supported by the Countryside Commission, brought together woodland owners, users, workers and smaller markets. Thus, for example, in Suffolk and Sussex, modified forms of traditional management were restored to ancient coppices by a mixture of advice, practical help, subsidy and identification of markets.

Several problems were encountered when restoring coppice management. Inevitably, the stool density was low after a period of neglect, for the weaker stools had been killed in competition with their vigorous neighbours. Stools which had not been cut for decades had lost their vigour and often sprouted weakly after cutting. Ash sprouts, for example, commonly spread laterally, as if they had been flattened by an elephant. Very few oaks had regenerated to form potential standards. However, the principal problem in most woods remains the increase in deer, which congregate in freshly cut patches and can browse the coppice sprouts so heavily that growth is checked and the stool is eventually killed. Another problem has been the copious growth of bramble on many sites, which overarches the weaker stools and greatly inhibits natural regeneration. Bramble often develops under the neglected coppice canopy before cutting, and its subsquent vigour may be enhanced by eutrophication.

Although many coppices have grown well enough after cutting to provide a closed canopy within 5–7 years, special measures have been necessary in perhaps a majority of sites to ensure adequate regrowth. Fencing has been commonplace, including dead hedging with brushwood from the cut coppice. Brushwood has been stacked over the individual stools. Deer culls have been necessary to give at least temporary relief. Planting of oaks and other native trees has been necessary where growth and stocking has been unsatisfactory. In the future it may be necessary to coppice in large, rectangular patches and to place each year's cut next to last year's in order to mitigate deer browsing inexpensively (R. Putman, personal communication).

21.7.2 Pollarding

Pollards and stub trees have been characteristic of hedges, wood-pastures and the margins of coppices, but in recent decades their numbers have declined, very few new pollards have been started and most existing pollards have not been lopped. Exceptionally within woodlands, re-pollarding started at Burnham Beeches from 1954, initially fatally, but latterly with considerable success. Old pollard hornbeam and oak were cut again at Hatfield Forest in about 1975, with

mixed results. Meanwhile, pollard hornbeams were still cut in rotation at Hatch Park (Kent), osiers were still commonly cut on lowland floodplains, and hedgerow trees along farmland byeways were still often lopped.

During the 1970s, some boundary pollards were cut again, when conservationists brought woodland nature reserves under renewed management. Pollard restoration brought mixed results, however, for the branches were heavy and the moribund trees resprouted weakly or not at all. Recognising the difficulties and the need to pool experience, Mitchell (1989) was commissioned to review pollarding as a system and technique. The results were brought together at the Burnham Beeches meeting (Read, 1991) and there is now an encouraging level of interest and competence in maintaining and renewing the important British population of pollard trees. The NCC ensured that the Hatch Park pollards continued to be cut.

21.7.3 Grazing in upland woods and wood-pastures

Deer, sheep and, to a much lesser extent, cattle have been part of upland woods since the late 19th century. Woods which were regenerated in the 18th and 19th centuries became pastures with trees, providing shelter and some feed, especially valuable during hard weather. Under this regime tree regeneration virtually ceased, the underwood was shaded out and the ground vegetation became a closely-cropped sward of grasses, small herbs, bitten-down bilberry and a rich mixture of bryophytes. Saplings, shrubs and many woodland herbs were largely restricted to inaccessible gulleys and rock outcrops.

In these circumstances, woodland management is possible only if grazing is eliminated, or at least controlled at low levels for a decade or two. This creates an opportunity for some regeneration of the shrub layer, for saplings in canopy gaps to grow beyond check by browsing and, if the fence is set beyond the wood margin, some woodland expansion. Exclosures were generally successful in releasing seedlings (Sykes and Horrill, 1985), but sometimes failed if seedlings did not germinate shortly after exclosure. However, exclosure also initiated a succession in the ground

Trees and shrubs	No regeneration due to competition from dense ground vegetation	Creation of regeneration niches	Loss of seedlings. Damage to saplings	Loss of saplings. Severe tree browsing	Barking of mature trees. Loss of shrub layer	Creation of parkland or moorland
Higher plants	Reduced diversity dominated by a few vigorous species	Reduction in vigorous species. Increase in diversity	Reduction in vegetation structure. Increase in grazing tolerant species	Loss of plant diversity, particularly of grazing sensitive species	Loss of cover and damage due to trampling. Bare ground.	Impoverishment due to net loss of nutrients from the system
Lower plants	Reduced cover and diversity due to competition from higher plants	Increase in cover of ground dwelling species as competition from higher plants reduced		Damage to ground dwelling species due to trampling.	Reduction of drought sensitive bryophytes	Increase in epiphytic lichens associated with parkland
Small mammals	High small mammal populations, a few species predominate	Increase in diversity as structural diversity increases	Reduction in small mammal populations as ground vegetation structure simplified		Reductions of populations through competition for food	Loss of diversity and abundance. Species of open ground predominate
Birds	Favouring birds of dense shrub layers	Increase in diversity as structural diversity increases	Increase in species favouring low shrub cover	Loss of ground nesting birds due to poor concealment	Loss of species dependent on berrybearing shrubs	Reduction in raptors dependent on small mammals
Invertebrates	High populations of phytophilous species	Increase in diversity as sward structure diversified	Increase in dung utilizing species	Decline in woodland species		Increase in parkland moorland species

No grazing ——————————————————⬤▬▬▬▬▶ High grazing intensity

Fig. 21.11 A summary of the impact of increased grazing intensity on the flora and fauna of woodland. The shaded zone defines the optimum for nature conservation, which occurs at low, rather than zero, grazing levels. Reproduced from Mitchell and Kirby (1990), with permission.

vegetation as a few vigorous species (e.g. *Vaccinium myrtillis*, *Deschampsia flexuousa*) grew strongly and displaced the low-growing bryophytes and herbs (Mitchell and Kirby, 1990).

Recent conservation management has therefore turned somewhat from excluding all grazing to controlling grazing at levels which permit some regeneration (Mitchell and Kirby, 1990) (Fig. 21.11). At Creag Megaidh, for example, a sharp reduction in the red deer herds throughout the reserve has allowed the geriatric alder and birch woods to regenerate prolifically.

Recent study of regeneration in the New Forest (Morgan, 1991) shows how ancient wood-pastures functioned and may indicate how they can be managed in the future. Even under the recent regime of heavy grazing, regeneration has been possible in gaps protected by windfall branches, amongst holly scrub and in patches which appear marginal to the groups of deer and horses. Viewed on a long perspective, regeneration in grazed woods would be slow

and patchy, but provided natural processes are not too modified (e.g. by removal of fallen trees), planting may not be necessary.

21.7.4 High forest management of broadleaves

Many ancient woods which have been coppiced for centuries are changing into broadleaved high forest. Some coppices have been singled and thinned until they take the form of high forest. Far more have been clear-felled and replaced by plantations of oak, beech, ash, cherry and other broadleaves. Many other woods have been neglected, so the coppice has grown dense and tall.

Both plantation forestry and long-term neglect tend to discriminate against hazel and other low-growing or minor constituents of the original coppice. In neglected coppices the hitherto stable underwood composition is destabilised. Any remaining standard oaks are subjected to heavy side-shade, which kills their lower branches. Stand structure inevitably changes, and the age–class distribution of the wood as a whole becomes unbalanced. Gap creation rates, which were 6–10%/yr in worked coppices, decline to almost zero for decades. Permanent open spaces (rides) are reduced or totally overshaded.

These changes have continued during the 1980s, but a comprehensive review by Mitchell and Kirby (1989) has made us more aware of the effects on fauna and flora. The change from coppice to high forest has changed soils, fauna and flora, but the exact chain of cause and effect is difficult to determine because so many factors change in step. Moreover, many changes have been observed casually and understood intuitively, but not examined by rigorous scientific research.

Putting all the sources together, the most important effects of converting coppice to high forest appear to be as follows:

(1) Permanent open spaces (mainly rides) become heavily shaded in neglected woods, but remain open in woods which have been felled and replanted as high forest. However, as plantations mature, all the rides in a wood become shaded simultaneously, unless ride margins are cleared. The flora and fauna of rides, especially butterflies, decline in many woods.

(2) Temporary open spaces (clear-fell compartments) vanish for decades in both neglected woods and (after planting has been completed) in replanted woods. This reduces the incidence of the gap-phase flora and fauna and, through the break in continuity of gap-phase habitat in small woods, may be permanently impoverishing. The thicket-stage fauna is also depleted, e.g. dormice.

(3) Soils in plantation high forest are more disturbed and compacted than they had been under coppice, but more organic matter accumulates, including occasional large logs.

(4) The flora changes from dominance by spring-growing and flowering species (such as bluebell) towards a greater representation of summer-growing species. Accumulation of litter enhances the representation of nitrophilous species. Bramble accumulates under dappled shade at the end of high forest rotations, where deer were not abundant.

(5) In the greater structural complexity of mature high forest stands, some aspects of invertebrate diversity increase, e.g. leaf mines and spiders in derelict hazel coppice (Sterling and Hambler, 1988).

During the 1980s a new interest developed in irregular, or continuous-cover forestry. The new climate of opinion, in which broadleaved woodlands were to be retained and managed as such, led to constructive thinking beyond the mere planting, thinning and clear-felling of even-aged stands. Selection forestry with a mixture of species has long been practised in continental Europe, but in Britain only a few individuals who were regarded as courageous innovators by conventional foresters, gave it a try (e.g. Hutt, 1974). With the publication of Evans' (1984) guideline and Matthews' (1989) review of silvicultural systems the new interest burgeoned into a 'Continuous Cover Forestry Group' with links to European foresters such as Dusan Mlincek (1991).

Although conservationists have constantly argued for greater use of natural regeneration in high forest restocking, planting remains prevalent. Newbold and Goldsmith (1981) reviewed the literature on the natural regeneration of beech and oak, but reached no firm conclusions.

21.7.5 Treatment of exotic trees and shrubs

Conservationists have long been antipathetic to introduced trees and shrubs. The common justification has been that such trees support a lower diversity of wildlife than native trees, particularly of invertebrates (Southwood, 1961, updated by Kennedy and Southwood, 1984). More fundamentally, the aim of nature conservation is to conserve natural conditions, especially the remnants of original–natural conditions, in which introduced trees have no part.

The 1980s have seen a revisionist and more discriminating approach to introductions. Increasingly, each species has been treated on its merits and disadvantages. On this basis, *Rhododendron* has been recognised as an unmitigated threat, whereas sycamore has been partially rehabilitated. *Rhododendron* smothers ground vegetation, excludes tree regeneration and supports little wildlife. Much thought has been given to its control (Evans and Becker, 1988), and thick infestations have been eliminated from NNRs such as Dinnet Oakwood and Coedydd Maentwrog.

Sycamore has few damaging effects on ground vegetation and supports rich epiphytic assemblages (Harding and Rose, 1986; Taylor, 1985). Although it is invasive in many woodland types, it appears to be capable of eventually reaching an equilibrium with ash (Waters and Savill, 1992). Sycamore has been established long enough to dominate many mature stands, and beneath these it generally regenerates poorly. Despite being a favourite target of grey squirrels, sycamore grows good timber and is favoured by owners with more commercial objectives. It is also resistant to drought and therefore more likely to survive the lack of aftercare which characterises roadside planting schemes. Accordingly, a concensus is now developing (Boyd, 1992) whereby sycamore will be controlled as part of mixed stands, accepted for planting in most new woods but eliminated from strict reserves and ancient woods where it is not already well established, and not introduced to ancient woods which it has not already reached.

21.8 Upland forest management

Until recently ecologists largely ignored the upland conifer plantations created since 1919. Although relict mires and other rich semi-natural habitats were retained within some plantations, the new forests were regarded as artificial and biologically poor substitutes for the long-established semi-natural moorlands and grasslands they replaced (NNC, 1986).

During the last decade the lead given by Mark Hill and Dorian Moss in the 1970s on research into the ecological development of plantation forests was followed by several considerable investigations into the ecology of the established forests and their wildlife, and methods of enriching them as wildlife habitats. At the same time, ecologists also investigated the overall impact of afforestation on the uplands. Research and survey was used to demonstrate the

importance of open moorland, mire and montane habitats (Stroud *et al.*, 1987; Ratcliffe and Thompson, 1988) to dispute the effects of afforestation beyond the forest fence (Stroud and Reed, 1986; Avery, 1989; Avery and Leslie, 1990) and to document the influence of plantations on sediment output to fresh waters (Soutar, 1989), watercourses (Newson, 1985), fish (Stoner and Gee, 1985) and aquatic fauna (Ormerod *et al.*, 1987).

21.8.1 Open space habitats

Open space habitats have been recognised as the focus of habitat diversity within plantation forests (Goldsmith, 1981; Smith and Charman, 1988; Good *et al.*, 1990). They comprise (i) pre-plantation habitats which were retained within the forest fence, notably the riversides and larger mires, the grasslands of forest smallholdings, high altitude moorland and outcrops, and (ii) the open ground of rides, roads and beneath power lines, which is maintained mostly as a necessary part of forest management. As the plantations have reached maturity, so (iii) clear-fell compartments have also become increasingly prominent.

The pre-plantation habitats surviving within the forest fence have not remained constant. Changes to grazing and burning regimes have allowed successional changes in raised mires (Chapman and Rose, 1991) and heaths, which have mostly impoverished the flora. Correspondingly, the ride and road vegetation has developed as a mosaic of grassland, heath and ruderal habitats (Goldsmith, 1981). Clear fells have been vegetated by a mixture of species surviving from the preceeding moorland and woodland species which have colonised during the first rotation (Hill, 1979). Birch, rowan and other broadleaved trees have invaded strongly in some forests.

Streamsides were commonly overplanted. Some became almost as bare of ground vegetation as the cores of planted compartments, but open spaces survived along the larger and more incised watercourses. Relieved of persistent grazing, riparian zones developed into mosaics of overgrown heathland and acid grassland, regenerating residual woodland assemblages and herb-rich fens and neutral grasslands.

21.8.2 Mature forest habitats

As the new plantations have matured, it has become increasingly apparent that the fauna and flora associated with plantations at the end of their commercial rotation and beyond begin to develop old-growth features. The ground vegetation returns, especially on brown earth soils and beneath pine and larch (Hill, 1979). The bird diversity increases (Currie and Bamford, 1982). Goshawks have taken up residence in end-of-rotation stands in Kielder Forest (Petty, 1989). Little is known about invertebrates, but the increasing availability of dead trees in older plantations may eventually allow conifer saproxylic species to spread.

Recognising this potential, Ratcliffe and Petty (1986), Peterken (1987) and Peterken *et al.* (1992) have suggested that old-growth should be designed into plantation forests as long-rotation stands centred on protected sites in valleys. Most such stands would be managed on a continuous cover basis in most forests, but adaptations would be necessary in, for example, forests dominated by the light-demanding Scots pine. A few old-growth stands would be retained with minimal intervention, where deadwood habitats might develop and the 'natural' ecology of non-native trees could be studied.

21.8.3 Upland forest ecology at the landscape scale

As the forests have matured, so too have ecologists' attitudes towards them, and the larger forests have increasingly been regarded as ecosystems on a landscape scale. The model forest (Peterken, 1993c) (Fig. 21.12) comprises tesserae of plantations of different ages, separated by networks of rides and roads, containing a scatter of isolated moorland patches, mires, pools and other semi-natural habitats, linked along a dendroid network of watercourses. This perspective extends beyond the forest fence, where the heathland enclaves on the higher levels march with the moorland above the forest and the scattered broadleaves along the lower reaches of streams within forests link with streamside habitats below the forest.

The landscape-scale perspective is particularly important when considering the control and conservation of species which roam over the whole forest (Ratcliffe, 1989). Deer are essentially woodland animals and now that red deer have become established within extensive mature forests, they have grown larger than the deer that remained on the moorlands. Within the forest they require both shelter in the plantations and feed from rides, glades and other open spaces. In summer they move freely between forests and higher ground.

Recent studies have shown how birds use the whole forest. Goshawks appear to select mature plantations on high ground for nest sites, with a good view over their hunting grounds in the valleys (Petty, 1989). Blackcock use a variety of habitats for feeding, nesting, roosting and resting, and this usage varies with the seasons (Cayford and Jones, 1989). If plantations can be designed to accommodate this variety within the 1.5km the birds range from their lek, then these birds can thrive in upland plantation forests. Nightjars use heathland and clear fells within some forests if they are present, but have failed to take full advantage of the extensive habitats open to them in much of the uplands. We knew before 1980 how song birds adjusted to the age, structure and composition of plantations, but Bibby *et al.* (1985) demonstrated the importance of clear fells as foci of bird diversity which rotate around the whole forest as compartments are harvested and restocked.

A large-scale perspective is also valuable for less mobile species, for populations move over time in response to afforestation and changes in forest management. We need to know a good deal more about the spread of plants into and around plantations, and about the changes in invertebrate assemblages after afforestation (Coulson, 1988; Good *et al.*, 1990) if we are fully to understand the ecology of upland plantations.

North American foresters and forest ecologists manage forests on a large scale. Their accumulated insights provide an inspiration for biodiversity development in upland plantation forests (Thomas, 1979; Hunter, 1990).

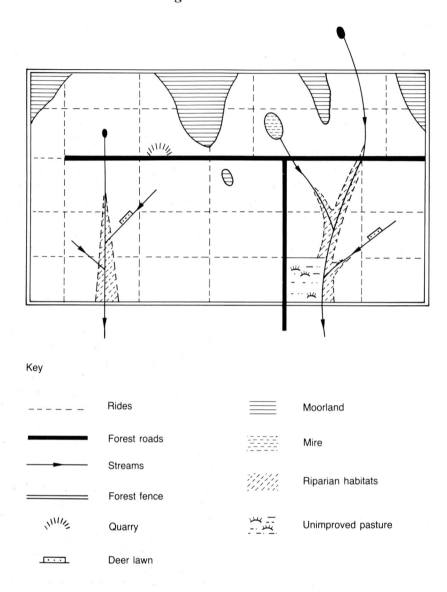

Key

– – – –	Rides	≡≡≡	Moorland
▬▬▬	Forest roads	⋯⋯	Mire
───▶	Streams	//////	Riparian habitats
════	Forest fence		Unimproved pasture
⸜⸝⸍	Quarry		
⌐·····⌐	Deer lawn		

Fig. 21.12 The generalised form of many upland plantation forests. The forest lies between moorland on higher ground and farmland below. Enclaves of moorland within the forest fence interdigitate with streams and habitats associated with drainage. Most streams arise within the plantations, but some drain from higher ground. The width of the riparian strips increases along the length of the streams. Farmland within the forest fence is usually directly linked to farmland outside. Deer lawns have been cut along streamsides. The regular network of rides and roads is broken by natural topographical features.

21.8.4 Changes in forest management

A feature of the 1980s has been the diversification of forests in a concerted attempt to improve them as habitats and amenities. Guidelines have been promulgated by the Forestry Commission on *Forests and Water* (revised 1991), *Forest Landscape Design* (1989) and *Forest Nature Conservation* (1990). The landscaping guidelines, which were based on concepts of visual harmony within and beyond forests and an interpretation of 'naturalness' in visual terms (Lucas, 1991), have been the main force behind forest redesign, but most of the landscaping principles have ecological analogues which should ensure that wildlife will benefit as well. Implementation is well advanced in some forests, such as Kielder (MacIntosh, 1989) and this has undoubtedly brought considerable benefits (Good *et al.*, 1990). Petty and Avery's (1990) account of forest birds was one of many publications dealing with particular species and groups within forests.

The principle features of change have been:

(1) Clearance of streamsides to leave wide strips of grassland, marsh and groups of native trees, together with glades for deer cut into plantations.

(2) Ride treatments, such as scalloping margins to create wide, sheltered bays, and cutting back mature plantations to allow rides to remain open and develop margins of scrub.

(3) Retention of open space habitats and some attempt at restoring past damage, e.g. by blocking drains from bogs.

(4) Broadleaved planting and retentions towards a target of 5% broadleaves in all forests.

(5) Diversification of age–class distribution, so that forests have a balanced age structure which can be maintained indefinitely.

(6) Retention of a few plantations beyond the commercial optimum for habitat, landscaping and as a backdrop to amenities such as picnic sites.

(7) Shaping of clear fells to produce irregular shapes and more edge habitats.

(8) Better links with surroundings, by means of irregular marginal shapes and contiguity between habitats inside and outside forests.

21.8.5 Pattern of forests

Throughout the 1980s the major conflict between forestry and nature conservation has been about the extent and distribution of afforestation (Smart and Andrews, 1985; NCC, 1986). It was generally agreed by the nature conservation protagonists that more forests were desirable, but not on existing important unwooded habitats, and preferably with native trees. The general principles of forest distribution were in dispute. Whilst some wanted integration of forestry and farming (McPhillimy, 1989), which implies a small-scale patchwork of forest patches scattered over most landscapes, others (Peterken, 1987) have argued for a large-scale patchwork of large forests and large forest-free uplands.

These issues were brought to a head in the Flow Country (Stroud *et al.*, 1987), where extensive afforestation was threatening to destroy a large wilderness of mire, moorland and important bird habitats. After much rancour — and at the cost, some claim, of breaking up the NCC — a compromise was reached on a landscape-scale zonation of large forests and forest-free areas. This led to a new approach for forestry planning, entitled 'Indicative Strategies', which identifies at a broad scale those areas where forestry must be constrained, and those areas where further forests are acceptable (Goodstadt, 1991).

Another recent development has been the trend to move forestry 'down the hill'. Changes in agricultural policy have made it possible for forestry to break out from its moorland fastnesses onto marginal agricultural land. Much of the new broadleaved planting during the first three years of the Broadleaves Policy was afforestation on the upland margins, though the absolute amounts were not large.

21.9 Creating new native woodlands

Little attention was paid to the expansion of native woodlands whilst the emphasis was on conserving existing woods and restraining the anarchic spread of upland conifer plantations. Such expansion was regarded as desirable but postponable, so scarce resources were concentrated in saving ancient woods from destruction. Incentives were available from

Tree species	1	2	3	4	5	6	7	8	9	10	W	D	H	A	N	X
Main trees																
Acer campestre, field maple				L	o	o	o	o	o	o		d	h		n	
Carpinus betulus, hornbeam						o	o						h	a	n	
Fagus sylvatica, beech						o	o	o				d		a	n	x
Fraxinus excelsior, ash	o	o	o	o	o	o	o	o	o	o	w	d	h		n	x
Pinus sylvestris, Scots pine		L										d		a		x
Populus nigra, black poplar					L	L	L	L	L		w				n	
Populus canescens, grey poplar						o	o	o	o	o	w				n	
Quercus petraea, sessile oak		o	o	o	L	L	L	L	L	o		d		a	n	x
Quercus robur, pedunculate oak		o	o	o	o	o	o	o	o	o	w		h	a	n	x
Tilia cordata, small-leaved lime				L	L	L	L	L	L	L		d*	h	a	n*	
Tilia platyphyllos, large-leaved lime				L	L	L							h		n	
Ulmus glabra, wych elm	o	o	o	o	o	o	o	o	o	o			h		n	x
Short-lived trees																
Alnus glutinosa, alder	o	o	o	o	o	o	o	o	o	o	w				n	
Betula pendula, silver birch	o	o	o	o	o	o	o	o	o	o		d	h	a	n	x
Betula pubescens, downy birch	o	o	o	o	o	o	o	o	o	o	w	d	h	a	n	x
Populus tremula, aspen	o	o	o	o	o	o	o	o	o	o		d*	h	a	n	x
Prunus avium, gean		o	o	o	o	o	o	o	o	o			h		n	
Salix alba, white willow			o	o	o	o	o	o	o	o	w				n	x
Salix caprea, goat willow	o	o	o	o	o	o	o	o	o	o	w		h		n	x
Salix cinerea, grey willow	o	o	o	o	o	o	o	o	o	o	w			a	n	x
Salix fragilis, crack willow			o	o	o	o	o	o	o	o	w				n	x
Salix pentandra, bay willow			o	o							w				n	
Salix purpurea, purple willow			o	o	o	o	o	o	o	o	w				n	x
Salix triandra, almond willow				o	o	o	o	o			w				n	
Salix viminalis, osier	o	o	o	o	o	o	o	o	o	o	w				n	
Subordinate trees and large shrubs																
Crataegus laevigata, hawthorn					L	L	L	L					h		n	
Crataegus monogyna, hawthorn	o	o	o	o	o	o	o	o	o	o		d	h	a	n	x
Corylus avellana, hazel	o	o	o	o	o	o	o	o	o	o		d	h		n	x
Ilex aquifolium, holly	o	o	o	o	o	o	o	o	o	o		d		a	n	
Malus sylvestris, crab apple			o	o	o	o	o	o	o	o		d	h		n	
Prunus padus, bird cherry	o	o	o	o	L						w				n	
Prunus spinosa, blackthorn	o	o	o	o	o	o	o	o	o	o	w	d	h		n	x
Sorbus aria, whitebeam		L		L			L	L				d		a*	n	x
Sorbus aucuparia, rowan	o	o	o	o	o	o	o	o	o	o		d		a		x
Sorbus torminalis, wild service					L	L	L	L	L	L			h	a	n	
Taxus baccata, yew				L			L	L				d			n	

*aspen: dry, acid soils only in zones 1–3

*lime: dry, neutral–alkaline soils only in zone 4

*whitebeam: acid, light soils only in zone 7; other soils indicated also appropriate for zone 7

Fig. 21.13 (opposite and above) A guide to the selection of locally native tree species for the creation of new native woodlands. The map and table summarise guidance given by Soutar and Peterken (1989). Great Britain is divided into ten zones, using the natural limits to the ranges of beech, hornbeam, bird cherry, small-leaved lime and Scots pine, in so far as complex distribution patterns can be resolved into simple lines. The species which are appropriate for each zone (o) and those which are appropriate with qualifications (L) are indicated. The species/zone combinations marked (L) indicate that only strictly local stock should be used, since these species in these zones have special biogeographical significance.

Most species will grow on medium-textured, moderately well drained, neutral soils, but are more particular on extreme soils. The table therefore also indicates which species are appropriate for particular soils:

- W: wet soils, swampy ground and the soils of floodplains, where flooding is seasonal;
- D: light, dry soils, typically heathy sands and sandy loams;
- H: heavy soils, mostly poorly-drained clays and clay loams;

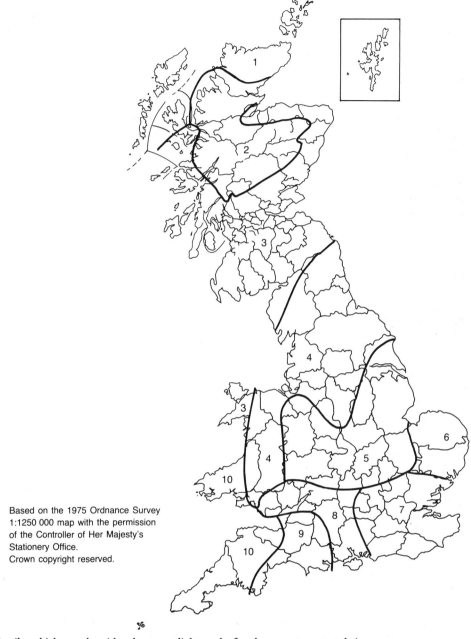

Based on the 1975 Ordnance Survey
1:1250 000 map with the permission
of the Controller of Her Majesty's
Stationery Office.
Crown copyright reserved.

- A: acid soils, which may be either heavy or light, and often have peaty accumulations;
- N: alkaline and neutral soils, typically those over chalk and limestone;
- X: exposed sites.

Examples:
(1) If the site available for planting lies on the middle Severn floodplain (zone 5), the planter would be presented with the following choice of main trees: ash, black poplar, grey poplar and pedunculate oak, but any black poplar planted should be raised from local stock.
(2) If the site available is a dry, heathy moor on lower Deeside (zones 2/3 boundary), the choice would be limited to sessile oak amongst the main trees, but silver birch, downy birch, aspen, hawthorn, holly, and rowan would qualify from other groups. Scots pine would be appropriate if local stock were used in the vicinity of a native pinewood.

about 1980 for planting Scots pine close to existing native pinewoods, but otherwise an expansion of native woodland was regarded as a conservationist's pipe-dream. During the 1980s, however, the spread of interest in native woodland conservation combined with changes in agricultural policy to create an opportunity for more afforestation with native species. The initial interest in expanding the native pinewoods has broadened into, amongst others, the Central Scotland Woodlands Project, Trees for Scotland, Community Forests, the New National Forest and the Farm Woodland Scheme. Under most of these schemes new planting is by no means restricted to native species, but the general aim is to include a substantial proportion of native trees. The Woodland Trust has created over 400ha of new woodland, often with the assistance of members of the public on tree planting weeks. The level of enthusiasm generated is epitomised by Leicester, where every schoolchild was out planting trees in national Tree Week, December 1992.

New native woodlands can develop from four sources: (i) natural succession on unwooded ground; (ii) planting appropriate native trees on bare ground; (iii) removing plantations of non-native tree species from ancient woods and allowing native trees to replace them; and (iv) planting or allowing natural succession of native tree species within existing upland plantation forests. All four processes have gathered pace during the 1980s.

Perhaps the most impressive of the long-established native woodland restoration schemes was that started in 1961 on the inner Hebridean island of Rum (Wormell, 1977). Moorland was ploughed and planted with a mixture of mainly broadleaved trees, selecting the tree and shrub species which were known from pollen analysis to have been present (or which were still present in surviving wooded fragments), and using local stock wherever possible. By 1987, 813 500 saplings of 28 species had been planted, the older plantations had developed closed canopies and the woodland invertebrate fauna was steadily increasing and diversifying. For 20 years the plantations had been established by ploughing through moorland and mires indiscriminately and using Lodgepole pine as a nurse, but in the 1980s a more sensitive approach was initiated, involving less ploughing, leaving more open space and removing the Lodgepole nurses.

Today, the nature conservation preference is to create new native woodland by natural regeneration and thus conserve local genotypes (Soutar and Spencer, 1991). This is practicable where new woods develop as expansions from existing woods or from scattered trees along streamsides and rivers, but planting will generally be necessary elsewhere, if only because people like to take positive action, rather than passively wait for trees to colonise. Accordingly, guidance was given on species selection (Soutar and Peterken, 1989) to ensure that only those species which are native to the district and suitable for the site are planted (Fig. 21.13).

In upland plantations a majority of the broadleaves introduced or retained in pursuit of the 5% quota was native species. In ancient woods, conifers underplanted below oaks were removed from Dalavich Wood (Kirby and May, 1989) and many other western oakwoods. In some important lowland ancient woods, such as Chalkney Wood, the original coppice mixture had not been completely exterminated by plantations, and measures were initiated to restore parts to native woodland.

21.10 A new relationship between forestry and nature conservation

In 1980 one could fairly claim that forestry and nature conservation were a long way apart. Public disquiet at the loss of broadleaved woodlands generated a great deal of critical media interest. Despite strong representations from conservation organisations, the Forestry Commission would not regard 'so-called' ancient woodlands as important. Their wildlife research branch was almost wholly concerned with pest control. Outside the controls and grants of official forestry an alternative forestry had sprung up, based on less intensive management, use of native species and a minor revival of coppicing and woodmanship in a form modified for a modern social and economic context.

The 1982 Loughborough conference on *Broadleaves in Britain* was a watershed. Thereafter, the

Forestry Commission, in consulation with many others, developed new policies which eventually emerged in 1985 as the Broadleaves Policy. By then the Wildlife and Conservation Research Branch had been revitalised as a group to study and advise on conserving wildlife as well as controlling pests. All this was assisted by the Wildlife and Countryside (Amendment) Act 1985, which conferred on the Forestry Commission a duty to conserve wildlife.

Since 1985 official forestry has developed even further towards accepting the validity of alternative forestry in a proportion of (mainly ancient) woods, notably with the introduction in 1992–3 of Special Management Grants and guidance for managing semi-natural woodlands and combined with special promotions, such as the Anglian Woodland Project. The distinction between woodland conservation and forestry has been broken down further by the implementation of wildlife-considerate policies in a large number of woods and forests where amenity, landscape and public access are factors.

Whilst these developments are excellent for nature conservation in woodlands, they raise new issues for nature conservationists. What is the main purpose of woodland nature reserves (WNRs) if wildlife is reasonably well protected in most woodlands outside nature reserves? Have the woodland National Nature Reserves (WNNRs) a special role, or have they become just another mechanism for protecting native woods from adverse changes which, on current policies, are unlikely to take place? What is the point of Sites of Special Scientific Interest in woodlands when forestry policy is based on the recognition of ancient woodlands as a special category? Even if SSSIs still have a role in woodlands, is there any justification for selecting them to represent typical woodland types, when these will be conserved in any case by forestry policies of wider application?

My answers to these questions are clear. WNRs are undoubtedly necessary, not least because forestry for other objectives will not provide the full range of desirable conditions (Peterken, 1991). The WNNRs would be justified as a separate category if they were primarily used as reference sites, places to test and demonstrate the effectiveness of management for nature conservation, and for long-term, ecological research into both these areas. These are the functions which NNRs were originally meant to fulfil. The most valuable woodlands for nature conservation should still be scheduled as SSSIs, because this mechanism provides, through specialist surveys, a high awareness of what is valuable and, through site-specific agreements, management which is designed to retain and enhance those values. SSSI management agreements still provide greater protection than the Forestry Authority's Codes, which remain voluntary.

One could argue that, after some legislative and administrative adjustments, the Forestry Commission could equally well oversee WNNRs and woodland SSSIs. These special woods could then be more firmly integrated into the national forestry strategy, rather than remain as a separate and competing function. Such a rearrangement could be made to work, but unfortunately it would create as many problems as it would solve. The nature conservation functions would be split between Forest Enterprise and Forestry Authority. Years would be needed to build up the necessary staffing and experience. The ethos of the Forestry Authority would have to change still further (which is, perhaps, a good argument for giving them the job!). Privatisation of all or part of the Forestry Commission would jeopardise the whole idea. Outside forestry, pressure for the far less desirable transfer of non-woodland NNRs and SSSIs to agricultural authorities might become irresistible.

The 1980s saw a substantial response by forestry to pressures for nature conservation, and a substantial broadening of interest in the recreational and aesthetic values of woodlands. Specialist nature conservationists and the statutory nature conservation agencies allowed themselves to become increasingly marginalised in woodland nature conservation, partly because they developed the curious opinion that the problems between forestry and nature conservation had somehow been solved, as if they formed an issue which could be settled, rather than a relationship which must be nurtured. In the 1990s these agencies would do well to reconsider their role in the new circumstances and return to basics with renewed conviction.

References

Adamson, R.S. (1912), An ecological study of a
    Cambridgeshire woodland, *J. Linn Soc. (Bot.)*, **40**,
    339–87.

Adamson, R.S. (1921), The woodlands of Ditcham
    Park, Hampshire, *J. Ecol.*, **9**, 114–219.

Allison, H. and Peterken, G.F. (1985), Changes in the
    number of non-woodland trees in Britain since
    1945, *Arboric. J.* **9**, 259–69.

Anderson, M.C. (1964), Studies of the woodland light
    climate. II. Seasonal variation in the light climate, *J.
    Ecol.*, **52**, 643–63.

Anderson, M.L. (1967), *A History of Scottish Forestry*,
    Nelson, London.

Applebaum, S. (1972), Roman Britain, In: *The Agrarian
    History of England and Wales I. A.D. 43–1042*
    (Finberg, H.P.R., ed.), Cambridge University Press,
    Cambridge.

Avery, B.W. (1964), *The soils and land use of the district
    around Aylesbury and Hemel Hempstead. Memoir of the
    Soil Survey of Great Britain, England and Wales*,
    HMSO, London.

Avery, B.W. (1973), Soil classification in the Soil
    Survey of England and Wales, *J. Soil Sci.*, **24**,
    324–8.

Avery, M.I. (1989), Effects of upland afforestation on
    some birds of the adjacent moorlands, *J. Appl. Ecol.*,
    26, 957–66.

Avery, M. and Leslie, R. (1990), *Birds and Forestry*,
    Poyser, London.

Backmeroff, C.E. and Peterken, G.F. (1989), Long-
    term changes in the woodlands of Clairinsh, Loch
    Lomond, *Trans. Bot. Soc. Edin.*, **45**, 253–97.

Bagley, J.J. and Rowley, P.B. (1966), *A Documentary
    History of England I. 1066–1540*, Penguin,
    Harmondsworth.

Baker, A.R.H. and Butlin, R.A., eds. (1973), *Studies of
    Field Systems in the British Isles*, Cambridge
    University Press, Cambridge.

Baker, C.A., Moxey, P.A. and Oxford, P.M. (1978),
    Woodland continuity and change in Epping Forest,
    Field Studies, **4**, 645–69.

Ball, D.F. and Stevens, P.A. (1981), The role of
    'ancient' woodlands in conserving 'undisturbed' soils
    in Britain, *Biol. Conserv.*, **19**, 163–76.

Ball, S.G. (1989), The Nature Conservancy Council's
    invertebrate site register, In: *Utilisation des Inventaires
    d'Invertebres pour l'Identification et la Surveillance
    d'Espaces de grand Intret Faunistique* Beufort, F. de
    and Maurin, H., eds. pp. 31–4. Museum National
    d'Histoire Naturelle, Paris.

Barber, K.E. (1975), Vegetational history of the New
    Forest: a preliminary note, *Proc. Hants Field Club
    Archeol. Soc.*, **30**, 5–8.

Barfield, T. ed. (1984), *A Herefordshire Woodland Survey*,
    Herefordshire and Radnorshire Nature Trust, Hereford.

Barkham, J.P. and Norris J.M. (1970), Multivariate
    procedures in an investigation of vegetation and soil
    relations of two beech woodlands, Cotswold Hills,
    England, *Ecology*, **51**, 630–9.

Barr, C.J. (1990), Mapping the changing face of
    Britain, *Geographical Magazine*, **62**(10), 44–7.

Bazeley, M.L. (1921), The extent of the English forest
    in the thirteenth century, *Trans. Roy. Hist. Soc., 4th
    Ser.*, **4**, 140–72.

Beevor, H. (1925), Norfolk woodlands from the evidence of contemporary chronicles, *Quart. J. For.*, **19**, 87–110.

Bennett, K.D. (1989), A provisional map of forest types of the British Isles 5000 years ago, *J. Quart. Sci.*, **4**, 141–4.

Bibby, C.J., Phillips, B.N. and Seddon, A.J.E. (1985), Birds of restocked conifer plantations in Wales, *J. Appl. Ecol.*, **22**, 619–33.

Birks, H.H. (1970), Studies in the vegetational history of Scotland I. A pollen diagram from Abernethy Forest, Inverness-shire, *J. Ecol.*, **58**, 827–46.

Birks, H.J.B. (1973), *Past and Present Vegetation of the Isle of Skye*, Cambridge University Press, Cambridge.

Birks, H.J.B., Deacon, J. and Peglar, S. (1975), Pollen maps for the British Isles 5000 years ago, *Proc. Roy. Soc., London, B*, **189**, 87–105.

Birks, H.J.B. (1982), Mid-Flandrian forest history of Roudsea Wood National Nature Reserve, *New Phytol.*, **90**, 339–54.

Birks. H.J.B. (1989), Holocene isochrone maps and patterns of tree-spreading in the British Isles, *J. Biogeog.*, **16**, 503–40.

Birks H.J.B. (1993), Quaternary paleoecology and vegetation science – current contributions and possible future developments, *Review of Paleobotany and Palynology*, in press.

Birrell, J.R. (1962), *The Honour of Tutbury in the fourteenth and fifteenth centuries*, M.A. Thesis, University of Birmingham.

Birse, E.L. and Robertson, J.S. (1976), Plant communities and soils of the lowland southern upland regions of Scotland, *Soil Survey of Scotland*, Aberdeen.

Blackwood, J.W. and Tubbs, C.R. (1970), A quantitative survey of chalk grassland in England, *Biol. Conserv.*, **3**, 1–5.

Boer, P.J. den (1990), Density limits and survival of local populations in 64 carabid species with different powers of dispersal, *J. Evol. Biol.*, **3**, 19–48.

Bond, J. (1981), Woodstock Park under the Plantagenet kings: the exploitation and the use of wood in a medieval deer park, *Arboric. J.*, **5**, 201–13.

Bormann, F.H. and Likens, G.E. (1979), *Pattern and Process in a Forested Ecosystem*, Springer, New York.

Bosshard, W., ed. (1978) Der Wald in der europäischen Landschaft, *Mitt. Eidgen, Anst. forstl Versuchs.*, **54**, 363–564.

Boycott, A.E. (1934), The habitats of land *Mollusca* in Britain, *J. Ecol.*, **22**, 1–38.

Boyd, J.M. (1992), Sycamore – a review of its status in conservation in Great Britain, *Biologist*, **39**, 29–31.

Boys, J. (1794), *General View of the Agriculture of the County of Kent*, Board of Agriculture, London.

Brandon, P.F. (1963), *The Commonlands and Wastes of Sussex*, PhD Thesis, University of London.

Brandon, P.F. (1969), Medieval clearances in the East Sussex weald, *Trans. Inst. Br. Geog.*, **48**, 135–53.

Brandon, P.F. (1974), *The Sussex Landscape*, Hodder and Stoughton, London.

Braun-Blanquet, J. and Tüxen, R. (1952), Irische Pflanzengesellschaften. *Veröff. Geobot. Inst. Rübel Zürich*, **25**, 224–421.

Brown, A.H.F. and Oosterhuis, L. (1981) The role of buried seeds in coppice woods, *Biol. Conserv.* **21**, 19–38.

Buckland, P.C. (1979), *Thorne Moors: a paleoecological study of a bronze age site*, Birmingham University Geography Department, Occasional Papers, 8.

Buckland, P.C. and Kenward, H.K. (1973) Thorne Moor: a paleo-ecological study of a Bronze Age site, *Nature*, **241**, 405–6.

Buckley, G.P. ed. (1992) *Ecology and Management of Coppice Woodlands*, Chapman & Hall, London.

Buckley, G.P. and Knight, D.G. (1989) The feasibility of woodland reconstruction, In: *Biological Habitat Reconstruction* (Buckley, G.P., ed), pp. 171–88, Belhaven Press, London.

Budd, J.T.C. (1989), National Countryside Monitoring Scheme, In: *Rural Information for Forward Planning*. ITE Symposium 21 (Bunce, R.G.H. and Barr, C.J., eds), Institute of Terrestrial Ecology, Grange-over-Sands.

Bunce, R.G.H. (1982, 1989) *A Field Key for Classifying British Woodland Vegetation, Parts 1 and 2*, Institute of Terrestrial Ecology, Cambridge.

Bunce, R.G.H. and Jeffers, J.N.R., eds. (1977), *Native Pinewoods of Scotland*, Institute of Terrestrial Ecology, Cambridge.

Burgess, R.L. and Sharpe, D.M., eds, (1981), *Forest Island Dynamics in Man-Dominated Landscapes*, Ecological Studies 41, Springer, New York.

Cantor, L.M. and Hatherly, J. (1979), The medieval parks of England, *Geography*, **64**, 71–85.

Cantor, L.M. and Wilson, J.D. (1969), The mediaeval deer-parks of Dorset IX. *Proc. Dorset Nat. Hist. Archaeol. Soc.*, **91**, 196–205.

Cayford, J. and Jones, P.H. (1989), Black grouse in Wales, *RSPB Conserv. Rev.*, **3**, 79–81.

Centre for Agricultural Strategy (1980), Strategy for the UK forest industry, CAS Report, no. 6, Reading.

Chapman, S.B. and Rose, R.J. (1991). Changes in the vegetation at Coom Rigg Moss National Nature Reserve within the period 1958–86, *J. Appl. Ecol.*, **28**, 140–53.

Chilterns Standing Conference (1970), *A Plan for the Chilterns*.

Christy, M. (1924), The hornbeam (*Carpinus betulus L.*) in Britain, *J. Ecol.*, **12**, 39–94.

Clapham, A.R., Tutin, T.G. and Warburg, E.F. (1962), *Flora of the British Isles*, Cambridge University Press, Cambridge.

Clarke, R.T. and McCulloch, J.S.G. (1975), Recent work on the comparison of the effects of alternative uses (coniferous forest, upland pastures) on water catchment behaviour, *Conservation and Land Drainage Conference* (Water Space Amenity Commission), 55–63.

Clements, F.E. (1910), *The Life History of Lodgepole Burn Forests*, Bulletin 79, USDA Forest Service, Washington.

Cmd. 7122 (1947), *Conservation of Nature in England and Wales*, HMSO, London.

Coleburn, P. (1983), *Hampshire's Countryside Heritage, 2, Ancient Woodland*, Hampshire County Council, Winchester.

Cooke, R.J. (1992), *Phase 2 Woodland NVC Surveys 1988–1991 in England and Wales*, unpublished report to English Nature.

Corbet, G.B. and Southern, H.N., eds. (1977), *The Handbook of British Mammals* (2nd edn.), Blackwells, Oxford.

Coulson, J.C. (1988), The structure and importance of invertebrate communities on peatlands and moorlands, and effects of environmental and management changes, In: *Ecological Change in the Uplands*. Special Publication 7 Usher, M.B. and Thompson, D.B.A., (eds), pp. 365–81, British Ecological Society Blackwell, Oxford.

Crocker, J. (1981), *Charnwood Forest: a changing landscape*, Loughborough Naturalists' Club, Loughborough.

Crocker, R.L. and Major, J. (1955), Soil development in relation to vegetation and surface age at Glacier Bay, Alaska, *J. Ecol.*, **43**, 427–48.

Cronon, W. (1983), *Changes in the Land*, Hill and Wang, New York.

Crowson, R.A. (1962), Observations on Coleoptera in Scottish oak woods, *The Glasgow Naturalist*, **18**, 177–95.

Crowson, R.A. (1972), Observations on Coleoptera in Eaves Wood, Silverdale, Lancs., *Entomologist's Monthly Magazine*, **108**, 176.

Crowther, R.E. (1977), Short-rotation coppice – time for a revival. *Forestry and British Timber*, **6**, 42–3.

Crowther, R.E. and Evans, J. (1986), *Coppice*,

Forestry Commission leaflet 83, H.M.S.O., London.

Currie, F.A. and Bamford, R. (1982), The value to birdlife of retaining small conifer stands beyond normal felling age within forests, *Quart. J. For.*, **76**, 153–60.

Darby, H.C. (1951), The clearing of the English woodlands, *Geography*, **36**, 71–83.

Darby, H.C. (1977), *Domesday England*, Cambridge University Press, Cambridge.

Darby, H.C. and Terrett, J.B. (1954), *The Domesday Geography of Midland England*, Cambridge University Press, Cambridge.

Day, S.P. (1991), Post-glacial vegetational history of the Oxford region, *New Phytol.*, **119**, 445–70.

Dent, J. (1962), *The Quest for Nonsuch*, Hutchinson, London.

Dewar, H.S.L. (c. 1926), The field archaeology of Doles, *Papers and Proc. Hants Field Club and Archaeol. Soc.*, **10**, 118–26.

Diamond, J.M. (1975), The island dilemma: lessons of modern biogeographic studies for the design of natural reserves, *Biol. Conserv.*, **7**, 129–46.

Dimbleby, G.W. (1962), The development of British heathlands and their soils, *Oxford Forestry Memoir*, 23.

Dimbleby, G.W. (1965), Post-glacial changes in soil profiles, *Proc. Roy. Soc., B*, **161**, 355–62.

Dimbleby, G.W. and Gill, J.M. (1955), The occurrence of podzols under deciduous woodland in the New Forest, *Forestry*, **28**, 95–106.

Doing, H. (1975), Subdivision of the alliance Quercion robori-petraeae into Vaccinio-Quercion and Violo-Quercion, In: *La Vegetation des Forêts Caducifoliees Acidiphiles* (Gehu, J., ed.), pp. 73–87, Cramer, Vaduz.

Donaldson, J. (1974), *General View of the Agriculture of the County of Northampton*, Board of Agriculture, London.

Donkin, R.A.(1960), The Cistercian settlement and the English royal forests, *Citeaux*, **11**, 39–55.

Duffy, D.C. and Meier, A.J. (1992), Do Appalachian herbaceous understories ever recover from clear-cutting?, *Cons. Biol.*, **6**, 196–201.

Duffey, E., ed. (1974), *Grassland Ecology and Wildlife Management*, Chapman & Hall, London.

Dzwonko, Z. and Loster, S. (1988), Species richness of small woodlands on the western Carpathian foothills, *Vegetatio*, **76**, 15–27.

Edgell, M.C.R. (1969), Vegetation of an upland ecosystem: Cader Idris, Merionethshire, *J. Ecol.*, **57**, 335–59.

Edlin, H.L. (1949), *Woodland Crafts in Britain*, Batsford, London.

Edwards, M.E. (1986), Disturbance histories of four Snowdonia woodlands and their relation to Atlantic bryophyte distributions. *Biol. Cons.*, **37**, 301–20.

Ellenberg, H. (1988), *Vegetation ecology of central Europe*, (4th ed.), Cambridge University Press, Cambridge.

Ellenberg, H. and Klötzli, F. (1972), Waldgesellschaften und Waldstandorte der Schweiz, *Mitt. schweiz. Anst. forstl. Versuchsw.*, **48**, 587–930.

Ellenberg, H. (1990), Eutrophication as a significant background problem for European wildlife, In: *Rapeseed 00 and Intoxication of Wild Animals*, (Askew, M.F., ed.), pp. 117–30, Commission of the European Communities, Luxembourg.

Elton, C.S. (1966), *The Pattern of Animal Communities*, Methuen, London.

Essex, S.J. (1984), The use of the annual agricultural census returns to measure change in the area of woodland, *East Midlands Geographer*, **8**, 159–62.

Evans, C.E. and Becker, D. (1988), Rhododendron control on RSPB reserves, *RSPB Conserv. Rev.*, **2**, 54–6.

Evans, J. (1984), *Silviculture of Broadleaved Woodland*, Forestry Commission Bulletin 62. H.M.S.O., London.

Evelyn, J. (1664), *Sylva, or a Discourse of Forest-trees* (4th edn.), publ. 1706, Doubleday, London.

Eversley, Lord (1910), *Commons, Forests and Footpaths* (revised edn.), Cassell, London.

Ewen, A.H. and Prime, C.T. (1975), *Ray's Flora of Cambridgeshire*, Wheldon and Wesley, Hitchen.

Fairey, J. (1815), *General View of the Agriculture of Derbyshire*, Board of Agriculture, London.

Falinski, J.B. (1986), Vegetation dynamics in temperate lowland primeval forests, *Geobotany*, **8**, Junk, Dordrecht.

Falkengren-Grerup, V. (1986), Soil acidification and vegetation changes in deciduous forest in southern Sweden, *Oecologia (Berlin)*, **70**, 339–47.

Fasham, P.J. (1983), Fieldwork in and around Micheldever Wood, Hampshire 1973– 1980, *Proc. Hamps. Field Club Arch. Soc.*, **39**, 5–45.

Fell, A. (1908), *The Early Iron Industry of Furness and District*, Hume Kitchen, Ulverston.

Finberg, H.P.R. (1972), *The Early Charters of the West Midlands* (2nd edn.), Leicester University Press, Leicester.

Fitzrandolph, H.E. and Hay, M.D. (1926), *The Rural Industries of England and Wales I. Timber and Underwood Industries and Some Village Workshops*, Clarendon Press, Oxford.

Fleming, S.C. (1983), *An examination of post-war change in woodlands of the farmed area of eastern Hampshire*, PhD thesis, University of Reading.

Fletcher, A. ed. (1982), *Survey and Assessment of Epiphytic Lichen Habitats*, unpublished report to NCC by British Lichen Society. CST Report 384.

Flower, N. (1977), *An Historical and Ecological Study of Inclosed and Uninclosed Woods in the New Forest, Hampshire*, PhD Thesis, University of London.

Ford, E.D., Malcolm, D.C. and Atterson, J., eds. (1979), *The Ecology of Even-aged Forest Plantations*, Institute of Terrestrial Ecology, Cambridge.

Ford, E.D. and Newbould, P.J. (1977), The biomass and production of ground vegetation and its relation to tree cover through a deciduous woodland cycle, *J. Ecol.*, **65**, 201–12.

Forestry Commission (1974), Annual Report for 1973–1974, 10–13.

Forestry Commission (1977), *The Wood Production Outlook in Britain*, Edinburgh.

Forestry Commission (1978), *The Place of Forestry in England and Wales*, Edinburgh.

Forestry Commission (1984), *Census of Woodlands and Trees, 1979–1982*, Great Britain, Forestry Commission, Edinburgh.

Forman, R.T.T. and Godron, M. (1986), *Landscape Ecology*, Wiley, New York.

Foster, D.R., Zebryk, T., Schoonmaker, P. and Lezberg, A. (1992), Post-settlement history of human land-use and vegetation dynamics of a *Tsuga canadensis* (hemlock) woodlot in central New England, *J. Ecol.*, **80**, 773–86.

Francis, J., Morton, A.J. and Boorman, L.A. (1992). The establishment of ground flora species in recently planted woodland, *Asp. Appl. Biol.*, **29**, 171–8.

Franklin, J.F., Cromack, K., Denison, W. *et al.* (1981), *Ecological Characteristics of Old-growth Douglas-fir Forests*. General Technical Report PNW-118. USDA Forest Service.

Fuller, R.J. (1992). Effects of coppice management on woodland breeding birds, In: *Ecology and Management of Coppice Woodlands* (Buckley, G.P., ed.), pp. 169–92, Chapman & Hall, London.

Game, M. and Peterken, G.F. (1984), Nature reserve selection strategies in the woodlands of central Lincolnshire, England. *Biol. Conserv.*, **29**, 157–81.

Garfitt, J.E. (1977), Irregular silviculture in the service of amenity, *Quart. J. For.*, **71**, 82–5.

Garland, S.P. (1983). Beetles as primary woodland indicators. *Sorby Record*, **22**, 3–38.

Garthwaite, P.F. (1977), Management and marketing of hardwoods in S.E. England, *Quart. J. For.*, **71**, 67–77 and 144–50.

Gehu, J., ed. (1975), *Colloques Phytosociologiques III. La Vegetation des Forêts Caducifoliees Acidiphiles*, Cramer, Vaduz.

Gibson, C.W.D. (1988), The distribution of 'ancient woodland' plant species among areas of different history in Wytham Woods, Oxfordshire, In: *Woodland Conservation and Research in the Clay Vale of Oxfordshire and Buckinghamshire*. Research and Survey in Nature Conservation 15 (Kirby, K.J. and Wright, F.J. eds.), pp. 32–40, Nature Conservancy Council, Peterborough.

Gilmour, J.S.L. and Walters, S.M. (1964), Philosophy and classification, *Vistas Bot.*, **4**, 1–22.

Gimmingham, C.H. (1972), *Ecology of Heathlands*, Chapman & Hall, London.

Godwin, H. (1962), Vegetational history of the Kentish chalk downs as seen at Wingham and Frogholt, *Veröff. Geobot. Inst. Rübel Zürich*, **37**, 83–99.

Godwin, H. (1968), Studies of the post-glacial history of British vegetation XV. Organic deposits of Old Buckenham Mere, Norfolk, *New Phytol.*, **67**, 95–107.

Godwin, H. (1975), *The History of the British Flora* (2nd edn.), Cambridge University Press, Cambridge.

Godwin, H. and Deacon, J. (1974), Flandrian history of oak in the British Isles, In: *The British Oak: its History and Natural History* (M.G. Morris and F.H. Perring, eds.), pp. 51–61, Classey, Faringdon.

Godwin, H., Clowes, D.R. and Huntley, B. (1974), Studies in the ecology of Wicken Fen V. Development of fen carr, *J. Ecol.*, **62**, 197–214.

Goldsmith, F.B., ed. (1981), *The Afforestation of the Uplands: the botanical interest of areas left unplanted*, Discussion Papers on Conservation, 35. University College, London.

Goldsmith, F.B. (1991a), The selection of protected areas. In: *The Scientific Management of Temperate Communities for Conservation*. British Ecological Society symposium 31 (Spellerberg, I.F., Goldsmith, F.B. and Morris, M.G., eds.), pp. 273–91, Blackwell, Oxford.

Goldsmith, F.B. ed. (1991b) *Monitoring for Conservation and Ecology*, Chapman & Hall, London.

Good, J.E.G., Williams, T.G., Wallace, H.L. *et al.* (1990), *Nature Conservation in Upland Conifer Forests*. Report to the Forestry Commission and the Nature Conservancy Council, Institute of Terrestrial Ecology, Bangor.

Goodfellow, S. and Peterken, G.F. (1981), A method for survey and assessment of woodlands for nature conservation using maps and species lists: the example of Norfolk woodlands, *Biol Conserv.*, **21**, 177–95.

Goodier, R. and Bunce, R.G.H. (1977), The native pinewoods of Scotland: the current state of the resource, In: *Native Pinewoods of Scotland* (Bunce, R.G.H. and Jeffers, J.N.R., eds.), pp. 78–87, Institute of Terrestrial Ecology, Cambridge.

Goodstadt, V. (1991), Indicative forestry strategies – the Scottish experience, In: *The Right Trees in the Right Place*. Seminar Proceedings 4, (Mollan, C. and Maloney, M. eds.), pp. 108–15, Royal Dublin Society, Dublin.

Gordon, G.P. (1911), Primitive woodland and plantation types in Scotland, *Trans. Roy. Scot. Arboric. Soc.*, **24**, 153–77.

Green, B.H. and Pearson, M.C. (1968), The ecology of Wybunbury Moss, Cheshire I. The present vegetation and some physical, chemical and historical factors controlling its nature and distribution, *J. Ecol.*, **56**, 245–67.

Greig, J. (1982), Past and present lime woods of Europe, In: *Archaeological aspects of woodland ecology*, British Archaeological Report S. 146 (Bell, M. and Limbrey, S. eds.), pp. 23–55, Oxford.

Grieve, I.C. (1978), Some effects of the plantation of conifers on a freely drained lowland soil, Forest of Dean, UK, *Forestry*, **51**, 21–8.

Grimm, E.C. (1984), Fire and other factors controlling the Big Woods vegetation of Minnesota in the mid-nineteenth century, *Ecol. Mono.*, **54**, 291–311.

Gustafsson, L. and Hällingbäck, T. (1988), Bryophyte flora and vegetation of managed and virgin coniferous forests in south-west Sweden, *Biol. Cons.*, **44**, 283–300.

Hammond, P.M. (1974), Changes in the British Coleopterous fauna, In: *The Changing Flora and Fauna of Britain* (D.L. Hawksworth, ed.), pp. 323–69, Academic Press, London.

Hannon, G.E. and Bradshaw, R.H.W. (1989), Recent vegetation dynamics on two Connemara lake islands, western Ireland, *J. Biogeog.*, **16**, 75–81.

Harding, P.T. and Rose, F. (1986), *Pasture-Woodlands in Lowland Britain*, Institute of Terrestrial Ecology, Abbots Ripton.

Harkness, C.E. (1982), *Surveys of Moorland and Roughland Change, No. 12, Mapping Changes in the Extent of Woodland in Upland Areas*, Department of Geography, University of Birmingham, Birmingham.

Harmon, M.E., Franklin, J.F., Swanson, F.J. *et al.* (1981), Ecology of coarse woody debris in temperate ecosystems, *Adv. Ecol. Res.*, **15**, 133–302.

Harper, J.L. (1977), *Population Biology of Plants*, Academic Press, London.

Harris, L.D. (1984), *The Fragmented Forest*, University of Chicago Press, Chicago.

Hart, C.E. (1966), *Royal Forest. A History of Dean's Woods as Producers of Timber*, Clarendon Press, Oxford.

Harvey, J.H. (1974), The stocks held by early nurseries, *Agric. Hist. Rev.*, **22**, 18–35.

Heliövaara, K and Väisänen, R. (1984), Effects of modern forestry on northwestern European forest invertebrates: a synthesis, *Acta Forestalia Fennica*, **189**.

Helliwell, D.R. (1972), Notes on Forestry Commission Census data, 1947 and 1965–7, Merlewood R and D Paper 31, Merlewood Research Station, Grange-over-Sands, Cumbria.

Helliwell, D.R. (1973), Priorities and values in nature conservation, *J. Env. Man.*, **1**, 85–127.

Helliwell, D.R. (1974), Discount rates in land-use planning, *Forestry*, **47**, 147–52.

Hermy, M. and Stieperaere, H. (1981), An indirect gradient analysis of the ecological relationships between ancient and recent riverine woodlands to the south of Bruges (Flanders, Belgium), *Vegetatio*, **44**, 43–9.

Hiley, W.E. (1954), *Woodland management*, Faber, London.

Hill, D.A., Lambton, S., Proctor, I. and Bullock, I. (1991), Winter bird communities in woodland in the Forest of Dean, England, and some implications of livestock grazing, *Bird Study*, **38**, 57–70.

Hill, D., Taylor, S., Thaxton, R. and Amphlet, A. (1990), Breeding bird communities of native pine forest, Scotland, *Bird Study*, **37**, 133–41.

Hill, M.O. (1979), The development of a flora in even-aged plantations, In: *The Ecology of Even-aged Forest Plantations* (Ford, E.D., Malcolm, D.C. and Atterson, J. eds.), Institute of Terrestrial Ecology, Cambridge.

Hill, M.O. and Jones, E.W. (1978), Vegetation changes resulting from afforestation of rough grazings in Caeo Forest, South Wales, *J. Ecol.*, **66**, 433–56.

Hill, M.O., Bunce, R.G.H. and Shaw, M.W. (1975), Indicator species analysis, a divisive polythetic method of classification, and its application to a study of native pinewoods in Scotland, *J. Ecol.*, **63**, 597–613.

Hobson, P.M. (1988), Methven Wood (Almondbank, Perth): the history of its management, *Scot. For.*, **42**, 104–12.

Hooper, M.D. (1970), The size and surroundings of nature reserves, *Symp. Br. Ecol. Soc.*, **11**, 555–61.

Hooper, M.D. (1973), History, In: *Monks Wood. A Nature Reserve Record* (Steele, R.C. and Welch, R.C., eds.), pp. 22–35, Natural Environment Research Council, Huntingdon.

Hopkinson, J.W. (1927), Studies on the vegetation of Nottinghamshire I. The ecology of the Bunter Sandstone, *J. Ecol.*, **15**, 130–71.

Horn, H.S. (1974), The ecology of secondary succession, *Ann. Rev. Ecol. System.*, **5**, 25–37.

Horrill, A.D., Sykes, J.M. and Idle, E.T. (1975), The woodland vegetation of Inchcailloch, Loch Lomond, *Trans. Proc. Bot. Soc. Edinburgh*, **42**, 307–34.

Hoskins, W.G. (1955), *The Making of the English Landscape*, Hodder and Stoughton, London.

Hoskins, W.G. (1968), History of common land and common rights, *Report of the Royal Commission on Common Land, 1955–58*, Cmnd. 462, Appendix II, pp. 149–66, HMSO, London.

Hunter, F.A. (1977), Ecology of pinewood beetles, In: *Native Pinewoods of Scotland* (Bunce, R.G.H. and Jeffers, J.N.R., eds.), pp. 42–55, Institute of Terrestrial Ecology, Cambridge.

Hunter, M.L. (1990), *Wildlife, Forests and Forestry*, Prentice Hall, Eaglewood Cliffs (NJ).

Hunting Surveys (1986), *Monitoring Landscape Change*, Hunting Surveys and Consultants Ltd, Borehamwood.

Hutt, P.A. (1974), Bradford Plan continuous cover forestry: an experiment in uneven-aged forestry, *Timber Grower*, **53**, 26–36.

Insley, H. ed. (1988), *Farm Woodland Planning*, Forestry Commission Bulletin 80, H.M.S.O., London.

Iversen, J. (1960), Problems of the early post-glacial forest development in Denmark, *Danmarks Geol. Unders. IV Series*, **4**, No. 3.

Iversen, J. (1964), Retrogressive vegetational succession in the post-glacial, *J. Ecol.* (Supplement), **52**, 59–70.

Jobling, J. and Pearce, M.L. (1977), Free growth of oak, *Forestry Commission Record 113*.

Jones, E.W. (1945), The structure and reproduction of the virgin forest of the north temperate zone, *New Phytol.*, **44**, 130–48.

Jones, E.W. (1959), Biological flora of the British Isles. *Quercus L, J. Ecol.*, **47**, 169–222.

Jones, E.W. (1961), British forestry in 1790–1813, *Quart. J. For.*, **55**, 36–40 and 131–8.

Jones, E.W. (1983), Review of Peterken (1981), *Forestry*, **56**, 87–91.

Kelly, D.L. (1981), The native forest vegetation of Killarney, south-west Ireland: an ecological account, *J. Ecol.*, **69**, 437–72.

Kelly, D. and Moore, J.J. (1975), A preliminary sketch of the Irish acidophilous oakwoods, In: *La Vegetation des Forêts Caducifoliées Acidiphiles* (Gehu, J., ed.), Cramer, Vaduz.

Kennedy, C.E.J. and Southwood, T.R.E. (1984), The number of species of insect associated with British trees: a re-analysis, *J. Anim. Ecol.*, **53**, 455–78.

King, P.I., ed. (1954), *The Book of William Morton*, Northamptonshire Record Society, 16.

Kipling, C. (1974), Some documentary evidence on woodland in the vicinity of Windermere, *Trans. Cumberland Westmorland Antiq. Arch. Soc., New Series*, **74**, 65–88.

Kirby, K.J. (1986), Forest and woodland evaluation, In: *Wildlife Conservation Evaluation* (Usher, M.B. ed.), pp. 201–21, Chapman & Hall, London.

Kirby, K.J. (1988), *A Woodland Survey Handbook*, Research and Survey in Nature Conservation, 11. Nature Conservancy Council, Peterborough.

Kirby. K.J. and May, J. (1989), The effects of enclosure, conifer planting and the subsequent removal of conifers in Dalavich oakwood (Argyll), *Scot. For.*, **43**, 280–8.

Kirby, K.J., Webster, S.D. and Antczak, A. (1991), Effects of forest management on stand structure and the quantity of fallen dead wood: some British and Polish examples, *For. Ecol. Manag.*, **43** 167–74.

Kirby, K.J., Bines, T., Burn, A. *et al.* (1986), seasonal and observer differences in vascular plant records from British woodlands, *J. Ecol.*, **74**, 123–31.

Kirby, P. (1992), *Habitat Management for Invertebrates: a practical handbook*, Royal Society for the Protection of Birds, Sandy.

Klötzli, F. (1979), Eichen-, Edellaub- und Bruchwälder der Britischen Inseln, *Schweiz. Z. Forst.*, **121**, 329–66.

Koop, H. (1989), *Forest Dynamics*, Springer, Berlin.

Kopecký, K. and Hejný, S. (1973), Neue syntaxonomische Auffassung der Gesellschaften ein- bis zweijähriger Pflanzen der *Galio-Urticetea* in Böhmen, *Folia Geobot. Phytotax., Praha*, **8**, 49–66.

Köstler, J. (1956), *Silviculture* (M.L. Anderson, translator), Oliver and Boyd, Edinburgh and London.

Lane, A. and Tait, J. (1990), *Practical Conservation. Woodlands*, Hodder and Stoughton, London.

Langdale-Brown, I. (1980), *Lowland Agricultural Habitats (Scotland), Air Photo Analysis of Change*, CSD Report 332, Nature Conservancy Council, Edinburgh.

Le Duc, M.G., Sparks, T.H. and Hill, M.O. (1992), Predicting potential plant colonisers of new woodland plantations, *Asp. Appl. Biol.*, **29**, 41–8.

Lemée, G. (1985), Role des arbres intolerants a l'ombrage dans la dynamic d'une hêtraie naturelle (Forêt de Fontainebleau), *Acta Oecologia*, **6**, 3–20.

Likens, G.E. ed. (1989), *Long-term Studies in Ecology*, Springer, New York.

Lindsay, J.M. (1974), *The Use of Woodland in Argyllshire and Perthshire Between 1650 and 1850*, PhD Thesis, University of Edinburgh.

Lindsay, J.M. (1975a), The history of oak coppice in Scotland, *Scot. For.*, **29**, 87–95.

Lindsay, J.M. (1975b), Charcoal iron smelting and its fuel supply; the example of Lorn furnace, Argyllshire, 1753–1876, *J. Hist. Geog.*, **1**, 283–98.

Lindsay, J.M. (1977), Forestry and agriculture in the Scottish Highlands 1700–1850: a problem in estate management, *Agric. Hist. Rev.*, **25**, 23–36.

Linnard, W. (1982), *Welsh Woods and Forests: history and utilisation*, National Museum of Wales, Cardiff.

Locke, G.M.L. (1970), *Census of Woodlands 1965–67*, Forestry Commission, HMSO, London.

Locke, G.M.L. (1976), The place of forestry in Scotland, *For. Comm. Res. Dev. Paper 113*.

Lonsdale, D., Hickman, I.T., Mobbs, I.D. and Matthews, R.W. (1989), A quantitative analysis of beech health and pollution across southern Britain, *Naturwissenschaften*, **76**, 571–3.

Lorimer, C.G. (1977), The presettlement forest and natural disturbance cycle of northeastern Maine, *Ecology*, **58**, 139–48.

Lorimer, C.G. (1980), Age structure and disturbance history of a southern Appalachian virgin forest, *Ecology*, **61**, 1169–84.

Lorrain-Smith, R. (1969), The economy of the private woodland in Great Britain, *Commonwealth Forestry Institute, Oxford*, paper 40.

Lorrain-Smith, R. (1973), Systems of management for timber production and wildlife conservation in lowland forestry, *Quart. J. For.*, **67**, 67–80.

Lucas, O.W.R. (1991), *The Design of Forest Landscapes*, Oxford University Press, Oxford.

Macarthur, R.H. and Wilson, E.O. (1967), *The Theory of Island Biogeography* (Monographs on Population Biology no. 1), Princeton University Press.

Macintosh, R. (1989), Forest design: Kielder Forest restructuring, *Timber Grower, Autumn edition*, 19–20.

Mackney, D. (1961), A podzol development sequence in oakwoods and heath in central England, *J. Soil. Sci*, **12**, 23–40.

McVean, D.N. (1956), Ecology of *Alnus glutinosa* (L.) Geartn, *J. Ecol.*, **44**, 321–30.

McVean, D.N. (1964), Woodland and scrub, In: *The Vegetation of Scotland* (Burnett, J.H., ed.), pp. 144–67, Oliver and Boyd, Edinburgh and London.

McVean, D.N. and Ratcliffe, D.A. (1962), *Plant Communities of the Scottish Highlands*, HMSO, London.

Malcolm, D.C. (1979), The future development of even-aged plantations: silvicultural implications, In: *The Ecology of Even-aged Forest Plantations* Ford, E.D., Malcolm, D.C. and Atterson, J., eds.), pp. 481–504, Institute of Terrestrial Ecology, Cambridge.

Malcolm, D.C., Evans, J. and Edwards, P.N. (1982), *Broadleaves in Britain*, Institute of Chartered Foresters, Edinburgh.

Manners, J.E. (1974), *Country Crafts Today*, David and Charles, Newton Abbot.

Manners, J.G. and Edwards, P.J. (1986), Death of old beech trees in the New Forest, *Proc. Hamps. Field Club Arch. Soc.*, **42**, 155–6.

Mansfield, A. (1952), *The Historical Geography of the Woodlands of the Southern Chilterns, 1600–1947*, MSc Thesis, University of London.

Margules, C.R. and Usher, M.B. (1981), Criteria used in assessing wildlife conservation potential: a review, *Biol. cons.*, **21**, 79–109.

Marks, J.B. (1942), Land use and plant succession in Coon valley, Wisconsin, *Ecol. Mono.*, **12**, 113–33.

Marquiss, M., Newton, I. and Ratcliffe, D.A. (1978), The decline of the raven, *Corvus corax*, in relation to afforestation in southern Scotland and northern England, *J. Appl. Ecol.*, **15**, 129–44.

Marren, P. (1990), *Britain's Ancient Woodland. Woodland Heritage*, David and Charles, Newton Abbot.

Marren, P. (1992), *The Wild Woods. A regional guide to Britain's ancient woodland*, David and Charles, Newton Abbot.

Marshall, W. (1788), *The Rural Economy of Yorkshire*, T. Cadgell, London.

Martin, M.H. and Pigott, C.D. (1975), Soils, In: *Hayley Wood. Its History and Ecology* (Rackham, O., ed.) pp. 61–71, Cambridge and Isle of Ely Naturalists' Trust, Cambridge.

Maser, C., Tarrant, R.F., Trappe, J.M. and Franklin, J.F. eds. (1988) *From the Forest to the Sea: a story of fallen trees*, General Technical Report PNW-GTR-229, USDA Forest Service.

Mather, A.S. (1971), Problems of afforestation in north Scotland, *Trans. Inst. Br. Geog.*, **54**, 19–32.

Mather, A.S. (1978), Patterns of afforestation in Britain since 1945, *Geography*, **50**, 157–66.

Matthews, J.D. (1955), The influence of weather on the frequency of beech mast years in England, *Forestry*, **28**, 107–16.

Matthews, J.D. (1989), *Silvicultural Systems*, Clarendon Press, Oxford.

McPhillimy, D. (1989), *Conservation in forests. Case studies of good practice in Scotland.* Countryside Commission for Scotland, Battleby.

Mensing, S.A. (1988), *Blue oak (Quercus douglasi) regeneration in the Tehachapi Mountains, Keen County, California*, MA Thesis, University of California, Berkeley.

Merton, L.F.H. (1970), The history and status of the woodlands of the Derbyshire limestone, *J. Ecol.*, **58**, 723–44.

Miles, J. (1978), The influence of trees on soil properties, *Ann. Report Inst. Terr. Ecol. 1977*, pp. 7–11.

Miller, F.R., Savill, P.S. and Kirk, C.F. (1988), *The Broadleaved Woodlands Policy in Privately Owned Woodlands*, Oxford Forestry Institute, Oxford.

Mitchell, F.J.G. (1988), The vegetational history of the Killarney oakwoods, SW Ireland: evidence from fine spatial resolution pollen analysis, *J. Ecol.*, **76**, 415–36.

Mitchell, F.J.G. and Kirby, K.J. (1990), The impact of large herbivores on the conservation of semi-natural woods in the British uplands, *Forestry*, **63**, 333–53.

Mitchell, P.L. (1989), Repollarding large neglected pollards: a review of current practice and results. *Arboric. J.*, **13**, 125–42.

Mitchell, P.L and Kirby, K.J. (1989), *Ecological Effects of Forestry Practices in Long-established Woodland and their Implications for Nature Conservation*, Occasional Paper 39, Oxford Forestry Institute, Oxford.

Mlinsek, D. (1991), The holistic concept of natural productive forests, *Quart. J. For.*, **85**, 8–11.

Moore, N.W. (1962), The heaths of Dorset and their conservation, *J. Ecol.*, **50**, 369–91.

Moore, N.W. (1969), Experience with pesticides and the theory of conservation, *Biol. Conserv.*, **1**, 201–7.

Moore, N.W. and Hooper, M.D. (1975), On the number of bird species in British woods, *Biol. Conserv.*, **8**, 239–50.

Moore, P.D. (1974), Prehistoric human activity and blanket peat initiation on Exmoor, *Nature*, **250**, 439–41.

Morgan, R.K. (1991), The role of protective understorey in the regeneration of a heavily browsed woodland, *Vegetatio*, **92**, 119–32.

Moss, C.E. (1907), *Geographical Distribution of Vegetation in Somerset: Bath and Bridgwater district*, Royal Geographical Society, London.

Moss, C.E. (1913), *Vegetation of the Peak District*, Cambridge University Press, Cambridge.

Moss, C.E., Rankin, W.M. and Tansley, A.G. (1910), The woodlands of England, *New Phytol.*, **9**, 113–49.

Moss, D. (1978), Song-bird populations in forestry plantations, *Quart. J. For.*, **72**, 5–14.

Muller, R.N. (1982), Vegetation patterns in the mixed mesophytic forests of eastern Kentucky, *Ecology*, **63**, 1901–17.

Nash, R. (1973), *Wilderness and the American Mind* (revised edn.), Yale University Press, New Haven and London.

Nature Conservancy Council (1976), *Nature Conservation and Agriculture*, Nature Conservancy Council, London.

Nature Conservancy Council (1984), *Nature Conservation in Great Britain*, Nature Conservancy Council, Peterborough.

Nature Conservancy Council (1986), *Nature Conservation and Afforestation in Britain*, Nature Conservancy Council, Peterborough.

Nature Conservancy Council (1989), *Guidelines for Selection of Biological SSSIs*, Nature Conservancy Council, Peterborough.

Neilson, N. (1928), The cartulary and terrier of the priory of Bilsington, Kent, *Records of the Social and Economic History of England and Wales*, **7**, British Academy, Oxford.

Newbold, A.J. and Goldsmith, F.B. (1981), *The Regeneration of Oak and Beech: a literature review*, Discussion Papers in Conservation 33, University College, London.

Newson, M. (1985), Forestry and water in the uplands of Britain: the background of hydrological research and options for harmonious land-use, *Quart. J. For.*, **79**, 113–20.

Nicholls, P.H. (1972), On the evolution of a forest landscape, *Trans. Inst. Br. Geog.*, **56**, 57–76.

Nichols, G.E. (1913), The vegetation of Connecticut. II, Virgin forests, *Torreya*, **13**, 199–215.

Nisbet, J. (1905), Forestry, In: *Victoria History of the Counties of England. Surrey 2*, pp. 561–78, Constable, Westminster.

Nisbet, J. and Lascelles, G.W. (1903), Forestry, In: *Victoria History of the Counties of England. Hampshire and Isle of Wight 2*, pp. 409–70, Constable, Westminster.

Nisbet, J. and Vellacott, C.H. (1907), Forestry, In: *Victoria History of the Counties of England, Gloucester 2*, 263–86, Constable, Westminster.

Noirfalise, A. and Vanesse, R. (1975), *Conséquences de la Monoculture des Conifères pour la Conservation des Sols et pour le Bilan Hydrologique*, Association des Espaces Verts, Brussels.

Oberdorfer, E. (1957), Süddeutsche Pflanzengesellschaften, *Pflanzensociologie*, **10**, Gustav Fischer, Jena.

Oberdorfer, E., ed. (1967), Systematische Übersicht der westdeutschen Phaneragamen- und Gefässkryptogamen-Gesellschaftern, *Schrift. Vegetations.*, **2**, 7–62.

Oberdorfer, E. (1970), *Pflanzensoziologische Excursionsflora für Süddeutschland und die angrenzenden Gebiete*, Ulmer, Stuttgart.

Ogley, R. (1987), *In the Wake of the Hurricane*, Froglets, Brasted Chart.

Olsen, J.S. (1958), Rates of succession and soil changes on Southern Lake Michigan sand dunes, *Bot. Gaz.*, **119**, 125–70.

Ormerod, S.J., Mawle, G.W. and Edwards, R.W. (1987). The influence of forest on aquatic fauna, In: *Environmental Aspects of Plantation Forestry in Wales.* ITE Symposium 22, (Good, J.E.G., ed.), 37–49, Institute of Terrestrial Ecology, Grange-over-Sands.

O'Sullivan, P.E. (1973a), Land use changes in the Forest of Abernethy, Inverness-shire (1750–1900 A.D.), *Scot. Geog. Mag.*, **89**, 95–106.

O'Sullivan, P.E. (1973b), Pollen analysis of mor humus layers from a native Scots pine ecosystem, interpreted with surface samples, *Oikos*, **24**, 259–72.

O'Sullivan, P.E. (1977), Vegetation history and the native pinewoods, In: *Native Pinewoods of Scotland* (Bunce, R.G.H. and Jeffers. J.N.R., eds.), Institute of Terrestrial Ecology, Cambridge.

Ovington, J.D. (1964), The ecological basis of the management of woodland nature reserves in Great Britain, *J. Ecol.* (Supplement), **52**, 29–37.

Packham, J.R. (1978), Biological Flora of the British Isles. *Oxalis acetosella* L., *J. Ecol.*, **66**, 669–93.

Packham, J.R. (1983), Biological Flora of the British Isles. *Lamiastrum galeobdolon* (L.) Ehrend & Polatschek, *J. Ecol.*, **71**, 975–97.

Packham, J.R. and Cohn, E.V.J. (1990), Ecology of the woodland field layer. *Arboric. J.*, **14**, 357–71,

Paul, C.R.C. (1975), The ecology of Mollusca in ancient woodland 1. The fauna of Hayley Wood, Cambridgeshire, *J. Conchol.*, **28**, 301–27.

Paul, C.R.C. (1978), The ecology of Mollusca in ancient woodland 3. Frequency of occurrence in west Cambridgeshire woods, *J. Conchol.*, **29**, 295–300.

Peglar, S. (1979), A radiocarbon-dated pollen diagram from Loch of Winless, Caithness, north-east Scotland, *New Phytol.*, **82**, 245–63.

Pennington, W., Howarth, E.Y., Bonny, A.P. and Lishman, J.P. (1972), Lake sediments in northern Scotland, *Phil. Trans. Roy. Soc., B*, **246**, 191–294.

Perring, F.H., Radford, G.L. and Peterken, G.F. (1973), Reserve recording. Monks Wood Experimental Station, Huntingdon, Cambridgeshire.

Peterken, G.F. (1966), Mortality of holly (*Ilex aquifolium*) seedlings in relation to natural regeneration in the New Forest, *J. Ecol.*, **54**, 259–69.

Peterken, G.F. (1969a), Development of vegetation in Staverton Park, Suffolk, *Field Studies*, **3**, 1–39.

Peterken, G.F. (1969b), An event record for nature reserves, *J. Devon Trust Nat. Conserv.*, **21**, 920–8.

Peterken, G.F. (1972), Conservation coppicing and the coppice crafts, *Quart. J. Devon Trust Nat. Conserv.*, **4**, 157–64.

Peterken, G.F. (1974a), A method for assessing woodland flora for conservation using indicator species, *Biol. Conserv.*, **6**, 239–45.

Peterken, G.F. (1974b), Developmental factors in the management of British woodlands, *Quart. J. For.*, **68**, 141–9.

Peterken, G.F. (1976), Long-term changes in the woodlands of Rockingham Forest and other areas, *J. Ecol.*, **64**, 123–46.

Peterken, G.F. (1977a), Habitat conservation priorities in British and European woodlands, *Biol. Conserv.*, **11**, 223–36.

Peterken, G.F. (1977b), General Management principles for nature conservation in British woodlands, *Forestry*, **50**, 27–48.

Peterken, G.F. (1981), Wood anemone in central Lincolnshire: an ancient woodland indicator? *Trans. Linc. Natural. Union*, **20**, 78–82.

Peterken, G.F. (1983a), Wild service-tree in central Lincolnshire, *Trans. Linc. Natural. Union*, **20**, 158–62.

Peterken, G.F. (1983b), Woodland surveys can mislead. *New Scientist*, **100**, 802–3.

Peterken, G.F. (1987), Natural features in the management of upland conifer forests. *Proc. Roy. Soc. Edin.*, **93B**, 223–34.

Peterken, G.F. (1991), Ecological issues in the management of woodland nature reserves, In: *The Scientific Management of Temperate Communities for Conservation*, (Spellerberg, I.F., Goldsmith, F.B. and Morris, M.G., eds.), pp. 245–72, Blackwell, Oxford.

Peterken, G.F. (1992), Coppices in the lowland landscape, In: *Ecology and Management of Coppice Woodlands* (Buckley, G.P. ed.), pp. 3–17, Chapman & Hall, London.

Peterken, G.F. (1993a), Long-term floristic development of woodland on former agricultural land in Lincolnshire, England, In: *Ecological effects of afforestation* (Watkins, C. ed.), pp. 31–43, CAB International, Wallingford.

Peterken, G.F. (1993b), Long-term studies in forest nature reserves, *Proceedings of the Symposium on European Forest Reserves*, Wageningen, May 1992, Pudoc, Wageningen (in press).

Peterken, G.F. (1993c), Conifer forests, In: *Wildlife and Conservation Management Handbook* (Ratcliffe, P.R. ed.), Forestry Commission, Edinburgh (in press).

Peterken, G.F. and Allison, H. (1989), *Woods, Trees and Hedges: a review of changes in the British countryside*, Focus on Nature Conservation 22. Nature Conservancy Council, Peterborough.

Peterken, G.F. and Backmeroff, C.E. (1988), *Long-term Monitoring in Unmanaged Woodland Nature Reserves*, Research and Survey in Nature Conservation, 9. Nature Conservancy Council, Peterborough.

Peterken, G.F. and Game, M. (1981), Historical factors affecting the distribution of *Mercurialis perennis* in central Lincolnshire, *J. Ecol.*, **69**, 781–96.

Peterken, G.F. and Game, M. (1984), Historical factors affecting the number and distribution of vascular plant species in the woodlands of central Lincolnshire, *J. Ecol.*, **72**, 155–82.

Peterken, G.F. and Harding, P.T. (1974), Recent changes in the conservation value of woodlands in Rockingham Forest, *Forestry*, **47**, 109–28.

Peterken, G.F. and Harding, P.T. (1975), Woodland conservation in eastern England: comparing the effects of changes in three study areas. *Biol. Conserv.*, **8**, 279–98.

Peterken, G.F. and Hubbard, J.C.E. (1972), The shingle vegetation of southern England: the holly wood on Holmstone Beach, Dungeness, *J. Ecol.*, **60**, 547–72.

Peterken, G.F. and Hughes, F. (1990), The changing lowlands, In: *Britain's Changing Environment from the Air* (Bayliss-Smith, T. and Owens, S. eds.), pp. 48–76, Cambridge University Press, Cambridge.

Peterken, G.F. and Jones, E.W. (1987), Forty years of change in Lady Park Wood: the old growth stands, *J. Ecol.*, **75**, 477–512.

Peterken, G.F. and Jones, E.W. (1989), Forty years of change in Lady Park Wood: the young growth stands, *J. Ecol.*, **77**, 401–29.

Peterken, G.F. and Stace, C.E. (1987), Structure and development of the Black Wood of Rannoch, *Scot. For.*, **41**, 29–44.

Peterken, G.F. and Tubbs, C.R. (1965), Woodland regeneration in the New Forest, Hampshire, since 1650, *J. Appl. Ecol.*, **2**, 159–70.

Peterken, G.F. and Welsh, R.C., eds. (1975), *Bedford Purlieus: Its History, Ecology and Management*, Monks Wood Experimental Station Symposium, 7.

Peterken, G.F., Ausherman, D., Buchenau, M. and Forman, R.T.T. (1992), Old-growth conservation within British upland conifer plantations, *Forestry*, **65**, 127–44.

Pettit, P.A.J. (1968), The royal forests of Northamptonshire: a study of the economy 1558–1714, Northamptonshire Record Society, 23.

Petty, S.J. (1989), *Goshawks: their status, requirements and management*, Forestry Commission Bulletin No. 90. HMSO London.

Petty, S.J. and Avery, M.I. (1990), *Forest Bird Communities*, Occasional Paper 26. Forestry Commission, Edinburgh.

Pickett, S.T.A. and Thompson, J.N. (1978), Patch dynamics and the design of nature reserves, *Biol. Conserv.*, **13**, 27–37.

Pickett, S.T.A. and White, P.S. (1985), *The Ecology of Natural Disturbance and Patch Dynamics*, Academic Press, Orlando.

Pigott, C.D. (1969), The status of *Tilia cordata* and *T. platyphyllos* on the Derbyshire limestone. *J. Ecol.*, **57**, 491–504.

Pigott, C.D. (1975), Natural regeneration of *Tilia cordata* in relation to forest-structure in the forest of Bialowieza, Poland, *Phil. Trans. Roy. Soc., B*, **270**, 151–79.

Pigott, C.D. (1977), The scientific basis of practical conservation: aims and methods of conservation, *Proc. Roy. Soc. Lond., B*, **197**, 59–68.

Pigott, C.D. and Huntley, J.P. (1978–82), Factors controlling the distribution of *Tilia cordata* at the northern limits of its geographical range, *New Phytol.*, **81**, 429–41; **84**, 145–64; **87**, 817–39.

Pigott, M.E. and Pigott, C.D. (1959), Stratigraphy and pollen analysis of Malham Tarn and Tarn Moss, *Field Studies*, **1**, 1–18.

Plaisance, G. (1963), *Guide des forêts de France*. La Nef, Paris.

Plomer, W., ed. (1964), *Kilvert's Diary 1870–1879*, Jonathan Cape, London.

Pollard, E. (1973), Hedges VII. Woodland relic hedges in Huntingdon and Peterborough, *J. Ecol.*, **61**, 343–52.

Pollard, E. (1977), A method for assessing changes in the abundance of butterflies, *Biol Conserv.*, **12**, 115–34.

Pollard, E. (1979), Population ecology and change in range of the white admiral butterfly *Lagoda camilla* L. in Britain, *Ecol. Entomol.*, **4**, 61–74.

Pollard, E., Hooper, M.D. and Moore, N.W. (1974), *Hedges*, Collins, London.

Poore, M.E.D. (1955), The use of phytosociological methods in ecological investigations, parts I, II and III, *J. Ecol.*, **43**, 226–44, 245–69 and 606–51.

Poore, M.E.D. (1956a), The use of phytosociological methods in ecological investigations IV. General discussion of phytosociological problems, *J. Ecol.*, **44**, 28–50.

Poore, M.E.D. (1956b), The ecology of Woodwalton Fen, *J. Ecol.*, **44**, 455–92.

Poore, M.E.D. (1962), The method of successive approximation in descriptive ecology, *Adv. Ecol. Res.*, **1**, 35–68.

Price, C. (1976), Blind alleys and open prospects in forest economics, *Forestry*, **49**, 99–107.

Prince, H.C. (1958), Parkland in the English landscape, *Amat. Hist.*, **3**, 332–49.

Proctor, M.C.F., Spooner, G.M. and Spooner, M.F. (1980), Changes in Wistman's Wood, Dartmoor: photographic and other evidence, *Rep. Trans. Devon. Ass. Advmt. Sci.*, **112**, 43–79.

Pryor, F. (1991), *English Heritage Book of Flag Fen*, Prehistoric Fenland Centre, Peterborough.

Pryor, F., French, C. and Taylor, M. (1985), An interim report on excavations at Etton, Maxey, Cambridgeshire, 1982–1984, *The Antiquaries Journal*, **65**.

Pyatt, D.G. and Craven, M.M. (1979), Soil changes under even-aged plantations, In: *The Ecology of Even-aged Forest Plantations* (Ford, E.D., Malcolm, D.C. and Atterson, J., eds.), Institute of Terrestrial Ecology, Cambridge.

Pyne, S.J. (1982), *Fire in America: a cultural history of wildland and rural fire*, Princeton University Press, Princeton.

Rackham, O. (1969), Knapwell Wood, *Nature in Cambridgeshire*, **12**, 25–31.

Rackham, O. (1971), Historical studies and woodland conservation, *Symp. Br. Ecol. Soc.*, **11**, 563–80.

Rackham, O. (1972), Grundle House: on the quantities of timber in certain East Anglian buildings in relation to local supplies, *Vernacular Architecture*, **3**, 3–8.

Rackham, O. (1974), The oak tree in historic times, In: *The British Oak: its History and Natural History* (Morris, M.G. and Perring, F.H., eds.), pp. 62–79, Classey, Faringdon.

Rackham, O. (1975), *Hayley Wood. Its History and Ecology*, Cambridge and Isle of Ely Naturalists' Trust, Cambridge.

Rackham, O. (1976), *Trees and Woodland in the British Landscape*, Dent. London. (Revised ed. pub. 1990)

Rackham, O. (1977a), Neolithic woodland management in the Somerset levels: Garvin's Walton Heath, and Rowland's tracks, *Somerset Levels Papers*, **3**, 65–71.

Rackham, O. (1977b), Hedgerow trees: their history, conservation and renewal, *Arboric. J.*, **3**, 169–77.

Rackham, O. (1978), Archaeology and land use history, In: Epping Forest – the natural aspect? (D. Corke, ed.), *Essex Naturalist*, **2**, 16–57.

Rackham, O. (1979), Neolithic woodland management in the Somerset Levels: Sweet Track I, *Somerset Levels papers*, **5**, 59–61.

Rackham, O. (1980), *Ancient Woodland*, Arnold, London.

Rackham, O. (1986a), *The Woods of South-east Essex*, Rochford District Council, Rochford.

Rackham, O. (1986b), The ancient woods of Norfolk, *Trans. Norwich Natural. Soc.*, **27**, 161–77.

Rackham, O. (1986c), *The History of the Countryside*, Dent, London.

Rackham, O. (1989), *The Last Forest. The Story of Hatfield Forest*, Dent, London.

Rackham, O. (1992), Gamlingay Wood, *Nature in Cambridgeshire*, **34**, 3–15.

Radley, J. (1961), Holly as a winter feed, *Agric. Hist. Rev.*, **9**, 89–92.

Ratcliffe, D.A. (1968), An ecological account of Atlantic bryophytes in the British Isles, *New Phytol.*, **67**, 365–439.

Ratcliffe, D.A., ed. (1977), *A Nature Conservation Review*, Cambridge University Press, Cambridge.

Ratcliffe, D.A. and Thompson, D.B.A. (1988), The British uplands: their ecological character and international significance, In: *Ecological Change in the Uplands*, British Ecological Society Special Publication 7, (Usher, M.B. and Thompson, D.B.A., eds.), pp. 9–36, Blackwell, Oxford.

Ratcliffe, P.R. (1989), The control of red and sika deer populations in commercial forests, In: *Mammals as Pests* (Putman, R.J., ed.), pp. 98–115, Chapman & Hall, London.

Ratcliffe, P.R. (1992), The interaction of deer and vegetation in coppice woods, In: *Ecology and Management of Coppice Woodlands* (Buckley, G.P., ed.), pp. 233–45, Chapman & Hall, London.

Ratcliffe, P.R. and Petty, S.J. (1986), The management of commercial forests for wildlife, In: *Trees and Wildlife in the Scottish Uplands* (Jenkins, D. ed.), pp. 177–87, Institute of Terrestrial Ecology, Abbots Ripton.

RCHM (Royal Commission on Historical Monuments, England) (1975), *An Inventory of the Historical Monuments in the County of Northampton I. Archaeological Sites in North-east Northamptonshire*, HMSO, London.

Read, H.J., ed., (1991), *Pollard and Veteran Tree Management*, Corporation of the City of London, Burnham Beeches.

Reade, M.G. (translator) (1969), Silviculture of selection forest (Ammon, W., author), *Quart. J. For.*, **58**, 197–210.

Reed, J.L. (1954), *Forests of France*, Faber, London.

Řehák, J. (1968), Srovnávaci výzkum v Boubinském pralese, *Lesnická práce*, **47**, 206–12.

Richard, P.W. (1952), *The Tropical Rain Forest*, Cambridge University Press, Cambridge.

Richens, R.H. (1958), Studies on *Ulmus* II. The village elms of Cambridgeshire, *Forestry*, **31**, 132–46.

Richens, R.H. (1983), *Elm*, Cambridge University Press, Cambridge.

Rixon, P. (1975), History and former woodland management, In: *Bedford Purlieus: Its History, Ecology and Management* (Peterken, G.F. and Welch, R.C., eds.), pp. 14–38, Monks Wood Experimental Station. Huntingdon.

Roberts, A.J., Russell, C., Walker, G.J. and Kirby, K.J. (1992), Regional variation in the origin, extent and composition of Scottish woodland. *Bot. J. Scotl.*, **46**, 167–89.

Roden, D. (1968), Woodland and its management in the medieval Chilterns, *Forestry*, **41**, 59–71.

Rodwell, J.S. (1991), *British Plant Communities. 1. Woodlands and Scrub*, Cambridge University Press, Cambridge.

Röhrig, E. and Ulrich, B., eds. (1991) *Ecosystems of the World, 7. Temperate Deciduous Forests*, Elsevier, Amsterdam.

Rose, F. (1974), The epiphytes of oak, In: *The British Oak: Its History and Natural History* (Morris, M.G. and Perring, F.H., eds.), pp. 250–73, Classey, Faringdon.

Rose, F. (1976), Lichenological indicators of age and environmental continuity in woodlands, In: *Lichenology: Progress and Problems*, (Systematics Association Special Volume 8) (Brown, D.H., Hawksworth, D.L. and Bailey, R.H., eds.), pp. 279–307, Academic Press, London.

Rose, F. (1992), Temperate forest management: its effects on bryophyte and lichen floras and habitats, In: *Bryophytes and Lichens in a Changing Environment* (Bales, J.W. and Farmer, A.M., eds.), pp. 211–33, Clarendon Press, Oxford.

Rose, F. and James, P.W. (1974), Regional studies on the British lichen flora I. The corticolous and lignicolous species of the New Forest, Hampshire, *Lichenologist*, **6**, 1–72.

Runkle, J.R. (1981), Gap regeneration in some old-growth forests of the eastern United States, *Ecology*, **62**, 1041–51.

Runkle, J.R. (1982), Patterns of disturbance in some old-growth mesic forests of eastern North America, *Ecology*, **63**, 1533–46.

Russell, D. (1992), *A Review of the Management of National Trust Trees and Woodlands*. National Trust, Cirencester.

Rybníčková, E. (1974), *Die Entwicklung der Vegetation und Flora im südlichen Teil der Böhmisch–Mährischen Höhe während des Spätglaziels und Holozäns*, Academia Verlag des Tchechoslowakischen, Praha.

Ryle, G.B. (1969), *Forest Service. The First Forty-five Years of the Forestry Commission in Great Britain*, David and Charles, Newton Abbot.

Rymer, L. (1980), Recent woodland history of North Knapdale, Argyllshire, Scotland, *Scot. For.*, **34**, 244–56.

Salbitano, F., ed. (1988), *Human Influence on Forest Ecosystems Development in Europe*, Pitagora Editrice, Bologna.

Salisbury E.J. (1916), The oak–hornbeam woods of Hertfordshire II. The *Quercus robur–Carpinus* woods, *J. Ecol.*, **4**, 88–117.

Salisbury, E.J. (1918), The oak–hornbeam woods of Hertfordshire III and IV. The *Quercus sessiliflora–Carpinus* woods, and comparative studies, *J. Ecol.*, **6**, 14–52.

Salisbury, E.J. (1924), The effects of coppicing as illustrated by the woods of Hertfordshire, *Trans. Herts. Nat. Hist. Soc.*, **18**, 1–21.

Salisbury, E.J. and Tansley, A.G. (1921), The durmast oakwoods (*Querceta sessiliflorae*) of the Silurian and Malvernian strata near Malvern, *J. Ecol.*, **9**, 19–38.

Saunders, D.A., Arnold, G.W., Burbridge, A.A. and Hopkins, A.J.M. eds. (1987), *Nature Conservation: the role of remnants of native vegetation*, Surrey Beatty, Chipping Norton.

Sernander, R. (1936), The primitive forests of Granskar and Fiby: a study of the part played by storm-gaps and dwarf trees in the regeneration of the Swedish spruce forest, *Acta Phytogeographica Suecica* **8**, 1–232.

Sharp, L. (1975), Timber, science and economic reform in the seventeenth century, *Forestry*, **48**, 51–86.

Sheail, J. (1976), *Nature in Trust*, Blackie, Glasgow and London.

Sheppard, J.A. (1973), Field systems of Yorkshire, In: *Studies of Field Systems in the British Isles* (Baker, A.R.H. and Butlin, R.A. eds.), pp. 45–187, Cambridge University Press, Cambridge.

Shimwell, D.W. (1971), *The Description and Classification of Vegetation*, Sidgwick and Jackson, London.

Shirley, E.P. (1867), *Some Account of English Deer Parks*, John Murray, London.

Siccama, T.G. (1971), Presettlement and present forest vegetation in northern Vermont with special reference to Chittendon County, *The American Midland Naturalist*, **85**, 153–72.

Silver, T. (1990), *A New Face on the Countryside. Indians, Colonists, and Slaves in South Atlantic Forests, 1500-1800*, Cambridge University Press, Cambridge.

Simberloff, D. (1987), The spotted owl fracas; mixing academic, applied and political ecology, *Ecology*, **68**, 766–72.

Sinker, C.A. (1962), The north Shropshire meres and mosses: a background for ecologists, *Field Studies*, **1**, 101–38.

Smart, N. and Andrews, J. (1985) *Birds and Broadleaves Handbook*, Royal Society for Protection of Birds, Sandy.

Smith, R. (1900), Botanical survey of Scotland, *Scot. Geog. Mag.*, **16**, 385–416, 441–67.

Smith, R.S. and Charman, D.J. (1988), The vegetation of upland mires within conifer plantations in Northumberland, northern England, *J. Appl. Ecol.*, **25**, 579–94.

Soutar, R.G. (1989), Afforestation and sediment yields in British fresh waters, *Soil Use Manag.*, **5**, 82–6.

Soutar, R.G. and Peterken, G.F. (1989), Regional lists of native trees and shrubs for use in afforestation schemes, *Arboric. J.*, **13**, 33–43.

Soutar, R.G. and Spencer, J.W. (1991), The conservation of genetic variation in Britain's native trees, *Forestry*, **64**, 1–12.

Southwood, T.R.E. (1961), The number of species of insects associated with various trees, *J. Anim. Ecol.*, **30**, 1–8.

Speight, M.C.D. (1989), *Sparoxylic Invertebrates and their Conservation*, Council of Europe, Strasbourg.

Spellerberg, I.F. (1991a), Biogeographical basis of conservation, In: *The Scientific Management of Temperate Communities for Conservation*. British Ecological Society Symposium 31 (Spellerberg, I.F., Goldsmith, F.B. and Morris, M.G., eds.), pp. 293–322, Blackwell, Oxford.

Spellerberg, I.F. (1991b), *Monitoring Ecological Change*, Cambridge University Press, Cambridge.

Spencer, J. (1990), Indications of antiquity. Some observations on the nature of plants associated with ancient woodland. *British Wildlife*, **2**, 90–100.

Spencer, J.W. and Kirby, K.J. (1992), An inventory of ancient woodland for England and Wales, *Biol. Conserv.*, **62**, 77–93.

Standish, A. (1616), New directions of experience ... for the increasing of timber and fire-wood, with the least waste and losse of ground ...

Steele, R.C. (1972a), *Wildlife Conservation in Woodlands*, Forestry Commission Booklet 29, pp. 21–8.

Steele, R.C., ed. (1972b), *Lowland Forestry and Wildlife Conservation*, Monks Wood Experimental Station Symposium, 6.

Steele, R.C. (1975), Forests and wildlife, *Phil. Trans. Roy. Soc., B*, **271**, 163–78.

Steele, R.C. and Peterken, G.F. (1982), Management objectives for broadleaved woodland – conservation, In: *Broadleaves in Britain* (Malcolm, D.C., Evans, J. and Edwards, P.N., eds.), pp. 91–103, Institute of Chartered Foresters, Edinburgh.

Steele, R.C. and Schofield, J.M. (1973), Conservation and management, In: *Monks Wood. A Nature Reserve Record* (Steele, R.C. and Welch, R.C., eds.), pp. 296–335, Natural Environment Research Council, Huntingdon.

Steele, R.C. and Welch, R.C., eds. (1973), *Monks Wood. A Nature Reserve Record*, Natural Environment Research Council, Huntingdon.

Sterling, P.H. and Hambler, C. (1988), Coppicing for conservation: do hazel communities benefit? In: *Woodland Conservation and Research in the Clay Vale of Oxfordshire and Buckinghamshire*. Research and Survey in Nature Conservation, 15 (Kirby, K.J. and Wright, F.J., eds.), pp. 69–80, Nature Conservancy Council, Peterborough.

Stern, R.C. (1973), Growing of hardwoods for pulpwood, *Scot. For.*, **27**, 11–16.

Steven, H.M. and Carlisle, A. (1959), *The Native Pinewoods of Scotland*, Oliver and Boyd, Edinburgh and London.

Stevens, P.A. (1975), Geology and Soils, In: *Bedford Purlieus: Its History, Ecology and Management* (Peterken, G.F. and Welch, R.C., eds.), pp. 43–64, Monks Wood Experimental Station, Huntingdon.

Stevenson, J.F. (1990), How ancient is the woodland of Mugdock? *Scot. For.*, **44**, 161–72.

Stevenson, W. (1813), *General View of the Agriculture of the County of Surrey*, Board of Agriculture, London.

Stewart, G.G. (1978), Inter-relations between agriculture and forestry in the uplands of Scotland: a forestry view, *Scot. For.*, **32**, 153–81.

Stone, T. (1793), *General View of the Agriculture of the County of Huntingdon*, Board of Agriculture, London.

Stoner, J.H. and Gee, A.S. (1985), Effects of forestry on water quality and fish in Welsh rivers and lakes. *J. Inst. Water Eng. Scien.*, **38**, 323–30.

Strayer, D., Glitzenstein, J.S., Jones, C.G. *et al.* (1986), *Long-term Ecological Studies: an illustrated account of their design, operation, and importance to ecology*, Occasional Publications 2. Institute of Ecosystem Studies, New York.

Streeter, D. (1974), Ecological aspects of oak woodland conservation, In: *The British Oak: Its History and Natural History* (Morris, M.G. and Perring, F.H., eds.), pp. 341–54, Classey, Farringdon.

Stroud, D. and Glue, D. (1991) *Britain's birds in 1989/90: the conservation and monitoring review*. British Trust for Ornithology/Nature Conservancy Council, Thetford.

Stroud, D.A. and Reed, T.M. (1986), The effect of plantation proximity on moorland breeding waders, *Bull. Wader Study Grp*, **46**, 25–8.

Stroud, D.A., Reed, T.M., Pienkowski, M.W. and Lindsay, R.A. (1987), *Nirds, Bogs and Forestry. The peatlands of Caithness and Sutherland*, Nature Conservancy Council, Peterborough.

Stubbs, A.E. (1972), Wildlife conservation and dead wood, *Quart. J. Devon Trust Nat. Conserv.*, **4**, 169–82.

Summerhayes, V.S., Cole, L.W. and Williams, P.H. (1924), The vegetation of the unfelled portions of Oxshott Heath and Esher Common, Surrey, *J. Ecol.*, **12**, 287–306.

Sumner, H. (1926), J. Norden's survey of medieval coppices in the New Forest. A.D. 1609, *Papers Proc. Hants Field Club Archaeol. Soc.*, **10**, 95–117.

Sykes, J.M. and Horrill, A.D. (1985), Natural regeneration in a Caledonian pinewood: progress after eight years of enclosure at Coille Coire Chuilc, Perthshire. *Arboric. J.*, **9**, 13–24.

Sykes, J.M., Horrill, A.D. and Mountford, M.D. (1983), Use of visual cover assessments as quantitative estimators of some British woodland taxa, *J. Ecol.*, **71**, 437–50.

Sylvester, D. (1969), *The Rural Landscape of the Welsh Borderland. A Study in Historical Geography*, Macmillan, London.

Tallis, J.H. and McGuire, J. (1972), Central Rossendale: the evolution of an upland vegetation I. The clearance of woodland, *J. Ecol.*, **60**, 721–37.

Tamm, C.O. (1956), Survival and flowering of perennial herbs, *Oikos*, **7**, 273–92.

Tansley, A.G., ed. (1911), *Types of British Vegetation*, Cambridge University Press, Cambridge.

Tansley, A.G. (1939), *The British Islands and their Vegetation*, Cambridge University Press, Cambridge.

Taylor, C.C. (1967), Whiteparish. A study of the development of a forest-edge parish, *Wiltshire Archaeol. Nat. Hist. Mag.*, **62**, 79–102.

Taylor, C.C. (1973), *The Cambridgeshire Landscape*, Hodder and Stoughton, London.

Taylor, C.C. (1983), *Village and Farmstead*, George Philip, London.

Taylor, L.R. (1988), Objective and experiment in long-term research, In: *Long-term Studies in Ecology*, (Likens, G.E., ed.), pp. 20–70, Springer, New York.

Taylor, N.W. (1985), *The sycamore in Britain – its natural history and value to wildlife*. Discussion Papers in Nature Conservation, 42. University College, London.

Terborgh, J. (1974), Preservation of natural diversity: the problem of extinction prone species, *Bioscience*, **24**, 715–22.

Thomas, J.A. (1974), *Factors influencing the numbers and distribution of the Brown Hairstreak,* Thecla betulae *L. (Lepidoptera: Lycaenidae) and the Black Hairstreak,* Strymonidia pruni *L. (Lepidoptera: Lycaenidae)*, PhD Thesis, University of Leicester.

Thomas, J.W., ed. (1979), *Wildlife Habitats in Managed Forests. The Blue Mountains of Oregon and Washington*, Agriculture Handbook No. 553, USDA Forest Service, Washington D.C.

Thomas, K. (1983), *Man and the Natural World. Changing Attitudes in England 1500–1800*, Penguin, Harmondsworth.

Thomas, R.C. (1988), Historical ecology of Bernwood Forest Nature Reserve 1900–1981, In: *Woodland Conservation and Research in the Clay Vale of Oxfordshire and Buckinghamshire*. Research and Survey in Nature Conservation 15 (Kirby, K.J. and Wright, F.J., eds.), pp. 20–9, Nature Conservancy Council, Peterborough.

Tittensor, R.M. (1970), History of Loch Lomond oakwoods, *Scot. For.*, **24**, 100–18.

Tittensor, R.M. (1978), A history of The Mens: a Sussex woodland common, *Sussex Archaeol. Coll.*, **116**, 347–74.

Tomialojc, L. (1991), Characteristics of old growth in the Bialowieza Forest, Poland, *Nat. Areas J.*, **11**, 7–18.

Troup, R.S. (1928), *Silvicultural Systems*, Clarendon Press, Oxford.

Tubbs, C.R. (1964), Early encoppicements in the New Forest, *Forestry*, **37**, 95–105.

Tubbs, C.R. (1968), *The New Forest: An Ecological History*, David and Charles, Newton Abbot.

Tubbs, C.R. (1986), *The New Forest*, Collins, London.

Turner, G.J. (1901), *Select Pleas of the Forest* (Selden Soc. Publ., 13), Quaritch, London.

Turner, J. (1965), A contribution to the history of forest clearance, *Proc. Roy. Soc., B*, **161**, 343–54.

Usher, M.B., ed. (1986), *Wildlife Conservation Evaluation*, Chapman & Hall, London.

Usher, M.B., Brown, A.C. and Bedford, S.E. (1992), Plant species richness in farm woodlands, *Forestry*, **65**, 1–13.

Väisänen, R.A., Järvinen, O. and Rauhala, P. (1986), How are extensive human-caused habitat alterations expressed on the scale of local bird populations in boreal forests? *Ornis Scandinavica*, **17**, 282–92.

Vancouver, C. (1795), *General View of the Agriculture of the County of Essex*, Board of Agriculture, London.

Vancouver, C. (1810), *General View of the Agriculture of Hampshire Including the Isle of Wight*, Board of Agriculture, London.

Vancouver, C. (1813), *General View of the Agriculture of the County of Devon*, Board of Agriculture, London.

Verboom, J., Schotman, A., Opdam, P. and Metz, J.A.J. (1991), European nuthatch metapopulations in a fragmented agricultural landscape, *Oikos*, **61**, 149–56.

Walker, G.J. and Kirby, K.J. (1987), An historical approach to woodland conservation in Scotland, *Scot. For.*, **41**, 87–98.

Ward, L.K. (1974), Ecological characteristics and classification of scrub communities, In: *Grassland Ecology and Wildlife Management* (Duffey, E., ed.), pp. 124–62, Chapman & Hall, London.

Ward, S.D. and Evans, D.F. (1976), Conservation assessment of British limestone pavements based on floristic criteria, *Biol. Conserv.*, **9**, 217–33.

Wardle, P. (1959), The regeneration of *Fraxinus excelsior* in woods with a field layer of *Mercurialis perennis*, *J. Ecol.*, **47**, 483–97.

Warren, M.S. (1987), The ecology and conservation of the Heath Fritillary butterfly, *Mellicta athalia*, *J. Appl. Ecol.*, **24**, 467–513.

Warren, M.S. and Key, R.S. (1991), Woodlands: past, present, and potential for insects, In: *The Conservation of Insects and their Habitats* (Collins, N.M. and Thomas, J.A. eds.), pp. 155–211, Academic Press, London.

Waters, T.L. and Savill, P.S. (1992), Ash and sycamore regeneration and the phenomenon of their alternation, *Forestry*, **65**, 417–33.

Watkins, C. (1983), *Woodlands in Nottinghamshire since 1945: a study of changing distribution, type and use*, PhD thesis, University of Nottingham.

Watkins, C. (1984), The use of Forestry Commission censuses for the study of woodland change, *J. Hist. Geog.*, **10**, 396–406.

Watkins, C. (1985), Sources for the assessment of British woodland change in the twentieth century, *Appl. Geog.*, **5**, 151–66.

Watkins, C. (1990), *Britain's Ancient Woodland. Woodland Management and Conservation*, David and Charles, Newton Abbot.

Watt, A.S. (1923), On the ecology of British beechwoods with special reference to their regeneration I. The causes of failure of natural regeneration of the beech (*Fagus silvatica* L.), *J. Ecol.*, **11**, 1–48.

Watt, A.S. (1924), On the ecology of British beechwoods with special reference to their regeneration II. The development and structure of beech communities on the Sussex Downs, *J. Ecol.*, **12**, 145–204.

Watt, A.S. (1925), On the ecology of British beechwoods with special reference to their regeneration II (cont.). The development and structure of beech communities on the Sussex Downs, *J. Ecol.*, **13**, 27–73.

Watt, A.S. (1926), Yew communities of the South Downs, *J. Ecol.*, **14**, 282–316.

Watt, A.S. (1934), The vegetation of the Chiltern Hills, with special reference to the beechwoods and their seral relationships, *J. Ecol.*, **22**, 230–70 and 445–507.

Watt, A.S. (1947), Pattern and process in the plant community, *J. Ecol.*, **35**, 1–22.

Way, L.J.U. (1913), An account of the Leigh Woods, in the parish of Long Ashton, County of Somerset, *Trans. Bristol Gloucestershire Archaeol. Soc.*, **36**, 55–102.

Webb, D.A. (1954), Is the classification of plant communities either possible or desirable? *Bot. Tidsskr.*, **51**, 362–70.

Westhof, V. and Den Held, A.J. (1969), *Plantengemeenschappen in Nederland*, Thieme, Zutphen.

Wheeler, P.T. (1984), A survey of woodland change in Nottinghamshire 1920–1980, *East Midlands Geographer*, **8**, 134–47.

Whitbread, A.M. (1991a), *When the Wind Blew. Life in Our Woods After the Great Storm of 1987*, Royal Society for Nature Conservation.

Whitbread, A.M. (1991b), *Research on the Ecological Effects on Woodland of the 1987 Storm.* Research and Survey in Nature Conservation No. 40. Nature Conservancy Council, Peterborough.

White, F.B.W. (1898), *The Flora of Perthshire*, Blackwood, Edinburgh.

White, P.S. (1979), Pattern, process and natural disturbance in vegetation, *Botan. Rev.*, **45**, 229–99.

Whitney, G.G. (1987), An ecological history of the Great Lakes forest of Michigan. *J. Ecol.*, **75**, 667–84.

Whitney, G.G. and Foster, D.R. (1988), Overstorey composition and age as determinants of the understorey flora of woods of central New England. *J. Ecol.*, **76**, 867–76.

Whittaker, J. (1892), *A Descriptive List of the Deer-parks and Paddocks of England*, Ballantyne, Hansen, London.

Wightman, A.D. ed. (1992), *A Forest for Scotland*, Scottish Wildlife and Countryside Link, Perth.

Wightman, W.R. (1968), The pattern of vegetation in the Vale of Pickering area c. 1300 A.D., *Trans. Inst. Br. Geog.*, **45**, 125–42.

Wilcox, H.A. (1933), *The Woodlands and Marshlands of England*, Liverpool University Press, Liverpool.

Williamson, K. (1964), Bird census work in woodland, *Bird Study*, **11**, 1–22.

Wilson, M. (1911), Plant distribution in the woods of north-east Kent, *Ann. Bot.*, **25**, 857–902.

Woodroffe-Peacock, E.A. (1918), A fox-covert study, *J. Ecol.*, **6**, 110–25.

Woods, K.S. (1949), *The Rural Industries round Oxford*, Clarendon Press, Oxford.

Woodward, S.F. (1992), *Swithland Wood. A Study of its History and Vegetation*, Leicester Museums, Arts and Records Service, Leicester.

Wormell, P. (1977), Woodland insect population changes on the Isle of Rhum in relation to forest history and woodland restoration, *Scot. For.*, **31**, 13–36.

Wright, D.F. (1977), A site evaluation scheme for use in the assessment of potential nature reserves, *Biol. Conserv.*, **11**, 293–305.

Yates, E.M. (1960), History in a map, *Geog. J.*, **126**, 32–51.

Yelling, J.A. (1968), Common land and enclosure in east Worcestershire, 1540–1870, *Trans. Inst. Br. Geog.*, **45**, 157–68.

Young, B. (1981), *Habitat Changes in West Berkshire*, Unpublished report to Nature Conservancy Council.

Site index

Site index

Site index

Subject index

Subject index

Changes in woodland area 36–39, 85, 261, 297, 325, 327
Chases 15
Chestnut woods, in classification 25, 141, 146, 177–178, 319
Choice of species 271–273
Classifications
    continental phytosociological 185–190
    for management guides 332
    types of 109
Clear felling 66–67, 71, 74, 280
Clearance of woodland 6, 9, 30–31, 34–39, 261–264, 325, 327
    effects of wildlife 261–264
    of ancient woods 37–38
    prehistoric 6, 9, 30
    recent 36–37, 261–264, 327–329
Coed Cymru 332, 333
Coleoptera 64–65
Colonising ability of plants 54, 63, 65, 92–99, 270–271, 304, 311–312, 314, 318–319
Common bird census 218
Commons, wooded 13, 24
Community Forests 333, 344
Composition of Atlantic forests 7–8
Composition of underwood 21–22, 25, 27, 33–34, 57–59, 114, 314–315
Conifer plantations 33, 44, 67–69, 77–78, 223, 272, 280, 316, 325, 338–341
    long-rotations 339
    open space habitats 84, 338–339, 341
Conservation areas 256–259, 308–309
Conifers, removal from native woods 344
Conservation (cf. nature conservation) 194–195
Continuous variation in plant communities 110
Convallaria majalis 51–52
Coppice
    changes in compartments 26
    cyclical effects on flora 52–56
    decline of 27–29, 307
    in forests 15–16
    in parks 14
    management of 17–30, 33, 284–289, 334–335
    origins of 4, 24, 56
Cyclical changes in forests 5–6, 182–183
Czechoslovakia, woods in 3–4, 229, 304–307

Dead wood 312, 316–317
Deer 14–17, 19, 33, 194, 318, 335–336, 339
Deer parks 13–15, 31, 62

Deschampsia caespitosa
    distribution in relation to soils 50
    distribution in Bedford Purlieus 217
Deschampsia flexuosa, distribution in relation to soils 50
Design principles for reserves 264
Distribution maps of stand types, sources for 119
Distribution of nature conservation value 240–241
Disturbance, natural 5, 315–316
Disturbed forms of stand types 176–177
Diversity, management for 273, 275
Documentary sources for woodland history 208
Domesday book, woods in 24, 82, 205
Dormant seed in soil 55, 76
Drought, effects of 315
Dryopteris aemula 52, 311

Elms 314
Endymion non-scriptus
    distribution in central Lincolnshire 95–96
    indicator of ancient woods in Norfolk 49
Epiphytes 46, 60–61, 273, 311, 316, 321
Essex woods, coppice rotations in 18th century 21
European woods and forests 3–6, 8, 66, 71–73, 183, 185–190, 228–229, 303–309, 320
    changes in 304
    conservation needs 307–308
    distribution 303–305
    management of 305–307
Eutrophication of woodlands 318, 327
Evaluation for nature conservation 231–249, 317–318, 321–322
Evaluation of particular features
    disturbance 238
    glades 234
    historical record 237
    research needs 238, 322–324
    rides 234–235
    size 235
    species 232
    structure 233–234
    surroundings 237–238
    time factors 235–237
    water bodies 235
    woodland types 232–233, 244–246
Evaluation, examples of woodland 238–240, 282
Event records 299–230, 279–280
Exclosures for regeneration 335–336
Extent of woodland types 296–298
Extinction-prone species 103, 200–201, 204, 282

Farm Woodland Scheme 333, 344

370